STUDENT SOLUTIONS MANUAL

John Garlow

BEGINNING AND INTERMEDIATE ALGEBRA

John Tobey Jeffrey Slater

Prentice Hall

Upper Saddle River, NJ 07458

Executive Editor: Karin E. Wagner
Supplements Editor: Elizabeth Covello
Editorial Assistant: Heather Balderson
Assistant Managing Editor: John Matthews
Production Editor: Wendy A. Perez
Supplement Cover Manager: Paul Gourhan
Supplement Cover Designer: PM Workshop Inc.
Manufacturing Buyer: Ilene Kahn

Prentice-Hall, Inc.
Upper Saddle River, NJ 07458

Printed in the United States of America

10 9 8 7 6 5 4 3 2 1

ISBN 0-13-092414-8

Pearson Education Ltd., *London*
Pearson Education Australia Pty. Ltd., *Sydney*
Pearson Education Singapore, Pte. Ltd.
Pearson Education North Asia Ltd., *Hong Kong*
Pearson Education Canada, Inc., *Toronto*
Pearson Educacíon de Mexico, S.A. de C.V.
Pearson Education—Japan, *Tokyo*
Pearson Education Malaysia, Pte. Ltd.
Pearson Education, *Upper Saddle River, New Jersey*

CONTENTS

Chapter 1

Pretest Chapter 1

1. $(-3)-(-6)=(-3)+(+6)=3$

2. $12-16+3-14=12+(-16)+3+(-14)$
$=-15$

3. $(-7)+(-11)=-18$

4. $7(-13)=7+13=20$

5. $(-7)(-2)(+3)(-1)=-(7)(2)(3)(1)=-42$

6. $\dfrac{-\frac{2}{3}}{-\frac{1}{4}}=-\dfrac{2}{3}+-\dfrac{1}{4}=\left(-\dfrac{2}{3}\right)\left(-\dfrac{4}{1}\right)=\dfrac{8}{3}$

7. $-2.3(-4)=9.2$

8. $20\div(-5)=-4$

9. $(-2)^4=(-2)(-2)(-2)(-2)=16$

10. $4^3=(4)(4)(4)=64$

11. $\left(-\dfrac{2}{3}\right)^3=\left(-\dfrac{2}{3}\right)\left(-\dfrac{2}{3}\right)\left(-\dfrac{2}{3}\right)=-\dfrac{8}{27}$

12. $-4^4=-(4)(4)(4)(4)=-256$

13. $6\cdot 2+8\cdot 3-4\cdot 2=12+24-8=28$

14. $3(4)^3+18\div 3-1=3(64)+6-1$
$=192+6-1=197$

15. $-\dfrac{1}{6}\left(\dfrac{1}{2}\right)+\dfrac{3}{4}\div\dfrac{1}{12}$
$=-\dfrac{1}{12}+9=-\dfrac{1}{12}+\dfrac{108}{12}=\dfrac{107}{12}=8\dfrac{11}{12}$

16. $1.22-4.1(1.4)+(-3.3)^2$
$=1.22-5.74+10.89$
$=6.37$

17. $-2x(3x-2xy+z)$
$=-2x(3x)+(-2x)(-2xy)+(-2x)(z)$
$=-6x^2+4x^2y-2xz$

18. $-(3x-4y-12)=-3x+4y+12$

19. $5x^2-3xy-6x^2y-8xy$
$=5x^2-6x^2y+(-3-8)xy$
$=5x^2-6x^2y-11xy$

20. $11x-3y+7-4x+y=7x-2y+7$

21. $3(2x-5y)-(x-8y)$
$=6x-15y-x+8y$
$=5x-7y$

22. $2y(x^2+y)-3(x^2y-5y^2)$
$=2x^2y+2y^2-3x^2y+15y^2$
$=-x^2y+17y^2$

23. If $x=-2$, then $3x^2-5x-4$
$=3(-2)^2-5(-2)-4$
$=3(4)-5(-2)-4$
$=12+10-4=18$

24. $ab+3a^2-5b=(3)(-1)+3(3)^2-5(-1)$
$=-3+3(9)+5$
$=-3+27+5$
$=29$

25. If $F=77$, then
$C=\frac{5}{9}(F-32)$
$=\frac{5}{9}[77-32]=\frac{5}{9}(45)=25^0 C$

26. $3[(x-y)-(2x-y)]-3(2x+y)$
$=3[x-y-2x+y]-6x-3y$
$=3[x]-6x-3y$
$=-3x-6x-3y$
$=-9x-3y$

27. $2x^2-3x[2x-(x+2y)]$
$=2x^2-3x(2x-x-2y)$
$=2x^2-3x(x-2y)$
$=2x^2-3x^2+6xy$
$=-x^2+6xy$

1.1 Exercises

	Number	WN	RaN	IR	ReN
1.	23	x	x		x
3.	π				x
5.	$-6.666...$		x		x
7.	$-2.3434...$		x		x
9.	$\sqrt{2}$			x	x

11. $-20,000$

13. Down $2\frac{3}{8} \Rightarrow -2\frac{3}{8}$

15. Rise 7^0 F $\Rightarrow +7$

17. Opposite of $\frac{3}{4}$ is $-\frac{3}{4}$

19. Opposite of -2.73 is 2.73

21. $|-1.3|=1.3$

23. $\left|\frac{5}{6}\right|=\frac{5}{6}$

25. $-6+(-5)=-11$

27. $\left(-\frac{1}{3}\right)+\left(+\frac{2}{3}\right)=\frac{1}{3}$

29. $-17+(-14)=-31$

27. $(+35)+(+20)=55$

29. $(-15)+(-16)=-31$

31. $\left(-\frac{2}{13}\right)+\left(-\frac{5}{13}\right)=-\frac{7}{13}$

33. $(-1.5)+(-2.3)=-3.8$

35. $(+0.6)+(-0.2)=0.4$

37. $(-12)+(-13)=-25$

39. $\left(-\frac{2}{5}\right)+\left(+\frac{3}{7}\right)=\left(-\frac{14}{35}\right)+\left(+\frac{15}{35}\right)=\frac{1}{35}$

41. $-8+5+(-3)=-3+(-3)=-6$

43. $(-3)+(+8)+(+5)+(-7)$
$=(+5)+(+5)+(-7)$
$=(10)+(-7)=3$

45. $21+(-4)=17$

47. $-98+103=5$

49. $-\frac{3}{5}+\frac{2}{3}=-\frac{9}{15}+\frac{10}{15}=\frac{1}{15}$

51. $-\frac{2}{3}+\left(-\frac{1}{4}\right)=-\frac{8}{12}+\left(-\frac{3}{12}\right)=-\frac{11}{12}$

53. $5.7+(-9.1)=-3.4$

55. $27+(-11)+(-4)=16+(-4)=12$

57. $18+(-39)+25+(-3)$
$=-21+25+(-3)$
$=4+(-3)=1$

59. $23.17+5.03+(-11.81)$
$=28.2+(-11.81)=16.39$

61. $-12+7=-5$
-5^0 F

63. $-85+(-180)=-265$
-265 feet

65. Account balance:
$643.85-185.50=\$458.35$
Short: $475-458.35=\$16.54$

67. $-258+(-32)+150=-290+150=-140$
He still owes $140.

69. $30+14+(-12)+(-18)+8=22$
$22,000,000

71. $-18+?=10$
$-18+28=10$
$?=28$

73. This year:
$\frac{328}{800}=0.41=41\%$
Next year:
$56\%\times 800=0.56\times 800=448$ kittens
Increase: $448-328=120$ kittens

1.2 Exercises

1. $-8-(-3)=-8+3=-5$

3. $(+20)-(+46)=(+20)+(-46)=-26$

5. $-14-(-3)=-14+3=-11$

7. $(-52)-(-60)=(-52)+(+60)=8$

9. $(0)-(-5)=(0)+(+5)=5$

11. $(+15)-(+20)=(+15)+(-20)=-5$

13. $(-18)-(-18)=(-18)+(+18)=0$

15. $-11-(-8)=-11+8=-3$

17. $(-0.6)-(+0.3)=(-0.6)+(-0.3)=-0.9$

19. $(+2.64)-(-1.83)=(+2.64)+(+1.83)$
$=4.47$

21. $2.3-(-1.4)=2.3+1.4=3.7$

23. $\left(+\frac{3}{4}\right)-\left(-\frac{3}{5}\right)=\left(+\frac{15}{20}\right)+\left(+\frac{12}{20}\right)$
$=\frac{27}{20}=1\frac{7}{20}$

25. $\left(-\frac{3}{4}\right)-\left(+\frac{5}{6}\right)=\left(-\frac{9}{12}\right)+\left(-\frac{10}{12}\right)$
$=-\frac{19}{12}=-1\frac{7}{12}$

27. $34-87=34+(-87)=-53$

29. $-67-32=(-67)+(-32)=-99$

31. $2.3-(-4.8)=2.3+(+4.8)=7.1$

33. $8-\left(-\frac{3}{4}\right)=8+\frac{3}{4}=8\frac{3}{4}$

35. $\frac{3}{5}-(-8)=\frac{3}{5}+(+8)=\frac{3}{5}+\frac{40}{5}=\frac{43}{5}=8\frac{3}{5}$

37. $-\frac{3}{10}-\frac{3}{4}=\left(-\frac{6}{20}\right)+\left(-\frac{15}{20}\right)$
$=-\frac{21}{20}=-1\frac{1}{20}$

39. $-135-(-126.5)=-135+(+126.5)=-8.5$

41. $0.0067-(-0.0432)=0.0067+(+0.0432)$
$=0.0499$

43. $\frac{1}{5}-6=\frac{1}{5}+\left(-\frac{30}{5}\right)=-\frac{29}{5}=-5\frac{4}{5}$

45. $-0.0023-6=-0.0023+(-6)=-6.0023$

47. $-2-(-9)=-2+(+9)=7$

49. $-35-(13)=-35+(-13)=-48$

51. $7+(-6)-(+3)=7+(-6)+(-3)$
$=1+(-3)=-2$

53. $-10+6-(-15)=-10+6+(+15)$
$=-4+(+15)=11$

55. $7+(-42)-27=7+(-42)+(-27)$
$=-35+(-27)=-62$

57. $-21-(-36)-8=-21+(+36)+(-8)$
$=15+(-8)=7$

59. $42-(-30)-65-(-11)+20$
$=42+(+30)+(-65)+(+11)+20$
$=72+(-65)+(+11)+20$
$=7+(+11)+20$
$=18+20=38$

61. $112-(-37)=112+(+37)=\$149$

63. $-45-(-29)=-45+(+29)$
$=-16^{0}\text{ F}$

Cumulative Review Problems

65. $-37+16=-21$

67. $-3+(-6)+(-10)=-9+(-10)$
$=-19$

69. $8-(-15)=8+(+15)$
$=23^{0}\text{ F}$

1.3 Exercises

1. To multiply two real numbers, multiply the absolute values. The sign of the result is positive if both numbers have the same sign, but negative if the two numbers have opposite signs.

3. $-8\times 3=-24$

5. $0(-12)=0$

7. $16\times 1.5=24$

9. $(-1.32)(-0.2)=0.264$

11. $(0.7)(-2.5)=-1.75$

13. $\left(\frac{12}{5}\right)(-10)=\left(\frac{12}{5}\right)\left(-\frac{10}{1}\right)=-24$

15. $\left(-\frac{4}{9}\right)\left(-\frac{3}{5}\right)=\frac{4}{15}$

17. $(12)\div(-4)=\frac{12}{-4}=-3$

19. $0\div(-15)=0$

21. $-45\div 9=\frac{-45}{9}=-5$

23. $(240)\div(-15)=\frac{240}{-15}=-16$

25. $(-0.6)\div 0.3=\frac{-0.6}{0.3}=-2$

27. $-6.3\div 0.7=\frac{-6.3}{0.7}=-9$

29. $(-7.2)\div 8=\frac{-7.2}{8}=-0.9$

31. $\left(-\frac{1}{5}\right)\div\left(\frac{2}{3}\right)=\left(-\frac{1}{5}\right)\left(\frac{3}{2}\right)=-\frac{3}{10}$

33. $\left(\frac{5}{7}\right)\div\left(-\frac{2}{3}\right)=\frac{5}{7}\cdot\left(-\frac{28}{3}\right)=-\frac{20}{3}=-6\frac{2}{3}$

35. $\frac{12}{-\frac{2}{5}} = 12 \div \left(-\frac{2}{5}\right) = \left(\frac{12}{1}\right)\left(-\frac{5}{2}\right) = -30$

37. $\frac{-\frac{3}{8}}{-\frac{2}{3}} = \left(-\frac{3}{8}\right) \div \left(-\frac{2}{3}\right) = \left(-\frac{3}{8}\right)\left(-\frac{3}{2}\right) = \frac{9}{16}$

39. $\frac{\frac{7}{15}}{-\frac{3}{5}} = \frac{7}{15}\left(-\frac{5}{3}\right) = -\frac{7}{9}$

41. $(-1)(-2)(-3)(4) = -(1)(2)(3)(4) = -24$

43. $-2(-2)(2)(-1)(-3) = 24$

45. $(-3)(-2)\left(\frac{1}{3}\right)(-4)(2)$
$= -(3)(2)\left(\frac{1}{3}\right)(4)(2) = -16$

47. $(-3)(-0.03)(100)(-2)$
$= -(3)(0.03)(100)(2) = -18$

49. $\left(-\frac{4}{5}\right)\left(-\frac{6}{7}\right)\left(-\frac{1}{3}\right) = -\left(\frac{24}{35}\right)\left(\frac{1}{3}\right) = -\frac{8}{35}$

51. $\left(-\frac{3}{4}\right)\left(-\frac{7}{15}\right)\left(-\frac{8}{21}\right)\left(-\frac{5}{9}\right)$
$= +\left(\frac{3}{4}\right)\left(\frac{7}{3\cdot 5}\right)\left(\frac{4\cdot 2}{3\cdot 7}\right)\left(\frac{5}{9}\right) = \frac{2}{27}$

53. $(-20) \div 5 = \frac{-20}{5} = -4$

55. $5 + (-7) = -2$

57. $8 - (-9) = 8 + (9) = 17$

59. $6(-12) = -72$

61. $(-37) \div 37 = \frac{-37}{37} = -1$

63. $? - 100 = -70$
$30 - 100 = -70$
There was \$30 in the account.

65. $\frac{15,768}{48} = \$328.50$

67. $5(4) = +20$, gained 20 yards

69. $-10(7) = -70$, lost 70 yards

71. Total $= -70 + 90 = 20$, gained 20 yards

73. Additional large-losses $= -15(3)$
$= -45$
Avoided medium-gains $= -(+15)(4)$
$= -60$
Fewer yards gained $= 45 + 60$
$= 105$ yards

Cumulative Review Problems

75. $(-13) + (-39) + (-20) = (-52) + (-20)$
$= -72$

77. $(-37) - (51) = (-37) + (-51) = -88$

1.4 Exercises

1. $4^4 = (4)(4)(4)(4) = 256$

3. A negative number to an odd power is negative.

5. $(-2)^4 = (-2)(-2)(-2)(-2) = 64$
   $-2^4 = -(2)(2)(2)(2) = -64$

7. $3^3 = 27$

9. $3^4 = 81$

11. $7^3 = 343$

13. $(-3)^3 = -27$

15. $(-2)^6 = 64$

17. $-5^3 = -(5)(5)(5) = -125$

19. $\left(\frac{1}{4}\right)^2 = \frac{1}{16}$

21. $\left(\frac{2}{5}\right)^3 = \frac{8}{125}$

23. $(0.9)^2 = 0.81$

25. $(0.2)^4 = 0.0016$

27. $(-16)^2 = 256$

29. $-16^2 = -256$

31. $(6)(6)(6)(6)(6) = 6^5$

33. $(w)(w) = w^2$

35. $(x)(x)(x)(x) = x^4$

37. $(3q)(3q)(3q) = (3q)^3 = 3^3q^3$

39. $5^3 + 6^2 = 125 + 36 = 161$

41. $(-3)^3 - (8)^2 = -27 - 64 = -91$

43. $5^3 - (-3)^3 = 125 - 9 = 116$

45. $(-4)^3(-3)^2 = (-64)(9) = -576$

47. $8^2(-2)^3 = 64(-8) = -512$

49. $4^{12} = 16,777,216$

51. $x^7 = 128,\ x = 2$

Cumulative Review Problems

53. $(-11) + (-13) + 6 + (-9) + 8$
    $= -24 + 6 + (-9) + 8$
    $= -18 + (-9) + 8$
    $= -27 + 8$
    $= -19$

55. $-17 - (-9) = -17 + 9 = -8$

57. $\left(\frac{2}{5}\right)\left(-\frac{1}{4}\right)\left(\frac{10}{11}\right)\left(-\frac{3}{4}\right)$
    $= +\left(\frac{2}{5}\right)\left(\frac{1}{2\cdot 2}\right)\left(\frac{2\cdot 5}{11}\right)\left(\frac{3}{4}\right) = \frac{3}{44}$

1.5 Exercises

1. $3(4)+6(5)$

3. $3(4)+6(5)=12+6(5)=18(5)=90$

   (b) $3(4)+6(5)=12+30=42$

5. $(2-5)^2 \div 3\times 4=(-3)^2 \div 3\times 4$

   $=9\div 3\times 4$

   $=3\times 4$

   $=12$

7. $8-2^3 \cdot 5+3=8-8\cdot 5+3$

   $=8-40+3=-29$

9. $4+27\div 3\cdot 2-8=4+9\cdot 2-8$

   $=4+18-8=14$

11. $8-5(2)^3 \div(-8)=8-5(8)\div(-8)$

    $=8-40\div(-8)$

    $=8+5$

    $=13$

13. $3(5-7)^2-6(3)=3(-2)^2-6(3)$

    $=3(4)-6(3)=12-18=-6$

15. $5\cdot 6-(3-5)^2+8\cdot 2=5\cdot 6-(-2)^2+8\cdot 2$

    $=5\cdot 6-4+8\cdot 2$

    $=30-4+16=42$

17. $\frac{1}{2}\div\frac{2}{3}+6\cdot\frac{1}{4}=\frac{1}{2}\cdot\frac{3}{2}+\frac{6}{1}\cdot\frac{1}{4}$

    $=\frac{3}{4}+\frac{6}{4}=\frac{9}{4}=2\frac{1}{4}$

19. $0.8+0.3(0.6-0.2)^2$

    $=0.8+0.3(0.4)^2$

    $=0.8+0.3(0.16)$

    $=0.8+0.048=0.848$

21. $\frac{3}{4}\left(-\frac{2}{5}\right)-\left(-\frac{3}{5}\right)=-\frac{3}{10}+\frac{6}{10}=\frac{3}{10}$

23. $-6.3-(-2.7)(1.1)+(3.3)^2$

    $=-6.3+2.97+10.89=7.56$

25. $\left(\frac{1}{2}\right)^3+\left(\frac{1}{4}\right)-\left(\frac{1}{6}-\frac{1}{12}\right)-\frac{2}{3}\cdot\left(\frac{1}{4}\right)^2$

    $=\left(\frac{1}{2}\right)^3+\frac{1}{4}-\left(\frac{1}{12}\right)-\frac{2}{3}\cdot\left(\frac{1}{4}\right)^2$

    $=\frac{1}{8}+\frac{1}{4}-\frac{1}{12}-\frac{2}{3}\cdot\frac{1}{16}$

    $=\frac{1}{8}+\frac{1}{4}-\frac{1}{12}-\frac{1}{24}$

    $=\frac{6}{24}=\frac{1}{4}$

Cumulative Review Problems

27. $(0.5)^3=0.125$

29. $-1^{20}=-1$

31. $15\div 2\times(6)=45$ ounces

1.6 Exercises

1. A var*iable* is a symbol used to represent an unknown number.

3. We are multiplying 4 by x by x. We know from the definition of exponents that x multiplied by x is x^2, this gives us an answer of $4x^2$.

5. $a(b-c)=ab-ac$ works:
$$=a(b-c)=a[b+(-c)]$$
$$=ab+(a)(-c)$$
$$=ab+(-ac)$$
$$=ab-ac$$

7. $-(x+4y)=-x-4y$

9. $-3(2a-5b)$
$$=-3(2a)-3(-5b)$$
$$=-6a+15b$$

11. $2(4x+y-2)$
$$=2(4x)+2(y)+2(-2)$$
$$=8x+2y-4$$

13. $-3x(-2x+3y-z)$
$$=(-3x)(-2x)+(-3x)(3y)+(-2x)(-z)$$
$$=6x^2-9xy+2xz$$

15. $2x(4x-y-6)$
$$=2x(4x)+2x(-y)+2x(-6)$$
$$=8x^2-2xy-12x$$

17. $-5(3x+9-7y)$
$$=(-5)(3x)+(-5)(9)+(-5)(-7y)$$
$$=-15x-45+35y$$

19. $\frac{1}{4}(x^2+2x-8)$
$$=\frac{1}{4}x^2+\frac{1}{4}(2x)+\frac{1}{4}(-8)$$
$$=\frac{x^2}{4}+\frac{x}{2}-2$$

21. $\frac{2}{3}(-27a^4+9a^2-21)$
$$=\frac{2}{3}(-27a^4)+\frac{2}{3}(9a^2)+\frac{2}{3}(-21)$$
$$=-18a^4+6a^2-14$$

23. $\frac{y}{3}(3y-4x-6)$
$$=\frac{y}{3}(3y)+\frac{y}{3}(-4x)+\frac{y}{3}(-6)$$
$$=y^2-\frac{4xy}{3}-2y$$

25. $3a(2a+b-c-4)$
$$=3a(2a)+3ab+3a(-c)+3a(-4)$$
$$=6a^2+3ab-3ac-12a$$

27. $(5x+1)(-4)=5x(-4)+1(-4)=-20x-4$

29. $3x(4x-5y-6)$
$$=3x(4x)+3x(-5y)+3x(-6)$$
$$=12x^2-15xy-18x$$

31. $(4a-2b-1)(-ab)$
$$=4a(-ab)+(-2b)(-ab)+(-1)(-ab)$$
$$=-4a^2b+2ab^2+ab$$

33. $(-2x+y-3)(4xy)$
$= (-2x)(4xy)+y(4xy)+(-3)(4xy)$
$= -8x^2y+4xy^2-12xy$

35. $2.5(1.5a^2-3.5a+2.0)$
$= (2.5)(1.5a^2)$
$+ (2.5)(-3.5a)+(2.5)(2.0)$
$= 3.75a^2-8.75a+5$

37. $-0.9q(2.1q-0.2r-0.8s)$
$= (-0.9q)(2.1q)+(0.9q)(-0.2r)$
$+ (-0.9q)(-0.8s)$
$= -1.89q^2+0.8qr+0.72qs$

39. $700(12x+8y)=700(12x)+700(8y)$
$= 8400x+5600y$ square feet

41. $3x(1500-4y)=3x(1500)+3x(-4y)$
$= 4500x-12xy$ square feet

Cumulative Review Problems

43. $-18+(-20)+36+(-14)$
$= -38+36+(-14)$
$= -2+(-14)=-16$

45. $-27-(-41)=-27+41=14$

47. $3^3 \div (2-11)-2\cdot 5 = 3^3 \div (-9)-2\cdot 5$
$= 27\div(-9)-2\cdot 5$
$= -3-10$
$= -13$

1.7 Exercises

1. A term is a number, a variable, or a product of numbers and variables.

3. The two terms $5x$ and $-8x$ are like terms because they both have the variable x with the exponent of one.

5. The only like terms are $7xy$ and $-14xy$ because the other two have different exponents even though they have the same variables.

7. $-14b^2-11b^2=(-14-11)b^2=-25b^2$

9. $10x^4+8x^4+7x^2=(10+8)x^4+7x^2$
$= 18x^4+7x^2$

11. $2ab+1-6ab-8=(2-6)ab+1-8$
$= -4ab-7$

13. $1.3x-2.6y+5.8x-0.9y$
$= (1.3+5.8)x+(-2.6-0.9)y$
$= 7.1x-3.5y$

15. $1.6x-2.8y-3.6x-5.9y$
$= (1.6-3.6)x+(-2.8-5.9)y$
$= -2x-8.7y$

17. $\frac{1}{2}x^2 - 3y - \frac{1}{3}y + \frac{1}{4}x^2$
$= \left(\frac{1}{2} + \frac{1}{4}\right)x^2 + \left(-3 - \frac{1}{3}\right)y$
$= \left(\frac{2}{4} + \frac{1}{4}\right)x^2 + \left(-\frac{9}{3} - \frac{1}{3}\right)y$
$= \frac{3}{4}x^2 - \frac{10}{3}y$

19. $\frac{1}{3}x - \frac{2}{3}y - \frac{2}{5}x + \frac{4}{7}y$
$= \left(\frac{1}{3} - \frac{2}{5}\right)x + \left(-\frac{2}{3} + \frac{4}{7}\right)y$
$= \left(\frac{5}{15} - \frac{6}{15}\right)x + \left(-\frac{14}{21} + \frac{12}{21}\right)y$
$= -\frac{1}{15}x - \frac{2}{21}y$

21. $3p - 4q + 2p + 3 + 5q - 21$
$= (3+2)p + (-4+5)q + 3 - 21$
$= 5p + q - 18$

23. $5x^2y - 10xy^6 + 6xy^2 - 7xy^2$
$= 5x^2y - 10xy^6 + (6-7)xy^2$
$= 5x^2y - 10xy^6 - xy^2$

25. $2ab + 5bc - 6ac - 2ab$
$= (2-2)ab + 5bc - 6ac$
$= 5bc - 6ac$

27 $2x^2 - 3x - 5 - 7x + 8 - x^2$
$= (2-1)x^2 + (-3-7)x - 5 + 8$
$= x^2 - 10x + 3$

29. $2y^2 - 8y + 9 - 12y^2 - 8y + 3$
$= (2-12)y^2 + (-8-8)y + 9 + 3$
$= -10y^2 - 16y + 12$

31. $ab + 3a - 4ab + 2a - 8b$
$= (1-4)ab + (3+2)a - 8b$
$= -3ab + 5a - 8b$
$= 5a - 3ab - 8b$

33. $3(x+y) - 5(-2y+3x)$
$= 3x + 3y + 10y - 15x$
$= -12x + 13y$

35. $2x(x-3y) - 4(-3x^2 - 2xy)$
$= 2x^2 - 6xy + 12x^2 + 8xy$
$= 14x^2 + 2xy$

37. $-7(7xy - 11y^2) - 2y(-2x + 3y)$
$= -21xy + 33y^2 + 4xy - 6y^2$
$= -17xy + 27y^2$

39. $7(3-x) - 6(8-13x)$
$= 21 - 7x - 48 + 78x$
$= 71x - 27$

41. $2(7x-2) + 2(3x+4) = 14x - 4 + 6x + 8$
$= 20x + 4$ meters

43. $2(4a-5) + 2(3a+8) + 2(9a+2)$
$= 8a - 10 + 6a + 16 + 18a + 4$
$= 32a + 10$ feet

45. $\left(-\frac{5}{3}\right)\left(\frac{1}{2}\right) = -\frac{5}{6}$

47. $\left(\frac{5}{7}\right)\div\left(-\frac{14}{3}\right)=\left(\frac{5}{7}\right)\left(-\frac{3}{14}\right)=-\frac{15}{98}$

1.8 Exercises

1. If $x = 3$, then $-2x+1$
$= -2(3)+1 = -6+1 = -5$

3. If $x = -9$, then $\frac{2}{3}x - 5 = \frac{2}{3}(-9) - 5$
$= -6 - 5 = -11$

5. If $x = \frac{1}{2}$, then $5x+10$
$= 5\left(\frac{1}{2}\right)+10 = \frac{5}{2}+10 = \frac{25}{2} = 12\frac{1}{2}$

7. If $x = 7$, then $2-4x = 2-4(7)$
$= 2-28 = -26$

9. If $x = -2$, then $x^2 - 3x = (-2)^2 - 3(-2)$
$= 4+6 = 10$

11. If $x = -1$, then $3x^2 = 3(-1)^2 = 3$

13. If $x = 2$, then $-3x^3 = -3(2)^3 = -3(8)$
$= -24$

15. If $x = -\frac{3}{4}$, then $9x+13$
$= 9\left(-\frac{3}{4}\right)+13 = -\frac{27}{4}+\frac{52}{4} = \frac{25}{4}$

17. If $x = -3$, then $2x^2+3x = 2(-3)^2+3(-3)$
$= 2(9)-9 = 18-9 = 9$

19. If $x = 3$, then $(2x)^2 + x$
$= [2(3)]^2 + 3$
$= [6]^2 + 3 = 36+3 = 39$

21. If $x = -2$, then $2-(-x)^2$
$= 2-[-(-2)]^2 = 2-(2)^2 = 2-4 = -2$

23. If $x = -3$, then
$7x+(2x)^2 = 7(-3)+[2(-3)]^2$
$= -21+(-6)^2 = -21+36 = 15$

25. If $x = 3$, then $x^2-7x+3 = (3)^2-7(3)+3$
$= 9-21+3 = -9$

27. If $x = -4$, then $\frac{1}{2}x^2 - 3x + 9$
$= \frac{1}{2}(-4)^2 - 3(-4) + 9$
$= \frac{1}{2}(16)+12+9$
$= 8+12+9 = 29$

29. If $x = 4$ and $y = -1$, then $2x^2-3xy+2y$
$= 2(4)^2-3(4)(-1)+2$
$= 2(16)+12-2$
$= 32+12-2 = 42$

31. If $a = 3$, $b = 2$, and $c = -4$, then
$a^2-2ab+2c^2 = 3^2-2(3)(2)+2(-4)^2$
$= 9-12+2(16)$
$= 9-12+32 = 29$

33. $A = ab$, $b = 92$, $a = 54$
$A = (92)(54) = 4968$ square feet

35. $A = s^2$

$$\text{Decrease} = A_{\text{old}} - A_{\text{new}}$$
$$= (23)^2 - (20)^2$$
$$= 529 - 400$$
$$= 129 \text{ square millimeters}$$

37. $A = \frac{1}{2}a(b_1 + b_2),\ a = 9,\ b_1 = 17,\ b_2 = 20$

$$A = \frac{1}{2}(9)(20+17) = \frac{9(37)}{2}$$
$$\approx 166.5 \text{ square inches}$$

39. $A = \frac{1}{2}ab,\ a = 14,\ b = 19$

$$A = \frac{1}{2}(14)(19) = 133 \text{ square feet}$$

41. $A = \pi r^2 = (3.14)(5)^2 = (3.14)(25)$

$$= 78.5 \text{ square meters}$$

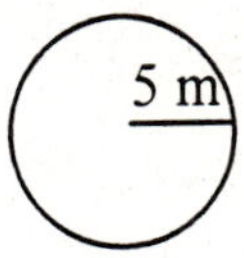

43. $\text{F} = \frac{9}{5}\text{C} + 32$

$$= \frac{9}{5}(-10) + 32$$
$$= -18 + 32 = 14$$
$$14^0\text{F}$$

45. $A = \frac{1}{2}\pi r^2,\ r = 15$

$$A \approx \frac{1}{2}(3.14)(15)^2 = 353.25 \text{ square inches}$$
$$\text{Cost} = (0.85)(353.25) = \$300.26$$

47. $\text{F} = \frac{9}{5}\text{C} + 32$

$$F = \frac{9}{5}(-60) + 32 = -108 + 32 = -76$$
$$F = \frac{9}{5}(-30) + 32 = -54 + 32 = -22$$
$$-76^0 F \text{ to } -22^\circ F$$

49. $k = 1.61r$

$$20 = 1.61r$$
$$r = \frac{20}{1.61} = 12.4$$

Approximately 12.4 miles

Cumulative Review Problems

51. $3(x-2y) - (x^2 - y) - (x - y)$

$$= 3x - 6y - x^2 + y - x + y$$
$$= -x^2 + 2x - 4y$$

Putting Your Skills to Work

1. Weight in $kg = 0.45(125) = 56.25$

Height in $m = 0.0254(5 \cdot 12 + 4)$

$$= 0.0254(64) = 1.6256$$
$$\text{BMI} = \frac{56.25}{(1.6256)^2} = 21.3$$

2. Weight in $kg = 0.45(185) = 83.25$
   Height in $m = 0.0254(5 \cdot 12 + 11)$
   $= 0.0254(71) = 1.8034$
   $\text{BMI} = \dfrac{83.25}{(1.8034)^2} = 25.6$
   Yes, just slightly; BMI = 25.6.

3. Weight in $kg = 0.45(95) = 42.75$
   Height in $m = 1.6$
   $\text{BMI} = \dfrac{42.75}{(1.6)^2} = 16.7$
   Yes, BMI = 16.7.

4. $\text{BMI} = \dfrac{0.45p}{m^2}$

1.9 Exercises

1. $-3x - 2y = -(3x + 2y)$

3. To simplify expressions with grouping symbols, we use the *distributive* property.

5. $6x - 3(x - 2y) = 6x - 3x + 6y$
   $= 3x + 6y$

7. $2(a + 3b) - 3(b - a)$
   $= 2a + 6b - 3b + 3a$
   $= 5a + 3b$

9. $5[3 + 2(x - 26) + 3x]$
   $= 5[3 + 2x - 52 + 3x]$
   $= 5(5x - 49)$
   $= 25x - 245$

11. $7x^2 + [2(3 - x) - 3x^2]$
    $= 7x^2 + [6 - 2x - 3x^2]$
    $= 7x^2 + 6 - 2x - 3x^2$
    $= 4x^2 - 2x + 6$

13. $2(x - 2y) - [3 - 2(x - y)]$
    $= 2x - 4y - (3 - 2x + 2y)$
    $= 2x - 4y - 3 + 2x - 2y$
    $= 4x - 6y - 3$

15. $5[3a - 2a(3a + 6b) + 6a^2]$
    $= 5[3a - 6a^2 - 12ab + 6a^2]$
    $= 5[3a - 12ab]$
    $= 15a - 60ab$

17. $x(x^2 + 2x - 3) - 2(x^3 + 6)$
    $= x^3 + 2x^2 - 3x - 2x^3 - 12$
    $= -x^3 + 2x^2 - 3x - 12$

19. $3a^2 - 4[2b - 3b(b + 2)]$
    $= 3a^2 - 4(2b - 3b^2 - 6b)$
    $= 3a^2 - 4(-4b - 3b^2)$
    $= 3a^2 + 16b + 12b^2$

21. $6b - \{5a - 2[a + (b - 2a)]\}$
    $= 6b - \{5a - 2[a + b - 2a]\}$
    $= 6b - \{5a - 2[-a + b]\}$
    $= 6b - \{5a + 2a - 2b\}$
    $= 6b - \{7a - 2b\}$
    $= 6b - 7a + 2b$
    $= -7a + 8b$

23. $-4\{3a^2-2[4a^2-(b+a^2)]\}$
$=-4[3a^2-2(4a^2-b-a^2)]$
$=-4[3a^2-2(3a^2-b)]$
$=-4(3a^2-6a^2+2b)$
$=-4(-3a^2+2b)$
$=12a^2-8b$

25. Attempts $=(4)(6)(365)=8760$
Successful $=2.5\%(8760)$
$=(.025)(8760)$
$=219$
Unsuccessful $=8760-219=8541$

Cumulative Review Problems

27. If $C=36.4$, then
$F=1.8C+32$
$=1.8(36.4)+32=65.52+32=97.52^0\text{F}$

29. $A=\frac{1}{2}a(b_1+b_2)$,
$a=400,\ b_1=800, b_2=700$
$A=\frac{1}{2}(400)(800+700)=200(1500)$
$=300{,}000$ sq. ft.
Cost $=55(300{,}000)=\$16{,}500{,}000$

Putting Your Skills to Work

1. $\frac{167.6}{100}(200)=\$335.20$

2. $\frac{220.5}{100}(10{,}000)=\$22{,}050.00$

3. $\frac{59.8}{188.4}(1884)=\$598.00$

4. $\frac{50.7}{148.5}(1485)=\$507.00$

5. Medical care in Boston

6. $\frac{165.8}{131.0}\approx 1.27$

Chapter 1 Review Problems

1. $(-6)+(-2)=-8$

2. $(-12)+(+7.8)=-4.2$

3. $(+5)+(-2)+(-12)=(+3)+(-12)=-9$

4. $(+3.7)+(-1.8)=1.9$

5. $\left(+\frac{1}{2}\right)+\left(-\frac{5}{6}\right)=\left(+\frac{3}{6}\right)+\left(-\frac{5}{6}\right)=-\frac{2}{6}=-\frac{1}{3}$

6. $\left(-\frac{3}{11}\right)+\left(-\frac{1}{22}\right)=\left(-\frac{6}{22}\right)+\left(-\frac{1}{22}\right)$
$=-\frac{7}{22}$

7. $\left(+\frac{3}{4}\right)+\left(-\frac{1}{12}\right)+\left(-\frac{1}{2}\right)$
$=\left(+\frac{9}{12}\right)+\left(-\frac{1}{12}\right)+\left(-\frac{6}{12}\right)=\frac{2}{12}=\frac{1}{6}$

8. $\left(-\frac{4}{15}\right)+\left(+\frac{12}{5}\right)+\left(-\frac{2}{3}\right)$
$=\left(-\frac{4}{15}\right)+\left(+\frac{36}{15}\right)+\left(-\frac{10}{15}\right)=\frac{22}{15}$

9. $(+5)-(-3)=(+5)+(+3)=8$

10. $(-2)-(-15)=(-2)+(+15)$
$=13$

11. $(-30)-(+3)=(-30)+(-3)$
$=-33$

12. $(+8)-(-1.2)=(+8)+(+1.2)$
$=9.2$

13. $\left(-\frac{7}{8}\right)+\left(-\frac{3}{4}\right)=\left(-\frac{7}{8}\right)+\left(-\frac{6}{8}\right)$
$=-\frac{13}{8}=-1\frac{5}{8}$

14. $\left(-\frac{3}{14}\right)+\left(+\frac{5}{7}\right)=\left(-\frac{3}{14}\right)+\left(+\frac{10}{14}\right)$
$=\frac{7}{14}=\frac{1}{2}$

15. $-20.8-1.9=-20.8+(-1.9)=-22.7$

16. $-151-(-63)=-151+63=-88$

17. $-5+(-2)-(-3)=-5+(-2)+(+3)$
$=(-7)+(+3)=-4$

18. $6-(-4)+(-2)+(8)=6+4+(-2)+(8)$
$=18+(-2)$
$=16$

19. $(-16)+(-13)=-29$

20. $(-11)-(-12)=-11+12=1$

21. $87\div(-29)=-3$

22. $-5(-6)+4(-3)=30+(-12)=18$

23. $\frac{-24}{-\frac{3}{4}}=\left(\frac{-24}{1}\right)\left(-\frac{4}{3}\right)=32$

24. $\left(-\frac{1}{2}\right)\div\frac{3}{4}=\left(-\frac{1}{2}\right)\left(\frac{2\cdot 2}{3}\right)=-\frac{2}{3}$

25. $\left(\frac{5}{7}\right)\div\left(-\frac{5}{25}\right)=\frac{5}{7}\cdot\left(-\frac{25}{5}\right)$
$=-\frac{25}{7}$ or $-3\frac{4}{7}$

26. $(-6)(3)(4)=(-18)(4)$
$=-72$

27. $(-1)(-2)(-3)(-5)=(1)(2)(3)(5)=30$

28. $(-5)\left(-\frac{1}{2}\right)(4)(-3)=\left(\frac{5}{2}\right)(-12)$
$=-30$

29. $-\frac{4}{3}+\frac{2}{3}+\frac{1}{6}=-\frac{2}{3}+\frac{1}{6}=-\frac{4}{6}+\frac{1}{6}$
$=-\frac{3}{6}=-\frac{1}{2}$

30. $-\frac{6}{7}+\frac{1}{2}+\left(-\frac{3}{14}\right)=-\frac{12}{14}+\frac{7}{14}+\left(-\frac{3}{14}\right)$
$=-\frac{5}{14}+\left(-\frac{3}{14}\right)$
$=-\frac{8}{14}$
$=-\frac{4}{7}$

31. $(-3)(-2)(-5)=-(3)(2)(5)=-30$

32. $-6+(-2)-(-3)=-6+(-2)+3$
$=-8+3$
$=-5$

33. $3.5(-2.6)=-9.1$

34. $(-5.4)\div 4(-6)=0.9$

35. $5-(-3.5)+1.6=5+3.5+1.6=10.1$

36. $-8+2-(-4.8)=-6+4.8$
$=-1.2$

37. $17+3.4+(-16)+(-2.5)=20.4+(-18.5)$
$=1.9$

38. $37+(-44)+12.5+(-6.8)$
$=49.5+(-50.8)$
$=-1.3$

39. $(-8)(3)=-24$; 24 yards lost

40. $-34+12=-22$
-22° F

41. $6895-(-468)=6895+468=7363$ feet

42. $+1\frac{1}{2}-3\frac{1}{4}+2-2\frac{1}{2}=3\frac{1}{2}+\left(-5\frac{3}{4}\right)$
$=-2\frac{1}{4}$
$2\frac{1}{4}$ point loss

43. $(-3)^4=(-3)(-3)(-3)(-3)=81$

44. $(-2)^7=-128$

45. $(-5)^4=-625$

46. $\left(\frac{2}{3}\right)^3=\frac{8}{27}$

47. $-9^2=-81$

48. $(0.6)^2=0.36$

49. $\left(\frac{5}{6}\right)^2=\left(\frac{5}{6}\right)\left(\frac{5}{6}\right)=\frac{25}{36}$

50. $\left(\frac{3}{4}\right)^3=\left(\frac{3}{4}\right)\left(\frac{3}{4}\right)\left(\frac{3}{4}\right)$
$=\frac{27}{64}$

51. $(5)(-4)+(3)(-2)^3$
$=(5)(-4)+(3)(-8)$
$=-20+(-24)$
$=-44$

52. $20-(-10)-(-6)+(-5)-1$
$=20+10+6+(-5)-1$
$=36-6$
$=30$

53. $(7-9)^3+(-6)(-2)+(-3)$
$=(-2)^3+(-6)(-2)+(-3)$
$=-8+(-6)(-2)+(-3)$
$=-8+(12)+(-3)=1$

54. $5(3x-7y)=5(3x)+5(-7y)=15x-35y$

55. $2x(3x-7y+4)$
$=2x(3x)+2x(-7y)+2x(4)$
$=6x^2-14xy+8x$

56. $-(7x^2-3x+11)$
$=-1(7x^2)+(-1)(-3x)+(-1)(11)$
$=-7x^2+3x-11$

57. $(2xy+x-y)(-3y)$
$=(2xy)(-3y)+x(-3y)-y(-3y)$
$=-6xy^2-3xy+3y^2$

58. $3a^2b-2bc+6bc^2-8a^2b-6bc^2+5bc$
$=(3-8)a^2+(-2+5)bc+(6-6)bc^2$
$=-5a^2b+3bc$

59. $9x+11y-12x-15y=-3x-4y$

60. $4x^2-13x+7-9x^2-22x-16$
$=(4-9)x^2+(-13-22)x+7-16$
$=-5x^2-35x-9$

61. $-x+\frac{1}{2}+14x^2-7x-1-4x^2$
$=(14-4)x^2+(-7-1)x+\frac{1}{2}-1$
$=10x^2-8x-\frac{1}{2}$

62. If $x=-7$, then $7x-6=7(-7)-6$
$=-49-6=-55$

63. If $x=8$, then $7-\frac{3}{4}x=7-\frac{3}{4}(8)$
$=7-6=1$

64. $x^2+3x-4=(-3)^2+3(-3)-4$
$=9-9-4$
$=-4$

65. If $x=2$, then
$-3x^2-4x+5=-3(2)^2-4(2)+5$
$=-3(4)-4(2)+5$
$=-12-8+5=-15$

66. $-3x^3-4x^2+2x+6$
$=-3(-2)^3-4(-2)^2+2(-2)+6$
$=-3(-8)-4(4)-4+6$
$=24-16-4+6$
$=30-20$
$=10$

67. If $v=24$, $t=2$, and $a=32$, then
$vt-\frac{1}{2}at^2=24(2)-\frac{1}{2}(32)(2)^2$
$=48-16(4)=48-64=-16$

68. $\frac{nRT}{V} = \frac{16(-2)(4)}{-20}$
$= \frac{-32}{-5} = \frac{32}{5}$

69. If $p = 6000$, $r = 18\%$, and $t = \frac{3}{4}$, then
$I = prt = 6000(0.18)\left(\frac{3}{4}\right) = \$810$

70. $F = \frac{9C + 160}{5} = \frac{9(30) + 160}{5}$
$= \frac{270 + 160}{5} = \frac{430}{5} = 86$
$86^\circ F$

71. If $r = 15$, $A = \pi r^2 = 3.14(15)^2$
$= 3.14(225) = 706.5$ square meters
Cost $= (706.5 \text{ sq m})(\$3/\text{sq m})$
$= \$2119.50$

72. $P = 180S - R - C$
$= 180(56) - 300 - 1200$
$= 10{,}080 - 300 - 1200$
$= 8580$
$8580

73. $A = \frac{1}{2}a(b_1 + b_2)$,
$a = 200$, $b_1 = 700$, $b_2 = 300$
$A = \frac{1}{2}(200)(700 + 300)$
$= 100(1000)$
$= 100{,}000$ square feet
Cost $= 2(100{,}000) = \$200{,}000$

74. $A = \frac{1}{2}ab$, $a = 3.8$, $b = 5.5$
$A = \frac{1}{2}(3.8)(5.5) = 10.45$ square feet
Cost $= 66(10.45) = \$689.70$

75. $5x - 7(x - 6) = 5x - 7x + 42$
$= -2x + 42$

76. $3(x - 2) - 4(5x + 3)$
$= 3x - 6 - 20x - 12$
$= -17x - 18$

77. $2[3 - (4 - 5x)] = 2(3 - 4 + 5x)$
$= 2(-1 + 5x) = -2 + 10x$

78. $-3x[x + 3(x - 7)] = -3x(x + 3x - 21)$
$= -3x(4x - 21)$
$= -12x^2 + 63x$

79. $2xy^3 - 6x^3y - 4x^2y^2 + 3(xy^3 - 2x^2y - 3x^2y^2)$
$= 2xy^3 - 6x^3y - 4x^2y^2 + 3xy^3 - 6x^2y - 9x^2y^2$
$= (2 + 3)xy^3 - 6x^3y + (-4 - 9)x^2y^2 - 6x^2y$
$= 5xy^3 - 6x^3y - 13x^2y^2 - 6x^2y$

80. $-5(x + 2y - 7) + 3x(2 - 5y)$
$= -5x - 10y + 35 + 6x - 15xy$
$= x - 10y + 35 - 15xy$

81. $2\{x-3(y-2)+4[x-2(y+3)]\}$
$=2[x-3y+6+4(x-2y-6)]$
$=2(x-3y+6+4x-8y-24)$
$=2(5x-11y-18)$
$=10x-22y-36$

82. $-5\{2a-b[5a-b(3+2a)]\}$
$=-5[2a-b(5a-3b-2ab)]$
$=-5(2a-5ab+3b^2+2ab^2)$
$=-10a+25ab-15b^2-10ab^2$

83. $-3\{2x-[x-3y(x-2y)]\}$
$=-3[2x-(x-3xy+6y^2)]$
$=-3(2x-x+3xy-6y^2)$
$=-3(x+3xy-6y^2)$
$=-3x-9xy+18y^2$

84. $2\{3x+2[x+2y(x-4)]\}$
$=2[3x+2(x+2xy-8y)]$
$=2(3x+2x+4xy-16y)$
$=2(5x+4xy-16y)$
$=10x+8xy-32y$

Chapter 1 Test

1. $-3+(-4)+9+2=-7+9+2$
$=2+2=4$

2. $-0.6-(-0.8)=-0.6+(0.8)=0.2$

3. $(-8)(-12)=96$

4. $(-5)(-2)(7)(-1)=-(10)(7)(1)$
$=-(70)(1)=-70$

5. $(-12)\div(-3)=4$

6. $(-1.8)\div(0.6)=-3$

7. $(-4)^3=(-4)(-4)(-4)=-64$

8. $(1.3)^2=1.69$

9. $\left(\frac{2}{3}\right)^4=\left(\frac{2}{3}\right)\left(\frac{2}{3}\right)\left(\frac{2}{3}\right)\left(\frac{2}{3}\right)$
$=\frac{16}{81}$

10. $7-(6-9)^2+5(2)=7-(-3)^2+10$
$=7-(9)+10$
$=-2+10$
$=8$

11. $3(4-6)^3+12\div(-4)+2$
$=3(-2)^3+12\div(-4)+2$
$=3(-8)+12\div(-4)+2$
$=-24-3+2$
$=-25$

12. $-5x(x+2y-7)$
$=-5x(x)-5x(2y)-5x(-7)$
$=-5x^2-10xy+35x$

13. $-2ab^2(-3a-2b+7ab)$
$=-2ab^2(-3a)-2ab^2(-2b)-2ab^2(7ab)$
$=6a^2b^2+4ab^3-14a^2b^3$

14. $6ab - \frac{1}{2}a^2b + \frac{3}{2}ab + \frac{5}{2}a^2b$
$= \left(6 + \frac{3}{2}\right)ab + \left(-\frac{1}{2} + \frac{5}{2}\right)a^2b$
$= \left(\frac{12}{2} + \frac{3}{2}\right)ab + \frac{4}{2}a^2b$
$= \frac{15}{2}ab + 2a^2b$

15. $12a(a+b) - 4(a^2 - 2ab)$
$= 12a^2 + 12ab - 4a^2 + 8ab$
$= (12-4)a^2 + (12+8)ab$
$= 8a^2 + 20ab$

16. $3(2-a) - 4(-6-2a) = 6 - 3a + 24 + 8a$
$= 5a + 30$

17. $5(3x - 2y) - (x + 6y)$
$= 15x - 10y - x - 6y$
$= (15-1)x + (-10-6)y$
$= 14x - 16y$

18. $x^3 - 3x^2y + 2y - 5$
$= 3^3 - 3(3)^2(-4) + 2(-4) - 5$
$= 27 - 3(9)(-4) - 8 - 5$
$= 27 + 108 - 8 - 5$
$= 122$

19. If $x = -3$, then $3x^2 - 7x - 11$
$= 3(-3)^2 - 7(-3) - 11$
$= 37$

20. $2a - 3b$
$= 2(-4) - 3(-3)$
$= -8 + 9$
$= 1$

21. (60 miles/hr)(1.61 km/mile)
= 96.6 kilometers per hour

22. $A = \frac{1}{2}a(b_1 + b_2)$,
$a = 120,\ b_1 = 200,\ b_2 = 180$
$A = \frac{1}{2}(120)(200 + 180) = 60(380)$
$= 22,800$ square feet

23. $A = \frac{1}{2}ab,\ a = 6.8,\ b = 8.5$
$A = \frac{1}{2}(6.8)(8.5) = 28.9$ square feet
Cost $= 0.80(28.9) = \$23.12$

24. $A = \pi r^2,\ r = 12$
$A \approx 3.14(12)^2 = 452.16$ square feet

25. $-3\{a + b[3a - b(1-a)]\}$
$= -3[a + b(3a - b + ab)]$
$= -3(a + 3ab - b^2 + ab^2)$
$= -3a - 9ab + 3b^2 - 3ab^2$

26. $3\{[x - (5 - 2y)] - 4[3 + (6x - 7y)]\}$
$= 3[(x - 5 + 2y) - 4(3 + 6x - 7y)]$
$= 3(x - 5 + 2y - 12 - 24x + 28y)$
$= 3(-23x + 30y - 17)$
$= -69x + 90y - 51$

Chapter 2

Pretest Chapter 2

1. $$\begin{aligned} x-30&=-46\\ x-30+30&=-46+30\\ x&=-16 \end{aligned}$$

2. $$\begin{aligned} \frac{1}{4}x&=12\\ 4\left(\frac{1}{4}x\right)&=4(12)\\ x&=48 \end{aligned}$$

3. $$\begin{aligned} -5x&=20\\ \frac{-5x}{-5}&=\frac{20}{-5}\\ x&=-4 \end{aligned}$$

4. $$\begin{aligned} 7x+2&=30\\ 7x&=28\\ x&=4 \end{aligned}$$

5. $$\begin{aligned} 3x-8&=7x+6\\ 3x+(-3x)-8&=7x+(-3x)+6\\ -8&=4x+6\\ -8+(-6)&=4x+6+(-6)\\ -14&=4x\\ \frac{-14}{4}&=\frac{4x}{4}\\ -\frac{7}{2}&=x \end{aligned}$$

6. $$\begin{aligned} 5x+3-6x&=4-8x-2\\ -x+3&=-8x+2\\ 7x+3&=2\\ 7x&=-1\\ x&=-\frac{1}{7} \end{aligned}$$

7. $$\begin{aligned} 5.2x+0.9&=2.8x+3.3\\ 5.2x+(-2.8x)+0.9&=2.8x+(-2.8x)+3.3\\ 2.4x+0.9&=3.3\\ 2.4x+0.9+(-0.9)&=3.3+(-0.9)\\ 2.4x&=2.4\\ \frac{2.4x}{2.4}&=\frac{2.4}{2.4}\\ x&=1 \end{aligned}$$

8. $$\begin{aligned} 2(x-4)-2&=3(x-2)\\ 2x-8-2&=3x-6\\ 2x-10&=3x-6\\ 2x+(-2x)-10&=3x+(-2x)-6\\ -10&=x-6\\ -10+6&=x-6+6\\ -4&=x \end{aligned}$$

9. $$\begin{aligned} \frac{2}{5}x+\frac{1}{4}&=\frac{1}{2}x\\ 20\left(\frac{2}{5}x\right)+20\left(\frac{1}{4}\right)&=20\left(\frac{1}{2}x\right)\\ 8x+5&=10x\\ 8x+(-8x)+5&=10x+(-8x)\\ 5&=2x\\ \frac{5}{2}&=x \end{aligned}$$

10. $$\frac{2}{3}(x-2)+\frac{1}{2}=5-(3+x)$$
$$6\left[\frac{2}{3}(x-2)+\frac{1}{2}\right]=6\left[5-(3+x)\right]$$
$$4(x-2)+3=6(2-x)$$
$$4x-5=12-6x$$
$$10x-5=12$$
$$10x=17$$
$$x=\frac{17}{10}$$

11. $$x-\frac{3}{2}=-\frac{4}{3}-\frac{5}{6}$$
$$6x-6\left(\frac{3}{2}\right)=-6\left(\frac{4}{3}\right)-6\left(\frac{5}{6}\right)$$
$$6x-9=-8-5$$
$$6x-9=-13$$
$$6x-9+9=-13+9$$
$$6x=-4$$
$$\frac{6x}{6}=-\frac{4}{6}$$
$$x=-\frac{2}{3}$$

12. $$x-2=\frac{5}{7}(x+4)$$
$$7x-7(2)=7\left(\frac{5}{7}\right)(x+4)$$
$$7x-14=5(x+4)$$
$$7x-14=5x+20$$
$$7x+(-5x)-14=5x+(-5x)+20$$
$$2x-14=20$$
$$2x-14+14=20+14$$
$$2x=34$$
$$\frac{2x}{2}=\frac{34}{2}$$
$$x=17$$

13. (a) $C=\frac{5}{9}(F-32)$
$$\frac{9}{5}C=\frac{9}{5}\left(\frac{5}{9}\right)(F-32)$$
$$\frac{9}{5}C=F-32$$
$$\frac{9}{5}C+32=F-32+32$$
$$\frac{9}{5}C+32=F$$

(b) $C=-15^0$
$$\frac{9}{5}(-15)+32=F$$
$$-27+32=F$$
$$5^0=F$$

14. a. $I=Prt$
$$\frac{I}{Pt}=\frac{Prt}{Pt}$$
$$\frac{I}{Pt}=r$$

b. $r=\frac{240}{(2000)(2)}=0.06$

6%

15. $1<10$

16. $2>-3$

17. $-12>-13$

18. $0<0.9$

19. $$-2x+5\le 4-x+3$$
$$-2x+5\le -x+7$$
$$-2x+x+5\le -x+x+7$$
$$-x+5\le 7$$
$$-x+5+(-5)\le 7+(-5)$$
$$-x\le 2$$
$$x\ge -2$$

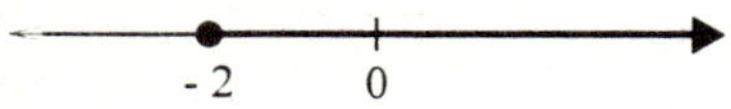

20. $$\frac{1}{2}(x+2)-1>\frac{x}{3}-5+x$$
$$6\left[\frac{1}{2}(x+2)-1\right]>6\left(\frac{x}{3}-5+x\right)$$
$$3(x+2)-6>2x-30+6x$$
$$3x+6-6>8x-30$$
$$3x>8x-30$$
$$-5x>-30$$
$$x<6$$

21. Let $x=$ second side

$3x-3=$ 1st side

$3x-8=$ 3rd side

$x+3x-3+3x-8=38$

$7x-11=38$

$7x=49$

$x=7$

$3x-3=18$

$3x-8=13$

1st side $=18$ feet

2nd side $=7$ feet

3rd side $=13$ feet

22. Let $x=$ weight of 1st package

$x-3\frac{1}{2}=$ weight of 2nd package

$x-2=$ weight of 3rd package

$x+x-3\frac{1}{2}+x-2=17$

$3x-5\frac{1}{2}=17$

$3x=22\frac{1}{2}$

$3x=\frac{45}{2}$

$x=\frac{45}{6}=7\frac{3}{6}=7\frac{1}{2}$

$x-3\frac{1}{2}=7\frac{1}{2}-3\frac{1}{2}=4$

$x-2=7\frac{1}{2}-2=5\frac{1}{2}$

1st package $=7\frac{1}{2}$ lb; 2nd package $=4$ lb

3rd package $=5\frac{1}{2}$ lb

23. Let $x=$ width

$2x+11=$ length

$P=2L+2W$

$88=2(2x+11)+2x$

$88=4x+22+2x$

$88=6x+22$

$66=6x$

$x=11$ feet: width

$2x+11=33$ feet: length

24. Let x = amount at 12%

$1000 - x$ = amount at 9%

$0.12x + 0.09(1000 - x) = 102$

$100[0.12x + 0.09(1000 - x)] = 100(102)$

$12 + 9(1000 - x) = 10,200$

$12x + 9000 - 9x = 10,200$

$3x + 9000 = 10,200$

$3x = 1200$

$x = 400$

$1000 - x = 1000 - 400 = 600$

\$400 at 12% and \$600 at 9%

25. Let x = original population

$0.11x$ = population growth

$x + 0.11x = 24,420$

$1.11x = 24,420$

$x = 22,000$

The population was 22,000

26. Let x = number of nickels

$x + 4$ = number of dimes

$(x+4)+3 = x+7$ = number of quarters

$0.05x + 0.10(x+4) + 0.25(x+7) = 3.35$

$100[0.05x + 0.10(x+4) + 0.25(x+7)]$

$= 100(3.35)$

$5x + 10(x+4) + 25(x+7) = 335$

$5x + 10x + 40 + 25x + 175 = 335$

$40x + 215 = 335$

$40x = 120$

$x = 3$

$x + 4 = 3 + 4 = 7$

$x + 7 = 3 + 7 = 10$

3 nickels, 7 dimes, 10 quarters

2.1 Exercises

1. When we use the equal sign, we indicate two expressions are equal in value.

3. The solution of an equation is a value of the variable that makes the equation true.

5. Answers may vary. A sample answer is "to isolate the variable".

7. $x + 12 = 18$

$x + 12 + (-12) = 18 + (-12)$

$x = 6$

Check: $6 + 12 \stackrel{?}{=} 18$

$18 = 18$

9. $15 = x + 9$

$15 + (-9) = x + 9 + (-9)$

$6 = x$

Check: $15 \stackrel{?}{=} 6 + 9$

$15 = 15$

11. $x - 11 = 5$

$x - 11 + 11 = 5 + 11$

$x = 16$

Check: $16 - 11 \stackrel{?}{=} 5$

$5 = 5$

13. $0 = x - 7$
$0 + 7 = x - 7 + 7$
$7 = x$

Check: $0 \stackrel{?}{=} 7 - 7$
$0 = 0$

15. $8 - 2 = x + 5$
$6 = x + 5$
$6 + (-5) = x + 5 + (-5)$
$1 = x$

Check: $8 - 2 \stackrel{?}{=} 1 + 5$
$6 = 6$

17. $18 - 2 + 3 = x + 19$
$19 = x + 19$
$19 + (-19) = x + 19 + (-19)$
$0 = x$

Check $18 - 2 + 3 \stackrel{?}{=} 0 + 19$
$19 = 19$

19. $x - 11 = -13$
$x - 11 + 11 = -13 + 11$
$x = -2$

Check: $-2 - 11 \stackrel{?}{=} -13$
$-13 = -13$

21. $-18 + x = 48$
$-18 + 18 + x = 48 + 18$
$x = 66$

Check: $-18 + 66 \stackrel{?}{=} 48$
$48 = 48$

23. $23 - 8 = x - 12$
$15 = x - 12$
$15 + 12 = x - 12 + 12$
$27 = x$

Check: $23 - 8 \stackrel{?}{=} 27 - 12$
$15 = 15$

25. $3 - 17 + 8 = 8 + x - 3$
$-6 = 5 + x$
$-6 + (-5) = 5 + (-5) + x$
$-11 = x$

Check: $3 - 17 + 8 \stackrel{?}{=} 8 + (-11) - 3$
$-6 = -6$

27. $-19 + x = 7,\ x \stackrel{?}{=} 12$
$-19 + 12 \stackrel{?}{=} 7$
$-7 = 7$
$x = 12$ is not the solution.
$-19 + x = 7$
$-19 + 19 + x = 7 + 19$
$x = 26$

29. $-16 + 5 \stackrel{?}{=} -5 - 6$
$-11 = -11$
$x = -5$ is the solution.

31. $-39 = x - 47, \quad x \stackrel{?}{=} -8$

$-39 \stackrel{?}{=} -8 - 47$

$-39 = -55$

$x = -8$ is not the solution.

$-39 = x - 47$

$-39 + 47 = x - 47 + 47$

$8 = x$

33. $x + 8 = 12 - 19 + 3, \quad x \stackrel{?}{=} -12$

$-12 + 8 \stackrel{?}{=} 12 - 19 + 3$

$-4 = -4$

$x = -4$ is the solution.

35. $-11 + x = -7$

$-11 + 11 + x = -7 + 11$

$x = 4$

37. $27 = 5 + x$

$27 + (-5) = 5 + (-5) + x$

$22 = x$

39. $3.6 + 1.2 = x + 1.3$

$4.8 = x + 1.3$

$4.8 + (-1.3) = x + 1.3 + (-1.3)$

$3.5 = x$

41. $4.2 + x = 1.3$

$4.2 + (-4.2) + x = 1.3 + (-4.2)$

$x = -2.9$

43. $x + \frac{1}{3} = \frac{2}{3}$

$x + \frac{1}{3} + \left(-\frac{1}{3}\right) = \frac{2}{3} + \left(-\frac{1}{3}\right)$

$x = \frac{1}{3}$

45. $\frac{2}{5} + x = \frac{1}{2} - \frac{3}{10}$

$\frac{4}{10} + x = \frac{5}{10} - \frac{3}{10}$

$\frac{4}{10} + x = \frac{2}{10}$

$\frac{4}{10} + \left(-\frac{4}{10}\right) + x = \frac{2}{10} + \left(-\frac{4}{10}\right)$

$x = -\frac{2}{10}$

$x = -\frac{1}{5}$

47. $\frac{7}{12} - \frac{2}{3} = x - \frac{5}{4}$

$\frac{7}{12} - \frac{8}{12} = x - \frac{15}{12}$

$-\frac{1}{12} = x - \frac{15}{12}$

$-\frac{1}{12} + \frac{15}{12} = x - \frac{15}{12} + \frac{15}{12}$

$\frac{14}{12} = x$

$\frac{7}{6} = x$

$1\frac{1}{6} = x$

49. $$7\frac{1}{8} = -20 + x$$
$$7\frac{1}{8} + 20 = -20 + 20 + x$$
$$27\frac{1}{8} = x$$

51. $$1.6 - 5x - 3.2 = -2x + 5.6 + 4x - 8x$$
$$-5x - 1.6 = -6x + 5.6$$
$$-5x + 6x - 1.6 = -6x + 6x + 5.6$$
$$x - 1.6 = 5.6$$
$$x - 1.6 + 1.6 = 5.6 + 1.6$$
$$x = 7.2$$

53. $$x - 0.2314 = -4.0144$$
$$x - 0.2314 + 0.2314 = -4.0144 + 0.2314$$
$$x = -3.783$$

Cumulative Review Problems

55. $$y^2 + y - 12 - 3y^2 - 5y + 16$$
$$= (1-3)y^2 + (1-5)y - 12 + 16$$
$$= 2y^2 - 4y + 4$$

57. $$x - 49.99 = -35.07$$
$$x - 49.99 + 49.99 = -35.07 + 49.99$$
$$x = \$14.92 \text{ initial balance}$$

2.2 Exercises

1. Divide each side of $6x = -24$ by $\underline{6}$.

3. Multiply each side of $\frac{1}{7}x = -2$ by $\underline{7}$.

5. $$\frac{1}{7}x = 5$$
$$7\left(\frac{1}{7}x\right) = 7(5)$$
$$x = 35$$
Check: $\frac{1}{7}(35) \stackrel{?}{=} 5$
$$5 = 5$$

7. $$\frac{1}{3}x = -9$$
$$3\left(\frac{1}{3}x\right) = 3(-9)$$
$$x = -27$$
Check: $\frac{1}{3}(-27) \stackrel{?}{=} -9$
$$-9 = -9$$

9. $$\frac{x}{5} = 16$$
$$5\left(\frac{x}{5}\right) = 5(16)$$
$$x = 80$$
Check: $\frac{80}{5} \stackrel{?}{=} 16$
$$16 = 16$$

11. $$-3 = \frac{x}{5}$$
$$5(-3) = 5\left(\frac{x}{5}\right)$$
$$-15 = x$$
Check: $-3 \stackrel{?}{=} \frac{-15}{5}$
$$-3 = -3$$

13. $12x = 48$

$\frac{12x}{12} = \frac{48}{12}$

$x = 4$

Check: $12(4) \stackrel{?}{=} 48$

$48 = 48$

15. $-16 = 6x$

$\frac{-16}{6} = \frac{6x}{6}$

$-\frac{8}{3} = x$

Check: $-16 \stackrel{?}{=} 6\left(-\frac{8}{3}\right)$

$-16 = -16$

17. $1.5x = 75$

$\frac{1.5x}{1.5} = \frac{75}{1.5}$

$x = 50$

Check: $1.5(50) \stackrel{?}{=} 75$

$75 = 75$

19. $-15 = -x$

$\frac{-15}{-1} = \frac{-x}{-1}$

$15 = x$

Check: $-15 \stackrel{?}{=} (-1)(15)$

$-15 = -15$

21. $-84 = 12x$

$\frac{-84}{12} = \frac{12x}{12}$

$-7 = x$

Check: $-84 \stackrel{?}{=} 12(-7)$

$-84 = -84$

23. $0.4x = 0.08$

$\frac{0.4x}{0.4} = \frac{0.08}{0.4}$

$x = 0.2$ or $\frac{1}{5}$

Check: $(0.4)(0.2) \stackrel{?}{=} 0.08$

$0.08 = 0.08$

25. $-3x = 21,\ x \stackrel{?}{=} 7$

$-3(7) \stackrel{?}{=} 21$

$-21 \neq 21$

$x = 7$ is not the solution.

$-3x = 21$

$\frac{-3x}{-3} = \frac{21}{-3}$

$x = -7$

27. $-11x = 66,\ x \stackrel{?}{=} -6$

$-11(-6) \stackrel{?}{=} 66$

$66 = 66$

$x = -6$ is the solution.

29. $-3y = 2.4$

$\frac{-3y}{-3} = \frac{2.4}{-3}$

$y = -0.8$

31. $-56 = -21t$

$\frac{-56}{-21} = \frac{-21t}{-21}$

$\frac{8}{3} = t$

Check: $-56 \stackrel{?}{=} -21\left(\frac{8}{3}\right)$

$-56 = -56$

33. $4.6y = -3.22$

$\frac{4.6y}{4.6} = \frac{-3.22}{4.6}$

$y = -0.7$

35. $4x + 3x = 21$

$7x = 21$

$\frac{7x}{7} = \frac{21}{7}$

$x = 3$

37. $2x - 7x = 20$

$-5x = 20$

$\frac{-5x}{-5} = \frac{20}{-5}$

$x = -4$

39. $-6x - 3x = -7$

$-9x = -7$

$\frac{-9x}{-9} = \frac{-7}{-9}$

$x = \frac{7}{9}$

41. $-\frac{2}{3} = -\frac{4}{7}x$

$-\frac{7}{4}\left(-\frac{2}{3}\right) = -\frac{7}{4}\left(-\frac{4}{7}x\right)$

$\frac{7}{6} = x$

43. $3.6172x = -19.026472$

$\frac{3.6172x}{3.6172} = \frac{-19.026472}{3.6172}$

$x = -5.26$

45. To solve an equation, we are performing steps to get an equivalent equation that has the same solution. Now $a = b$ and $a(0) = b(0)$ are not equivalent equations because they do not have the same solution. So we must have the requirement that when we multiply both sides of the equation by c, it is essential that c is nonzero.

Cumulative Review Problems

47. $(-6)(-8) + (-3)(2) = 48 - 6 = 42$

49. $5 + (2-6)^2 = 5 + (-4)^2$

$= 5 + 16$

$= 21$

51. $600 + 82 - 47 + 103 - 106$

$= 785 - 153$

$= \$632$

2.3 Exercises

1. $4x+13=21$

   $4x+13+(-13)=21+(-13)$

   $4x=8$

   $\frac{4x}{4}=\frac{8}{4}$

   $x=2$

   Check: $4(2)+13\stackrel{?}{=}21$

   $21=21$

3. $4x-11=13$

   $4x-11+11=13+11$

   $4x=24$

   $\frac{4x}{4}=-\frac{24}{4}$

   $x=6$

   Check: $4(6)-11\stackrel{?}{=}13$

   $13=13$

5. $4x-21=-91$

   $4x-21+21=-91+21$

   $4x=-70$

   $\frac{4x}{4}=-\frac{-70}{4}$

   $x=-17.5$

   Check: $4(-17.5)-21\stackrel{?}{=}-91$

   $-91=-91$

7. $-4x+17=-35$

   $-4x+17+(-17)=-35+(-17)$

   $-4x=-52$

   $\frac{-4x}{-4}=\frac{-52}{-4}$

   $x=13$

   Check: $-4(13)+17\stackrel{?}{=}-35$

   $-52+17\stackrel{?}{=}-35$

   $-35=-35$

9. $-15+2x=15$

   $-15+15+2x=15+15$

   $2x=30$

   $\frac{2x}{2}=\frac{30}{2}$

   $x=15$

   Check: $-15+2(15)\stackrel{?}{=}15$

   $-15+30\stackrel{?}{=}15$

   $15=15$

11. $\frac{1}{5}x-2=6$

    $\frac{1}{5}x-2+2=6+2$

    $\frac{1}{5}x=8$

    $5\left(\frac{1}{5}x\right)=5(8)$

    $x=40$

    Check: $\frac{1}{5}(40)-2\stackrel{?}{=}6$

    $6=6$

13. $\frac{1}{6}x+2=-4$

$\frac{1}{6}x+2+(-2)=-4+(-2)$

$\frac{1}{6}x=-6$

$6\left(\frac{1}{6}x\right)=6(-6)$

$x=-36$

Check: $\frac{1}{6}(-36)+2\stackrel{?}{=}-4$

$-6+2\stackrel{?}{=}-4$

$-4=-4$

15. $8x=48+2x$

$8x+(-2x)=48+2x+(-2x)$

$6x=48$

$\frac{6x}{6}=\frac{48}{6}$

$x=8$

Check: $8(8)\stackrel{?}{=}48+2(8)$

$64\stackrel{?}{=}48+16$

$64=64$

17. $-6x=-27+3x$

$-6x+(-3x)=-27+3x+(-3x)$

$-9x=-27$

$\frac{-9x}{-9}=\frac{-27}{-9}$

$x=3$

Check: $-6(3)\stackrel{?}{=}-27+3(3)$

$-18\stackrel{?}{=}-27+9$

$-18=-18$

19. $63-x=8x$

$63-x+x=8x+x$

$63=9x$

$\frac{63}{9}=\frac{9x}{9}$

$7=x$

Check: $63-7\stackrel{?}{=}8(7)$

$56=56$

21. $54-2x=-8x$

$54-2x+2x=-8x+2x$

$54=-6x$

$\frac{54}{-6}=\frac{-6x}{-6}$

$-9=x$

Check: $54-2(-9)\stackrel{?}{=}-8(-9)$

$54+18\stackrel{?}{=}72$

$72=72$

23. $5y+2=6y-6+y$

$5(4)+2\stackrel{?}{=}6(4)-6+4$

$20+2\stackrel{?}{=}24-2$

$22=22$

$y=4$ is the solution.

25. $9x+2-5x=-8+5x-2$

$9(-12)+2-5(-12)\stackrel{?}{=}-8+5(-12)-2$

$-108+2+60\stackrel{?}{=}-8-60-2$

$-46\neq-70$

$x=-12$ is not a solution.

$9x+2-5x=-8+5x-2$

$4x+2=5x-10$

$4x+(-4x)+2=5x+(-4x)-10$

$2=x-10$

$2+10=x-10+10$

$12=x$

27. $7y+21-5y=5y-7+y$

Left:

$2y+21=6y-7$

$2y+(-6y)+21=6y+(-6y)-7$

$-4y+21=-7$

$-4y+21+(-21)=-7+(-21)$

$-4y=-28$

$\frac{-4y}{-4}=\frac{-28}{-4}$

$y=7$

Right:

$2y+21=6y-7$

$2y+(-2y)+21=6y+(-2y)-7$

$21=4y-7$

$21+7=4y-7+7$

$28=4y$

$\frac{28}{4}=\frac{4y}{4}$

$7=y$

Neither approach is better.

29. $8-3x=7x+8$

$$8-3x+3x=7x+3x+8$$
$$8=10x+8$$
$$8-8=10x+8-8$$
$$0=10x$$
$$\frac{0}{10}=\frac{10x}{10}$$
$$0=x$$

31. $2x+5=4x-5$

$$2x+(-4x)+5=4x+(-4x)-5$$
$$-2x+5=-5$$
$$-2x+5+(-5)=-5+(-5)$$
$$-2x=-10$$
$$\frac{-2x}{-2}=\frac{-10}{-2}$$
$$x=5$$

33. $11y-8=9y-16$

$$11y-8+8=9y-16+8$$
$$11y=9y-8$$
$$11y-9y=9y-9y-8$$
$$2y=-8$$
$$\frac{2y}{2}=\frac{-8}{2}$$
$$y=-4$$

35. $9x-5+4x=7x+43-2x$

$$13x-5=5x+43$$
$$13x+(-5x)-5=5x+(-5x)+43$$
$$8x-5=43$$
$$8x-5+5=43+5$$
$$8x=48$$
$$\frac{8x}{8}=\frac{48}{8}$$
$$x=6$$

37. $6(x+2)=42$

$6x+12=42$

$6x+12-12=42-12$

$6x=30$

$\frac{6x}{6}=\frac{30}{6}$

$x=5$

Check: $6(5+2) \stackrel{?}{=} 42$

$6(7) \stackrel{?}{=} 42$

$42 = 42$

39. $4(2x+1) - 7 = 6 - 5$

$8x + 4 - 7 = 6 - 5$

$8x - 3 = 1$

$8x = 4$

$x = \frac{1}{2}$

Check: $4\left[2\left(\frac{1}{2}\right)+1\right] - 7 \stackrel{?}{=} 6 - 5$

$4(1+1) - 7 \stackrel{?}{=} 1$

$1 = 1$

41. $6(3-4x) + 17 = 8x - 3(2-3x)$

$18 - 24x + 17 = 8x - 6 + 9x$

$-24x + 35 = 17x - 6$

$41 = 41x$

$1 = x$

Check: $6[3-4(1)] + 17$

$\stackrel{?}{=} 8(1) - 3[2-3(1)]$

$6(3-4) + 17 \stackrel{?}{=} 8 - 3(2-3)$

$-6 + 17 \stackrel{?}{=} 8 + 3$

$11 = 11$

43. $3(x+0.2) = 2(x-0.3) + 4.3$

$3x + 0.6 = 2x - 0.6 + 4.3$

$3x + 0.6 = 2x + 3.7$

$3x - 2x + 0.6 = 2x - 2x + 3.7$

$x + 0.6 = 3.7$

$x + 0.6 - 0.6 = 3.7 - 0.6$

$x = 3.1$

Check: $3(3.1+0.2) \stackrel{?}{=} 2(3.1-0.3) + 4.3$

$3(3.3) \stackrel{?}{=} 2(2.8) + 4.3$

$9.9 = 9.9$

45. $3(x-2) + 2 = 2(x-4)$

$3x - 6 + 2 = 2x - 8$

$3x - 4 = 2x - 8$

$3x - 4 + 4 = 2x - 8 + 4$

$3x = 2x - 4$

$3x - 2 = 2x - 2x - 4$

$x = -4$

Check: $3(-4-2)+2 \stackrel{?}{=} 2(-4-4)$

$3(-6) + 2 \stackrel{?}{=} 2(-8)$

$-18 + 2 \stackrel{?}{=} -16$

$-16 = -16$

47. $3(x+0.2) - (2x+0.5) = 2(x+0.3) - 0.5$

$3x + 0.6 - 2x - 0.5 = 2x + 0.6 - 0.5$

$x + 0.1 = 2x + 0.1$

$x + 0.1 - 0.1 = 2x + 0.1 - 0.1$

$x = 2x$

$x - x = 2x - x$

$0 = x$

Check: $3(0+0.2)-[2(0)+0.5]$

$\stackrel{?}{=}2(0+0.3)-0.5$

$3(0.2)-(0+0.5)\stackrel{?}{=}2(0.3)-0.5$

$0.6-0.5\stackrel{?}{=}0.6-0.5$

$0.1=0.1$

49. $2(4x-x)+6=2(2x+x)+8-x$

$2(3x)+6=2(3x)+8-x$

$6x+6=6x+8-x$

$6x+6=5x+8$

$6x+6-6=5x+8-6$

$6x=5x+2$

$-5x+6x=-5x+5x+2$

$x=2$

Check: $2[4(2)-2]+6$

$\stackrel{?}{=}2[2(2)+2]+8-2$

$2(8-2)+6\stackrel{?}{=}2(4+2)+6$

$2(6)+6\stackrel{?}{=}2(6)+6$

$12+6\stackrel{?}{=}12+6$

$18=18$

51. $4x-3.1=5.3-3x$

$4x+3x-3.1=5.3-3x+3x$

$7x-3.1=5.3$

$7x-3.1+3.1=5.3+3.1$

$7x=8.4$

$\frac{7x}{7}=\frac{8.4}{7}$

$x=1.2$

53. $8-7z+2z=20+5z$

$8-5z=20+5z$

$8-5z+5z=20+5z+5z$

$8=20+10z$

$-20+8=-20+20+10z$

$-12=10z$

$\frac{-12}{10}=\frac{10z}{10}$

$-1.2=z$

55. $-6w-7=-3-8w$

$8w-6w-7=-3-8w+8w$

$2w-7=-3$

$2w-7+7=-3+7$

$2w=4$

$\frac{2w}{2}=\frac{4}{2}$

$w=2$

57. $4-7x-13=8x-3-5x$

$-7x-9=3x-3$

$-7x-3x-9=3x-3x-3$

$-10x-9=-3$

$-10x-9+9=-3+9$

$-10x=6$

$\frac{-10x}{-10}=\frac{6}{-10}$

$x=-\frac{6}{10}$

$x=-\frac{3}{5}$

59. $3x^2+4x-7=3x^2-5x+2$

$-3x^2+3x^2+4x-7=-3x^2+3x^2-5x+2$

$4x-7=-5x+2$

$5x+4x-7=5x-5x+2$

$$9x-7=2$$
$$9x-7+7=2+7$$
$$9x=9$$
$$\frac{9x}{9}=\frac{9}{9}$$
$$x=1$$

61. $$2.8x-0.9=5.2x-3.3$$
$$2.8x-0.9+0.9=5.2x-3.3+0.9$$
$$2.8x=5.2x-2.4$$
$$-5.2x+2.8x=-5.2x+5.2x-2.4$$
$$-2.4x=-2.4$$
$$\frac{-2.4x}{-2.4}=\frac{-2.4}{-2.4}$$
$$x=1$$

63. $$x-0.8x+4=2.6$$
$$0.2x+4=2.6$$
$$0.2x+4-4=2.6-4$$
$$0.2x=-1.4$$
$$\frac{0.2x}{0.2}=\frac{-1.4}{0.2}$$
$$x=-7$$

65. $$3x+2-1.7x=0.6x+31.4$$
$$1.3x+2=0.6x+31.4$$
$$1.3x+2-2=0.6x+31.4-2$$
$$1.3x=0.6x+29.4$$
$$-0.6x+1.3x=-0.6x+0.6x+29.4$$
$$0.7x=29.4$$
$$\frac{0.7x}{0.7}=\frac{29.4}{0.7}$$
$$x=42$$

67. $$3(2z-4)-4(z+5)=5(z-4)$$
$$6z-12-4z-20=5z-20$$
$$2z-32=5z-20$$
$$2z-2z-32=5z-2z-20$$
$$-32=3z-20$$
$$-32+20=3z-20+20$$
$$-12=3z$$
$$\frac{-12}{3}=\frac{3z}{3}$$
$$-4=z$$

69. $$-2.21x+8.65=3.69x-7.78$$
$$-2.21x+8.65-8.65=3.69x-7.78-8.65$$
$$-2.21x=3.69x-16.43$$
$$-3.69x-2.21x=-3.69x+3.69x-16.43$$
$$-5.9x=-16.43$$
$$\frac{-5.9x}{-5.9}=\frac{-16.43}{-5.9}$$
$$x=2.78$$

Cumulative Review Problems

71. $$2\{x-3[4+2(3+x)]\}$$
$$=2[x-3(4+6+2x)]$$
$$=2[x-3(10+2x)]$$
$$=2(x-30-6x)$$
$$=2(-5x-30)$$
$$=-10x-60$$

73. $W = \frac{1}{4}x + 5,\ L = \frac{5}{8}x - 1$

$$P = 2L + 2W$$
$$= 2\left(\tfrac{5}{8}x - 1\right) + 2\left(\tfrac{1}{4}x + 5\right)$$
$$= \tfrac{5}{4}x - 2 + \tfrac{2}{4}x + 10$$
$$= \tfrac{7}{4}x + 8$$
$$= 1\tfrac{3}{4}x + 8 \text{ feet}$$

2.4 Exercises

1. $\frac{1}{5}x + \frac{1}{10} = \frac{1}{2}$

$$10\left(\frac{1}{5}x\right) + 10\left(\frac{1}{10}\right) = 10\left(\frac{1}{2}\right)$$
$$2x + 1 = 5$$
$$2x = 4$$
$$x = 2$$

Check: $\frac{1}{5}(2) + \frac{1}{10} \stackrel{?}{=} \frac{1}{2}$

$$\frac{4}{10} + \frac{1}{10} \stackrel{?}{=} \frac{1}{2}$$
$$\frac{1}{2} = \frac{1}{2}$$

3. $\frac{2}{3}x = \frac{1}{3}x + \frac{1}{6}$

$$6\left(\frac{2}{3}x\right) = 6\left(\frac{1}{3}x\right) + 6\left(\frac{1}{6}\right)$$
$$4x = 2x + 1$$
$$2x = 1$$
$$x = \frac{1}{2}$$

Check: $\frac{2}{3}\left(\frac{1}{2}\right) \stackrel{?}{=} \frac{1}{3}\left(\frac{1}{2}\right) + \frac{1}{6}$

$$\frac{1}{3} \stackrel{?}{=} \frac{1}{6} + \frac{1}{6}$$
$$\frac{1}{3} = \frac{1}{3}$$

5. $\frac{x}{2} + \frac{x}{5} = \frac{7}{10}$

$$10\left(\frac{x}{2}\right) + 10\left(\frac{x}{5}\right) = 10\left(\frac{7}{10}\right)$$
$$5x + 2x = 7$$
$$7x = 7$$
$$\frac{7x}{7} = \frac{7}{7}$$
$$x = 1$$

Check: $\frac{1}{2} + \frac{1}{5} \stackrel{?}{=} \frac{7}{10}$

$$\frac{5}{10} + \frac{2}{10} \stackrel{?}{=} \frac{7}{10}$$
$$\frac{7}{10} = \frac{7}{10}$$

7. $20-\frac{1}{3}x=\frac{1}{2}x$

$6(20)-6\left(\frac{1}{3}x\right)=6\left(\frac{1}{2}x\right)$

$120-2x=3x$

$120=5x$

$24=x$

Check: $20-\frac{1}{3}(24)\stackrel{?}{=}\frac{1}{2}(24)$

$20-8\stackrel{?}{=}12$

$12=12$

9. $2+\frac{y}{2}=\frac{3y}{4}-3$

$4(2)+4\left(\frac{y}{2}\right)=4\left(\frac{3y}{4}\right)-4(3)$

$8+2y=3y-12$

$8=y-12$

$20=y$

Check: $2+\left(\frac{20}{2}\right)\stackrel{?}{=}\frac{3(20)}{4}-3$

$2+10\stackrel{?}{=}15-3$

$12=12$

11. $\frac{y-1}{2}=4-\frac{y}{7}$

$\frac{y}{2}-\frac{1}{2}=4-\frac{y}{7}$

$14\left(\frac{y-1}{2}\right)=14(4)-14\left(\frac{y}{7}\right)$

$7y-7=56-2y$

$9y-7=56$

$9y=63$

$y=7$

Check: $\frac{7-1}{2}\stackrel{?}{=}4-\frac{7}{7}$

$\frac{6}{2}\stackrel{?}{=}4-1$

$3=3$

13. $0.3x-2.2=3.2$

$10(0.3x-2.2)=10(3.2)$

$3x-22=32$

$3x=54$

$x=18$

Check: $0.3(18)-2.2\stackrel{?}{=}3.2$

$5.4-2.2\stackrel{?}{=}3.2$

$3.2=3.2$

15. $0.6x+5.9=3.8$

$10(0.6x)+10(5.9)=10(3.8)$

$6x+59=38$

$6x+59-59=38-59$

$6x=-21$

$\frac{6x}{6}=\frac{-21}{6}$

$x=-\frac{7}{2}=-3.5$

Check: $0.6(-3.5)+5.9\stackrel{?}{=}3.8$

$-2.1+5.9\stackrel{?}{=}3.8$

$3.8=3.8$

17. $\frac{1}{2}(y-2)+2=\frac{3}{8}(3y-4),\quad y\stackrel{?}{=}4$

$$\frac{1}{2}(4-2)+2\stackrel{?}{=}\frac{3}{8}(3\cdot 4-4)$$
$$1+2\stackrel{?}{=}\frac{3}{8}(8)$$
$$3=3$$

$y=4$ is a solution.

19. $\frac{1}{2}\left(y-\frac{1}{5}\right)=\frac{1}{5}(y+2),\quad y\stackrel{?}{=}\frac{5}{8}$

$$\frac{1}{2}\left(\frac{5}{8}-\frac{1}{5}\right)\stackrel{?}{=}\frac{1}{5}\left(\frac{5}{8}+2\right)$$
$$\frac{1}{2}\left(\frac{17}{40}\right)\stackrel{?}{=}\frac{1}{5}\left(\frac{21}{8}\right)$$
$$\frac{17}{80}\neq\frac{42}{80}$$

$y=\frac{5}{8}$ is not a solution.

21. $\frac{3}{4}(3x+1)=2(3-2x)+1$

$$4\left(\frac{3}{4}\right)(3x+1)=4(2)(3-2x)+4(1)$$
$$3(3x+1)=8(3-2x)+4$$
$$9x+3=24-16x+4$$
$$9x+3=-16x+28$$
$$9x+16x+3=-16x+16x+28$$
$$25x+3=28$$
$$25x+3-3=28-3$$
$$25x=25$$
$$\frac{25x}{25}=\frac{25}{25}$$
$$x=1$$

23. $0.7x-3.3=2.5-0.2x-5.8$

$$10(0.7x-3.3)=10(2.5-0.2x-5.8)$$
$$7x-33=25-2x-58$$
$$7x-33=-33-2x$$
$$9x=0$$
$$x=0$$

25. $0.3x-0.2(3-5x)=-0.5(x-6)$

$$10(0.3x)-10(0.2)(3-5x)=10(-0.5)(x-6)$$
$$3x-2(3-5x)=-5(x-6)$$
$$3x-6+10x=-5x+30$$
$$13x-6=-5x+30$$
$$13x+5x-6=-5x+5x+30$$
$$18x-6=30$$
$$18x-6+6=30+6$$
$$18x=36$$
$$\frac{18x}{18}=\frac{36}{18}$$
$$x=2$$

27. $-5(0.2x+0.1)-0.6=1.9$

$$-x-0.5-0.6=1.9$$
$$-x-1.1=1.9$$
$$-x=3$$
$$x=-3$$

29. $\frac{1}{3}(y+2)=3y-5(y-2)$

$$3\left(\frac{1}{3}\right)(y+2)=3(3y)-3(5)(y-2)$$
$$y+2=9y-15y+30$$
$$y+2=-6y+30$$
$$7y=28$$
$$y=4$$

31. $$\frac{1+2x}{5}+\frac{4-x}{3}=\frac{1}{15}$$
$$15\left(\frac{1+2x}{5}\right)+15\left(\frac{4-x}{3}\right)=15\left(\frac{1}{15}\right)$$
$$3+6x+20-5x=1$$
$$x+23=1$$
$$x=-22$$

33. $$\frac{1}{5}(x+3)=2x-3(2-x)-3$$
$$5\left(\frac{1}{5}\right)(x+3)=5(2x)-5(3)(2-x)-5(3)$$
$$x+3=10x-30+15x-15$$
$$x+3=25x-45$$
$$x-x+3=25x-x-45$$
$$3=24x-45$$
$$3+45=24x-45+45$$
$$48=24x$$
$$\frac{48}{24}=\frac{24x}{24}$$
$$2=x$$

35. $$\frac{1}{3}(x-2)=3x-2(x-1)+\frac{16}{3}$$
$$3\left(\frac{1}{3}\right)(x-2)=3(3x)-3(2)(x-1)+3\left(\frac{16}{3}\right)$$
$$x-2=9x-6x+6+16$$
$$x-2=3x+22$$
$$-2x=24$$
$$x=-12$$

37. $$\frac{3}{2}x+\frac{1}{3}=\frac{2x-3}{4}$$
$$\frac{3}{2}x+\frac{1}{3}=\frac{2x}{4}-\frac{3}{4}$$
$$12\left(\frac{3}{2}x\right)+12\left(\frac{1}{3}\right)=12\left(\frac{2x}{4}\right)-12\left(\frac{3}{4}\right)$$
$$18x+4=6x-9$$
$$12x=-13$$
$$13x=-6$$
$$x=-\frac{13}{12}\text{ or }-1\frac{1}{12}$$

39. $$0.8(x-3)=-5(2.1x+0.4)$$
$$0.8x-2.4=-10.5x-2$$
$$10(0.8x)-10(2.4)=10(-10.5x)-10(2)$$
$$8x-24=-105x-20$$
$$113x=4$$
$$x=\frac{4}{113}$$

41. $$\frac{1}{5}x+\frac{2}{3}+\frac{1}{15}=\frac{4}{3}x-\frac{7}{15}$$
$$15\left(\frac{1}{5}x\right)+15\left(\frac{2}{3}\right)+15\left(\frac{1}{15}\right)=15\left(\frac{4}{3}x\right)-15\left(\frac{7}{15}\right)$$
$$3x+10+1=20x-7$$
$$-17x=-18$$
$$x=\frac{18}{17}$$

43. $$-1+5(x-2)=12x+3-7x$$
$$-1+5x-10=5x+3$$
$$5x-11=5x+3$$
$$5x-5x-11=5x-5x+3$$
$$-11=3,\quad \text{No solution}$$

45. $$\begin{aligned} 9(x+3)-6 &= 24-2x-3+11x \\ 9x+27-6 &= 9x+21 \\ 9x+21 &= 9x+21 \\ 9x-9x+21 &= 9x-9x+21 \\ 21 &= 21 \end{aligned}$$

Infinite number of solutions

Cumulative Review Problems

47. $$\begin{aligned} \frac{3}{7}+1\frac{5}{10} &= \frac{3}{7}+\frac{15}{10} \\ &= \frac{30}{70}+\frac{105}{70} \\ &= \frac{135}{70} = 1\frac{65}{70} = 1\frac{13}{14} \end{aligned}$$

49. $$\begin{aligned} \left(2\frac{1}{5}\right)\left(6\frac{1}{8}\right) &= \left(\frac{11}{5}\right)\left(\frac{49}{8}\right) \\ &= \frac{539}{40} \\ &= 13\frac{19}{40} \end{aligned}$$

51. Population in 1985

$40+20\%(40)=48$

Population in 2000

$48+450\%(48)=264$ pairs

53. $A=\pi r^2$, $r=6$

$A \approx 3.14(6)^2 = 113.04$ square inches

Cost $=2(113.04)=\$226.08$

2.5 Exercises

1. Multiply each term by 5. Then add -160 to each side. Then divide each side by 9. We would obtain $\frac{5F-160}{9}=C$.

3. (a) $A=\frac{1}{2}bh$

$$\frac{2A}{h}=\frac{2\left(\frac{1}{2}bh\right)}{h}=b$$

$A=60\text{ m}^2$, $h=12$ m

$$b=\frac{2(60)}{12}=10$$

base $=10$ m

(b) $A=\frac{1}{2}bh$

$$\frac{2A}{b}=\frac{2\left(\frac{1}{2}bh\right)}{b}$$

$$\frac{2A}{b}=h$$

$A=88\ m^2$, $b=11\ m$

$$h=\frac{2(88)}{11}=16$$

height $=16\ m$

5. (a) $3x-5y=15$

$$3x+(-3x)-5y=-3x+15$$
$$-5y=-3x+15$$
$$\frac{-5y}{-5}=\frac{-3x+15}{-5}$$
$$y=\frac{-3x+15}{-5} \text{ or } y=\frac{3}{5}x-3$$

(b) $x=-5$

$$y=\frac{3}{5}(-5)-3=-6$$

7. $A=\frac{1}{2}bh$

$$2A=bh$$
$$\frac{2A}{h}=b$$

9. $I=\text{P}tr$

$$\frac{I}{rt}=P$$

11. $y=mx+b$

$$y-b=mx$$
$$\frac{y-b}{x}=m$$

13. $A=P(1+rt)$

$$A=P+Ptr$$
$$A-P=Ptr$$
$$\frac{A-P}{\text{Pr}}=t$$

15. $5x-6y=6$

$$-6y=-5x+6$$
$$y=\frac{5}{6}x-1$$

17. $y=-\frac{3}{4}x+9$

$$4y=-3x+36$$
$$4y-36=-3x$$
$$-\frac{4}{3}y+12=x$$

19. $ax+by=c$

$$by=c-ax$$
$$y=\frac{c-ax}{b}$$

21. $A=\pi r^2$

$$\frac{A}{\pi}=r^2$$

23. $S=\frac{1}{2}gt^2$

$$2S=2gt^2$$
$$\frac{2S}{t^2}=g$$

25. $S=2\pi rh+2\pi r^2$

$$S-2\pi r^2=2\pi rh+2\pi r^2-2\pi r^2$$
$$S-2\pi r^2=2\pi rh$$
$$\frac{S-2\pi r^2}{2\pi r}=\frac{2\pi rh}{2\pi r}$$
$$\frac{S-2\pi r^2}{2\pi r}=h$$

27. $V = \pi r^2 h$

$$\frac{V}{\pi r^2} = \frac{\pi r^2 h}{\pi r^2}$$

$$\frac{V}{\pi r^2} = h$$

29. $V = LWH$

$$\frac{V}{WH} = L$$

31. $V = \frac{1}{3}\pi r^2 h$

$$3V = \pi r^2 h$$

$$\frac{3V}{\pi h} = r^2$$

33. $P = 2L + 2W$

$$P - 2L = 2W$$

$$\frac{P - 2L}{2} = W$$

35. $c^2 = a^2 + b^2$

$$c^2 - b^2 = a^2$$

37. $F = \frac{9}{5}C + 32$

$$F - 32 = \frac{9}{5}C$$

$$\frac{5}{9}(F - 32) = C$$

39. $P = k\left(\frac{T}{V}\right)$

$$\frac{P}{k} = \frac{T}{V}$$

$$\frac{PV}{k} = T$$

41. $A = \frac{\pi r^2 S}{360}$

(a) $360A = \pi r^2 S$

$$\frac{360A}{\pi r^2} = r^2$$

43. $W = \frac{P - 2L}{2}$, $P = 5.8$, $L = 2.1$

$$W = \frac{5.8 - 2(2.1)}{2} = 0.8 \text{ miles}$$

45. $L = \frac{V}{WH}$, $V = 5940$, $W = 22$, $H = 9$

$$L = \frac{5940}{(22)(9)} = 30 \text{ feet}$$

47. (*a*) $V = 1100x + 7050$

$$V - 7050 = 1100x$$

$$\frac{V - 7050}{1100} = x$$

(b) $V = 25,750$

$$x = \frac{25,750 - 7050}{1100} = 17 \text{ years}$$

Year $= 1985 + 17 = 2002$

49. $I = Prt$

If $t \Rightarrow 2t,\ I \Rightarrow Pr(2t) = 2I$

I doubles if t doubles.

51. $A = \pi r^2$

If $r \Rightarrow 2r,\ A \Rightarrow \pi(2r)^2 = 4\pi r^2$

A is quadrupled if r doubles.

Cumulative Review Problems

53. $-8+7-3-(-12)+2$

$= -8+7+(-3)+12+2 = -11+21 = 10$

55. $3[2-2\{5+3(2-5)\}]$

$= 3[2-2\{5+3(-3)\}] = 3[2-2\{5-9\}]$

$= 3[2-2\{-4\}] = 3[2+8] = 3[10] = 30$

57. $\left(3\frac{1}{4}\right)(12{,}000) = \frac{13}{4}(12{,}000)$

$= 39{,}000$ square feet

Putting Your Skills to Work

1. Increase $= 24.9 - 21.6 = 3.3$ million tons

3. Decrease $= 20.1 - 18.2 = 1.9$ million tons

5. $V = 18.2 - 0.19x$

$V - 18.2 = -0.19x$

$\frac{V - 18.2}{-0.19} = x$

2.6 Exercises

1. $5 > -6$ is equivalent to $-6 < 5$.
Both statements imply that 5 is to the right of -6 on the number line.

3. $9 > -3$

5. $-4 < -2$

7. $\frac{3}{5} ? \frac{4}{7}$

$\frac{21}{35} ? \frac{20}{35}$

$\frac{3}{5} > \frac{4}{7}$

9. $-1.2 < 2.1$

11. $-\frac{13}{3} < -4$

13. $-\frac{5}{8} < -\frac{3}{5}$

15. $\frac{123}{4986} = 0.024669$

0.0247 is greater

17. $x \geq -6$

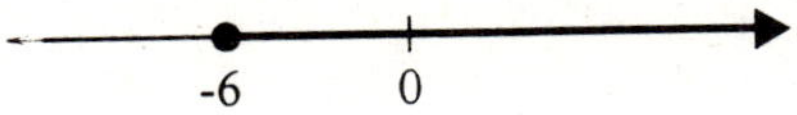

19. $x > 7$

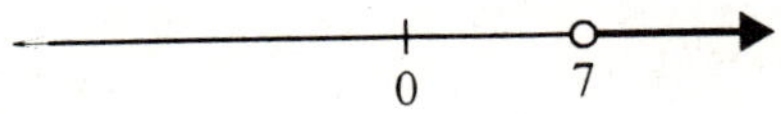

21. $x > \frac{3}{4}$

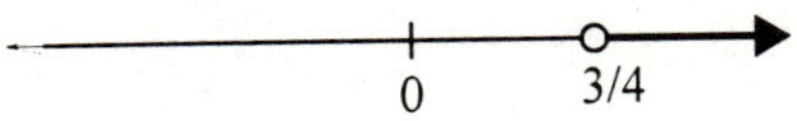

23. $x \leq -3.6$

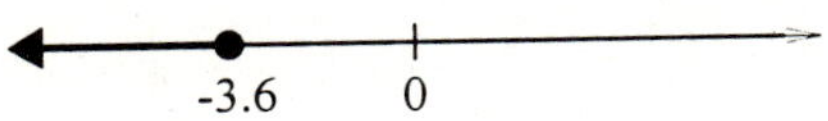

25. $25 < x$

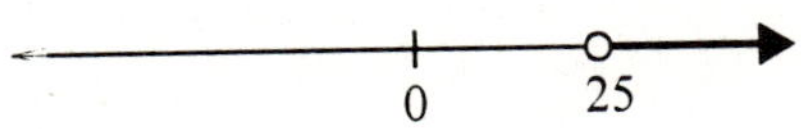

27. $x \geq -\frac{2}{3}$

29. $x < -20$

31. $x > 6.5$

33. V>580

35. $h \geq 37$

37. $h \geq 48$

39. $x \leq 2,\ x > -3,\ x < \frac{5}{2},\ x \geq -\frac{5}{2}$

$$-\frac{5}{2} \leq x \leq 2$$

Cumulative Review Problems

41. $\left(-\frac{3}{4}\right)^2 = \left(-\frac{3}{4}\right)\left(-\frac{3}{4}\right) = \frac{9}{16}$

43. $$\frac{x}{6} = 13$$
$$6\left(\frac{x}{6}\right) = 6(13)$$
$$x = 78$$

45. $p = 3500,\ r = 0.09,\ t = 5$
$I = prt = 3500(0.09)(5) = \$1575$

2.7 Exercises

1. $x + 7 \leq 4$
$x \leq -3$

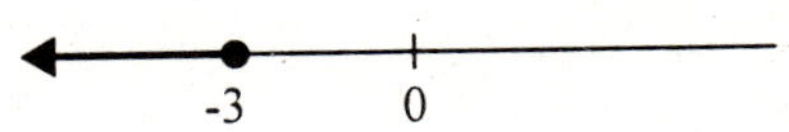

3. $-2x < 18$
$x > -9$

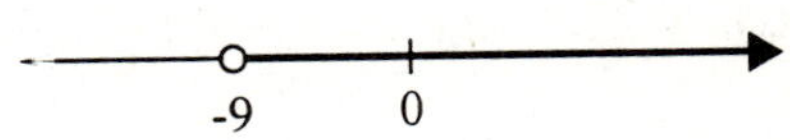

5. $\frac{1}{2}x \geq 4$
$x \geq 8$

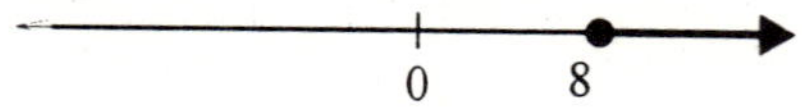

7. $2x - 3 < 4$
$$2x < 7$$
$$x < \frac{7}{2}$$

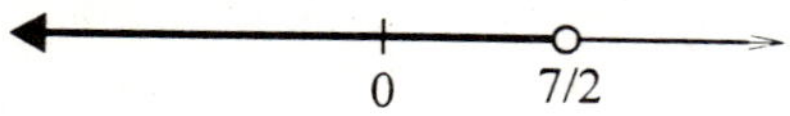

9. $-4 - 14x < 6 - 6x$
$$-8x < 10$$
$$x > -\frac{5}{4}$$

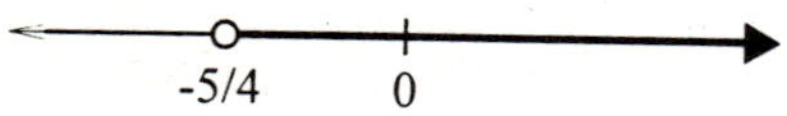

11. $\frac{5x}{6} - 5 > \frac{x}{6} - 9$
$$5x - 30 > x - 54$$
$$4x > -24$$
$$x > -6$$

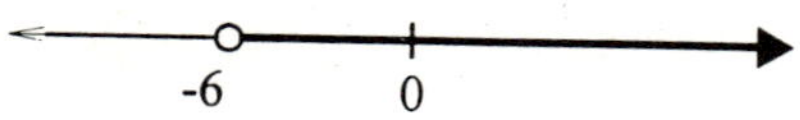

13. $2(3x+4) > 3(x+3)$
$$6x + 8 > 3x + 9$$
$$3x + 8 > 9$$
$$3x > 1$$
$$x > \frac{1}{3}$$

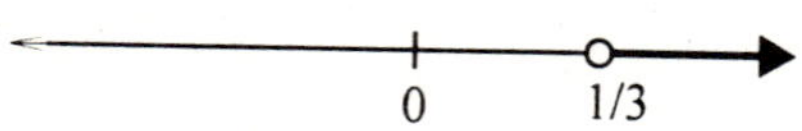

15. $5 > 3$
$$5 + (-2) > 3 + (-2)$$
$$3 > 1$$

Adding any number to both sides of an inequality does not reverse the direction.

17. $2x - 5 < 5x - 11$
$$-3x < -6$$
$$x > 2$$

19. $6x - 2 \geq 4x + 6$
$$2x \geq 8$$
$$x \geq 4$$

21. $0.3(x-1) < 0.1x - 0.5$
$$0.3x - 0.3 < 0.1x - 0.5$$
$$3x - 3 < x - 5$$
$$2x < -2$$
$$x < -1$$

23. $3 + 5(2-x) \geq -3(x+5)$
$$3 + 10 - 5x \geq -3x - 15$$
$$13 - 5x \geq -3x - 15$$
$$-2x + 13 \geq -15$$
$$-2x \geq -28$$
$$x \leq 14$$

25. $$\frac{x+6}{7}-\frac{6}{14}>\frac{x+3}{2}$$
$$14\left(\frac{x+6}{7}\right)-14\left(\frac{6}{14}\right)>14\left(\frac{x+3}{2}\right)$$
$$2(x+6)-6>7(x+3)$$
$$2x+12-6>7x+21$$
$$2x+6>7x+21$$
$$-5x>15$$
$$x<-\frac{15}{5}$$
$$x<-3$$

27. $$\frac{1}{6}-\frac{1}{2}(3x+2)<\frac{1}{3}\left(x-\frac{1}{2}\right)$$
$$1-3(3x+2)<2\left(x-\frac{1}{2}\right)$$
$$1-9x-6<2x-1$$
$$-9x-5<2x-1$$
$$-11x<4$$
$$x>-\frac{4}{11}$$

29. $$1.96x-2.58<9.36x+8.21$$
$$-7.40x<10.79$$
$$x>-1.46$$

31. $$\frac{75+83+86+x}{4}\geq 80$$
$$\frac{244+x}{4}\geq 80$$
$$244+x\geq 320$$
$$x\geq 76$$
Must get a 76 or higher.

33. $268+4x\geq 300$
$$4x\geq 32$$
$$x\geq 8$$
8 days or more.

Cumulative Review Problems

35. Width $=36+8=44$
Length $=78+8=86$
$P=2L+2W=2(86)+2(44)=260$ feet

37. $A=\pi r^2$, $r=\frac{9}{2}=4.5$
$A\approx 3.14(4.5)^2\approx 63.6$ square inches

2.8 Exercises

1. x = long length
$x-17$ = short length
$$x+x-17=47$$
$$2x=64$$
$$x=32$$
$$x-17=15$$
The long piece is 32 m long.
The short piece is 15 m long.

3. x = hours David worked
$x+15$ = hours Sarah worked
$x-5$ = hours Kate worked
$x+x+15+x-5=100$
$3x+10=100$
$3x=90$
$x=30$
$x+15=45$
$x-5=25$
David worked 30 hours
Sarah worked 45 hours
Kate worked 25 hours

5. x = height of Mt. Whitney
$x+5826$ = height of Mt. McKinley
$x-341$ = height of Mt. Oxford
$x+x+5826+x-341=48,967$
$3x+5485=48,967$
$3x=43,482$
$x=14,494$
$x+5826=20,320$
$x-341=14,153$
Mt. Whitney is 14,494 feet high
Mt. Mckinley is 20,320 feet high
Mt. Oxford is 14,153 feet high

7. x = width
$2x+50$ = length
$P=2L+2W$
$700=2(2x+50)+2x$
$700=4x+100+2x$
$700=6x+100$
$600=6x$
$100=x$
$2x+50=250$
width = 100 meters
length = 250 meters

9. w = width
$3w-35$ = length
$2w+2(3w-35)=190$
$2w+6w-70=190$
$8w=260$
$w=32.5$
$3w-35=62.5$
width = 32.5 cm
length = 62.5 cm

11. x = speed of jackal
$2x$ = speed of cheetah
$x+10$ = speed of elk
$x+2x+x+10=150$
$4x+10=150$
$4x=140$
$x=35$
$2x=70$
$x+10=45$

35 mi/hr is the speed of the jackal.

70 mi/hr is the speed of the cheetah.

45 mi/hr is the speed of the elk.

13. $x =$ length of longest side

$\frac{1}{2}x + 4 =$ shortest, $x - 3 =$ second longest

$x + x - 3 + \frac{1}{2}x + 4 = 46$

$2x + 2x - 6 + x + 8 = 92$

$5x + 2 = 92$

$5x = 90$

$x = 18$

$x - 3 = 18 - 3 = 15$

$\frac{1}{2}x + 4 = 9 + 4 = 13$

Longest side = 18 inches

Second longest side = 15 inches

Shortest side = 13 inches

15. $r = \frac{d}{t}$

(a) Over land: $r = \frac{684}{18} = 38$ mi/hr

(b) Over water: $r = \frac{1197}{21} = 57$ mi/hr

(c) Difference $= 57 - 38 = 19$ mi/hr

Faster over water

17. $x =$ length of side

$x + 2 =$ length of new side

$4(x + 2) = 5x - 3$

$4x + 8 = 5x - 3.$

$11 = x$

The original square was 11 m by 11 m.

Cumulative Review Problems

19. $-4x(2x^2 - 3x + 8) = -8x^3 + 12x^2 - 32x$

21. $-7x + 10y - 12x - 8y - 2$

$= (-7 - 12)x + (10 - 8)y - 2$

$= -19x + 2y - 2$

23.
$$\frac{2}{3}x - \frac{1}{6} = 2x + \frac{1}{2}$$
$$6\left(\frac{2}{3}x\right) - 6\left(\frac{1}{6}\right) = 6(2x) + 6\left(\frac{1}{2}\right)$$
$$4x - 1 = 12x + 3$$
$$4x - 4x - 1 = 12x - 4x + 3$$
$$-1 = 8x + 3$$
$$-1 - 3 = 8x + 3 - 3$$
$$-4 = 8x$$
$$\frac{-4}{8} = \frac{8x}{8}$$
$$-\frac{1}{2} = x$$

2.9 Exercises

1. $x =$ number of bags

$0.75x + 3(6.50) = 27.75$

$0.75x + 19.50 = 27.75$

$0.75x = 8.25$

$x = 11$

He filled 11 bags.

3. x = overtime hours

$6(40)+9x=303$

$240+9x=303$

$9x=63$

$x=7$

He needs 7 hours of overtime per week.

5. x = original price

$0.28x=100.80$

$x=360$

The original price was \$360.

7. x = Walter's profit

$x+0.12x$ = Jim's profit

$x+x+0.12x=9540$

$2.12x=9540$

$x=4500$

$x+0.12x=5040$

Walter's profit is \$4500

Jim's profit is \$5040.

9. x = investment

$0.09x$ = interest

$x+0.09x=6540$

$1.09x=6540$

$x=6000$

He invested \$6000.

11. x = amount earning 7%

$5000-x$ = amount earning 5%

$0.07x+0.05(5000-x)=310$

$0.07x+250-0.05x=310$

$0.02x+250=310$

$0.02x=60$

$x=3000$

$5000-x=2000$

They invested \$3000 at 7% and \$2000 at 5%.

13. x = amount invested at 12%

$400,000-x$ = amount invested at 8%

$0.12x+0.08(400,000-x)=38,000$

$0.12x+32,000-0.08x=38,000$

$0.04x+32,000=38,000$

$0.04x=6,000$

$x=150,000$

$400,000-x=250,000$

They invested \$150,000 at 12% and \$250,000 at 8%.

15. x = amount invested

$\frac{x}{2}$ = amount invested at 5%

$\frac{x}{3}$ = amount invested at 4%

$x-\frac{x}{2}-\frac{x}{3}=\frac{6x-3x-2x}{6}=\frac{x}{6}=$ amount invested at 3.5%

$0.05\left(\frac{x}{2}\right)+0.04\left(\frac{x}{3}\right)+(0.035)\left(\frac{x}{6}\right)=530$

$\left(\frac{0.05}{2}+\frac{0.04}{3}+\frac{0.035}{6}\right)x=530$

$\frac{53}{1200}x=530$

$x=12,000$

He invested \$12,000.

17. $x =$ number of quarters

    $x - 4 =$ number of nickels

    $0.25x + 0.05(x - 4) = 3.70$

    $25x + 5(x - 4) = 370$

    $25x + 5x - 20 = 370$

    $30x = 390$

    $x = 13$

    $x - 4 = 9$

    She has 13 quarters and 9 nickels.

19. $x =$ number of dimes

    $x + 3 =$ number of quarters

    $2(x + 3) = 2x + 6 =$ number of nickels

    $0.25(x + 3) + 0.10x + 0.05(2x + 6) = 3.75$

    $25(x + 3) + 10x + 5(2x + 6) = 375$

    $25x + +75 + 10x + 10x + 30 = 375$

    $45x + 105 = 375$

    $45x = 270$

    $x = 6$

    $x + 3 = 6 + 3 = 9$

    $2x + 6 = 12 + 6 = 18$

    He has 9 quarters, 6 dimes, and 18 nickels

21. $x =$ number of \$10 bills

    $2x =$ number of \$20 bills

    $x + 3 =$ number of \$100 bills

    $10x + 20(2x) + 100(x + 3) = 1500$

    $10x + 40x + 100x + 300 = 1500$

    $150x = 1200$

    $x = 8$

    $2x = 16$

    $x + 3 = 11$

    Eight \$10 bills, sixteen \$20 bills, eleven \$100 bills.

23. $x =$ measure of 1st angle

    $\frac{1}{2}x =$ measure of 2nd

    $x + 5 =$ measure of 3rd

    $x + \frac{1}{2}x + x + 5 = 180$

    $2x + x + 2x + 10 = 360$

    $5x = 350$

    $x = 70$

    $\frac{1}{2}x = 35$

    $x + 5 = 75$

    Angles measure 70^0, 35^0, and 75^0 respectively.

25. $x =$ amount invested at 7%

    $2969 - x =$ amount at 11%

    $0.07x + 0.11(2969 - x) = 273.47$

    $7x + 11(2969 - x) = 27,347$

    $7x + 32,659 - 11x = 27,347$

    $-4x = -5312$

    $x = 1328$

    $2969 - x = 1641$

    He invested \$1328 at 7% and \$1641 at 11%.

27. $x =$ miles driven
Golden Gate has the better bargain when the miles driven exceeds the number which makes their total charge equal to that of the West Suburban Agency.
$0.24x + 3(35) = 0.16x + 3(41)$
$24x + 10,500 = 16x + 12,300$
$8x = 1800$
$x = 225$
The salesperson must drive more than 225 miles.

Cumulative Review Problems

29. $5(3) + 6 \div (-2) = 15 + (-3) = 12$

31. If $x = -2$ and $y = 3$, then
$2x^2 + 3xy - 2y^2$
$= 2(-2)^2 + 3(-2)(3) - 2(3)^2$
$= 2(4) - 18 - 2(9)$
$= 8 - 18 - 18 = -28$

33. Discount $= 0.15(22,400) = 3360$
Discounted cost $= 22,400 - 3360$
$= \$19,040$
Average cost $= \dfrac{19,040}{200} = \$95.20/\text{part}$

Putting Your Skills to Work

1. $24.89(10) = 248.90$
Monthly payments will be \$248.90.
$4(12)(248.90) - 10,000 = 1947.2$
He will pay \$1947.20 in interest.

2. $47.07(3) = 141.21$
Monthly payments will be \$141.21.
$2(12)(141.21) - 3000 = 389.04$
She will pay \$389.04 in interest.

3. 5 years:
monthly payments: $19.8(15) = 297$
interest: $5(12)(297) - 15,000 = 2820$
4 years:
monthly payments: $23.95(15) = 359.25$
interest: $4(12)(359.25) - 15,000 = 2244$
difference in interest: $2820 - 2244 = 576$
He will pay \$576 more in interest.

4. $35,000 - 17,000 = 18,000$ to borrow
monthly payments for:
2 years: $48.96(18) = \$881.28$
3 years: $35.16(18) = \$632.88$
They will pick a time period of 3 years or more.

Chapter 2 Review Problems

1. $$5x+20=3x$$
$$5x+(-3x)+20=3x+(-3x)$$
$$2x+20=0$$
$$2x+20+(-20)=+(-20)$$
$$2x=-20$$
$$\frac{2x}{2}=\frac{-20}{2}$$
$$x=-10$$

2. $$7x+3=4x$$
$$-7x+7x+3=-7x+4x$$
$$3=-3x$$
$$\frac{3}{-3}=\frac{-3x}{-3}$$
$$x=-1$$

3. $$6-18x=4-17x$$
$$6-18x+18x=4-17x+18x$$
$$6=4+x$$
$$6-4=4-4+x$$
$$2=x$$

4. $$18-10x=63+5x$$
$$18-10x+10x=63+5x+10x$$
$$18=63+15x$$
$$18-63=63-63+15x$$
$$-45=15x$$
$$\frac{-45}{15}=\frac{15x}{15}$$
$$-3=x$$

5. $$6x-2(x+3)=5$$
$$6x-2x-6=5$$
$$4x-6=5$$
$$4x-6+6=5+6$$
$$4x=11$$
$$\frac{4x}{4}=\frac{11}{4}$$
$$x=\frac{11}{4}=2.75$$

6. $$1-2(6-x)=3x+2$$
$$1-12+2x=3x+2$$
$$2x-11=3x+2$$
$$2x-11+11=3x+2+11$$
$$2x=3x+13$$
$$-3x+2x=-3x+3x+13$$

7. $$x-(0.5x+2.6)=17.6$$
$$x-0.5x-2.6=17.6$$
$$0.5x-2.6=17.6$$
$$10(0.5x)-10(2.6)=10(17.6)$$
$$5x-26=176$$
$$5x-26+26=176+26$$
$$5x=202$$
$$\frac{5x}{5}=\frac{202}{5}$$
$$x=40.4$$

8. $$-0.2(x+1) = 0.3(x+11)$$
$$10[-0.2(x+1)] = 10[0.3(x+11)]$$
$$-2(x+1) = 3(x+11)$$
$$-2x-2 = 3x+33$$
$$-2x-2-33 = 3x+33-33$$
$$-2x-35 = 3x$$
$$2x-2x-35 = 2x+3x$$
$$-35 = 5x$$
$$\frac{-35}{5} = \frac{5x}{5}$$
$$-7 = x$$

9. $$3(x-2) = -4(5+x)$$
$$3x-6 = -20-4x$$
$$3x+4x-6 = -20-4x+4x$$
$$7x-6 = -20$$
$$7x-6+6 = -20+6$$
$$7x = -14$$
$$\frac{7x}{7} = \frac{-14}{7}$$
$$x = -2$$

10. $$\frac{1}{4}x = -16$$
$$4\left(\frac{1}{4}x\right) = 4(-16)$$
$$x = -64$$

11. $$y+37 = 26$$
$$y+37+(-37) = 26+(-37)$$
$$y = -11$$

12. $$6(8x+3) = 5(9x+8)$$
$$48x+18 = 45x+40$$
$$48x+18-18 = 45x+40-18$$
$$48x = 45x+22$$
$$-45x+48x = -45x+45x+22$$
$$3x = 22$$
$$x = \frac{22}{3} \text{ or } 7\frac{1}{3}$$

13. $$3(x-3) = 13x+21$$
$$3x-9 = 13x+21$$
$$3x+(-3x)-9 = 13x+(-3x)+21$$
$$-9 = 10x+21$$
$$-21-9 = 10x+21+(-21)$$
$$-30 = 10x$$
$$\frac{-30}{10} = \frac{10x}{10}$$
$$-3 = x$$

14. $$9x+10 = 3x+4$$
$$9x+10-10 = 3x+4-10$$
$$9x = 3x-6$$
$$-3x+9x = -3x+3x-6$$
$$6x = -6$$
$$\frac{6x}{6} = \frac{-6}{6}$$
$$x = -1$$

15.
$$\begin{aligned}24-3x&=4(x-1)\\24-3x&=4x-4\\24-3x+3x&=4x+3x-4\\24&=7x-4\\24+4&=7x-4+4\\28&=7x\\\frac{28}{7}&=\frac{7x}{7}\\4&=x\end{aligned}$$

16.
$$\begin{aligned}36&=9x-(3x-18)\\36&=9x-3x+18\\36&=6x+18\\36+(-18)&=6x+18+(-18)\\18&=6x\\\frac{18}{6}&=\frac{6x}{6}\\3&=x\end{aligned}$$

17.
$$\begin{aligned}2(3-x)&=1-(x-2)\\6-2x&=1-x+2\\6-2x+x&=3-x+x\\6-x&=3\\6+(-6)-x&=3+(-6)\\-x&=-3\\\frac{-x}{-1}&=\frac{-3}{-1}\\x&=3\end{aligned}$$

18.
$$\begin{aligned}4(x+5)-7&=2(x+3)\\4x+2x-7&=2x+6\\4x+13&=2x+6\\4x+13-13&=2x+6-13\\4x&=2x-7\\-2x+4x&=-2x+2x-7\\2x&=-7\\\frac{2x}{2}&=\frac{-7}{2}\\x&=-\frac{7}{2}\text{ or }-3\frac{1}{2}\end{aligned}$$

19.
$$\begin{aligned}0.9y+3&=0.4y+1.5\\10(0.9y)+10(3)&=10(0.4y)+10(1.5)\\9y+30&=4y+15\\9y+(-4y)+30&=4y+(-4y)+15\\5y+30&=15\\54y+30+(-30)&=15+(-30)\\5y&=-15\\\frac{5y}{5}&=\frac{-15}{5}\\y&=-3\end{aligned}$$

20.
$$\begin{aligned}7y-3.4&=1.3\\7y-3.4+3.4&=1.3+3.4\\7y&=14.7\\\frac{7y}{7}&=\frac{14.7}{7}\\y&=2.1\end{aligned}$$

21.
$$8(3x+5)-10=9(x-2)+13$$
$$24x+40-10=9x-18+13$$
$$24x+30=9x-5$$
$$24x+(-9x)+30=9x+(-9x)-5$$
$$15x+30=-5$$
$$15x+30+(-30)=-5+(-30)$$
$$15x=-35$$
$$\frac{15x}{15}=\frac{-35}{15}$$
$$x=\frac{-7}{3} \text{ or } -2\frac{1}{3}$$

22.
$$3=2x+5-3(x-1)$$
$$3=2x+5-3x+3$$
$$3=-x+8$$
$$3+(-8)=-x+8+(-8)$$
$$-5=-x$$
$$\frac{-5}{-1}=\frac{-x}{-1}$$
$$5=x$$

23.
$$-2(x-3)=-4x+3(3x+2)$$
$$-2x+6=-4x+9x+6$$
$$-2x+6=5x+6$$
$$-2x+6-6=5x+6-6$$
$$-2x=5x$$
$$2x-2x=2x+5x$$
$$0=7x$$
$$\frac{0}{7}=\frac{7x}{7}$$
$$0=x$$

24.
$$2(5x-1)-7=3(x-1)+5-4x$$
$$10x-2-7=3x-3+5-4x$$
$$10x-9=-x+2$$
$$10x+x-9=-x+x+2$$
$$11x-9=2$$
$$11x-9+9=2+9$$
$$11x=11$$
$$\frac{11x}{11}=\frac{11}{11}$$
$$x=1$$

25.
$$1=\frac{5x}{6}+\frac{2x}{3}$$
$$6=5x+4x$$
$$6=9x$$
$$\frac{2}{3}=x$$

26.
$$\frac{7x}{5}=5+\frac{2x}{5}$$
$$5\left(\frac{7x}{5}\right)=5(5)+5\left(+\frac{2x}{5}\right)$$
$$7x=25+2x$$
$$7x-2x=25+2x-2x$$
$$5x=25$$
$$\frac{5x}{5}=\frac{25}{5}$$
$$x=5$$

27. $$\frac{7x-3}{2}-4=\frac{5x+1}{3}$$
$$6\left(\frac{7x-3}{2}\right)-6(4)=6\left(\frac{5x+1}{3}\right)$$
$$3(7x-3)-24=2(5x+1)$$
$$21x-9-24=10x+2$$
$$21x-33=10x+2$$
$$21x+(-10x)-33=10x+(-10x)+2$$
$$11x-33=2$$
$$11x-33+33=2+33$$
$$11x=35$$
$$\frac{11x}{11}=\frac{35}{11}$$
$$x=\frac{35}{11}$$

28. $$\frac{3x-2}{2}+\frac{x}{4}=2+x$$
$$4\left(\frac{3x-2}{2}+\frac{x}{4}\right)=4(2+x)$$
$$2(3x-2)+x=8+4x$$
$$6x-4+x=8+4x$$
$$7x-4=4x+8$$
$$7x-4+4=4x+8+4$$
$$7x=4x+12$$
$$-4x+7x=-4x+4x+12$$
$$3x=12$$
$$\frac{3x}{3}=\frac{12}{3}$$
$$x=4$$

29. $$\frac{-3}{2}(x+5)=1-x$$
$$2\left(\frac{-3}{2}\right)(x+5)=2(1-x)$$
$$-3x-15=2-2x$$
$$-3x+3x-15=2-2x+3x$$
$$-15=2+x$$
$$-15+(-2)=2+(-2)+x$$
$$-17=x$$

30. $$\frac{-4}{3}(2x+1)=-x-2$$
$$3\left[\frac{-4}{3}(2x+1)\right]=3(-x-2)$$
$$-4(2x+1)=-3x-6$$
$$-8x-4=-3x-6$$
$$-8x-4+4=-3x-6+4$$
$$-8x=-3x-2$$
$$3x-8x=3x-3x-2$$
$$-5x=-2$$

31. $$\frac{1}{3}(x-2)=\frac{x}{4}+2$$
$$12\left(\frac{1}{3}\right)(x-2)=12\left(\frac{x}{4}\right)+12(2)$$
$$4(x-2)=3x+24$$
$$4x-8=3x+24$$
$$4x+(-3x)-8=3x+(-3x)+24$$
$$x-8=24$$
$$x-8+8=24+8$$
$$x=32$$

32. $$\frac{1}{5}(x-3)=20-\frac{x}{2}$$
$$10\left[\frac{1}{5}(x-3)\right]=10\left(2-\frac{x}{2}\right)$$
$$2(x-3)=20-5x$$
$$2x-6=20-5x$$
$$2x+5x-6=20-5x+5x$$
$$7x-6=20$$
$$7x-6+6=20+6$$
$$7x=26$$
$$\frac{7x}{7}=\frac{26}{7}$$
$$x=\frac{26}{7}$$

33. $$\frac{4}{5}+\frac{1}{2}x=\frac{1}{5}x+\frac{1}{2}$$
$$10\left(\frac{4}{5}\right)+10\left(\frac{1}{2}x\right)=10\left(\frac{1}{5}x\right)+10\left(\frac{1}{2}\right)$$
$$8+5x=2x+5$$
$$5x+(-2x)+8=2x+(-2x)+5$$
$$3x+8=5$$
$$3x+8+(-8)=5+(-8)$$
$$3x=-3$$
$$\frac{3x}{3}=\frac{-3}{3}$$
$$x=-1$$

34. $$3x+\frac{6}{5}-x=\frac{6}{5}x-\frac{4}{5}$$
$$2x+\frac{6}{5}=\frac{6}{5}x-\frac{4}{5}$$
$$5\left(2x+\frac{6}{5}\right)=5\left(\frac{6}{5}x-\frac{4}{5}\right)$$
$$10x+6=6x-4$$
$$10x+6-6=6x-4-6$$
$$10x=6x-10$$
$$-6x+10x=-6x+6x-10$$
$$4x=-10$$
$$\frac{4x}{4}=\frac{-10}{4}$$
$$x=-\frac{5}{2}\text{ or }-2\frac{1}{2}$$

35. $$\frac{10}{3}-\frac{5}{3}x+x=\frac{2}{9}+\frac{1}{9}x$$
$$9\left(\frac{10}{3}\right)-9\left(\frac{5}{3}x\right)+9x=9\left(\frac{2}{9}\right)+9\left(\frac{1}{9}x\right)$$
$$30-15x+9x=2+x$$
$$30-6x=2+x$$
$$30-6x+6x=2+x+6x$$
$$30=2+7x$$
$$30+(-2)=2+(-2)+7x$$
$$28=7x$$
$$\frac{28}{7}=\frac{7x}{7}$$
$$4=x$$

36. $$-\frac{8}{3}x-8+2x-5=-\frac{5}{3}$$
$$-\frac{8}{3}x-13+2x=-\frac{5}{3}$$
$$3\left(-\frac{8}{3}x-13+2x\right)=3\left(-\frac{5}{3}\right)$$
$$-8x-39+6x=-5$$
$$-2x-39+6x=-5$$
$$-2x-39+39=-5+39$$

$$-2x = 34$$
$$\frac{-2x}{-2} = \frac{34}{-2}$$
$$x = -17$$

37. $$\frac{1}{2} + \frac{5}{4}x = \frac{2}{5}x - \frac{1}{10} + 4$$
$$20\left(\frac{1}{2}\right) + 20\left(\frac{5}{4}x\right) = 20\left(\frac{2}{5}x\right) - 20\left(\frac{1}{10}\right) + 20(4)$$
$$10 + 25x = 8x - 2 + 80$$
$$10 + 25x = 8x + 78$$
$$10 + 25x + (-8x) = 8x + (-8x) + 78$$
$$10 + 17x = 78$$
$$10 + (-10) + 17x = 78 + (-10)$$
$$17x = 68$$
$$\frac{17x}{17} = \frac{68}{17}$$
$$x = 4$$

38. $$\frac{1}{6}x - \frac{2}{3} = \frac{1}{3}(x-4)$$
$$6\left(\frac{1}{6}x\right) - 6\left(\frac{2}{3}\right) = 6\left(\frac{1}{3}\right)(x-4)$$
$$x - 4 = 2x - 8$$
$$x + (-x) - 4 = 2x + (-x) - 8$$
$$-4 = x - 8$$
$$-4 + 8 = x - 8 + 8$$
$$4 = x$$

39. $$\frac{1}{2}(x-3) = \frac{1}{4}(3x-1)$$
$$4\left[\frac{1}{2}(x-3)\right] = 4\left[\frac{1}{4}(3x-1)\right]$$
$$2(x-3) = 3x - 1$$
$$2x - 6 = 3x - 1$$
$$2x - 6 + 1 = 3x - 1 + 1$$
$$2x - 5 = 3x$$
$$-2x + 2x - 5 = -2x + 3x$$
$$-5 = x$$

40. $$\frac{7}{12}(x-3) = -\frac{1}{3}x + 1$$
$$12\left(\frac{7}{12}\right)(x-3) = -12\left(\frac{1}{3}x\right) + 12(1)$$
$$7x - 21 = -4x + 12$$
$$7x + 4x - 21 = -4x + 4x + 12$$
$$11x - 21 = 12$$
$$11x - 21 + 21 = 12 + 21$$
$$11x = 33$$
$$\frac{11x}{11} = \frac{33}{11}$$
$$x = 3$$

41. $$\frac{1}{6} + \frac{1}{3}(x-3) = \frac{1}{2}(x+9)$$
$$6\left(\frac{1}{6}\right) + 6\left(\frac{1}{3}\right)(x-3) = 6\left(\frac{1}{2}\right)(x+9)$$
$$1 + 2(x-3) = 3(x+9)$$
$$1 + 2x - 6 = 3x + 27$$
$$2x - 5 = 3x + 27$$
$$2x - 2x - 5 = 3x - 2x + 27$$
$$-5 = x + 27$$
$$-5 - 27 = x + 27 - 27$$
$$-32 = x$$

42.
$$\frac{1}{7}(x+5)-\frac{6}{14}=\frac{1}{2}(x+3)$$
$$14\left(\frac{1}{7}\right)(x+5)-14\left(\frac{6}{14}\right)=14\left(\frac{1}{2}\right)(x+3)$$
$$2x+10-6=7x+21$$
$$2x+4=7x+21$$
$$2x+(-2x)+4=7x+(-2x)+21$$
$$4=5x+21$$
$$4+(-21)=5x+21+(-21)$$
$$-17=5x$$
$$\frac{-17}{5}=\frac{5x}{5}$$
$$-\frac{17}{5}=x$$

43.
$$-\frac{1}{3}(2x-6)=\frac{2}{3}(3+x)$$
$$3\left(-\frac{1}{3}\right)(2x-6)=3\left(\frac{2}{3}\right)(3+x)$$
$$-2x+6=6+2x$$
$$-2x+6=6+2x+2x$$
$$6=6+4x$$
$$6+(-6)=6+(-6)+4x$$
$$0=4x$$
$$\frac{0}{4}=\frac{4x}{4}$$
$$0=x$$

44.
$$\frac{7}{9}x+\frac{2}{3}=2+\frac{1}{3}x$$
$$9\left(\frac{7}{9}x+\frac{2}{3}\right)=9\left(2+\frac{1}{3}x\right)$$
$$7x+6=18+3x$$
$$7x+6-6=18-6+3x$$
$$7x=12+3x$$
$$7x-3x=12+3x-3x$$
$$4x=12$$
$$\frac{4x}{4}=\frac{12}{4}$$
$$x=3$$

45.
$$3x-y=10$$
$$3x+(-3x)-y=-3x+10$$
$$-y=-3x+10$$
$$\frac{-y}{-1}=\frac{-3x+10}{-1}$$
$$y=3x-10$$

46.
$$5x+2y+7=0$$
$$5x+2y=-7$$
$$2y=-5x-7$$
$$y=\frac{-5x-7}{2}$$

47.
$$A=P(1+rt)$$
$$A=P+\Pr t$$
$$A+(-P)=P+(-P)+\Pr t$$
$$A-P=\Pr t$$
$$\frac{A-P}{Pt}=\frac{\Pr t}{Pt}$$
$$\frac{A-P}{Pt}=r$$

48. $$A = 4\pi r^2 + 2\pi rh$$
$$A - 4\pi r^2 = 2\pi rh$$
$$\frac{A - 4\pi r^2}{2\pi r} = h$$

49. $$H = \frac{1}{3}(a + 2p + 3)$$
$$3H = 3\left(\frac{1}{3}\right)(a + 2p + 3)$$
$$3H = a + 2p + 3$$
$$3H + (-a) + (-3) = a + (-a) + 2p + 3 + (-3)$$
$$3H - a - 3 = 2p$$
$$\frac{3H - a - 3}{2} = \frac{2p}{2}$$
$$\frac{3H - a - 3}{2} = p$$

50. $$ax + by = c$$
$$ax - ax + by = c - ax$$
$$by = c - ax$$
$$\frac{by}{b} = \frac{c - ax}{b}$$
$$y = \frac{c - ax}{b}$$

51. $$H + 2d = 6c - 3d$$
$$H + 2d + 3d = 6c - 3d + 3d$$
$$H + 5d = 6c$$
$$H - H + 5d = 6c - H$$
$$5d = 6c - H$$
$$\frac{5d}{5} = \frac{6c - H}{5}$$
$$d = \frac{6c - H}{5}$$

52. $$H = \frac{3c + 2b}{4}$$
$$4H = 4\left(\frac{3c + 2b}{4}\right)$$
$$4H = 3c + 2b$$
$$4H - 3c = 3c - 3c + 2b$$
$$4H - 3c = 2b$$
$$\frac{4H - 3c}{2} = \frac{2b}{2}$$
$$\frac{4H - 3c}{2} = b$$

53. a. $$C = \frac{WRT}{1000}$$
$$1000C = WRT$$
$$\frac{1000C}{WR} = T$$

b. $$T = \frac{1000(0.36)}{(30)(0.002)} = 6000$$

54. a.
$$5x - 3y = 12$$
$$5x + (-5x) - 3y = -5x + 12$$
$$-3y = -5x + 12$$
$$\frac{-3y}{-3} = \frac{-5x + 12}{-3}$$
$$y = \frac{5}{3}x - 4$$

b. $x = 9$
$$y = \frac{5}{3}(9) - 4 = 15 - 4 = 11$$

55. a. $I = \frac{E}{R}$
$$RI = E$$
$$R = \frac{E}{I}$$

b. $R = \frac{100}{20} = 5$

56.
$$2 - 3x \le -5 + 4x$$
$$-2 + 2 - 3x \le -2 - 5 + 4x$$
$$-3x \le -7 + 4x$$
$$-3x - 4x \le -7 + 4x - 4x$$
$$-7x \le -7$$
$$\frac{-7x}{-7} \ge \frac{-7}{-7}$$
$$x \ge 1$$

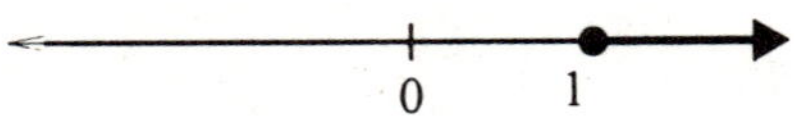

57. $2x - 3 + x > 5(x + 1)$
$$3x - 3 > 5x + 5$$
$$-2x > 8$$
$$x < -4$$

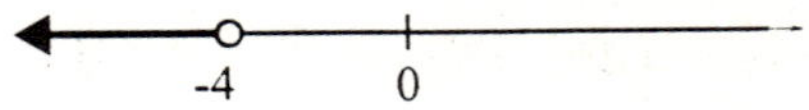

58.
$$-x + 4 < 3x + 16$$
$$-x + 4 - 4 < 3x + 16 - 4$$
$$-x < 3x + 12$$
$$-3x - x < -3x + 3x + 12$$
$$-4x < 12$$
$$\frac{-4x}{4} > \frac{12}{-4}$$
$$x > -3$$

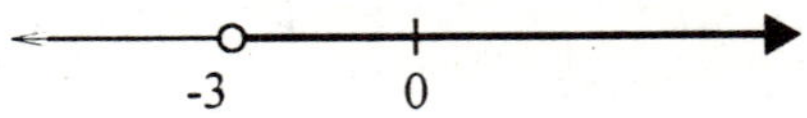

59. $4x \ge 2(12 - 2x)$
$$4x \ge 24 - 4x$$
$$8x \ge 24$$
$$x \ge 3$$

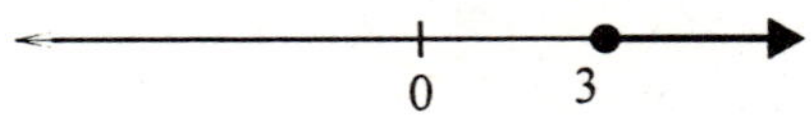

60.
$$5 - \frac{1}{2}x > 4$$
$$2\left(5 - \frac{1}{2}x\right) > 2(4)$$
$$10 - x > 8$$
$$-10 + 10 - x > -10 + 8$$
$$-x > -2$$
$$\frac{-x}{-1} < \frac{-2}{-1}$$
$$x < 2$$

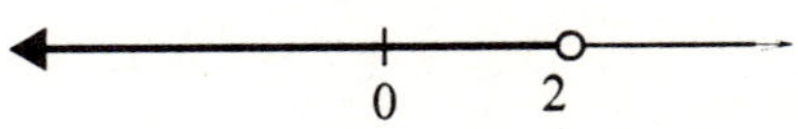

61. $2(x-1) \geq 3(2+x)$

$$2x-2 \geq 6+3x$$
$$-x \geq 8$$
$$x \leq -8$$

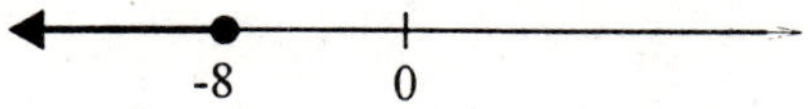

62. $3x+5-7x \leq -2x-1$

$$-4x+5 \leq -2x-1$$
$$-4x+5-5 \leq -2x-1-5$$
$$-4x \leq -2x-6$$
$$2x-4x \leq 2x-2x-6$$
$$-2x \leq -6$$
$$\frac{-2x}{-2} \geq \frac{-6}{-2}$$
$$x \geq 3$$

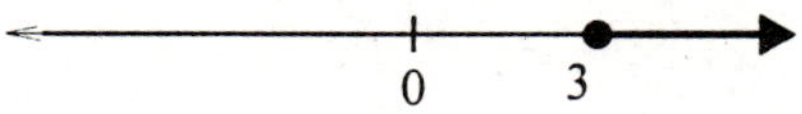

63. $-4x-14 < 4-2(3x-1)$

$$-4x-14 < 4-6x+2$$
$$-4x-14 < 6-6x$$
$$2x < 20$$
$$x < 10$$

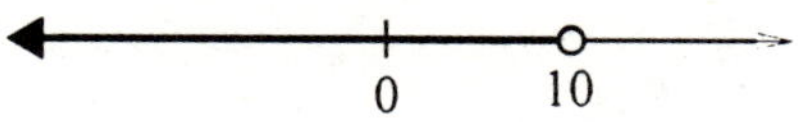

64. $\frac{1}{2}(2x+3) > 10$

$$2x+3 > 20$$
$$2x > 17$$
$$x > \frac{17}{2}$$

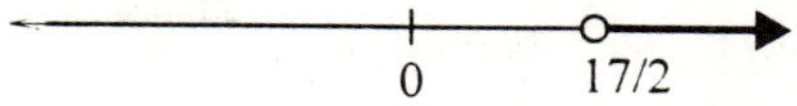

65. $\frac{1}{3}(x+2) \leq \frac{1}{2}(3x-5)$

$$6\left[\frac{1}{3}(x+2)\right] \leq 6\left[\frac{1}{2}(3x-5)\right]$$
$$2(x+2) \leq 3(3x-5)$$
$$2x+4 \leq 9x-15$$
$$2x+4+15 \leq 9x-15+15$$
$$2x+19 \leq 9x$$
$$-2x+2x+19 \leq -2x+9x$$
$$19 \leq 7x$$
$$\frac{19}{7} \leq \frac{7x}{7}$$
$$\frac{19}{7} \leq x \text{ or } x \geq \frac{19}{7}$$

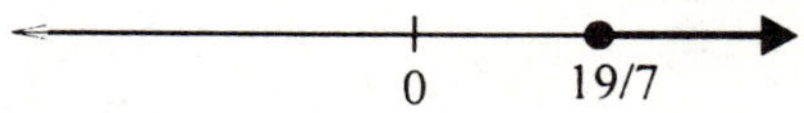

66. $4(2-x)-(-5x+1) \geq -8$

$$8-4x+5x-1 \geq -8$$
$$x+7 \geq -8$$
$$x \geq -15$$

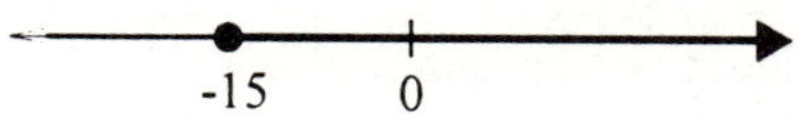

67. $5(1-x) < 3(x-1) - 2(3-x)$

$5 - 5x < 3x - 3 - 6 + 2x$

$5 - 5x < 5x - 9$

$5 + 9 - 5x < 5x - 9 + 9$

$14 - 5x < 5x$

$14 - 5x + 5x < 5x + 5x$

$14 < 10x$

$\frac{14}{10} < \frac{10x}{10}$

$\frac{7}{5} < x$ or $x > \frac{7}{5}$

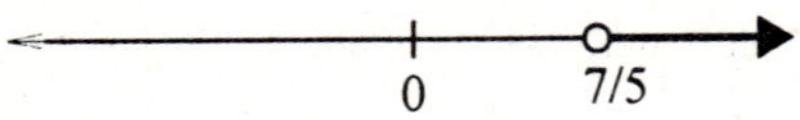

68. $70n \le 3220$

$n \le \frac{3220}{70}$

$n \le 46$

69. $85n \le 1445$

$n \le \frac{1445}{85}$

$n \le 17$

70. x = length of 1st side

$2x - 1$ = length of 2nd side

$x + 9$ = length of 3rd side

$x + 2x - 1 + x + 9 = 40$

$4x + 8 = 40$

$4x = 32$

$x = 8$

$2x - 1 = 15$

$x + 9 = 17$

The lengths are 8 yd, 15 yd, 17 yd, respectively.

71. 1st angle: x

2nd angle: $3x$

3rd angle: $2x - 12$

$x + 3x + 2x - 12 = 180$

$6x - 12 = 180$

$6x = 192$

$x = 32$

$3x = 3(32) = 96$

$2x - 12 = 2(32) - 12 = 52$

32^0, 96^0, and 52^0, respectively.

72. x = length of one piece

$\frac{3}{5}x$ = length of other

$x + \frac{3}{5}x = 50$

$5x + 3x = 250$

$8x = 250$

$x = 31.25$

$\frac{3}{5}x = 18.75$

The lengths are 31.25 yd and 18.75 yd.

73. x = Jon's salary

$3000+\frac{1}{2}x$ = Lauren's salary

$x+3000+\frac{1}{2}x=48,000$

$\frac{3}{2}x+3000=48,000$

$\frac{3}{2}x=45,000$

$x=30,000$

$3000+\frac{1}{2}x=3000+\frac{1}{2}(30,000)=18,000$

Jon: \$30,000; Lauren: \$18,000

74. x = number of kilowatt-hours

$25+0.15x=71.50$

$0.15x=46.50$

$x=310$

310 kilowatt-hours

75. x = number of miles driven

$0.25x+39(3)=187$

$0.25x+117=187$

$0.25x=70$

$x=280$

He drove 280 miles.

76. x = last year's salary

$x+0.09x=24,525$

$1.09x=24,525$

$x=22,500$

\$22,500

77. x = original price

$0.18x=36$

$x=200$

The original price was \$200.

78. x = amount invested at 12%

$9000-x$ = amount at 8%

$0.12x+0.08(9000-x)=1000$

$12x+8(9000-x)=100,000$

$12x+72,000-8x=100,000$

$4x=28,000$

$x=7000$

$9000-x=2000$

They invested \$7000 at 12% and \$2000 at 8%.

79. x = amount at 4.5%

$5000-x$ = amount at 6%

$0.045x+0.06(5000-x)=270$

$45x+60(5000-x)=270,000$

$45x+300,000-60x=270,000$

$-15x=-30,000$

$x=2000$

$5000-x=5000-2000=3000$

\$2000 at 4.5%; \$3000 at 6%

80. x = number of dimes

$x+3$ = number of quarters

$2(x+3)=2x+6$ = number of nickels

$0.05(2x+6)+0.10x+0.25(x+3)=3.75$

$5(2x+6)+10x+25(x+3)=375$

$10x+30+10x+25x+75=375$

$45x=270$

$x=6$

$x+3=6+3=9$

$2x+6=2(6)+6=18$

She has 18 nickels, 6 dimes, 9 quarters

81. $d-1=$ number of nickels

$d=$ number of dimes

$d+2=$ number of quarters

$0.05(d-1)+0.10d+0.25(d+2)=3.65$

$5(d-1)+10d+25(d+2)=365$

$5d-5+10d+25d+50=365$

$40d+45=365$

$40d=320$

$d=8$

$d-1=8-1=7$

$d+2=8+2=10$

7 nickels, 8 dimes, 10 quarters

Chapter 2 Test

1.
$$\begin{aligned} 3x+5.6&=11.6\\ 3x+5.6-5.6&=11.6-5.6\\ 3x&=6\\ \frac{3x}{3}&=\frac{6}{3}\\ x&=2 \end{aligned}$$

2.
$$\begin{aligned} 9x-8&=-6x-3\\ 9x+6x-8&=-6x+6x-3\\ 15x-8&=-3\\ 15x-8+8&=-3+8\\ 15x&=5\\ \frac{15x}{15}&=\frac{5}{15}\\ x&=\frac{1}{3} \end{aligned}$$

3.
$$\begin{aligned} 2(2y-3)&=4(2y+2)\\ 4y-6&=8y+8\\ 4y-6+6&=8y+8+6\\ 4y&=8y+14\\ -8y+4y&=-8y+8y+14\\ -4y&=14\\ \frac{-4y}{-4}&=\frac{14}{-4}\\ y&=-\frac{7}{2} \end{aligned}$$

4.
$$\begin{aligned} \frac{1}{7}y+3&=\frac{1}{2}y\\ 14\left(\frac{1}{7}y\right)+14(3)&=14\left(\frac{1}{2}y\right)\\ 2y+42&=7y\\ 2y-2y+42&=7y-2y\\ 42&=5y\\ \frac{42}{5}&=\frac{5y}{5}\\ 8\frac{2}{5}&=y \end{aligned}$$

5. $$\begin{aligned} 5(8-2x) &= -6x+8 \\ 40-10x &= -6x+8 \\ 40-10x+10x &= -6x+10x+8 \\ 40 &= 4x+8 \\ 40-8 &= 4x+8-8 \\ 32 &= 4x \\ \frac{32}{4} &= \frac{4x}{4} \\ 8 &= x \end{aligned}$$

6. $$\begin{aligned} 0.8x+0.18-0.4x &= 0.3(x+0.2) \\ 100(0.8x+0.18-0.4x) &= 100[0.3(x+0.2)] \\ 80x+18-40x &= 30(x+0.2) \\ 40x+18 &= 30x+6 \\ 40x+18-18 &= 30x+6-18 \\ 40x &= 30x-12 \\ -30x+40x &= -30x+30x-12 \\ 10x &= -12 \\ \frac{10x}{10} &= \frac{-12}{10} \\ x &= -\frac{6}{5} \text{ or } -1.2 \end{aligned}$$

7. $$\begin{aligned} \frac{2y}{3}+\frac{1}{5}-\frac{3y}{5}+\frac{1}{3} &= 1 \\ 15\left(\frac{2y}{3}\right)+15\left(\frac{1}{5}\right)-15\left(\frac{3y}{5}\right)+15\left(\frac{1}{3}\right) &= 15(1) \\ 10y+3-9y+5 &= 15 \\ y+8 &= 15 \\ y+8-8 &= 15-8 \\ y &= 7 \end{aligned}$$

8. $$\begin{aligned} 3-2y &= 2(3y-2)-5y \\ 3-2y &= 6y-4-5y \\ 3-2y &= y-4 \\ 3-2y+2y &= y+2y-4 \\ 3 &= 3y-4 \\ 3+3 &= 3y-4+4 \\ 7 &= 3y \\ \frac{7}{3} &= \frac{3y}{3} \\ \frac{7}{3} &= y \end{aligned}$$

9. $$\begin{aligned} 5(20-x)+10x &= 165 \\ 100-5x+10x &= 165 \\ 100+5x &= 165 \\ -100+100+5x &= -100+165 \\ 5x &= 65 \\ \frac{5x}{5} &= \frac{65}{5} \\ x &= 13 \end{aligned}$$

10. $$\begin{aligned} 2(x+75)+5x &= 1025 \\ 2x+150+5x &= 1025 \\ 7x+150 &= 1025 \\ 7x+150-150 &= 1025-150 \\ 7x &= 875 \\ x &= 125 \end{aligned}$$

11. $$-2(2-3x) = 76-2x$$
$$-4+6x = 76-2x$$
$$-76-4+6x = -76+76-2x$$
$$-80+6x = -2x$$
$$-80+6x-6x = -2x-6x$$
$$-80 = -8x$$
$$\frac{-80}{-8} = \frac{-8x}{-8}$$
$$10 = x$$

12. $$20-(2x+6) = 5(2-x)+2x$$
$$20-2x-6 = 10-5x+2x$$
$$-2x+14 = -3x+10$$
$$3x-2x+14 = 3x-3x+10$$
$$x+14 = 10$$
$$x+14-14 = 10-14$$
$$x = -4$$

13. $$2x-3 = 12-6x+3(2x+3)$$
$$2x-3 = 12-6x+6x+9$$
$$2x-3 = 21$$
$$2x-3+3 = 21+3$$
$$2x = 24$$
$$\frac{2x}{2} = \frac{24}{2}$$
$$x = 12$$

14. $$\frac{1}{3}x - \frac{3}{4}x = \frac{1}{12}$$
$$12\left(\frac{1}{3}x - \frac{3}{4}x\right) = 12\left(\frac{1}{12}\right)$$
$$4x - 9x = 1$$
$$-5x = 1$$
$$\frac{-5x}{-5} = \frac{1}{-5}$$
$$x = -\frac{1}{5}$$

15. $$\frac{3}{5}x + \frac{7}{10} = \frac{1}{3}x + \frac{3}{2}$$
$$30\left(\frac{3}{5}x + \frac{7}{10}\right) = 30\left(\frac{1}{3}x + \frac{3}{2}\right)$$
$$18x + 21 = 10x + 45$$
$$18x + 21 - 21 = 10x + 45 - 21$$
$$18x = 10x + 24$$
$$-10x + 18x = -10x + 10x + 24$$
$$8x = 24$$
$$\frac{8x}{8} = \frac{24}{8}$$
$$x = 3$$

16.
$$\frac{15x-2}{28}=\frac{5x-3}{7}$$
$$28\left(\frac{15x-2}{28}\right)=28\left(\frac{5x-3}{7}\right)$$
$$15x-2=4(5x-3)$$
$$15x-2=20x-12$$
$$15x-2+12=20x-12+12$$
$$15x+10=20x$$
$$-15x+15x+10=-15x+20x$$
$$10=5x$$
$$\frac{10}{5}=\frac{5x}{5}$$
$$2=x$$

17.
$$\frac{1}{3}(7x-1)+\frac{1}{4}(2-5x)=\frac{1}{3}(5+3x)$$
$$\frac{7x}{3}-\frac{1}{3}+\frac{1}{2}-\frac{5x}{4}=\frac{5}{3}+x$$
$$12\left(\frac{7x}{3}-\frac{1}{3}+\frac{1}{2}-\frac{5x}{4}\right)=12\left(\frac{5}{3}+x\right)$$
$$28x-4+6-15x=20+12x$$
$$13x+2=12x+20$$
$$13x+2-2=12x+20-2$$
$$13x=12x+18$$
$$-12x+13x=-12x+12x+18$$
$$x=18$$

18.
$$A=3w+2P$$
$$A-2P=3w$$
$$\frac{A-2P}{3}=w$$

19.
$$\frac{2w}{3}=4-\frac{1}{2}(x+6)$$
$$\frac{2w}{3}=4-\frac{1}{2}x-3$$
$$6\left(\frac{2w}{3}\right)=6(4)-6\left(\frac{1}{2}x\right)-6(3)$$
$$4w=24-3x-18$$
$$4w=3x+6$$
$$w=\frac{-3x+6}{4}$$

20.
$$A=\frac{1}{2}h(a+b)$$
$$2A=h(a+b)$$
$$2A=ha+hb$$
$$2A-hb=ha$$
$$\frac{2A-hb}{h}=a$$

21.
$$5ax(2-y)=3axy+5$$
$$10ax-5axy=3axy+5$$
$$10ax-5=8axy$$
$$\frac{10ax-5}{8ax}=y$$

22.
$$P=2L+2W$$
$$P-2L=2L-2L+2W$$
$$P-2L=2W$$
$$\frac{P-2L}{2}=\frac{2W}{2}$$
$$\frac{P-2L}{2}=W$$

23. $P = 120,\ L = 42$

$$W = \frac{P - 2L}{2} = \frac{120 - 2(42)}{2} = 18 \text{ feet}$$

24.
$$\begin{aligned} 4(x-1) &\ge 12x \\ 4x - 4 &\ge 12x \\ -4x + 4x - 4 &\ge -4x + 12x \\ -4 &\ge 8x \\ \frac{-4}{8} &\ge \frac{8x}{8} \\ -\frac{1}{2} &\ge x \end{aligned}$$

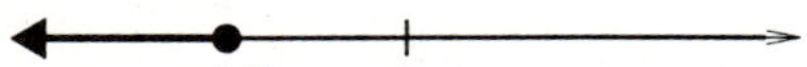

25.
$$\begin{aligned} 2 - 7(x+1) - 5(x+2) &< 0 \\ 2 - 7x - 7 - 5x - 10 &< 0 \\ -12x - 15 &< 0 \\ -12x - 15 + 15 &< 0 + 15 \\ -12x &< 15 \\ \frac{-12x}{-12} &> \frac{15}{-12} \\ x &> -\frac{5}{4} \end{aligned}$$

26.
$$\begin{aligned} 5 + 8x - 4 &< 2x + 13 \\ 8x + 1 &< 2x + 13 \\ 8x + 1 - 1 &< 2x + 13 - 1 \\ 8x &< 2x + 12 \\ -2x + 8x &< -2x + 2x + 12 \\ 6x &< 12 \\ \frac{6x}{6} &< \frac{12}{6} \\ x &< 2 \end{aligned}$$

27.
$$\begin{aligned} \frac{1}{4}x + \frac{1}{16} &\le \frac{1}{8}(7x - 2) \\ 16\left[\frac{1}{4}x + \frac{1}{16}\right] &\le 16\left[\frac{1}{8}(7x-2)\right] \\ 4x + 1 &\le 2(7x - 2) \\ 4x + 1 &\le 14x - 4 \\ 4x + 1 + 4 &\le 14x - 4 + 4 \\ 4x + 5 &\le 14x \\ -4x + 4x + 5 &\le -4x + 14x \\ 5 &\le 10x \\ \frac{5}{10} &\le \frac{10x}{10} \\ \frac{1}{2} &\le x \end{aligned}$$

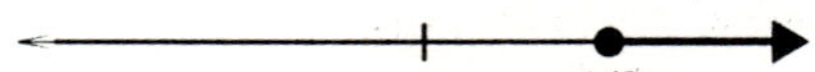

28. $\frac{2}{3}x = \text{1st side}$

$x = \text{2nd side}$

$x - 14 = \text{3rd side}$

$\frac{2}{3}x + x + x - 14 = 66$

$\frac{8}{3}x - 14 = 66$

$\frac{8}{3}x = 80$

$x = 30$

$\frac{2}{3}x = \frac{2}{3}(30) = 20$

$x - 14 = 30 - 14 = 16$

20 m, 30 m, and 16 m, respectively.

29. $w = \text{width}$

$2w + 7 = \text{length}$

$2w + 2(2w + 7) = 134$

$2w + 4w + 14 = 134$

$6w + 14 = 134$

$6w = 120$

$w = 20$

$2w + 7 = 2(20) + 7 = 47$

width = 20 m

length = 47 m

30. $2x = \text{1st pollutant}$

$x = \text{2nd pollutant}$

$0.75x = \text{3rd pollutant}$

$2x + x + 0.75x = 15$

$3.75x = 15$

$x = 4$

$2x = 2(4) = 8$

$0.75x = 0.75(4) = 3$

1st pollutant: 8ppm;

2nd pollutant: 4 ppm;

3rd pollutant: 3 ppm

31. $x = \text{number of months}$

$116x + 200 = 1940$

$116x = 1740$

$x = 15$

15 months

32. $x = \text{amount at 14\%}$

$4000 - x = \text{amount at 11\%}$

$0.14x + 0.11(4000 - x) = 482$

$100[0.14x + -.11(4000 - x)] = 100(482)$

$14x + 11(4000 - x) = 48,200$

$14x + 44,000 - 11x = 48,200$

$3x + 44,000 = 48,200$

$3x = 4200$

$x = 1400$

$4000 - x = 4000 - 1400 = 2600$

\$1400 at 14%; \$2600 at 11%

33. $2x$ = number of nickels

$x-1$ = number of dimes

x = number of quarters

$$0.05(2x)+0.10(x-1)+0.25(x)=3.50$$
$$5(2x)+10(x-1)+25x=350$$
$$10x+10x-10+25x=350$$
$$45x-10=350$$
$$45x=360$$
$$x=8$$
$$2x=2(8)=16$$
$$x-1=8-1=7$$

16 nickels; 7 dimes; 8 quarters

Cumulative Test for Chapters 1-2

1. $-\frac{2}{3}+-\frac{2}{7}=-\frac{14}{21}+-\frac{6}{21}=-\frac{20}{21}$

2. $(-3)(-5)(-1)(2)(-1)=(3)(5)(1)(2)(1)$
$=30$

3. $$\begin{array}{r} 0.41 \\ 2.5\overline{)1.025} \\ \underline{100} \\ 25 \\ \underline{24} \end{array}$$

4. $5ab-7ab^2-3ab-12ab^2+10ab-9ab^2$
$=(5-3+10)ab+(-7-12-9)ab^2$
$=12ab-28ab^2$

5. $(5x)^2=5^2(x)^2=25x^2$

6. $(3-5)^2-9\div(2-11)=(-2)^2-9\div(-9)$
$=4-9\div(-9)$
$=4+1$
$=5$

7. $2\{x+y[3-2x(1-4y)]\}$
$=2[x+y(3-2x+8xy)]$
$=2(x+3y-2xy+8xy^2)$
$=2x+6y-4xy+16xy^2$

8. $$4(7-2x)=3x-12$$
$$28-8x=3x-12$$
$$28-8x+8x=3x+8x-12$$
$$28=11x-12$$
$$28+12=11x-12+12$$
$$40=11x$$
$$\frac{40}{11}=\frac{11x}{11}$$
$$\frac{40}{11}=x \text{ or } 3\frac{7}{11}$$

9. $$\frac{1}{3}(x+5)=2x-5$$
$$3\left[\frac{1}{3}(x+5)\right]=3(2x-5)$$
$$x+5=6x-15$$
$$x+5+15=6x-15+15$$
$$x+20=6x$$
$$-x+x+20=-x+6x$$
$$20=5x$$
$$\frac{20}{5}=\frac{5x}{5}$$
$$4=x$$

10. $\frac{2y}{3}-\frac{1}{4}=\frac{1}{6}+\frac{y}{4}$

$8y-3=2+3y$

$5y=5$

$y=1$

11. $H=\frac{2}{3}(b+4a)$

$\frac{3H}{2}=b+4a$

$\frac{3H}{2}-4a=b$ or $b=\frac{3H-8a}{2}$

12. $I=\text{P}tr$

$\frac{I}{\text{Pr}}=t$

13. $A=\frac{ha}{2}+\frac{hb}{2}$

$2A=ha+hb$

$2A-hb=ha$

$\frac{2A-hb}{h}=a$ or $a=\frac{2A}{h}-b$

14. $\frac{1}{2}(x-5)\ge x-4$

$x-5\ge 2x-8$

$-x\ge -3$

$x\le 3$

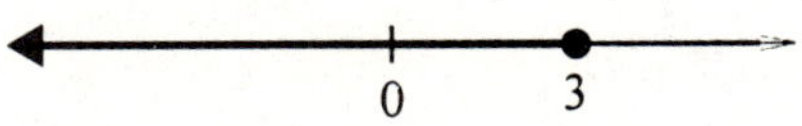

15. $4(2-x)>1-5x-8$

$8-4x>-7-5x$

$-8+8-4x>-8-7-5x$

$-4x>-15-5x$

$-4x+5x>-15-5x+5x$

$x>-15$

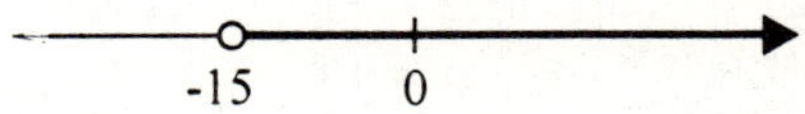

16. $x+\frac{5}{9}\le\frac{1}{3}+\frac{7}{9}x$

$9x+5\le 3+7x$

$2x\le -2$

$x\le -1$

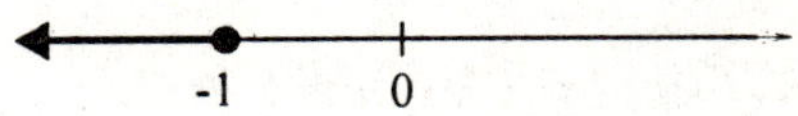

17. $4-16x\le 6-5(3x-2)$

$4-16x\le 6-15x+10$

$4-16x\le 16-15x$

$4-16x+16x\le 16-15x+16x$

$4\le 16+x$

$-16+4\le -16+16+x$

$-12\le x$

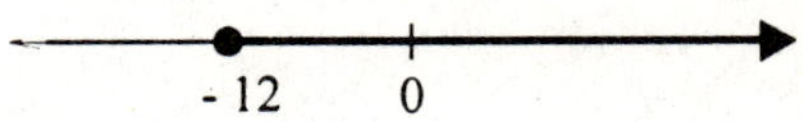

18. $\frac{0+82+89+87+x}{5}\ge 70$

$258+x\ge 5(70)$

$x\ge 350-258$

$x\ge 92$

He must receive a 92 or better.

19. w = width

    $3w + 11$ = length

    $P = 2L + 2W$

    $2(3w + 11) + 2w = 78$

    $6w + 22 + 2w = 78$

    $8w = 56$

    $w = 7$

    $3w + 11 = 32$

    length = 32 cm, width = 7 cm

20. x = amount invested at 15%

    $7000 - x$ = amount invested at 7%

    $0.15x + 0.07(7000 - x) = 730$

    $15x + 7(7000 - x) = 73,000$

    $15x + 49,000 - 7x = 73,000$

    $8x = 24,000$

    $x = 3000$

    $7000 - x = 4000$

    He invested \$3000 at 15% and \$4000 at 7%.

21. $x + 3$ = number of nickels

    $3x$ = number of dimes

    x = number of quarters

    $0.05(x + 3) + 0.10(3x) + 0.25(x) = 2.55$

    $5(x + 3) + 10(3x) + 25x = 255$

    $5x + 15 + 30x + 25x = 255$

    $60x + 15 = 255$

    $60x = 240$

    $x = 4$

    $x + 3 = 4 + 3 = 7$

    $3x = 3(4) = 12$

    7 nickels; 12 dimes; 4 quarters

Chapter 3

Pretest Chapter 3

1. $(-3, 1), (4, -4), (-2, -6)$

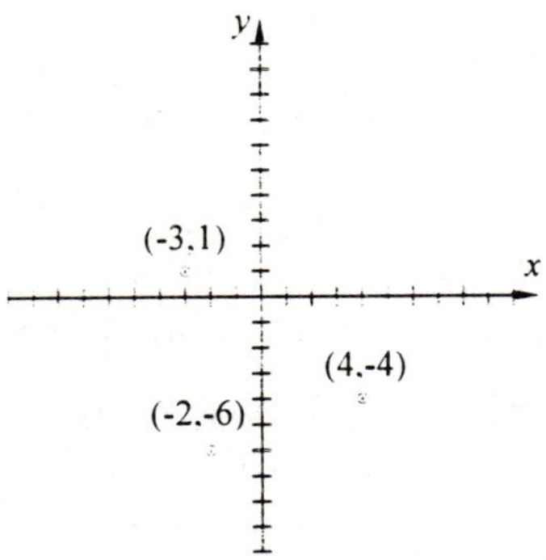

2. $A(-1, -2), B(3, -2), C(-1, 2)$

3. $y = -3x + 7$

   (a) $(-1, \ ), \ y = -3(-1) + 7 = 10$

   $(-1, 10)$

   (b) $(\ , 1), \ 1 = -3x + 7 \Rightarrow x = 2$

   $(2, 1)$

4. $5x - 2y = -10$

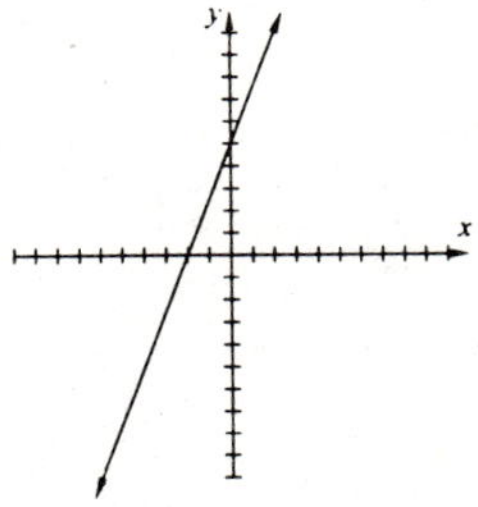

5. $y = 4$

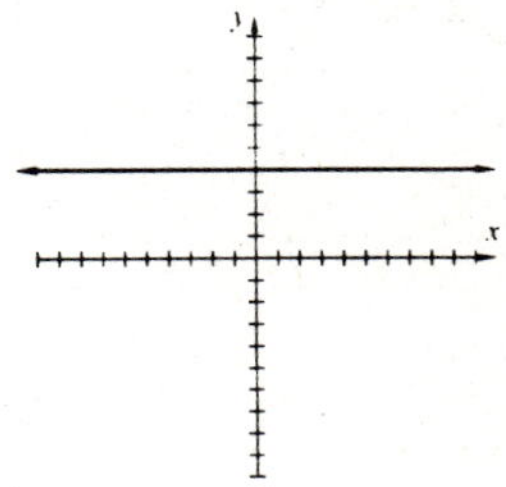

6. $(4, -2)$ and $(-4, 6)$

$$m = \frac{6-(-2)}{-4-4} = \frac{8}{-8} = -1$$

7. $4x - 3y - 7 = 0$

$$-3y = -4x - 7$$
$$y = \frac{4}{3}x + \frac{7}{3}$$
$$y = mx + b$$
$$m = \frac{4}{3}$$

8. (a) $y = mx + b$

   $y = 4x - 5$

   (b) $y = 4x - 5$

   $-4x + y = -5$

   $4x - y = 5$

9. $y = 2x + 1$

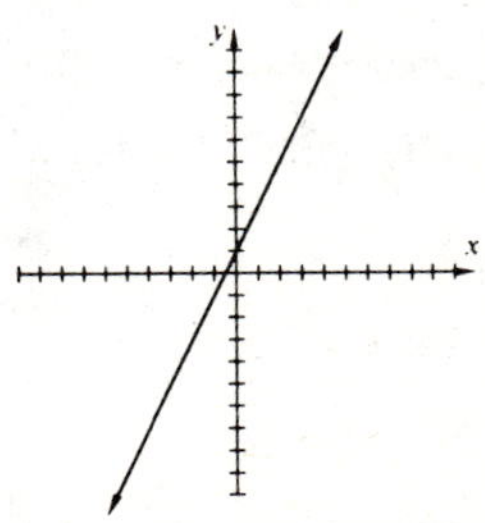

10. $y = mx + b$

$$y = -\frac{3}{2}x + 2$$

11. $m_1 = \frac{3}{5}$

(a) parallel: $m_2 = m_1 = \frac{3}{5}$

(b) perpendicular: $m_2 = -\frac{1}{m_1} = -\frac{5}{3}$

12. $y = mx + b$

$$-3 = \frac{2}{3}(-3) + b$$
$$-3 = -2 + b$$
$$-1 = b$$
$$y = \frac{2}{3}x - 1$$

13. $(-1, 6)$ and $(2, 3)$

$$m = \frac{y_2 - y_1}{x_2 - x_1} = \frac{3-6}{2-(-1)} = -1$$
$$y - y_1 = m(x - x_1)$$
$$y - 6 = -1[x - (-1)]$$
$$y = -x + 5$$

14. $y \geq 3x - 2$

Test point: $(0, 0)$

$0 \geq 3(0) - 2$

$0 \geq -2$ True

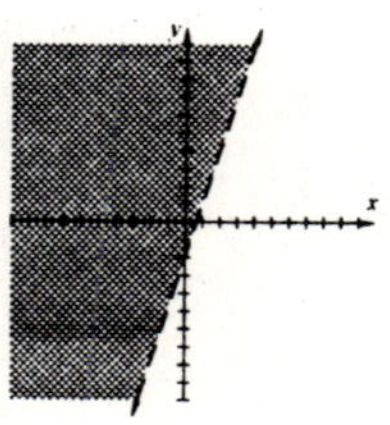

15. $4x + 2y < -12$

$$2x + y < -6$$

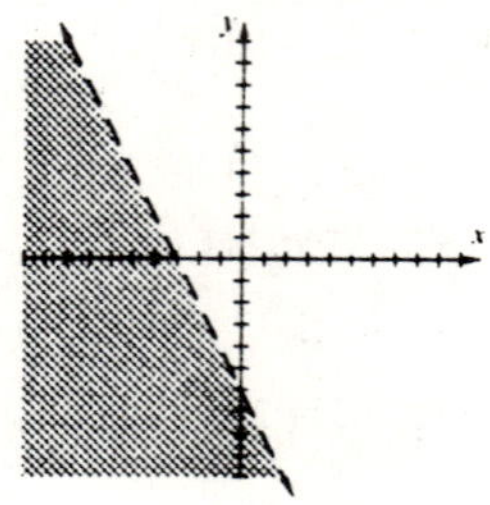

16. For each r there is one and only one C. It is a function.

17. Not a function, since a vertical line can cross the graph in more than one point.

18. $f(x) = 3x - 7$

(a) $f(-2) = 3(-2) - 7 = -13$

(b) $f(7) = 3(7) - 7 = 14$

19. $g(x) = 2x^2$

(a) $g(-6) = 2(-6)^2 = 72$

(b) $g\left(\frac{1}{2}\right) = 2\left(\frac{1}{2}\right)^2 = \frac{1}{2}$

3.1 Exercises

1.

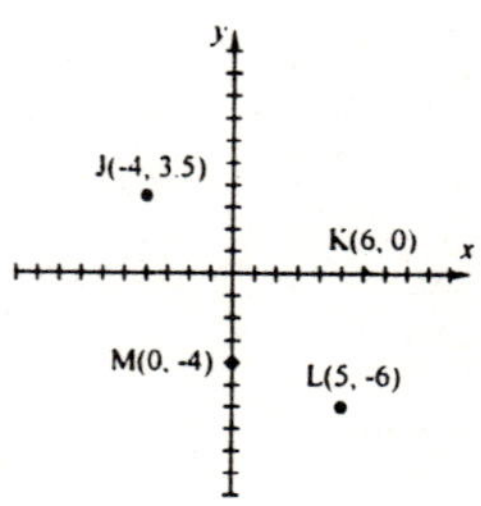

3. R: $(-3, -5)$

   S: $\left(-4\frac{1}{2}, 0\right)$

   X: $(3, -5)$

   Y: $\left(2\frac{1}{2}, 6\right)$

5. The x – coordinate of the origin is 0.

7. $(5, 1)$ is an ordered pair because the order is important. The graphs of $(5, 1)$ and $(1, 5)$ are different.

9. $(-5, -3), (-4, -4), (-4, -5), (-3, -3)$
   $(-2, -1), (-2, -2), (-1, -3), (0, -4)$
   $(0, -5), (1, -3)$

11. $y = 3x + 8$

    (a) $(0,\ ),\ y = 3(0) + 8 = 8$
    $(0, 8)$

    (b) $(4,\ ),\ y = 3(4) + 8 = 20$
    $(4, 20)$

13. $y = -2x + 3$

    (a) $(-6,\ ),\ y = -2(-6) + 3 = 15$
    $(-6, 15)$

    (b) $(3,\ ),\ y = -2(3) + 3 = -3$
    $(3, -3)$

15. $5x - 2y = 9$

    (a)
    $$5(7) - 2y = 9$$
    $$35 - 2y = 9$$
    $$-2y = -26$$
    $$y = 13$$
    $(7, 13)$

    (b)
    $$5x - 2(-7) = 9$$
    $$5x + 14 = 9$$
    $$5x = -5$$
    $$x = -1$$
    $(-1, -7)$

17. $-4x + 5y = -20$

    (a)
    $$-4(10) + 5y = -20$$
    $$-40 + 5y = -20$$
    $$5y = 20$$
    $$y = 4$$
    $(10, 4)$

    (b)
    $$-4x + 5(-8) = -20$$
    $$-4x - 40 = -20$$
    $$-4x = 20$$
    $$x = -5$$
    $(-5, -8)$

19. $4x + \frac{1}{3}y = 8$

    (a)
    $$4x + \frac{1}{3}(-12) = 8$$
    $$4x - 4 = 8$$
    $$4x = 12$$
    $$x = 3$$
    $(3, -12)$

19. (b) $4\left(\frac{3}{2}\right)+\frac{1}{3}y=8$

$6+\frac{1}{3}y=8$

$\frac{1}{3}y=2$

$y=6$

$\left(\frac{3}{2}, 6\right)$

21. D5

23. D1

25. C3

27. (a)

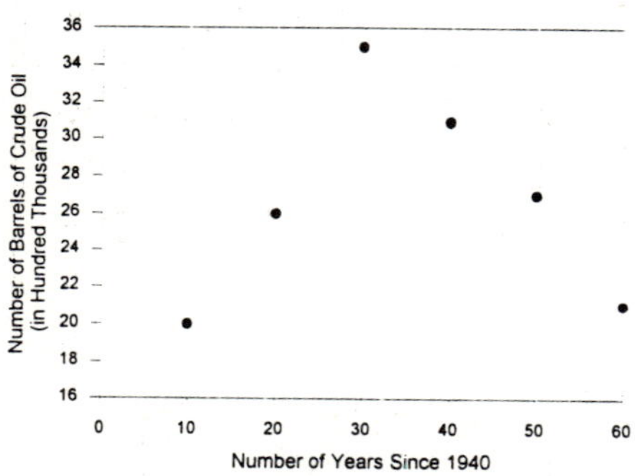

(b) The number of barrels of crude oil increased significantly from 1940 to 1970. From 1970 to 2000, it has decreased significantly.

Cumulative Review Problems

29. $x-3\le -5(x-6)$

$x-3\le -5x+30$

$x+5x-3\le -5x+5x+30$

$6x-3\le 30$

$6x-3+3\le 30+3$

$6x\le 33$

$\frac{6x}{6}\le\frac{33}{6}$

$x\le\frac{11}{2}$

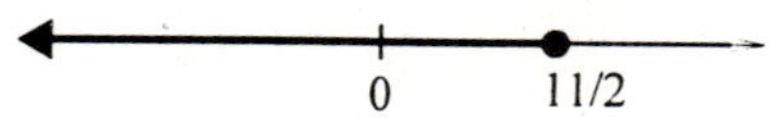

31. $x=$ the number of members it has now

$x=795+0.16(795)=922$

33. (a) Area $=30\times 22=660$ sq. ft.

$\frac{36,300}{660}=\$55$ per sq. ft.

(b) Area $=14\times 8=112$ sq. ft

Cost $=112(55)=\$6160$

3.2 Exercises

1. Replacing x by -2 and y by 5 in $2x+5y=0$ does not result in a true statement so $(-2, 5)$ is not a solution.

3. The x – intercept of a line is the point where the line crosses the <u>x - axis</u>.

5. $y = -2x + 1$

$(0, 1), (-2, 5), (1, -1)$

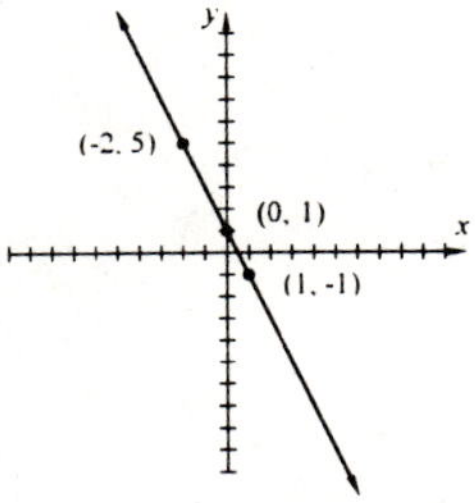

7. $y = x - 4$

$(0, -4), (2, -2), (4, 0)$

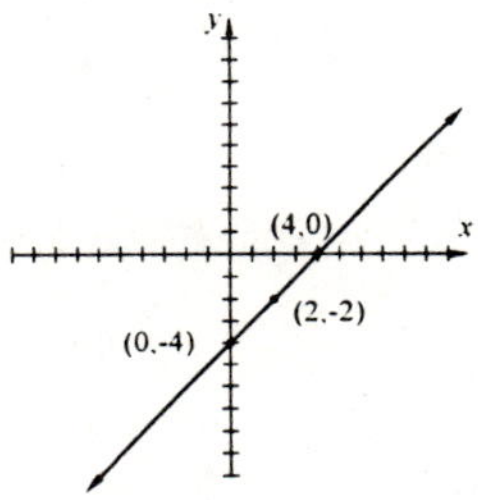

9. $y = 2x - 5$

x	y
0	−5
2	−1
4	3

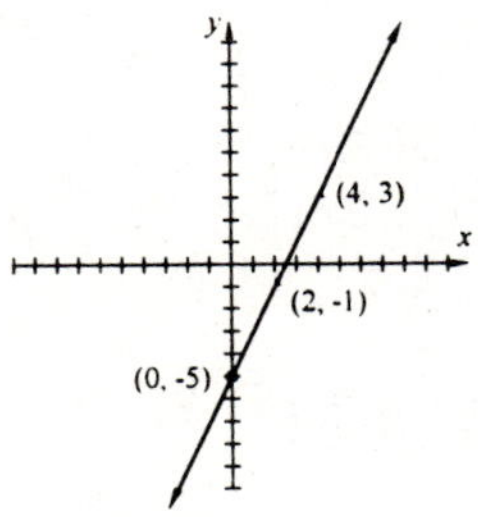

11. $y = 2x + 1$

x	y
−1	−1
0	1
1	3

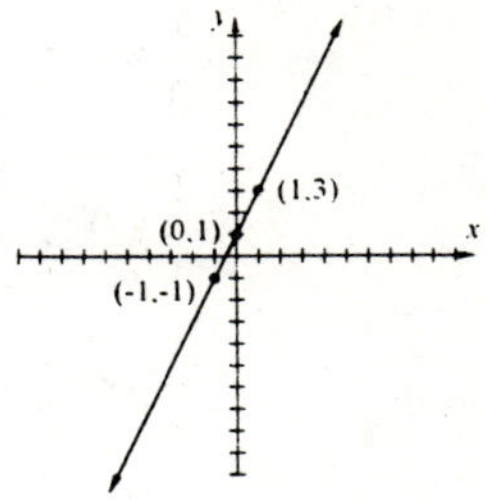

13. $2y - 5x = 0$

x	y
−2	−5
0	0
2	5

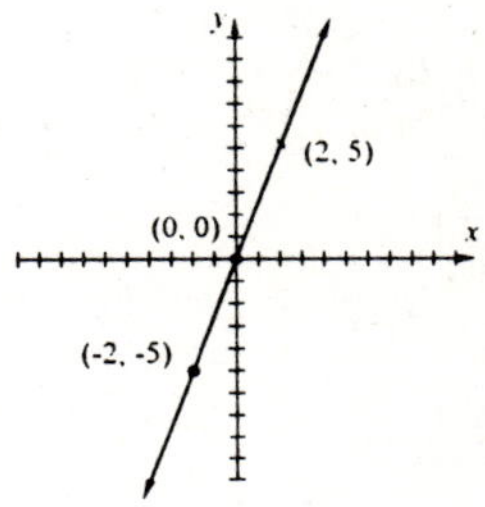

15. $y = -\frac{2}{3}x + 2$

x	y
0	2
3	0
6	−2

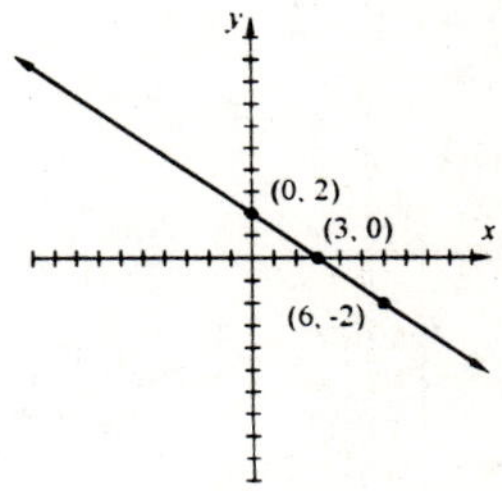

17. $y = 6 - 2x$

x	y
0	6
1	4
3	0

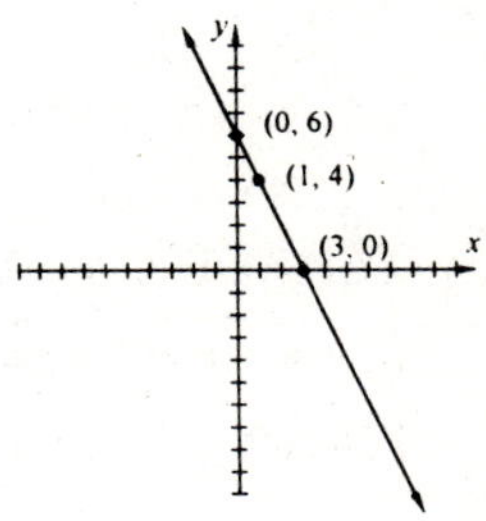

19. $x + 3 = 6y$

x	y
-3	0
0	$\frac{1}{2}$
3	1

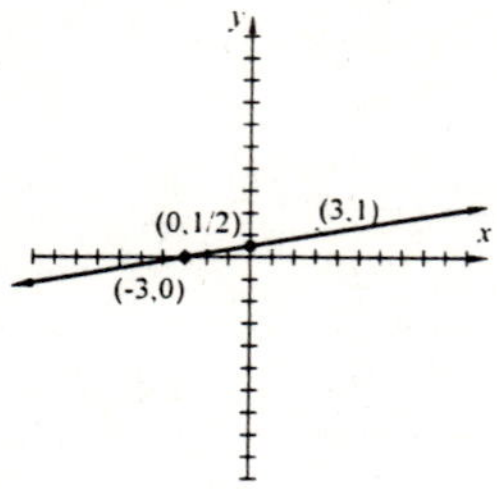

21. $3x + 2y = 6$

x	y
-2	6
0	3
2	0

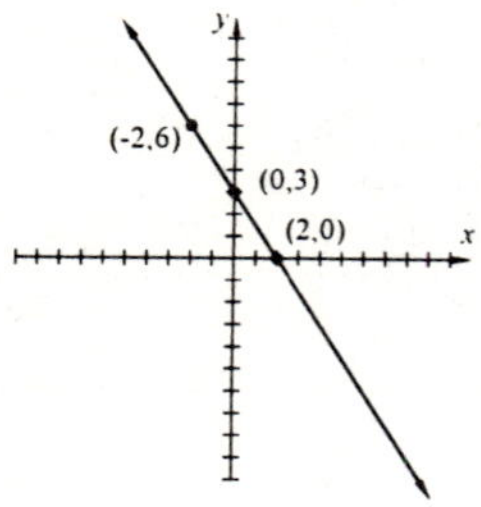

23. $2x + 5y - 2 = -12$

$2x + 5y = -10$

x	y
-5	0
0	-2
5	-4

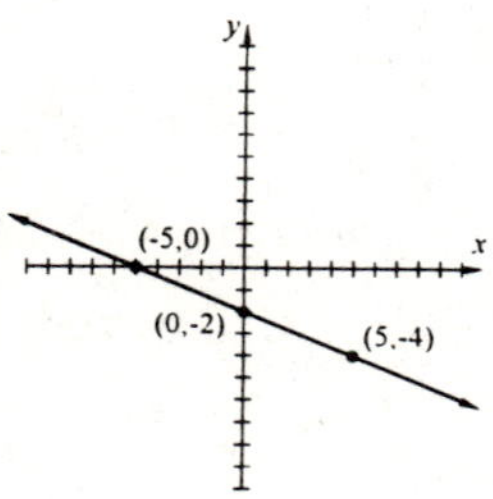

25. $3y + 1 = 7$

$3y = 6$

$y = 2$

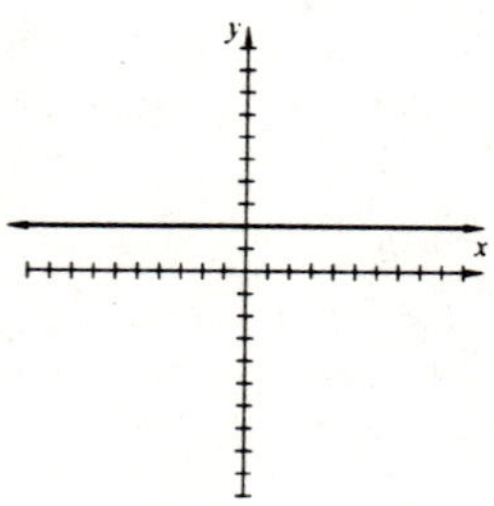

27. $C = 8m$

m	0	15	30	45	60	75
C	0	120	240	360	480	600

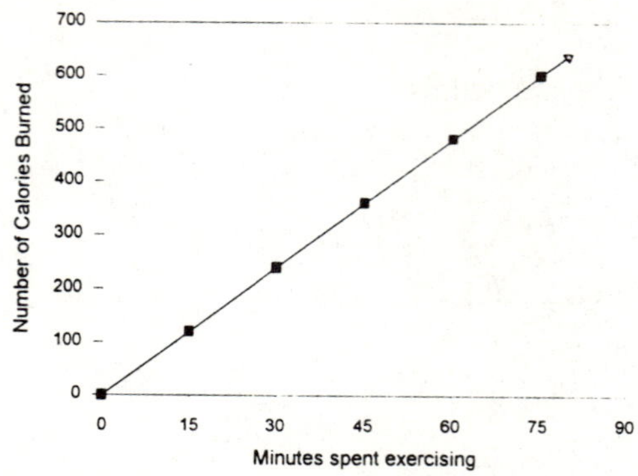

29. $C = 1.6t + 343$

t	0	4	8
C	343	349.4	355.8

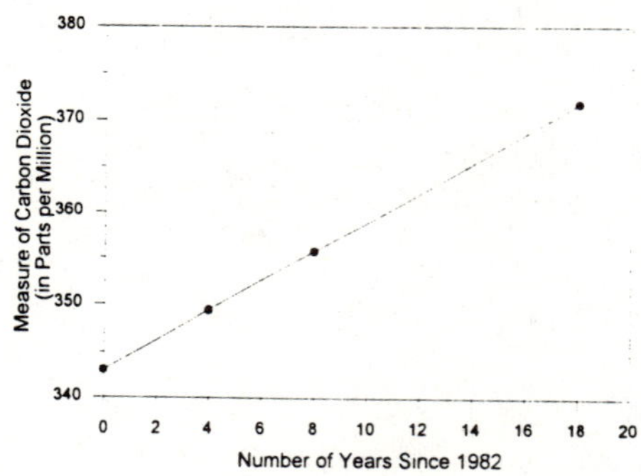

Cumulative Review Problems

31. $6 + (-2)^3 - (5 - 9) \div 2$

$= 6 + (-2)^3 - (-4) \div 2$

$= 6 + (-8) - (-4) \div 2$

$= 6 + (-8) - (-2)$

$= 0$

33. $x = \text{length}$

$\frac{x}{2} + 1 = \text{width}$

$$p = 2L + 2w$$
$$53 = 2x + 2\left(\frac{x}{2} + 1\right)$$
$$53 = 2x + x + 2$$
$$51 = 3x$$
$$17 = x$$
$$\frac{17}{2} + 1 = 9\frac{1}{2}$$

width $= 9\frac{1}{2}$ meters

Length = 17 meters

35. x = cost of mailing from Japan

$\frac{1}{2}x - 2$ = cost of mailing from U.S.

$$x + \frac{1}{2}x - 2 = 184$$
$$2x + x - 4 = 368$$
$$3x - 4 = 368$$
$$3x = 372$$
$$x = 124$$

The cost is \$1.24

3.3 Exercises

Use $m = \frac{y_2 - y_1}{x_2 - x_1}$ in Exercises 1-11.

1. $(6, 6)$ and $(9, 3)$

$$m = \frac{3-6}{9-6} = \frac{-3}{3} = -1$$

3. $(-2, 1)$ and $(3, 4)$

$$m = \frac{4-1}{3-(-2)} = \frac{3}{5}$$

5. $(-7, -4)$ and $(3, -8)$

$$m = \frac{-8-(-4)}{3-(-7)} = \frac{-4}{10} = -\frac{2}{5}$$

7. $(-3, 0)$ and $(0, -4)$

$$m = \frac{-4-0}{0-(-3)} = \frac{-4}{3} = -\frac{4}{3}$$

9. $\left(\frac{3}{4}, -4\right)$ and $(2, -8)$

$$m = \frac{-8-(-4)}{2-\frac{3}{4}} = \frac{-4}{\frac{5}{4}} = -\frac{16}{5}$$

11. You cannot find the slope of the line passing through $(5, -12)$ and $(5, -6)$ because division by zero is impossible.

Use $y = mx + b$ in Exercises 13-41.

13. $y = 8x + 9;\ m = 8,\ b = 9$

15. $y = -3x + 4,\ m = -3,\ b = 4$

17. $y = \frac{5}{6}x - \frac{2}{9};\ m = \frac{5}{6},\ b = -\frac{2}{9}$

19. $y = -6x;\ m = -6,\ b = 0$

21. $6x + y = \frac{4}{5}$

$$y = -6x + \frac{4}{5};\quad m = -6,\ b = \frac{4}{5}$$

23. $5x + 2y = 3$

$$y = -\frac{5}{2}x + \frac{3}{2};\quad m = -\frac{5}{2},\ b = \frac{3}{2}$$

25. $7x - 3y = 4$

$$y = \frac{7}{3}x - \frac{4}{3};\quad m = \frac{7}{3},\ b = -\frac{4}{3}$$

27. $m = \frac{3}{4},\ b = 2$

(a) $y = \frac{3}{4}x + 2$

(b) $4y = 3x + 8$

$3x - 4y = -8$

29. $m = 6,\ b = -3$

(a) $y = 6x - 3$

(b) $6x - y = 3$

31. (a) $y = -\frac{5}{4}x - \frac{3}{4}$

(b) $4y = -5x - 3$

$5x + 4y = -3$

33. $m = \frac{1}{2},\ b = -3$

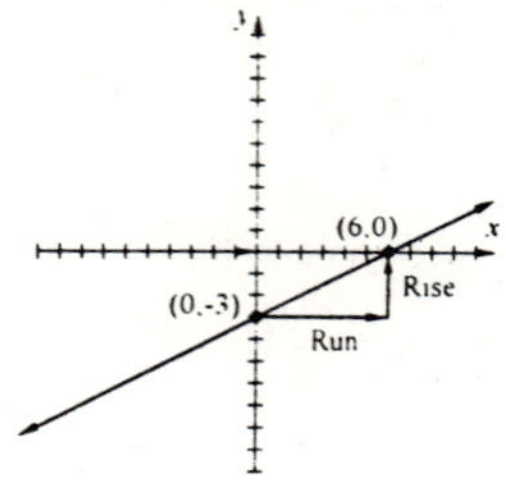

35. $m = -\frac{5}{3},\ b = 2$

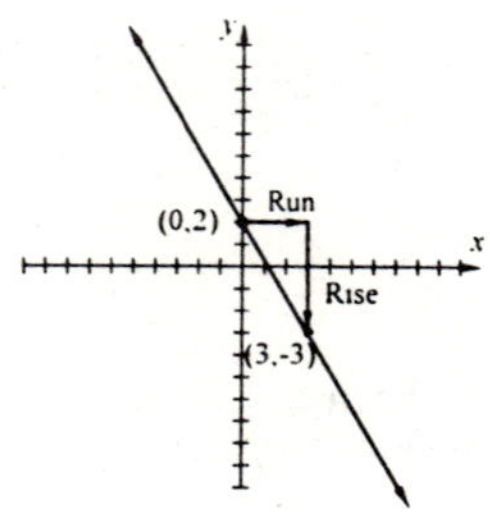

37. $y = \frac{3}{4}x - 5$

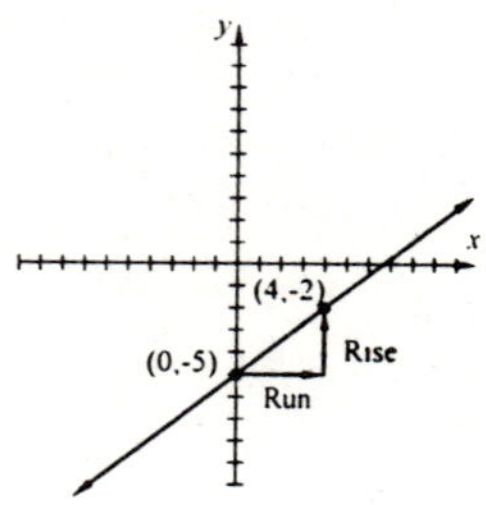

39. $y + 2x = 3$

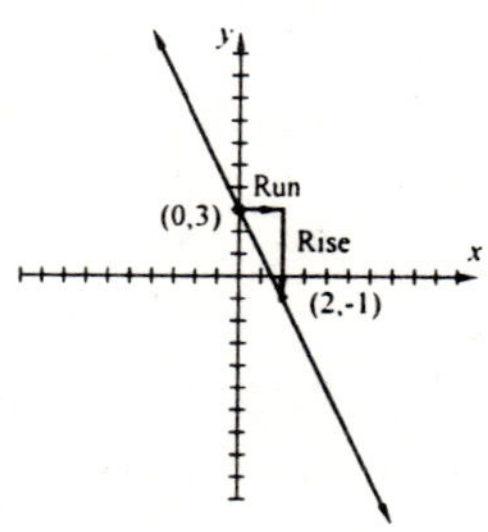

41. $y = 2x$

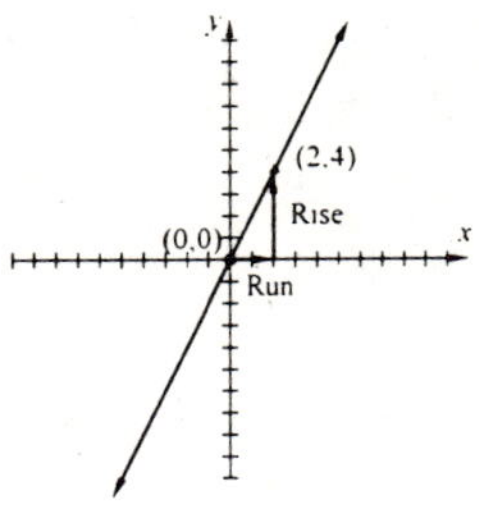

43. $m = \dfrac{13}{5}$

(a) $m_{11} = m = \dfrac{13}{5}$

(b) $m_{\perp} = -\dfrac{1}{m} = -\dfrac{5}{13}$

45. $m_1 = -\dfrac{5}{8}$

(a) $m_2 = m_2 = -\dfrac{5}{8}$

(b) $m_2 = -\dfrac{1}{m_1} = \dfrac{8}{5}$

47.

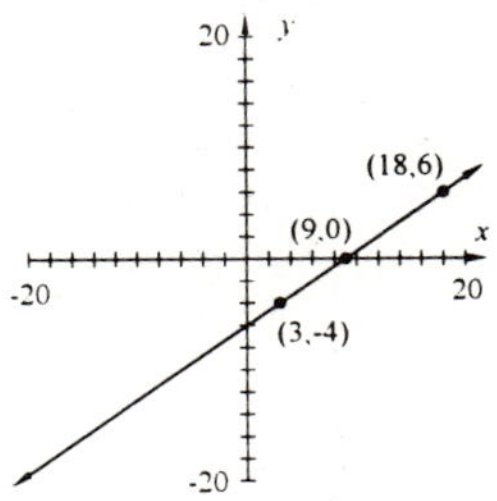

Yes

$$m = \frac{0-(-4)}{9-3} = \frac{4}{6} = \frac{2}{3}$$

From the graph, $b = -6$

$$y = \frac{2}{3}x - 6$$

49. $y = \dfrac{1}{10}(22x + 830)$

(a) $y = 2.2x + 83$

(b) $m = 2.2,\ b = 83$

(c) The slope is the increase in the number of civilians employed in the United States in millions for each year during the period from 1970 to 1990.

Cumulative Review Problems

51. $\dfrac{1}{4}x + 3 > \dfrac{2}{3}x + 2$

$$3x + 12 \cdot 3 > 4 \cdot 2x + 12 \cdot 2$$
$$3x + 36 > 8x + 24$$
$$-5x > -12$$
$$x < \frac{12}{5}$$

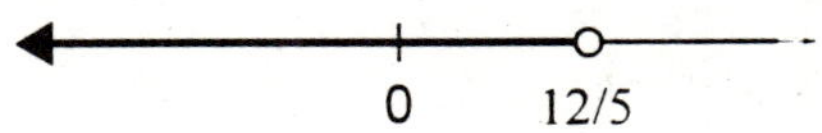

53. $7x - 2(x + 3) \le 4(x - 7)$

$$7x - 2x - 6 \le 4x - 28$$
$$x \le -22$$

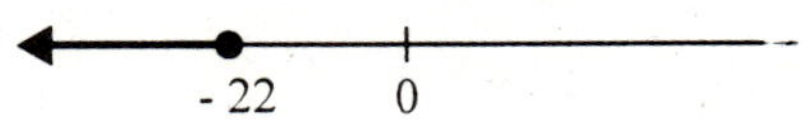

55. x = the number of counselors

$$\frac{3}{25} = \frac{x}{189}$$
$$25x = 3(189)$$
$$25x = 567$$
$$x \approx 22.68$$

A minimum of 23 counselors

Putting Your Skills to Work

Use $p = \frac{5}{11}d + 15$ in Exercises 1 - 5.

1. $d = 22$

$p = \frac{5}{11}(22) + 15 = 25 \text{ lb/in}^2$

2. $p = 35$

$35 = \frac{5}{11}(d) + 15$

$20 = \frac{5}{11}(d)$

$44 = d$

Depth is 44 feet

3. $(22, 25)$ and $(44, 35)$

$m = \frac{35-25}{44-22} = \frac{10}{22} = \frac{5}{11}$

4.

d	0	22	44
p	15	25	35

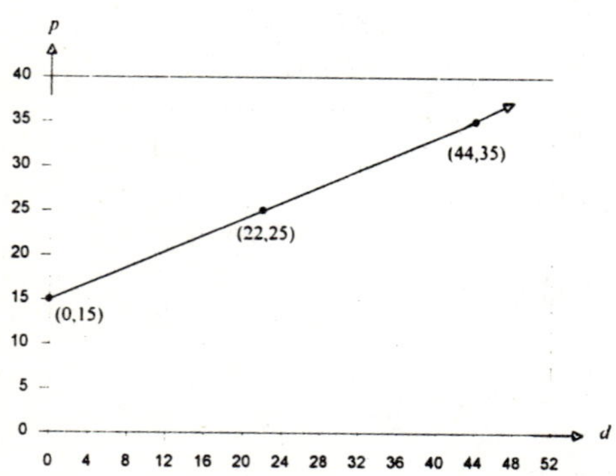

5. From any point on the graph, count over 5 in the positive x direction. Then you must count up 11 in the y – direction to get back to the line. This agrees.

6.

d	0	11	22	33	44
p	15	20	25	30	35

(a) $20 - 15 = 5$

(b) $11 - 0 = 11$

(c) $m = \frac{5}{11}$; Slope is the difference between the p values over the difference between the d values.

7. $(0, 18)$ and $(5, 22)$

$m = \frac{22-18}{5-0} = \frac{4}{5}$

8. $m = \frac{27.6-22}{12-5} = \frac{5.6}{7} = \frac{4}{5} = 0.8$

The two slopes are equivalent. The points lie on a line.

9. When $d = 0$, $p = 18$, so $b = 18$

$p = \frac{4}{5}d + 18$

10. $p = \frac{4}{5}(44) + 18$

$= 53.2$ pounds per square inch

The pressure is greater.

11. The new liquid is more dense because the pressure is greater at the same depth.

3.4 Exercises

Use $m = \dfrac{y_2 - y_1}{x_2 - x_1}$ and $y = mx + b$

as needed in Exercises 1-31.

1. $m = 4,\ (-3, 0)$

$0 = 4(-3) + b \Rightarrow b = 12$

$y = 4x + 12$

3. $m = -2,\ (4, 3)$

$3 = -2(4) + b \Rightarrow b = 11$

$y = -2x + 11$

5. $m = -3,\ \left(\frac{1}{2}, 2\right)$

$2 = -3\left(\frac{1}{2}\right) + b \Rightarrow b = \frac{7}{2}$

$y = -3x + \frac{7}{2}$

7. $m = -\frac{2}{5},\ (5, -3)$

$-3 = -\frac{2}{5}(5) + b \Rightarrow b = -1$

$y = -\frac{2}{5}x - 1$

9. $(3, -12)$ and $(-4, 2)$

$m = \dfrac{2-(-12)}{-4-3} = \dfrac{14}{-7} = -2$

$2 = -2(-4) + b \Rightarrow b = -6$

$y = -2x - 6$

11. $(2, -3)$ and $(-1, 6)$

$m = \dfrac{6-(-3)}{-1-2} = \dfrac{9}{-3} = -3$

$-3 = -3(2) + b \Rightarrow b = 3$

$y = -3x + 3$

13. $(3, 5)$ and $(-1, -15)$

$m = \dfrac{-15-5}{-1-3} = \dfrac{-20}{-4} = 5$

$5 = 5(3) + b \Rightarrow b = -10$

$y = 5x - 10$

15. $\left(1, \frac{5}{6}\right)$ and $\left(3, \frac{3}{2}\right)$

$m = \dfrac{\frac{3}{2} - \frac{5}{6}}{3-1} = \dfrac{\frac{4}{6}}{2} = \dfrac{1}{3}$

$\frac{5}{6} = \frac{1}{3}(1) + b \Rightarrow b = \frac{1}{2}$

$y = \frac{1}{3}x + \frac{1}{2}$

17. $b = 1,\ m = -\frac{2}{3}$

$y = -\frac{2}{3}x + 1$

19. $b = -4, \quad m = \frac{2}{3}$

    $y = \frac{2}{3}x - 4$

21. $b = 0, \quad m = 3$

    $y = 3x$

23. $b = -2, \quad m = 0$

    $y = -2$

25. $m = 0, \ (7, -2)$

    The line is horizontal so

    $y = -2$

27. $(0, 2)$, perpendicular to $y-$axis $\Rightarrow m = 0$

    $y = 2$

29. $(0, -4) \Rightarrow b = -4$

    Parallel to $y = \frac{3}{4}x + 2 \Rightarrow m_1 = \frac{3}{4}$

    $m_2 = m_1 = \frac{3}{4}$

    $y = \frac{3}{4}x - 4$

31. Perpendicular to $y = 2x - 9 \Rightarrow m_1 = 2$

    $m_2 = -\frac{1}{m_1} = -\frac{1}{2}$

    $y = mx + b, \ (2, 3)$

    $3 = -\frac{1}{2}(2) + b \Rightarrow b = 4$

    $y = -\frac{1}{2}x + 4$

33. $(0, 227)$ and $(10, 251)$

    $m = \frac{251 - 227}{10 - 0} = \frac{24}{10} = 2.4, \ b = 227$

    $y = 2.4x + 227$

Cumulative Review Problems

35. $\frac{x}{2} + \frac{1}{3} = 4\left(\frac{1}{2} + \frac{x}{24}\right), \ x \stackrel{?}{=} 5$

    $\frac{5}{2} + \frac{1}{3} \stackrel{?}{=} 4\left(\frac{1}{2} + \frac{5}{24}\right)$

    $\frac{17}{6} \stackrel{?}{=} 4\left(\frac{17}{24}\right)$

    $\frac{17}{6} = \frac{17}{6}$, $x = 5$ is the solution

37. 2nd week:

    Price $= 80 - 0.15(80) = \$68$

    3rd week:

    Price $= 68 - 0.1(68) = \$61.20$

39. Number shipped $= 35(60) = 2100$

    Percent defective $= \frac{104}{2100} \cdot 100\% \approx 4.95\%$

3.5 Exercises

1. Any convenient point not on the boundary line may be used as a test point.

3. $y > 2 - 3x$

   Test point: $(0, 0)$

   $0 > 2 - 3(0)$

   $0 > 2$ False

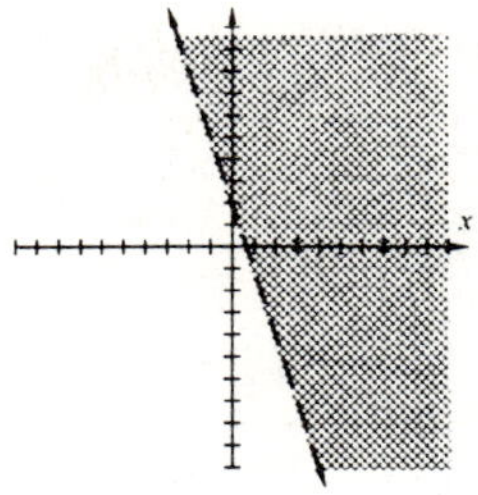

5. $2x - 3y < 6$

   Test point: $(0, 0)$

   $2(0) - 3(0) < 6$

   $0 < 6$ True

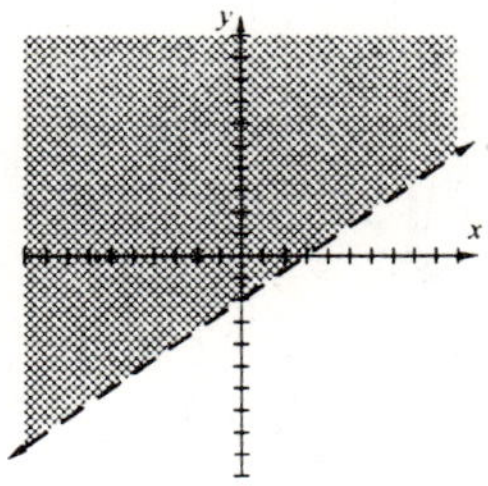

7. $2x - y \geq 3$

   Test point: $(0, 0)$

   $2(0) - (0) \geq 3$

   $0 \geq 3$ False

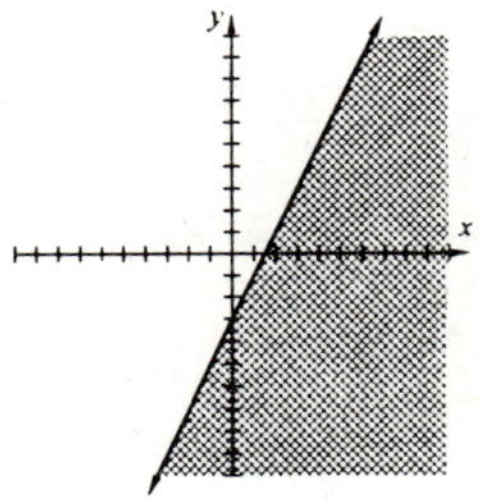

9. $y \geq 3x$

   Test point: $(1, 0)$

   $0 \geq 3(1)$

   $0 \geq 3$ False

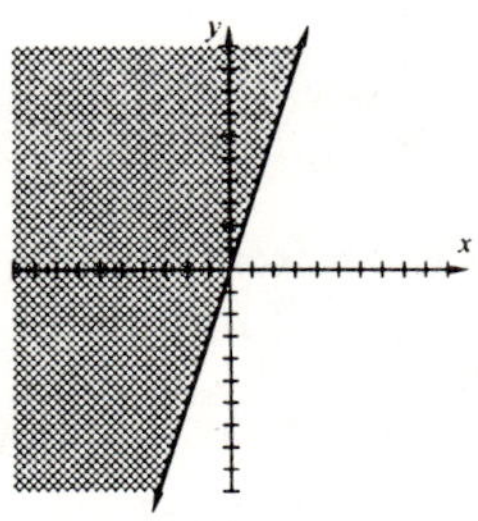

11. $y < -\frac{1}{2}x$

    Test point: $(1, 0)$

    $0 < -\frac{1}{2}(1)$

    $0 < -\frac{1}{2}$ False

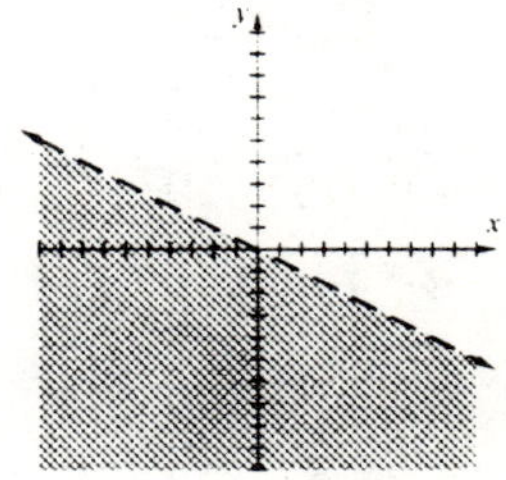

13. $x \geq 2$

    Test point: $(0, 0)$

    $0 \geq 2$ False

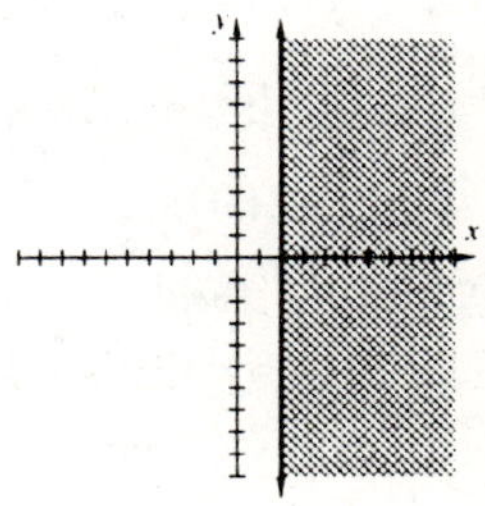

15. $3x - y + 1 \geq 0$

    Test point: $(0, 0)$

    $3(0) - (0) + 1 \geq 0$

    $1 \geq 0$ True

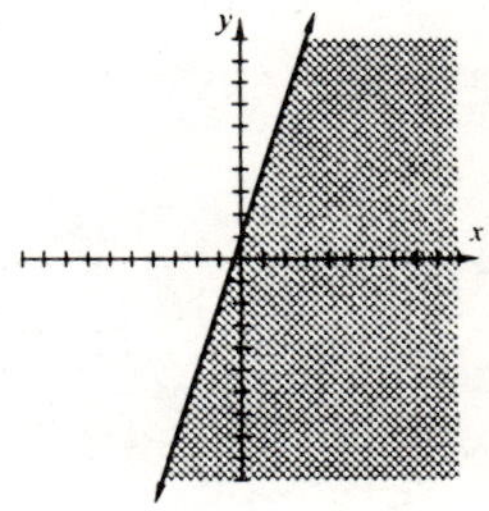

17. $2x > -3y$

    Test point: $(1, 1)$

    $2(1) > -3(1)$

    $2 > -3$ True

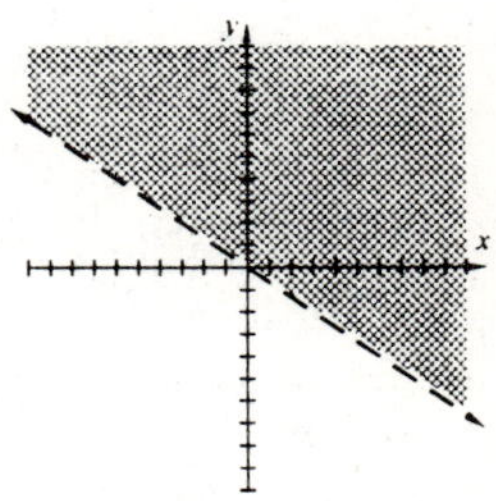

19. $2x > 3 - y$

Test point: $(0, 0)$

$2(0) > 3 - (0)$

$0 > 3$ False

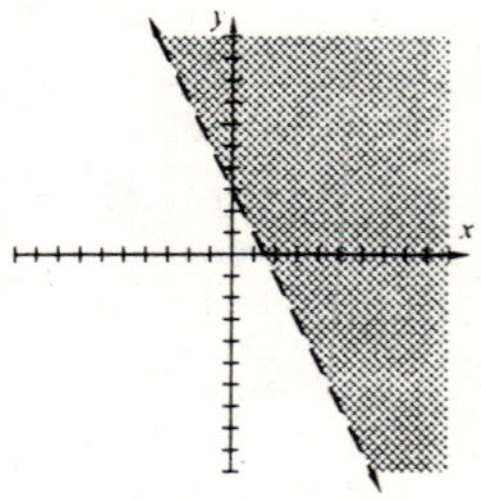

21. $x > -2y$

Test point: $(1, 0)$

$(1) > -2(0)$

$1 > 0$ True

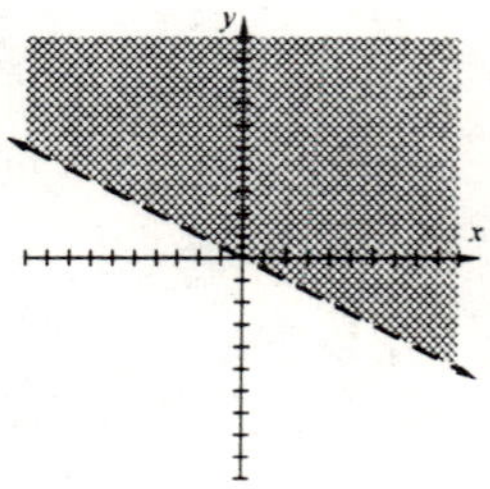

23. $3x + 3 > 5x - 3$

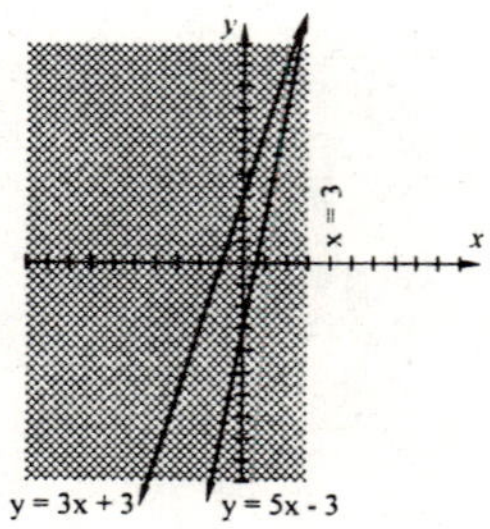

(a) $x < 3$

(b) $3x + 3 > 5x - 3$

$-2x > -6$

$x < 3$

(c) $3x + 3$ will be greater than $5x - 3$ for those values of x where the graph of $y = 3x + 3$ lies above the graph of $y = 5x - 3$.

Cumulative Review Problems

25. $m = \frac{y_2 - y_1}{x_2 - x_1}$, $(3, 0)$, $(6, -7)$

$m = \frac{-7 - 0}{6 - 3} = -\frac{7}{3}$

$y = mx + b$, $m = -\frac{7}{3}$, $(3, 0)$

$0 = -\frac{7}{3}(3) + b$

$0 = -7 + b$

$7 = b$

$y = -\frac{7}{3}x + 7$

27. x = the number of calls

$0.492(1000) = 492$ sales so far

$\frac{x + 492}{x + 1000} = 0.5$

$x + 492 = 0.5(x + 1000)$

$x + 492 = 0.5x + 500$

$0.5x = 8$

$x = 16$

She must make 16 more calls.

3.6 Exercises

1. You can describe a function using a table of values, an algebraic equation, or a graph.

3. The domain of a function is the set of possible values of the independent variable.

5. If a vertical line can intersect the graph more than once, the relation is not a function.

7. $\left\{\left(\frac{1}{2}, 5\right), \left(\frac{1}{4}, 10\right), \left(\frac{3}{4}, 6\right), \left(\frac{1}{2}, 6\right)\right\}$

   Domain = $\left\{\frac{1}{4}, \frac{1}{2}, \frac{3}{4}\right\}$

   Range = $\{5, 6, 10\}$

   Not a function

9. $\{(7.3, 1), (0, 8), (2, 1)\}$

   Domain = $\{0, 2, 7.3\}$

   Range = $\{1, 8\}$

   Function

11. $\{(5.6, 8), (5.8, 6), (6, 5.8), (5, 6)\}$

    Domain = $\{5, 5.6, 5.8, 6\}$

    Range = $\{5.8, 6, 8\}$

    Function

13. $\{(85, 3), (95, 11), (110, 15), (110, 20)\}$

    Domain = $\{85, 95, 110\}$

    Range = $\{3, 11, 15, 20\}$

    Not a function

15. $y = x^2 - 1$

x	y
−2	3
−1	0
0	−1
1	0
2	3

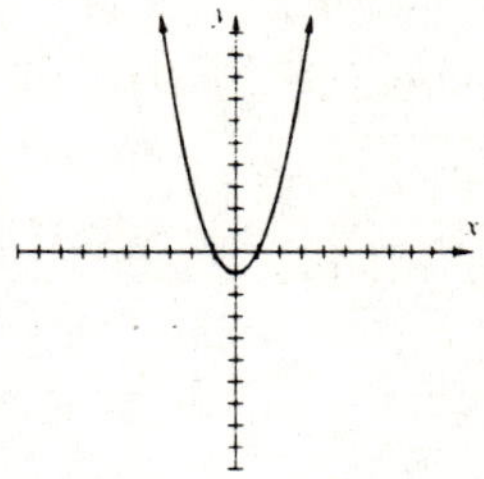

17. $y = \frac{1}{2}x^2$

x	y
−2	2
−1	0.5
0	0
1	0.5
2	2

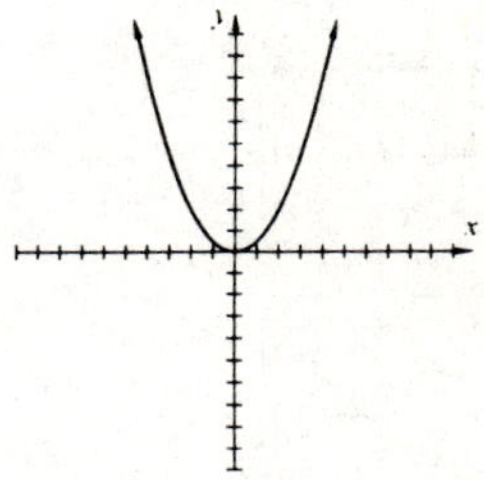

19. $x = \frac{1}{2}y^2$

x	y
2	−2
0.5	−1
0	0
0.5	1
2	2

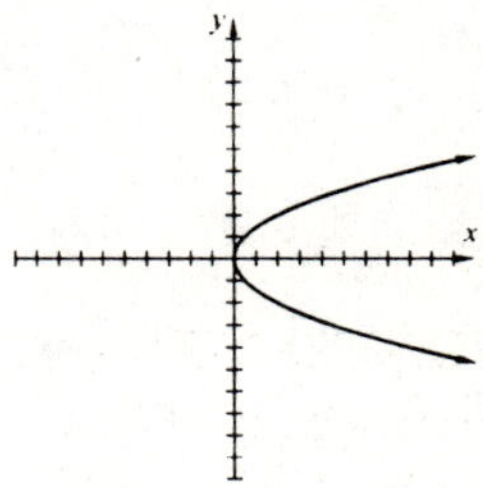

21. $x = 2y^2$

x	y
8	−2
2	−1
0	0
2	1
8	2

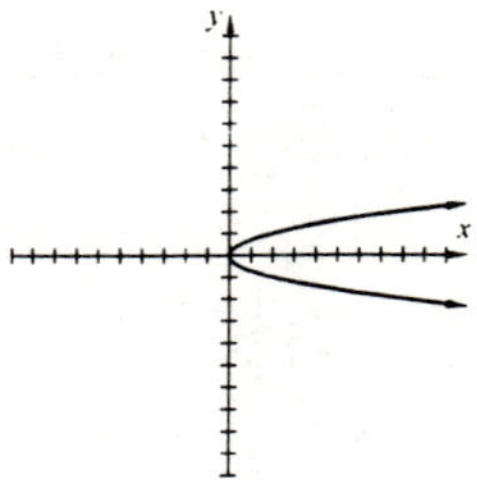

23. $y = -\frac{2}{x}$

x	y
−4	$\frac{1}{2}$
−2	1
−1	2
1	−2
2	−1
4	$-\frac{1}{2}$

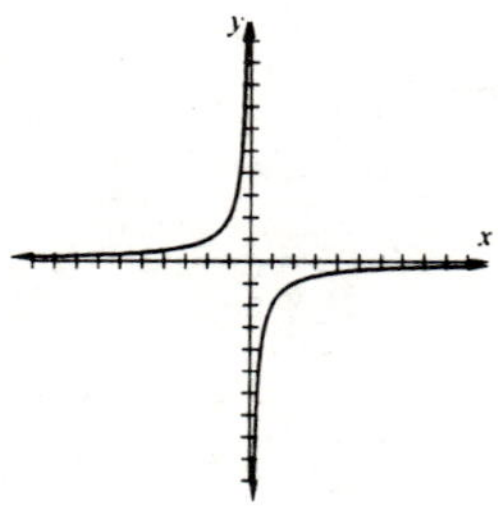

25. $y = -\frac{6}{x^2}$

x	y
−3	$-\frac{2}{3}$
−2	$-\frac{3}{2}$
−1	−6
1	−6
2	$-\frac{3}{2}$
3	$-\frac{2}{3}$

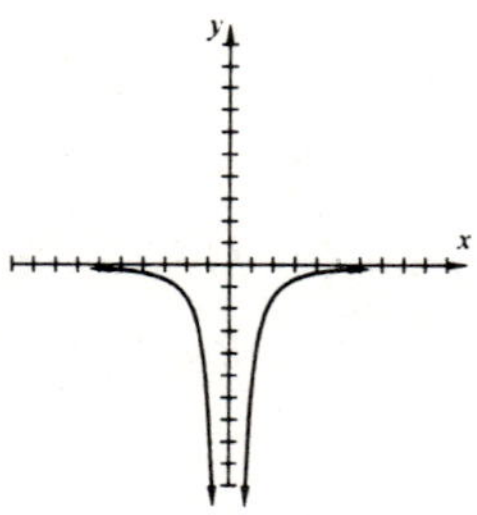

27. $x = (y-2)^2$

x	y
4	0
1	1
0	2
1	3
4	4

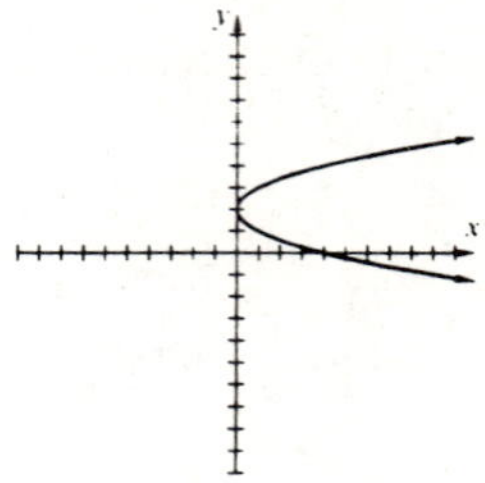

29. $x = \frac{2}{y+1}$

x	y
−2	$-\frac{1}{2}$
−1	−1
0	−2
1	2
2	1
3	$\frac{1}{2}$

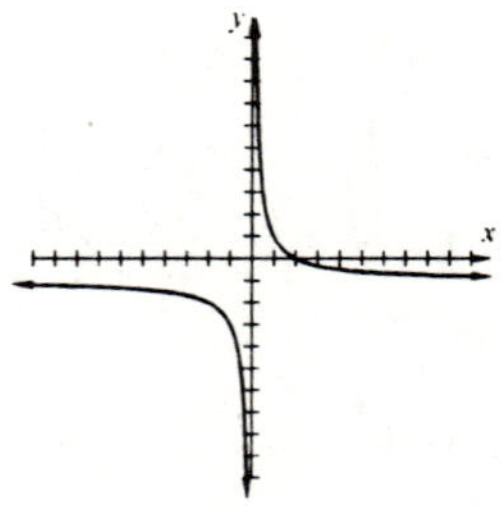

31. Function, passes the vertical line test

33. Not a function, fails the vertical line test

35. Function, passes the vertical line test

37. Not a function, fails the vertical line test

39. $f(x) = 3 - 5x$

(a) $f(-1) = 3 - 5(-1) = 8$

(b) $f(1) = 3 - 5(1) = -2$

(c) $f(2) = 3 - 5(2) = -7$

41. $f(x) = x^2 + 2x - 3$

(a) $f(-1) = (-1)^2 + 2(-1) - 3 = -4$

(b) $f(0) = (0)^2 + 2(0) - 3 = -3$

(c) $f(3) = (3)^2 + 2(3) - 3 = 12$

43. $f(x) = x - 2 + \frac{8}{x}$

(a) $f(-2) = -2 - 2 + \frac{8}{-2} = -8$

(b) $f(1) = 1 - 2 + \frac{8}{1} = 7$

(c) $f(4) = 4 - 2 + \frac{8}{4} = 4$

45. $f(1990) = 30\%$

47. $f(1985) - f(1990) = 40 - 30 = 10\%$

Between 1985 and 1990

49. $f(x) = -0.6x^2 + 22x + 411$

$f(0) = -0.6(0)^2 + 22(0) + 411 = 411$

$f(4) = -0.6(4)^2 + 22(4) + 411 = 489.4$

$f(10) = -0.6(10)^2 + 22(10) + 411 = 571$

$f(20) = -0.6(20)^2 + 22(20) + 411 = 611$

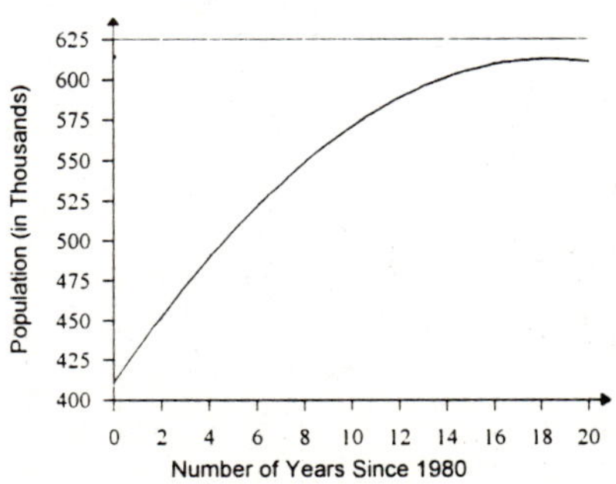

The curve slopes upward more steeply for smaller values of x. The growth rate is decreasing as x gets larger.

Cumulative Review Problems

51. If $x = 2$, and $y = -3$, then

$$\begin{aligned} x^2 + xy - 5 &= (2)^2 + (2)(-3) - 5 \\ &= 4 - 6 - 5 \\ &= -7 \end{aligned}$$

53.
$$\begin{aligned} 6x - 2y &= -10 \\ 6x - 6x - 2y &= -6x - 10 \\ -2y &= -6x - 10 \\ \frac{-2y}{-2} &= \frac{-6x}{-2} - \frac{10}{-2} \\ y &= 3x + 5 \end{aligned}$$

Putting Your Skills to Work

1. $2.47(60)(60)(24) = 213,408$ acres/day

$213,408(365) \approx 78,000,000$ acres/year

2. $1756(2.47) = 4337$ million acres

$= 4,337,000,000$ acres

3. $1992: 4337 - 2(78) = 4181$ million acres

$1995: 4337 - 5(78) = 3947$ million acres

4. t = number of years after 1990

$(0, 4337)$ and $(5, 3947)$

$m = \frac{3947 - 4337}{5 - 0} = -78, b = 4337$

$r(t) = -78t + 4337$

5.

t	0	5	10	15	20
r	4337	3947	3557	3167	2777

6.

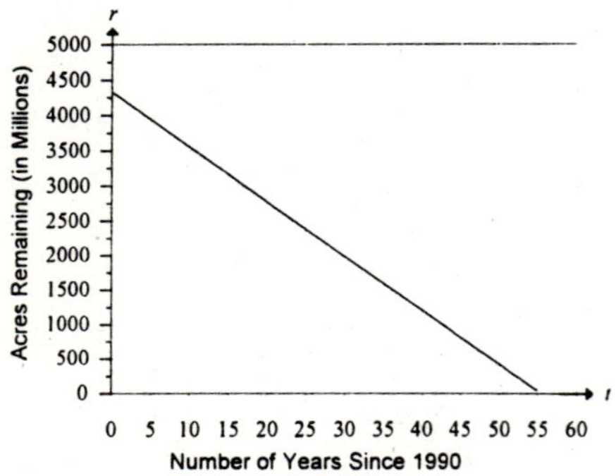

7. $t = 2035 - 1990 = 45$

$r(45) \approx 827$ million acres

8. $r(55) = 0$

There will be no rainforests after $1990 + 55 = 2045$.

Chapter 3 Review Problems

1. A: $(2, -3)$, B: $(-1, 0)$, C: $(3, 2)$, D: $(-2, -3)$

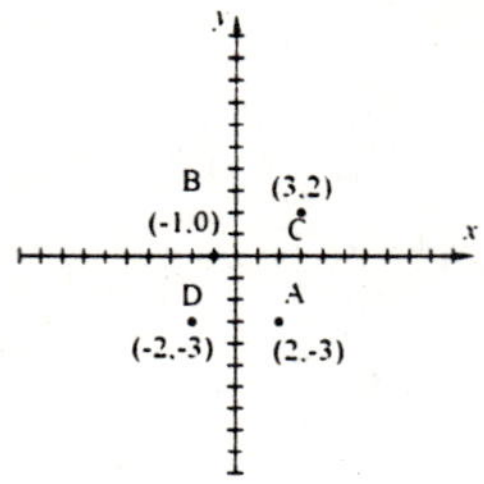

2. $E: (4, 4)$, $F: (0, 3)$, $G: (1, -4)$, $H: (-4, -1)$

3. $y = 3x - 5$

(a) $(0, \ )$: $y = 3(0) - 5 = -5$: $(0, -5)$

(b) $(3, \ )$: $y = 3(3) - 5 = 4$: $(3, 4)$

4. $2x + 5y = 12$

(a) $(1, \ )$: $2(1) + 5y = 12$

$5y = 10$

$y = 2$: $(1, 2)$

(b) $(\ , 4)$: $2x + 5(4) = 12$

$2x = -8$

$x = -4$: $(-4, 4)$

5. $x = 6$

(a) $(\ , -1)$: $x = 6$: $(6, -1)$

(b) $(\ , 3)$: $x = 6$: $(6, 3)$

6. $3y = 2x + 6$

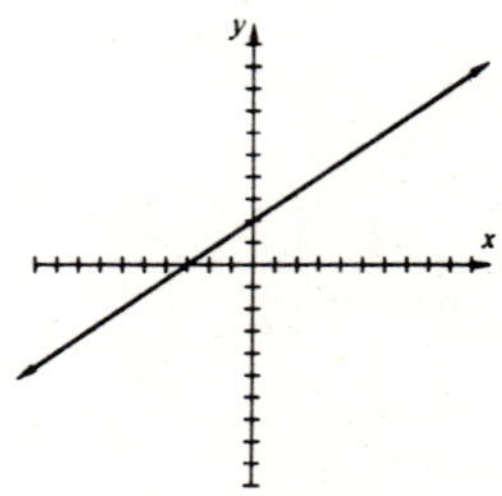

7. $5y + x = -15$

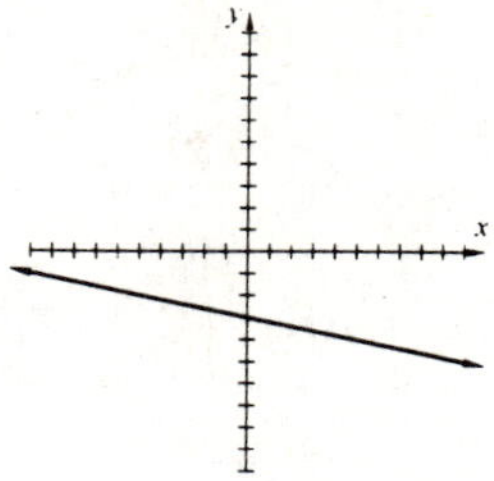

8. $3x + 5y = 15 + 3x$

$5y = 15$

$y = 3$

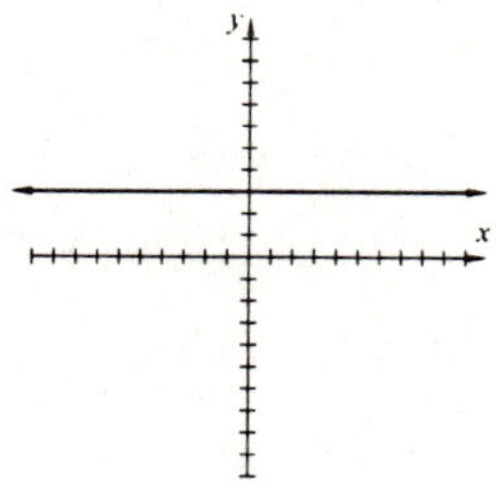

9. $(5, -3)$ and $\left(2, -\frac{1}{2}\right)$

$m = \frac{y_2 - y_1}{x_2 - x_1} = \frac{-\frac{1}{2} - (-3)}{2 - 5} = -\frac{5}{6}$

10. $9x - 11y + 15 = 0$

$9x + 15 = 11y$

$\frac{9}{11}x + \frac{15}{11} = y$

$m = \frac{9}{11},\ b = \frac{15}{11}$

11. $m = -\frac{1}{2},\ b = 3$

$y = mx + b$

$y = -\frac{1}{2}x + 3$

12. $m = -\frac{2}{3}$

$m_{\perp} = -\frac{1}{m} = \frac{3}{2}$

13. $y = -\frac{1}{2}x + 3$

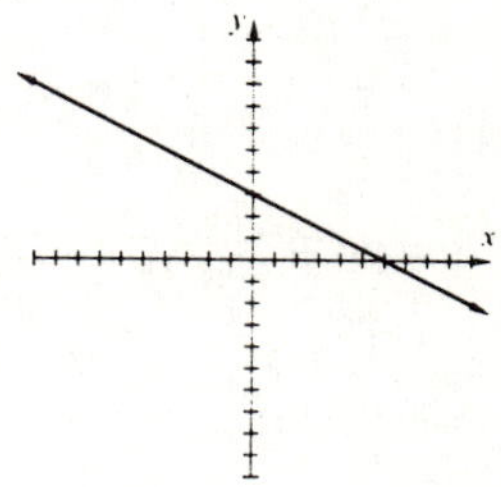

14. $2x - 3y = -12$

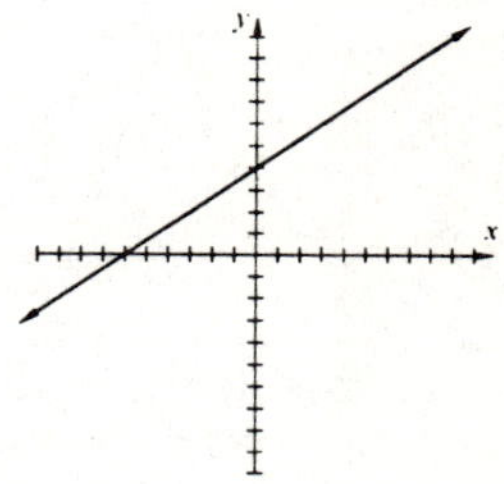

15. $5x + 2y = 20 + 2y$

$5x = 20$

$x = 4$

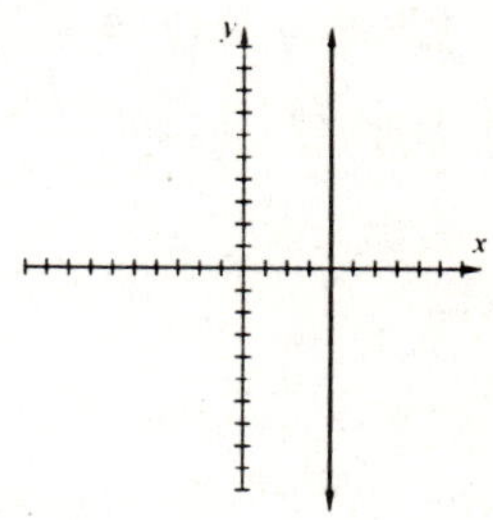

16. $(5, 6),\ m = 2$

$y = mx + b$

$6 = 2(5) + b \Rightarrow b = -4$

$y = 2x - 4$

17. $(3, -4)$, $m = -6$

$y = mx + b$

$-4 = -6(3) + b \Rightarrow b = 14$

$y = -6x + 14$

18. $(6, -3)$, $m = \frac{1}{3}$

$y = mx + b$

$-3 = \frac{1}{3}(6) + b \Rightarrow b = -5$

$y = \frac{1}{3}x - 5$

19. $(3, 7)$ and $(-6, 7)$

$m = \frac{y_2 - y_1}{x_2 - x_1} = \frac{7-7}{-6-3} = 0$

Horizontal line with $y = 7$

$y = 7$

20. $(0, -3)$ and $(3, -1)$

$m = \frac{-3-(-1)}{0-3} = \frac{2}{3}$, $b = -3$

$y = \frac{2}{3}x - 3$

21. $(0, 1)$ and $(1, -2)$

$m = \frac{-2-1}{1-0} = -3$, $b = 1$

$y = -3x + 1$

22. Vertical line: $x = -5$

23. $y < \frac{1}{3}x + 2$

Test point: $(0, 0)$

$0 < \frac{1}{3}(0) + 2$

$0 < 2$ True

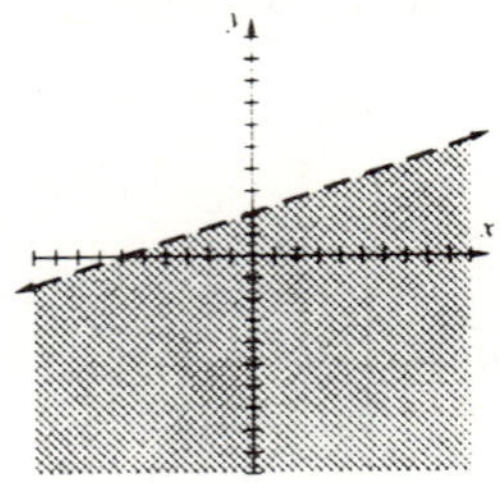

24. $3x + 2y \geq 12$

Test point: $(0, 0)$

$3(0) + 2(0) \geq 12$

$0 \geq 12$ False

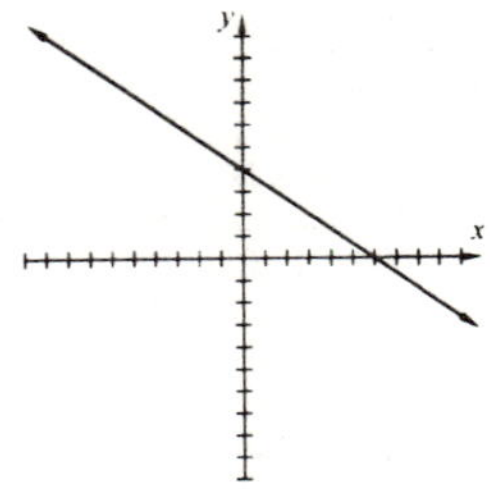

25. $y \geq -2$

Test point: $(0, 0)$

$0 \geq -2$ True

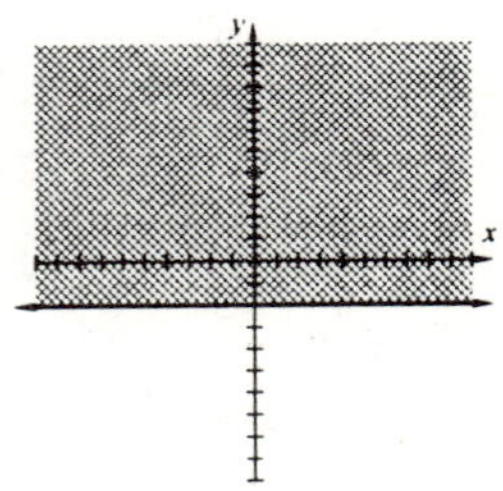

26. $\{(5, -6), (-6, 5), (-5, 5), (-6, -6)\}$

Domain $\{-6, -5, 5\}$

Range $\{-6, 5\}$

Not a function

27. $\{(3, -7), (-7, 3), (-3, 7), (7, -3)\}$

Domain $\{3, -7, -3, 7\}$

Range $\{-7, 3, 7, -3\}$

Function

28. Function, passes the vertical line test

29. Not a function, fails the vertical line test

30. Function, passes the vertical line test

31. $y = x^2 - 5$

x	y
−2	−1
−1	−4
0	−5
1	−4
2	−1

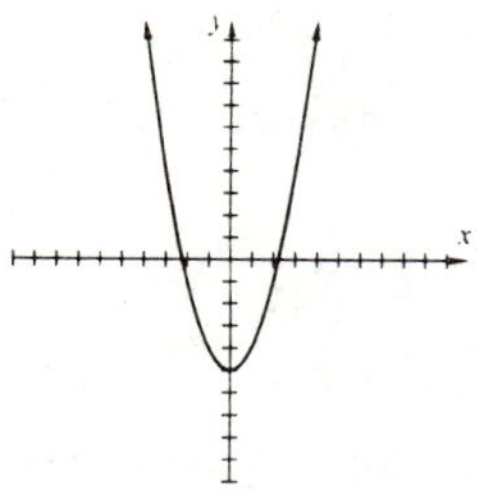

32. $x = y^2 + 3$

x	y
7	−2
4	−1
3	0
4	1
7	2

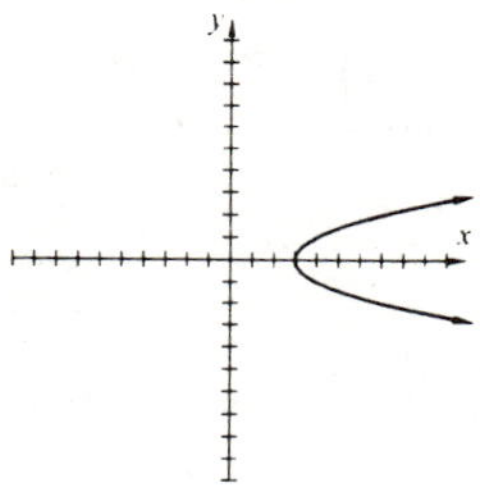

33. $y = (x-3)^2$

x	y
1	4
2	1
3	0
4	1
5	4

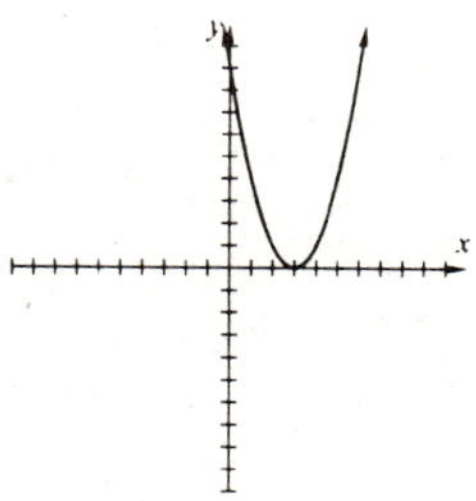

34. $f(x) = 7 - 6x$

(a) $f(0) = 7 - 6(0) = 7$

(b) $f(-4) = 7 - 6(-4) = 31$

35. $g(x) = -2x^2 + 3x + 4$

(a) $g(-1) = -2(-1)^2 + 3(-1) + 4 = -1$

(b) $g(3) = -2(3)^2 + 3(3) + 4 = -5$

36. $p(x) = \frac{-3}{x}$

(a) $p(-3) = \frac{-3}{-3} = 1$

(b) $p(0.5) = \frac{-3}{0.5} = -6$

37. $f(x) = \frac{2}{x+4}$

(a) $f(-2) = \frac{2}{-2+4} = 1$

(b) $f(6) = \frac{2}{6+4} = \frac{1}{5}$

38. $f(x) = \frac{x+6}{x-5}$

(a) $f(0) = \frac{0+6}{0-5} = -\frac{6}{5}$

(b) $f(-2) = \frac{-2+6}{-2-5} = -\frac{4}{7}$

Use $y = 150 + 0.15x$ in Exercises 39-44.

39. $y = 150 + 0.15(1000) = \$300$

40. $y = 150 + 0.15(3000) = \$600$

41. $y = 150 + 0.15x$
$y = mx + b$
$m = 0.15$

42. The slope indicates that the cost of the trip increases \$0.15 for each mile.

43. $540 = 150 + 0.15x$
$390 = 0.15x$
$x = 2600$ miles

44. $750 = 150 + 0.15x$
$600 = 0.15x$
$x = 4000$ miles

Use $y = 30 + 0.09x$ in Exercises 45-50,

45. $y = 30 + 0.09(2000) = \$210$

46. $y = 30 + 0.09(1600) = \$174$

47. $y = 0.09x + 30$
$y = mx + b,\ b = 30$
It tells us that if no electricity is used, the cost is a minimum of \$30.

48. $y = 0.09x + 30$
$y = mx + b,\ m = 0.09$
It tells us that the bill increases \$0.09 for each kilowatt-hour of use.

49. $147 = 0.09x + 30$
$117 = 0.09x$
$x = 1300$ kilowatt-hours

50. $246 = 0.09x + 30$
$216 = 0.09x$
$x = 2400$ kilowatt-hours

Chapter 3 Test

1. $B = (6, 1)$ $C = (-4, -3)$
$D = (-3, 0)$ $E = (5, -2)$

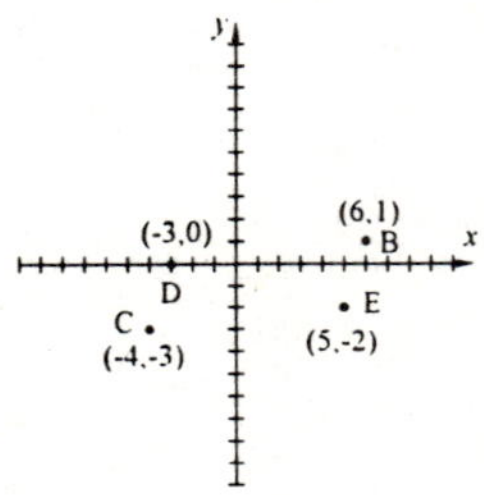

2. $6x - 3 = 5x - 2y$
$x + 2y = 3$

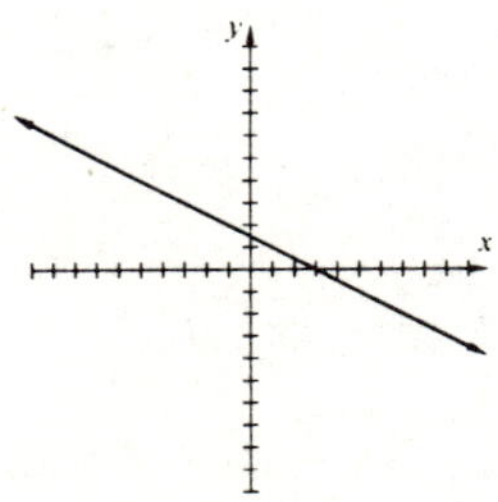

3. $12x - 3y = 6$

$4x - y = 2$

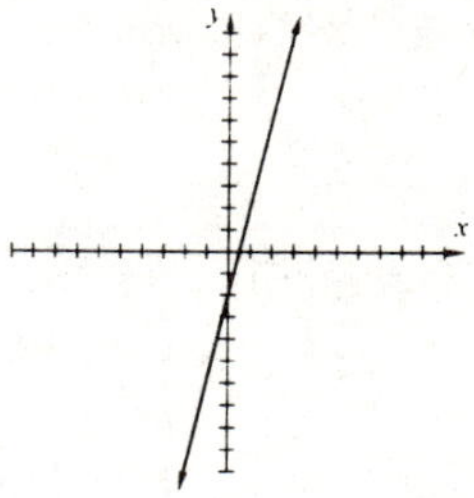

4. $y = \frac{2}{3}x - 4$

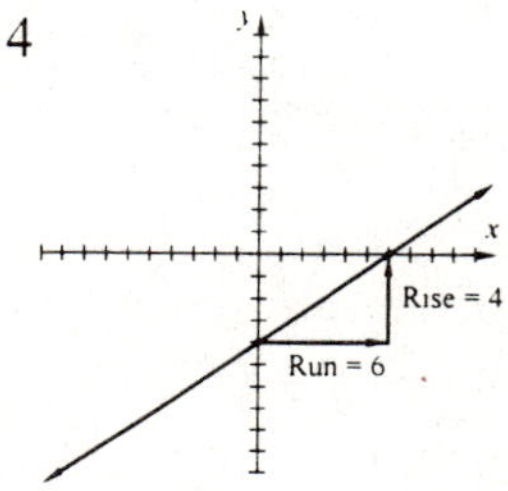

5. $3x + 2y - 5 = 0$

$2y = -3x + 5$

$y = -\frac{3}{2}x + \frac{5}{2}$

$m = -\frac{3}{2},\ b = \frac{5}{2}$

6. $(8, 6)$ and $(-3, -5)$

$m = \frac{-5-6}{-3-8} = \frac{-11}{-11} = 1$

7. $(4, -2),\ m = \frac{1}{2}$

$y = mx + b$

$-2 = \frac{1}{2}(4) + b \Rightarrow b = -4$

$y = \frac{1}{2}x - 4$ or $x - 2y = 8$

8. $(-3, 11)$ and $(-6, 11)$

$m = \frac{11-11}{-6-(-3)} = 0$

9. $(2, 5)$ and $(8, 3)$

$m = \frac{y_2 - y_1}{x_2 - x_1} = \frac{3-5}{8-2} = -\frac{1}{3}$

$y = mx + b$

$5 = -\frac{1}{3}(2) + b \Rightarrow b = \frac{17}{3}$

$y = -\frac{1}{3}x + \frac{17}{3}$ or $x + 3y = 17$

10. $(2, 7)$ and $(2, -2)$

$m = \frac{y_2 - y_1}{x_2 - x_1} = \frac{-2-7}{2-2} = \frac{-9}{0}$ undefined

Vertical line: $x = 2$

11. $4y \le 3x$

Test point: $(1,1)$

$4(1) \le 3(1)$

$4 \le 3$ False

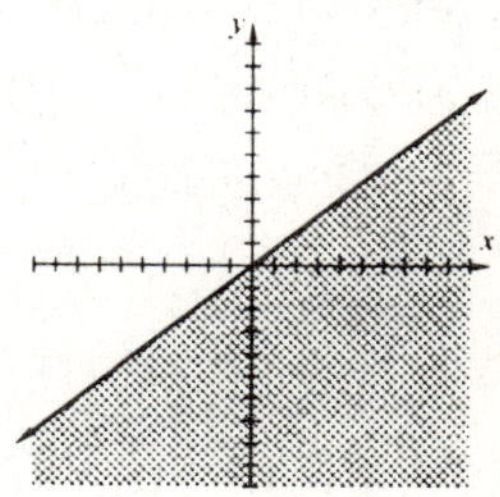

12. $-3x - 2y > 10$.

Test point: $(0, 0)$

$-3(0) - 2(0) > 10$

$0 > 10$ False

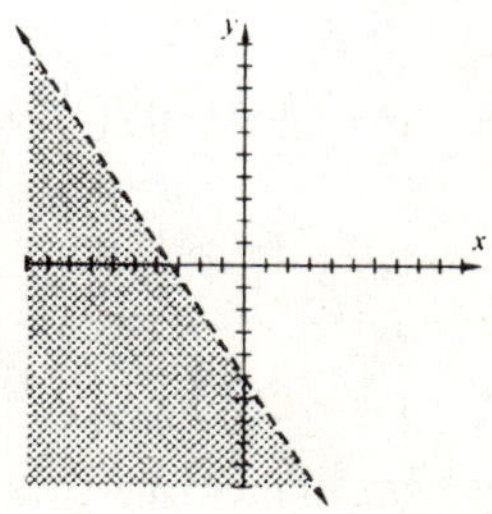

13. $\{(-8, 2), (3, 2), (5, -2)\}$
    This is a function because there is only one second element for each first element.

14. Function, passes the vertical line test

15. $y = 2x^2 - 3$

x	y
−2	5
−1	−1
0	−3
1	−1
2	5

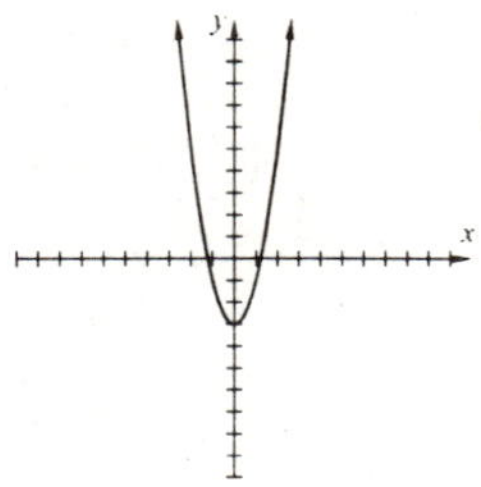

16. $f(x) = 2x^2 + 3x - 4$
    (a) $f(0) = 2(0)^2 + 3(0) - 4 = -4$
    (b) $f(-2) = 2(-2)^2 + 3(-2) - 4 = -2$

17. $g(x) = \frac{3}{x-4}$
    (a) $g(3) = \frac{3}{3-4} = -3$
    (b) $g(-11) = \frac{3}{-11-4} = \frac{3}{-15} = -\frac{1}{5}$

Cumulative Test for Chapters 1-3

1. $$\left(-\frac{1}{3}\right)\left(\frac{2}{5}\right)-\left(\frac{2}{5}\right)^2 = \left(-\frac{1}{3}\right)\left(\frac{2}{5}\right)-\left(\frac{4}{25}\right)$$
$$= \left(-\frac{2}{15}\right)-\left(\frac{4}{25}\right)$$
$$= \frac{-10-12}{75}$$
$$= -\frac{22}{75}$$

2. $x = 2^3 - 6 \div 3 - 1$
   $x = 8 - 2 - 1$
   $x = 5$

3. $-5x[3x + 2(x-6)] = -5x[3x + 2x - 12]$
   $= -5x[5x - 12]$
   $= -25x^2 + 60x$

5. $2 - 3(4 - x) = x - (3 - x)$
   $2 - 12 + 3x = x - 3 + x$
   $x = 7$

4. $p = 3500,\ 7 = 0.07,\ t = 5$
   $I = prt = (3500)(0.07)(5) = \$1225$

5. $2 - 3(4 - x) = x - (3 - x)$
   $2 - 12 + 3x = x - 3 + x$
   $x = 7$

6. $$\frac{3x-2}{4}-\frac{x}{3}=2$$
$$12\left(\frac{3x-2}{4}\right)-12\left(\frac{x}{3}\right)=12(2)$$
$$3(3x-2)-4x=24$$
$$9x-6-4x=24$$
$$5x-6=24$$
$$5x-6+6=24+6$$
$$5x=30$$
$$\frac{5x}{5}=\frac{30}{5}$$
$$x=6$$

7. $$A=lw+lh$$
$$A-lh=lw$$
$$\frac{A-lh}{l}=w$$

8. $3(x-2)\le 4x-7$
$$3x-6\le 4x-7$$
$$-x\le -1$$
$$x\ge 1$$

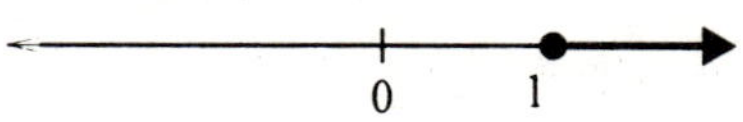

9. $h=$ height, $3h-3=$ length
$$2h+2l=p$$
$$2h+2(3h-3)=90$$
$$2h+6h-6=90$$
$$8h=96$$
$$h=12$$
$$3h-3=33$$
heigth $=12$ feet, length $=33$ feet

10. $(6, 8)$ and $(7, 11)$
$$m=\frac{y_2-y_1}{x_2-x_1}=\frac{11-8}{7-6}=\frac{3}{1}=3$$
$$y=mx+b$$
$$8=3(6)+b\Rightarrow b=-10$$
$$y=3x-10$$

11. $(7, -4)$ and slope is undefined
$$x=7$$

12. $(-2, 3)$, $m=\frac{1}{3}$
$$y=mx+b$$
$$3=\frac{1}{3}(-2)+b\Rightarrow b=\frac{11}{3}$$
$$y=\frac{1}{3}x+\frac{11}{3}$$

13. $(-8, -3)$ and $(11, -3)$
$$m=\frac{y_2-y_1}{x_2-x_1}=\frac{-3-(-3)}{11-(-8)}=0$$

14. $3x-7y=-2$
$$-7y=-3x-2$$
$$y=\frac{3}{7}x+\frac{2}{7}$$
$$y=mx+b$$
$$m=\frac{3}{7}$$

15. $y = \dfrac{2}{3}x - 4$

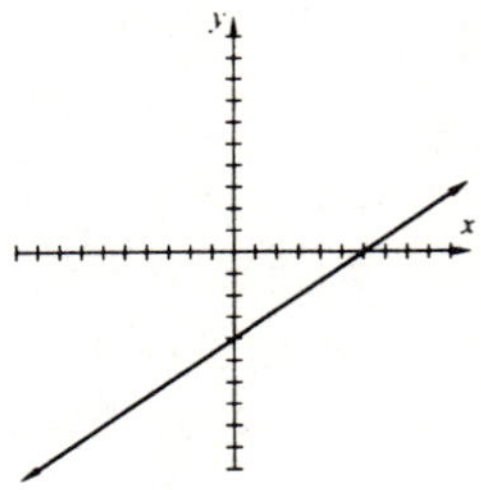

16. $3x + 8 = 5x$

$8 = 2x$

$4 = x$

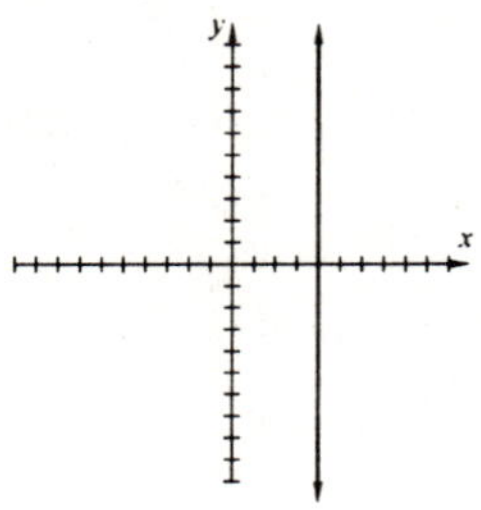

17. $2x + 5y \le -10$

Test point: $(0,\ 0)$

$2(0) + 5(0) \le -10$

$0 \le -10$ False

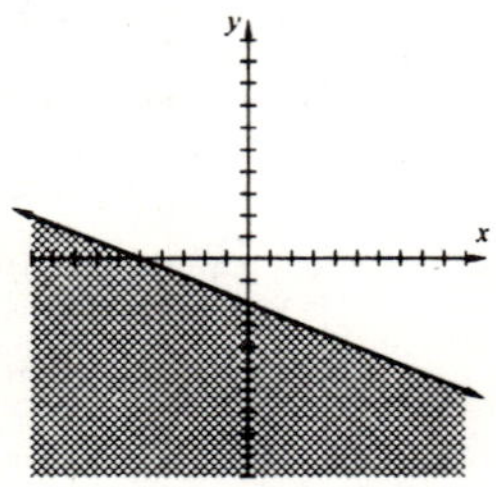

18. $x = y^2 + 3$

x	y
7	−2
4	−1
3	0
4	1
7	2

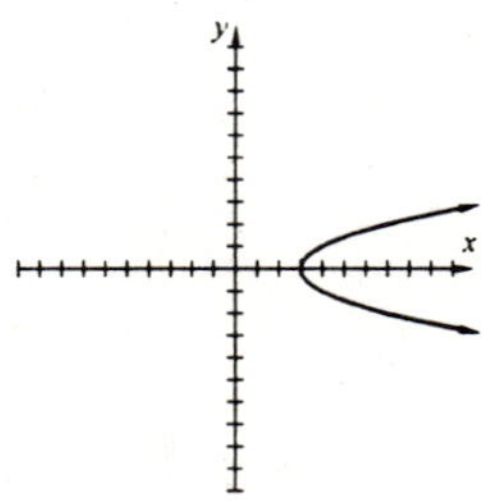

Not a Function

19. $\{(3,\ 10),\ (10,\ -3),\ (0,\ -3),\ (5,\ 10)\}$

This is a function

20. $f(x) = -2x^2 - 3x + 5$

(a) $f(4) = -2(4)^2 - 3(4) + 5 = -39$

(b) $f(-2) = -2(-2)^2 - 3(-2) + 5 = 3$

Chapter 4

Pretest Chapter 4

1. $5x-2y=27$ **(1)**
   $3x-5y=-18$ **(2)**
   Multiply **(1)** by 5, and **(2)** by -2 and add
   $$\begin{array}{r} 25x-10y=135 \\ -6x+10y=36 \\ \hline 19x=171 \\ x=9 \end{array}$$
   Substitute $x=9$ into **(1)**
   $$\begin{aligned} 5(9)-2y&=27 \\ 45-2y&=27 \\ 2y&=18 \\ y&=9 \end{aligned}$$
   The solution is $(9,9)$.

2. $7x+3y=15$ **(1)**
   $\frac{1}{3}x-\frac{1}{2}y=2$ **(2)**
   Multiply **(2)** by 6 to clear fractions
   $2x-3y=12$ **(3)**
   Add **(1)** and **(3)**
   $$\begin{array}{r} 7x+3y=15 \\ 2x-3y=12 \\ \hline 9x=27 \\ x=3 \end{array}$$
   Substitute $x=3$ into **(1)**
   $$\begin{aligned} 7(3)+3y&=15 \\ 3y&=-6 \\ y&=-2 \end{aligned}$$
   The solution is $(3,-2)$.

3. $2x=3+y$ **(1)**
   $3y=6x-9$ **(2)**
   Solve **(1)** for y
   $y=2x-3$ and substitute into **(2)**
   $$\begin{aligned} 3(2x-3)&=6x-9 \\ 6x-9&=6x-9 \\ 0&=0 \end{aligned}$$
   The equations are dependent and the system has an infinite number of solutions.

4. $2x+7y=-10$ **(1)**
   $5x+6y=-2$ **(2)**
   Multiply **(1)** by 5 and **(2)** by -2 and add
   $$\begin{array}{r} 10x+35y=-50 \\ -10x-12y=4 \\ \hline 23y=-46 \\ y=-2 \end{array}$$
   Substitute $y=-2$ into **(1)**
   $$\begin{aligned} 2x+7(-2)&=-10 \\ 2x-14&=-10 \\ 2x&=4 \\ x&=2 \end{aligned}$$
   The solution is $(2,-2)$.

5. $5x-2y+z=-1$ **(1)**
   $3x+y-2z=6$ **(2)**
   $-2x+3y-5z=7$ **(3)**
   Multiply **(1)** by 2 and add to**(2)**
   $$\begin{array}{r} 10x-4y+2z=-2 \\ 3x+y-2z=6 \\ \hline 13x-3y=4 \end{array}$$ **(4)**
   (continued)

5. (continued)

Multiply **(1)** by 5 and add to **(3)**

$$\begin{array}{r} 25x-10y+5z=-5 \\ \underline{-2x+3y-5z=7} \\ 23x-7y=2 \end{array} \quad \textbf{(5)}$$

Multiply **(4)** by 7 and **(5)** by -3 and add

$$\begin{array}{r} 91x-21y=28 \\ \underline{-69x+21y=-6} \\ 22x=22 \\ x=1 \end{array}$$

Substitute $x=1$ into **(5)**

$$\begin{aligned} 23(1)-7y&=2 \\ 7y&=21 \\ y&=3 \end{aligned}$$

Substitute $x=1,\ y=3$ into **(1)**

$$\begin{aligned} 5(1)-2(3)+z&=-1 \\ 5-6+z&=-1 \\ z&=0 \end{aligned}$$

The solution is $(1,3,0)$.

6. $x+y+2z=9$ **(1)**
$3x+2y+4z=16$ **(2)**
$2y+z=10$ **(3)**

Solve **(3)** for z and substitute into **(1)** and **(2)**

$$z=10-y$$

$$\begin{aligned} x+y+2(10-2y)&=9 \\ x+y+20-4y&=9 \\ x-3y&=-11 \quad \textbf{(4)} \end{aligned}$$

$$\begin{aligned} 3x+2y+4(10-2y)&=16 \\ 3x+2y+40-8y&=16 \\ 3x-6y&=-24 \quad \textbf{(5)} \end{aligned}$$

Multiply (4) by -2 and (5) and add to **(5)**

(continued)

6. (continued)

$$\begin{array}{r} -2x+6y=22 \\ \underline{3x+6y=-24} \\ x=-2 \end{array}$$

Substitute $x=-2$ into **(5)**

$$\begin{aligned} 3(-2)-6y&=-24 \\ -6y&=-18 \\ y&=3 \end{aligned}$$

Substitute $x=-2,\ y=3$ into **(1)**

$$\begin{aligned} -2+3+2z&=9 \\ 2z&=8 \\ z&=4 \end{aligned}$$

The solution is $(-2,3,4)$.

7. $x-2z=-5$ **(1)**
$y-3z=-3$ **(2)**
$2x-z=-4$ **(3)**

Solve **(3)** for z and substitute into **(1)**

$$z=2x+4$$

$$\begin{aligned} x-2(2x+4)&=-5 \\ x-4x-8&=-5 \\ -3x&=3 \\ x&=-1 \end{aligned}$$

Substitute $x=-1$ into **(3)**

$$\begin{aligned} 2(-1)-z&=-4 \\ -z&=-2 \\ z&=2 \end{aligned}$$

Substitute $z=2$ into **(2)**

$$\begin{aligned} y-3(2)&=-3 \\ y-6&=-3 \\ y&=3 \end{aligned}$$

The solution is $(-1,3,2)$.

8. Let $x =$ cost of shirt and $y =$ cost of pants.

$2x+3y=75$ **(1)**
$3x+5y=121$ **(2)**

Multiply **(1)** by 3 and **(2)** by -2 and add

$$\begin{array}{r} 6x+9y=225 \\ -6x-10y=-242 \\ \hline -y=-17 \\ y=17 \end{array}$$

Substitute $y=17$ into **(1)**

$$2x+3(17)=75$$
$$2x=24$$
$$x=12$$

The shirts cost \$12 and the pants cost \$17.

9. Let $x =$ the number of A packets, $y =$ the number of B packets, and $z =$ the number of C packets.

$3x+2y+2z=21$ **(1)**
$2x+4y+2z=22$ **(2)**
$4x+5y+z=26$ **(3)**

Solve **(3)** z and substitute into **(1)** and **(2)**

$$z=26-4x-5y$$
$$3x+2y+2(26-4x-5y)=21$$
$$3x+2y+52-8x-10y=21$$
$$5x+8y=31 \quad \textbf{(4)}$$
$$2x+4y+2(26-4x-5y)=22$$
$$2x+4y+52-8x-1y=22$$
$$6x+6y=30$$
$$x+y=5$$
$$y=5-x \quad \textbf{(5)}$$

Substitute $y=5-x$ into **(4)**

(continued)

9. (continued)

$$5x+8(5-x)=31$$
$$-3x=-9$$
$$x=3$$

Substitute $x=3$ into **(5)**

$y=5-3=2$. Substitute $x=3$, $y=2$ into **(3)**

$$4(3)+5(2)+z=26$$
$$z=4$$

She should use 3 of packet A, 2 of packet B, and 4 of packet C.

10. Let $x =$ the number of easy exercises and $y =$ the number of hard exercises.

$x+y=10$ **(1)**
$7x+10y=82$ **(2)**

Solve **(1)** for y and substitute into **(2)**

$$y=10-x$$
$$7x+10(10-x)=82$$
$$-3x=-18$$
$$x=6$$

Substitute $x=6$ into **(1)**

$$6+y=10$$
$$y=4$$

You can solve four hard exercises and 6 easy exercises.

11. $2x+2y\ge -4$

Test point: $(0,0)$

$$2(0)+2(0)\ge -4$$
$$0\ge -4 \text{ True}$$

$-3x+y\le 2$

Test point: $(0,0)$

$$-3(0)+0\le 2$$
$$0\le 2 \text{ True}$$

(continued)

11. (continued)

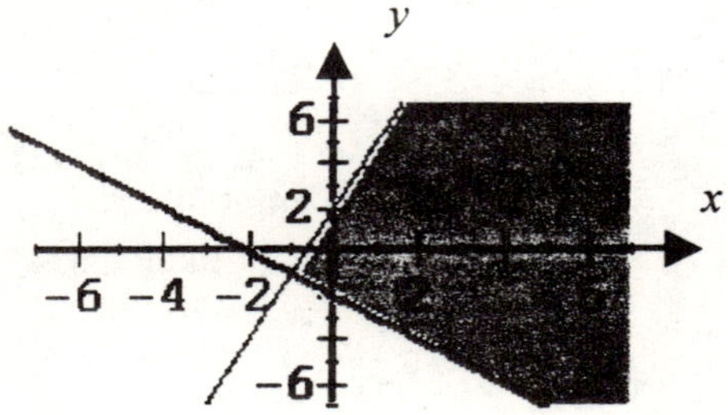

12. $x-3y<-6$

Test point: $(0,0)$

$0-3(0)<-6$

$0<-6$ False

$x<3$

Test point: $(0,0)$

$0<3$ True

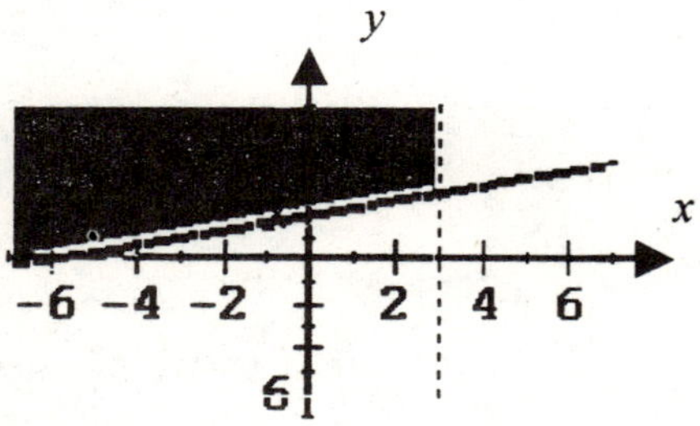

4.1 Exercises

1. There is no solution. There is no point (x,y) that satisfies both equations.

   The graph of such a system yields two parallel lines.

3. $4x+1=6-y$

   LHS: $4\left(\frac{3}{2}\right)+1=6+1=7$

   RHS: $6-(-1)=6+1=7$

   $2x-5y=8$

   $2\left(\frac{3}{2}\right)-5(-1)=3+5=8$

   $\left(\frac{3}{2},-1\right)$ is a solution.

5. $3x+y=2$

   $2x-y=3$

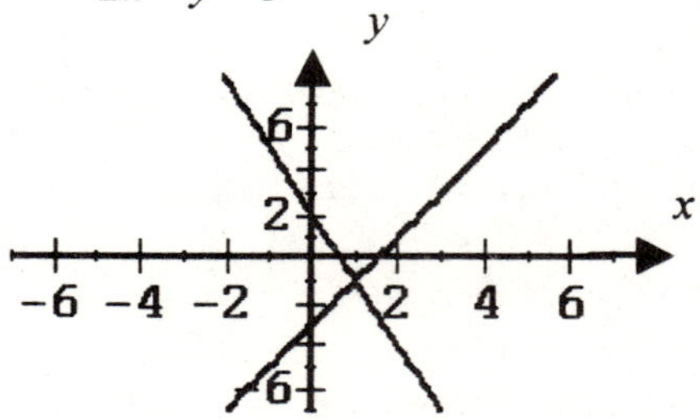

$(1,-1)$ is the solution.

7. $2x+3y=6$

   $2x+y=-2$

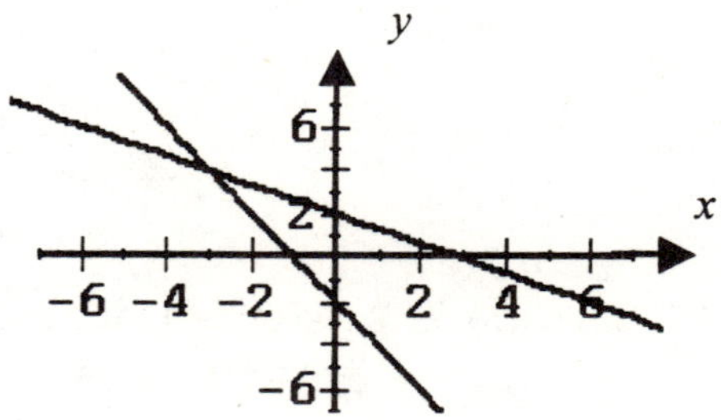

$(-3,4)$ is the solution.

9. $y = -x + 3$
   $3x + 3y = -2$

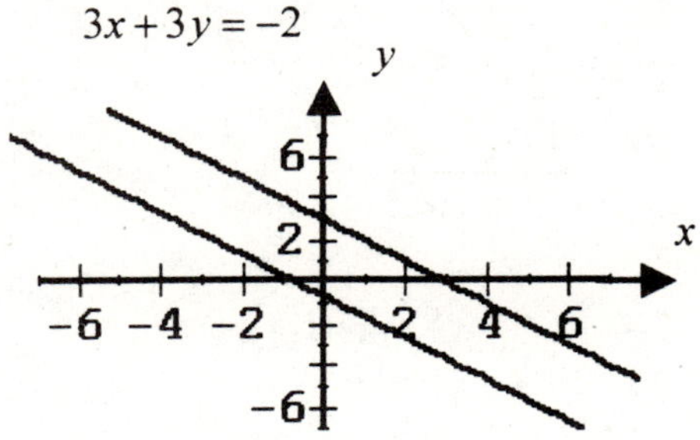

The system has no solution.

11. $3x + 2y = -17$ **(1)**
    $2x + y = 3$ **(2)**

Solve **(2)** for y and substitute into **(1)**

$$y = 3 - 2x$$
$$3x + 2(3 - 2x) = -17$$
$$3x + 6 - 4x = -17$$
$$x = 23$$

Substitute $x = 23$ into **(2)** and solve for y

$$2(23) + y = 3$$
$$46 + y = 3$$
$$y = -43$$
$$x = 23,\ y = -43$$

is the solution.

13. $-x + 3y = -8$ **(1)**
    $2x - y = 6$ **(2)**

Solve **(2)** for y and substitute into **(1)**

$$y = 2x - 6$$
$$-x + 3(2x - 6) = -8$$
$$-x + 6x - 18 = -8$$
$$5x = 10$$
$$x = 2$$

Substitute $x = 2$ into **(1)**

(continued)

13. (continued)

$$-2 + 3y = -8$$
$$3y = -6$$
$$y = -2$$

$x = 2,\ y = -2$ is the solution.

Check:

(1) $-2 + 3(-2) \stackrel{?}{=} -8$

$$-2 - 6 \stackrel{?}{=} -8$$
$$-8 = -8$$

(2) $2(2) - (-2) \stackrel{?}{=} 6$

$$4 + 2 \stackrel{?}{=} 6$$
$$6 = 6$$

15. $\frac{1}{5}x - \frac{1}{2}y = 1$ **(1)**
    $\frac{1}{5}x - 3y = -9$ **(2)**

Multiply **(2)** by 5 to clear fractions, solve for x, then substitute the result into **(1)**

$$x - 15y = -45$$
$$x = 15y - 45$$
$$\frac{1}{5}(15y - 45) - \frac{1}{2}y = 1$$
$$3y - 9 - \frac{1}{2}y = 1$$
$$6y - 18 - y = 2$$
$$5y = 20$$
$$y = 4$$

Substitute $y = 4$ into **(1)**

$$\frac{1}{5}x - \frac{1}{2}(4) = 1$$

(continued)

15. (continued)

$\frac{1}{5}x-2=1$

$\frac{1}{5}x=3$

$x=15$

$x=15,\ y=4$ is the solution.

Check:

(1) $\frac{1}{5}(15)-\frac{1}{2}(4)\stackrel{?}{=}1$

$3-2\stackrel{?}{=}1$

$1=1$

(2) $\frac{1}{5}(15)-3(4)\stackrel{?}{=}-9$

$3-12\stackrel{?}{=}-9$

$-9=-9$

17. $3(x-1)+2(y+2)=6$ **(1)**

$4(x-2)+y=2$ **(2)**

Solve **(2)** for y and substitute into **(1)**

$4x-8+y=2$

$y=10-4x$

$3x-3+2(10-4x+2)=6$

$3x-3+20-8x+4=6$

$-5x=-15$

$x=3$

Substitute $x=3$ into **(2)**

$4(3-2)+y=2$

$4+y=2$

$y=-2$

$x=3,\ y=-2$ is the solution.

19. $9x+2y=2$ **(1)**

$3x+5y=5$ **(2)**

(continued)

19. (continued)

Multiply **(2)** by -3 and add to **(1)**

$$\begin{aligned} 9x+2y&=2\\ -9x-15y&=-15\\ \hline -13y&=-13 \end{aligned}$$

$y=1$

Substitute $y=1$ into **(1)**

$9x+2(1)=2$

$9x+2=2$

$9x=0$

$x=0$

$x=0,\ y=1$ is the solution.

21. $6s-3t=1$ **(1)**

$5s+6t=15$ **(2)**

Multiply **(1)** by 2 and add to **(2)**

$$\begin{aligned} 12s-6t&=2\\ 5s+6t&=15\\ \hline 17s&=17 \end{aligned}$$

$s=1$

Substitute $s=1$ into **(2)**

$5(1)+6t=15$

$6t=10$

$t=\frac{5}{3}$

$s=1,\ t=\frac{5}{3}$ is the solution.

23. $\frac{7}{2}x+\frac{5}{2}y=-4$ **(1)**

$3x+\frac{2}{3}y=1$ **(2)**

Clear fractions

$7x+5y=-8$ **(1)**

$9x+2y=3$ **(2)**

(continued)

23. (continued)
Multiply **(1)** by 2 and **(2)** by -5 and add

$$\begin{array}{r} 14x+10y=-16 \\ -45x-10y=-15 \\ \hline -31x=-31 \end{array}$$

$$x=1$$

Substitute $x=1$ into **(2)**

$$9(1)+2y=3$$
$$2y=-6$$
$$y=-3$$

$x=1,\ y=-3$ is the solution.

25. $1.6x+1.5y=1.8$ **(1)**
$0.4x+0.3y=0.6$ **(2)**

Multiply **(2)** by -0.5 and add to **(1)**

$$\begin{array}{r} 1.6x+1.5y=1.8 \\ -2.0x-1.5y=-3.0 \\ \hline -0.4x=-1.2 \end{array}$$

$$x=3$$

Substitute $x=3$ into **(1)**

$$1.6(3)+1.5y=1.8$$
$$4.8+1.5y=1.8$$
$$1.5y=-3.0$$
$$y=-2$$

$x=3,\ y=-2$ is the solution.
Check:

(1) $1.6(3)+1.5(-2)\overset{?}{=}1.8$
$$4.8-3.0\overset{?}{=}1.8$$
$$1.8=1.8$$

(2) $0.4(3)+0.3(-2)\overset{?}{=}0.6$
$$1.2-0.6\overset{?}{=}0.6$$
$$0.6=0.6$$

27. $2x+y=4$ **(1)**
$\frac{2}{3}x+\frac{1}{4}y=2$ **(2)**

Solve **(1)** for y and substitute into **(2)**

$$y=4-2x$$
$$\frac{2}{3}x+\frac{1}{4}(4-2x)=2$$
$$8x+12-6x=24$$
$$2x=12$$
$$x=6$$

Substitute $x=6$ into **(1)**

$$2(6)+y=4$$
$$12+y=4$$
$$y=-8$$

Check:

(1) $2(6)+(-8)\overset{?}{=}4$
$$12-8\overset{?}{=}4$$
$$4=4$$

(2) $\frac{2}{3}(6)+\frac{1}{4}(-8)\overset{?}{=}2$
$$4-2\overset{?}{=}2$$
$$2=2$$

29. $0.2x=0.1y-1.2$ **(1)**
$2x-y=6$ **(2)**

Solve **(2)** for y and substitute into **(1)**

$$y=2x-6$$
$$0.2x=0.1(2x-6)-1.2$$
$$0.2x=0.2x-0.6-1.2$$
$$0=-1.8$$

This is an inconsistent system of equations and has no solution.

31. $5x-7y=12$ **(1)**
$-10x+14y=-24$ **(2)**

Multiply **(1)** by 2 and add to **(2)**

$$\begin{array}{r} 10x-14y=24 \\ -10x+14y=-24 \\ \hline 0=0 \end{array}$$

This is a dependent system of equations and has an infinite number of solutions.

33. Multiply both equations by 10 to clear decimals
$8x+9y=13$ **(1)**
$6x-5y=45$ **(2)**

Multiply **(1)** 5 and **(2)** by 9 and add

$$\begin{array}{r} 40x+45y=65 \\ 54x-45y=405 \\ \hline 94x=4760 \end{array}$$

$$x=5$$

Substitute $x=5$ into **(1)**

$$8(5)+9y=13$$
$$40+9y=13$$
$$9y=-27$$
$$y=-3$$

$x=5,\ y=-3$ is the solution.

35. Solve the first equation for a and substitute into the second equation.

$$a=\frac{4}{5}b-\frac{1}{5}$$

$$15\left(\frac{4}{5}b-\frac{1}{5}\right)-12b=4$$
$$12b-3-12b=4$$
$$-3=4$$

This is an inconsistent system of equations and has no solution.

37. Solve the first equation for y and substitute into the second equation.

$$y=14-\frac{3}{8}x$$

$$2x-\frac{7}{4}\left(14-\frac{3}{8}x\right)=18$$
$$8x-98+\frac{21}{8}x=72$$
$$\frac{85}{8}x=170$$
$$x=16$$

Substitute $x=16$ into the first equation

$$\frac{3}{8}(16)+y=14$$
$$6+y=14$$
$$y=8$$

$x=16,\ y=8$ is the solution.

39. (a) $y=200+50x$ for Old World Tile
$y=300+30x$ for Modern Bath

(b)

x	$y=300+30x$
0	300
4	420
8	540

x	$y=200+50x$
0	200
4	400
8	600

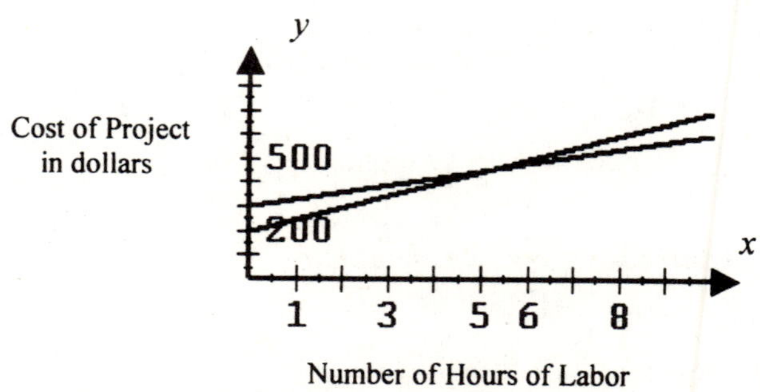

(continued)

39. (continued)
(c) From the graph the cost will be the same for 5 hours of installing new tile.
(d) From the graph Modern Bathroom Headquarters will cost less to remove old time and install new tile if the time needed to install the new tile is 6 hrs.

41. $y_1 = -1.7x + 3.8$
$y_2 = 0.7x - 2.1$

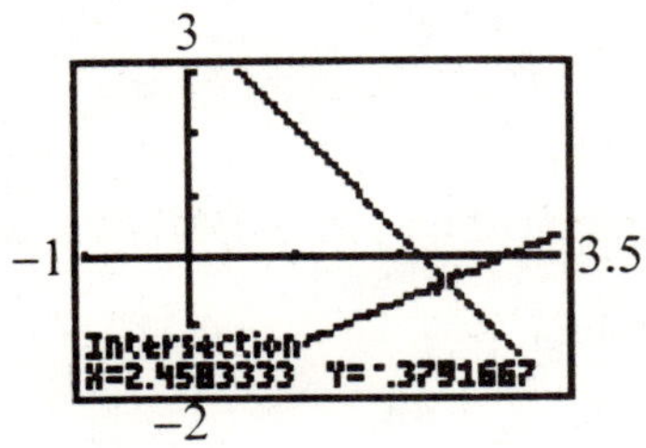

To the nearest hundredth, the point of intersection is $(2.46, -0.38)$.

43. Write $0.5x + 1.1y = 5.5$ as $y_1 = (5.5 - 0.5x)/1.1$ and $-3.1x + 0.9y = 13.1$ as $y_1 = (13.1 + 3.1x)/0.9$ for input to the calculator.

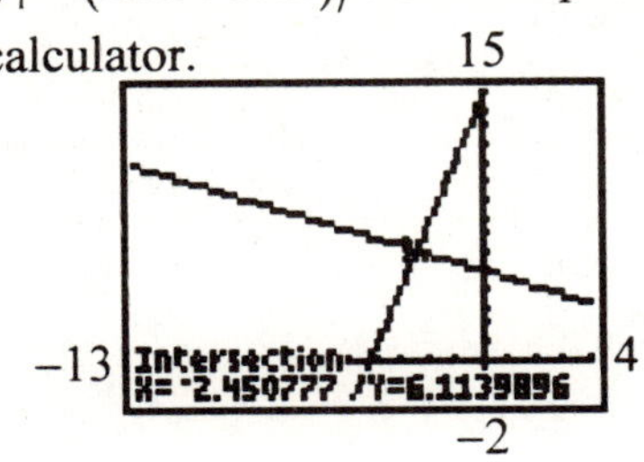

The point of intersection is $(2.45, 6.11)$.

Cumulative Review Problems

45. $(18 \text{ in})(27 \text{ in})\left(\dfrac{\text{ft}^2}{144 \text{ in}^2}\right)\left(\dfrac{\$8}{\text{ft}^2}\right)$
$= \$27$

Putting Your Skills to Work

1. $\dfrac{100{,}000 - 12{,}000}{2000 - 1900} = 880$ lost per year
$c(t) = 100{,}000 - 880t$ where t is the number of years since 1900.

2. $\dfrac{70{,}000 - 10{,}500}{2000 - 1970} \approx 1980$ lost per year
$r(t) = 70{,}000 - 1980t$ where t is the where t is the number of years since 1900.

3. $c(90) = 100{,}000 - 880(90) = 20{,}800$
$c(t) = 20{,}800 - 880t$ where now t is the number of years since 1990.

4. $r(20) = 70{,}000 - 1980(20) = 30{,}400$
$r(t) = 30{,}400 - 1980t$ where now t is the number of years since 1990.

5.

t	$c(t)$
0	20,800
5	16,400
10	12,000

t	$r(t)$
0	30,400
5	20,500
10	10,600

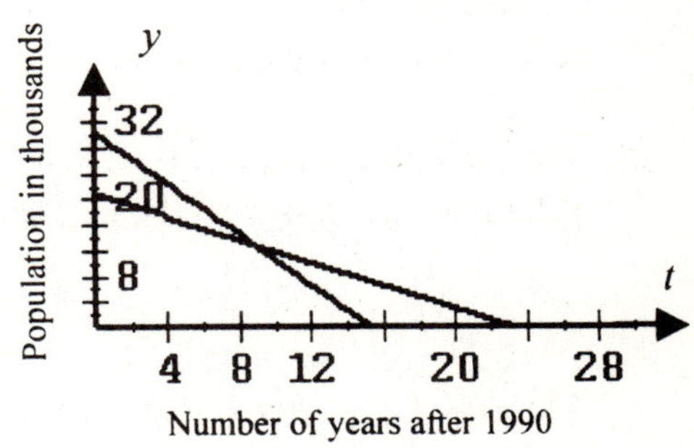

6. The slope of the rhino line is -1980, a decrease of 1980/yr. The slope of the cheetah line is -880, a decrease of 880/yr. Rhinos are decreasing faster.

7. $30,400-1980t=20,800-880t$

$$1100t=9600$$
$$t=8.7$$

The two populations will be equal 8.7 years after 1990. In the year 1999 the two populations were equal.

8. The cheetah will become extinct when
$20,800-880t=0$

$$880t=20,800$$
$$t\approx 23.6$$

$1990+23.6=2013.6$. The cheetah will become extinct sometime between 2013 and 2014. The rhino will become extinct when
$30,400-1980t=0$

$$1980t=30,400$$
$$t\approx 15.4$$

$1990+15.4=2005.4$. The rhino will become extinct sometime between 2005 and 2006.

4.2 Exercises

1. $2x-3y+2z=-7$ **(1)**
$x+4y-z=10$ **(2)**
$3x+2y+z=4$ **(3)**

(1)

$$2(2)-3(1)+2(-4)\stackrel{?}{=}-7$$
$$4-3-8\stackrel{?}{=}-7$$
$$-7=-7$$

(2)

$$2+4(1)-(-4)\stackrel{?}{=}10$$
$$2+4+4\stackrel{?}{=}10$$
$$10=10$$

(continued)

1. (continued)

(3)

$$3(2)+2(1)+(-4)\stackrel{?}{=}4$$
$$6+2-4\stackrel{?}{=}4$$
$$4=4$$

$(2,1,-4)$ is a solution.

3. $x+y+2z=0$ **(1)**
$2x-y-z=1$ **(2)**
$x+2y+3z=1$ **(3)**
Add **(1)** and **(2)**
$3x+z=1$ **(4)**
Add $2\cdot$**(2)** and **(3)**
$5x+z=3$ **(5)**
Subtract **(5)** from **(4)**
$-2x=-2$
$x=1$
Substitute $x=1$ into **(5)**
$5(1)+z=3$
$z=-2$
Substitute $x=1,\ z=-2$ into **(1)**
$1+y+2(-2)=0$
$1+y-4=0$
$y=3$
$x=1,\ y=3,\ z=-2$ is the solution.

5. $x+2y-3z=-11$ **(1)**
$-2x+y-z=-11$ **(2)**
$x+y+z=6$ **(3)**
Add **(2)** and **(3)**
$-x+2y=-5$ **(4)**
Add $3\cdot$**(3)** and **(1)**
$4x+5y=7$ **(5)**
Add $4\cdot$**(4)** and **(5)**
$13y=-13$
$y=-1$

(continued)

5. (continued)
Substitute $y = -1$ into **(5)**
$$4x + 5(-1) = 7$$
$$4x - 5 = 7$$
$$4x = 12$$
$$x = 3$$
Substitute $x = 3$, $y = -1$ into **(3)**
$$3 + (-1) + z = 6$$
$$2 + z = 6$$
$$z = 4$$
$x = 3$, $y = -1$, $z = 4$ is the solution.

7. $-4x + y - 3z = 2$ **(1)**
$5x - 3y + 4z = 1$ **(2)**
$3x - 2y + 5z = 1$ **(3)**
Add $3 \cdot$**(1)** and **(2)**
$-7x - 5z = 7$ **(4)**
Add $2 \cdot$**(1)** and **(3)**
$-5x - z = 5$ **(5)**
Add $-5 \cdot$**(5)** and **(4)**
$$18x = -18$$
$$x = -1$$
Substitute $x = -1$ into **(5)**
$$-5(-1) - z = 5$$
$$5 - z = 5$$
$$-z = 0$$
$$z = 0$$
Substitute $x = -1$ and $z = 0$ into **(1)**
$$-4(-1) + y - 3(0) = 2$$
$$4 + y = 2$$
$$y = -2$$
$x = -1$, $y = -2$, $z = 0$ is the solution.

9. $x + 4y - z = -5$ **(1)**
$-2x - 3y + z = 5$ **(2)**
$x - \frac{2}{3}y + z = \frac{11}{3}$ **(3)**(continued)

9. (continued)
Add **(1)** and **(3)**
$2x + \frac{10}{3}y = -\frac{4}{3}$ **(4)**
Add $2 \cdot$**(1)** and **(2)**
$$5y = -5$$
$$y = -1$$
Substitute $y = -1$ into **(4)**
$$2x + \frac{10}{3}(-1) = -\frac{4}{3}$$
$$6x - 10 = -4$$
$$6x = 6$$
$$x = 1$$
Substitute $x = 1$ and $y = -1$ into **(3)**
$$1 - \frac{2}{3}(-1) + z = \frac{11}{3}$$
$$3 + 2 + 3z = 11$$
$$3z = 6$$
$$z = 2$$
$x = 1$, $y = -1$, $z = 2$ is the solution.

11. $2x - 3y + 2z = -7$ **(1)**
$\frac{3}{2}x + y + \frac{1}{2}z = 2$ **(2)**
$x + 4y - z = 10$ **(3)**
Add **(1)** and $2 \cdot$**(3)**
$4x + 5y = 13$ **(4)**
Add **(3)** and $2 \cdot$**(2)**
$4x + 6y = 14$ **(5)**
Subtract **(5)** from **(4)**
$$-y = -1$$
$$y = 1$$
Substitute $y = 1$ into **(5)**
$$4x + 6(1) = 14$$
$$4x = 8$$
$$x = 2$$
(continued)

11. (continued)
Substitute $x = 2,\ y = 1$ into **(3)**
$2 + 4(1) - z = 10$
$6 - z = 10$
$-z = 4$
$z = -4$
$x = 2,\ y = 1,\ z = -4$ is the solution.

13. $a = 8 + 3b - 2c$ **(1)**
$4a + 2b - 3c = 10$ **(2)**
$c = 10 + b - 2a$ **(3)**
Substitute *a* from **(1)** into **(2)**
$4(8 + 3b - 2c) + 2b - 3c = 10$
$32 + 12b - 8c + 2b - 3c = 10$
$14b - 11c = -22$ **(4)**
Substitute *a* from **(1)** into **(3)**
$c = 10 + b - 2(8 + 3b - 2c)$
$c = 10 + b - 16 - 6b + 4c$
$5b - 3c = -6$ **(5)**
Add $5 \cdot$**(4)** and $-14 \cdot$**(5)**
$-13c = -26$
$c = 2$
Substitute $c = 2$ into **(5)**
$5b - 3(2) = -6$
$5b - 6 = -6$
$5b = 0$
$b = 0$
Substitute $b = 0,\ c = 2$ into **(1)**
$a = 8 + 3(0) - 2(2)$
$a = 8 - 4$
$a = 4$
$a = 4,\ b = 0,\ c = 2$ is the solution.

15. Multiply all three equations by 10 to clear decimals.
$2a + b + 2c = 1$ **(1)**
$3a + 2b + 4c = -1$ **(2)**
$6a + 11b + 2c = 3$ **(3)** (con't)

15. (continued)
Subtract **(3)** from **(1)**
$-4a - 10b = -2$
$2a + 5b = 1$ **(4)**
Add $-2 \cdot$**(1)** and **(2)**
$-a = -3$
$a = 3$
Substitute $a = 3$ into **(4)**
$2(3) + 5b = 1$
$6 + 5b = 1$
$5b = -5$
$b = -1$
Substitute $a = 3,\ b = -2$ into **(1)**
$2(3) + (-1) + 2c = 1$
$6 - 1 + 2c = 1$
$2c = -4$
$c = -2$
$a = 3,\ b = -1,\ c = -2$ is the solution.

17. $-x + 3y - 2z = 11$ **(1)**
$2x - 4y + 3z = -15$ **(2)**
$3x - 5y - 4z = 5$ **(3)**
Add $3 \cdot$**(1)** and **(3)**
$4y - 10z = 38$ **(4)**
Add $2 \cdot$**(1)** and **(2)**
$2y - z = 7$ **(5)**
Add **(4)** and $2 \cdot$**(5)**
$-8z = 24$
$z = -3$
Substitute $z = -3$ into **(5)**
$2y - (-3) = 7$
$2y + 3 = 7$
$2y = 4$
$y = 2$
Substitute $z = -3,\ y = 2$ into **(1)**
$-x + 3(2) - 2(-3) = 11$
(continued)

17. (continued)

$-x+6+6=11$

$-x+12=11$

$-x=-1$

$x=1$

$x=1,\ y=2,\ z=-3$ is the solution.

19. When a calculator is used it is convenient to keep all three equations together as the operations are performed.

$x-4y+4z=-3.72186$

$-x+3y-z=5.98115$

$2x-y+5z=7.93645$

Now perform two operations on the system; first, add the first equation to the second and add -2 times the first to the third. Note that this *does not* change the first equation but the second and third.

$x-4y+4z=-3.72186$

$-y+3z=2.25929$

$7y-3z=15.38017$

Add the second equation to the third

$x-4y+4z=-3.72186$

$-y+3z=2.25929$

$6y=17.63946$

From the third equation $y=2.93991$ which may be substituted into the second equation to give $z=1.73307$. Substituting these values for x and y into the first equation gives $x=1.10551$.

The solution, to five decimal places, is $x=1.10551,\ y=2.93991,\ z=1.73307$

21. $x+y=1$ **(1)**

$y-z=-3$ **(2)**

$2x+3y+z=1$ **(3)**

Solve **(1)** for y

$y=1-x$ **(4)**

Solve **(2)** for z

$z=y+3$ **(5)**

Substitute y from **(4)** into **(5)**

$z=1-x+3=4-x$ **(6)**

Substitute y from **(4)** and z from **(6)** into **(3)**

$2x+3(1-x)+4-x=1$

$2x+3-3x+4-x=1$

$-2x+7=1$

$-2x=-6$

$x=3$

Substitute $x=3$ into **(6)**

$z=4-3$

$z=1$

Substitute $x=3$ into **(4)**

$y=1-3$

$y=-2$

Substitute $y=-2,\ z=1$ into **(3)**

$2x+3(-2)+1=1$

$2x-6+1=1$

$2x=6$

$x=3$

$x=3,\ y=-2,\ z=1$ is the solution.

23. $-y+2z=1$ **(1)**

$x+y+z=2$ **(2)**

$-x+3z=2$ **(3)**

Add **(2)** and **(3)**

$y+4z=4$ **(4)**

Add **(1)** and **(4)**

$6z=5$

$z=\frac{5}{6}$ (continued)

23. (continued)

Substitute $z=\frac{5}{6}$ into **(4)**

$$y+4\left(\frac{5}{6}\right)=4$$
$$6y+20=24$$
$$6y=4$$
$$y=\frac{2}{3}$$

Substitute $z=\frac{5}{6},\ y=\frac{2}{3}$ into **(2)**

$$x+\frac{2}{3}+\frac{5}{6}=2$$
$$x=\frac{1}{2}$$

$x=\frac{1}{2},\ y=\frac{2}{3},\ z=\frac{5}{6}$ is the solution.

25. $x-2y+z=0$ **(1)**
$-3x-y=-6$ **(2)**
$y-2z=-7$ **(3)**

Multiply **(1)** by 2 and add to **(3)**
$2x-3y=-7$ **(4)**

Multiply **(2)** by -3 and add to **(4)**
$$11x=11$$
$$x=1$$

Substitute $x=1$ into **(2)**
$$-3-y=-6$$
$$y=3$$

Substitute $x=1,\ y=3$ into **(1)**
$$1-6+z=0$$
$$z=5$$

$x=1,\ y=3,\ z=5$ is the solution.

27. $\frac{a}{2}-b+c=8$ **(1)**
$\frac{3}{2}a+b+2c=0$ **(2)**
$a+c=2$ **(3)**

Add **(1)** and **(2)**
$2a+3c=8$ **(4)**
Subtract $3\cdot$**(3)** from **(4)**
$$-a=2$$
$$a=-2$$

Substitute $a=-2$ into **(3)**
$$-2+c=2$$
$$c=4$$

Substitute $a=-2,\ c=4$ into **(1)**
$$\frac{-2}{2}-b+4=8$$
$$-b=5$$
$$b=-5$$
$a=-2,\ b=-5,\ c=4$ is the solution.

29. $2x+y=-3$ **(1)**
$2y+16z=-18$ **(2)**
$-7x-3y+4z=6$ **(3)**

Solve **(1)** for x, $x=\frac{-3-y}{2}$ and **(2)** for z, $z=\frac{-18-2y}{16}$ and substitute into **(3)**

$$-7\left(\frac{-3-y}{2}\right)-3y+4\left(\frac{-18-2y}{16}\right)=6$$
$$\frac{21}{2}+\frac{7y}{2}-3y-\frac{18}{4}-\frac{y}{2}=6$$
$$6=6$$

The system of equations is a dependent system and has an infinite number of solutions.

31. $3x+3y-3z=-1$ **(1)**
$4x+y-2z=1$ **(2)**
$-2x+4y-2z=-8$ **(3)**
Subtract **(3)** from **(2)**
$6x-3y=9$ **(4)**
Multiply **(1)** by -2 and add to 3 times **(2)**
$6x-3y=5$ **(5)**
Comparing **(4)** and **(5)** gives $5=4$ which is false. This is an inconsistent system of equations and has no solution.

Cumulative Review Problems

33. $m=\frac{y_2-y_1}{x_2-x_1}=\frac{4-3}{1-(-2)}=\frac{1}{3}$
$y-y_1=m(x-x_1)$
$y-4=\frac{1}{3}(x-1)=\frac{1}{3}x-\frac{1}{3}$
$y=\frac{1}{3}x+\frac{11}{3}$

35. Let $c=$ the number of cattle purchased, $h=$ the number of horses purchased, and $s=$ the number of sheep purchased, then
$c+601=1.8(h+346)$
$s+545=1.74(h+346)$
where h can be any number. Let $h=0$, which gives
$c+601=1.8(346)=622.8$
$c=21.8$
$s+545=1.74(346)=602.04$
$s=57.04$
The rancher should by no horses, 22 cattle, and 57 sheep. After the purchase he will have 346 horses, 623 cattle and 602 sheep.

4.3 Exercises

1. Let $x=$ number of heavy equipment operators and $y=$ number of laborers.
$x+y=35,\ y=35-x$
$140x+90y=3950$
$140x+90(35-x)=3950$
$140x+3150-90x=3950$
$50x=800$
$x=16$
$y=35-x=35-16=19$

16 heavy equipment operators and 19 laborers were employed.

3. $x=$ number of tickets for regular coach seats
$y=$ number of tickets for sleeper car seats
$x+y=98$ **(1)**
$120x+290y=19{,}750$ **(2)**
Solve **(1)** for y and substitute into **(2).**
$y=98-x$
$120x+290(98-x)=19{,}750$
$120x+28{,}420-290x=19{,}750$
$-170x=-8670$
$x=51$
Substitute 51 for x in **(1)**
$51+y=98$
$y=47$
Number of coach tickets = 51
Number of sleeper tickets = 47

5. $x=$ number of managers with computer experience
$y=$ number of managers without computer experience

(continued)

5. (continued)

$2x + 5y = 190,\ y = \dfrac{190 - 2x}{5}$

$3x + 8y = 295$

$3x + 8\left(\dfrac{190 - 2x}{5}\right) = 295$

$15x + 1520 - 16x = 1475$

$-x = -45$

$x = 45$

$y = \dfrac{190 - 2(45)}{5} = 20$

The company should train 45 managers with computer experience and 20 managers without computer experience.

7. $x =$ number of packages of old fertilizer
$y =$ number of packages of new fertilizer

$50x + 65y = 3125,\ y = \dfrac{3125 - 50x}{65}$

$60x + 45y = 2925$

$60x + 45\left(\dfrac{3125 - 50x}{65}\right) = 2925$

$3900x + 140{,}625 - 2250x = 190{,}125$

$1650x = 49{,}500$

$x = 30$

$y = \dfrac{3125 - 50(30)}{65} = 25$

The farmer should use 30 packages of the old fertilizer and 25 packages of the new fertilizer.

9. $x =$ cost of one doughnut
$y =$ cost of one large coffee

$3x + 4y = 4.91$ **(1)** (continued)

9. (continued)

$5x + 6y = 7.59$ **(2)**

Multiply **(1)** by 5 and add to -3 times **(2)**

$$\begin{aligned} 15x + 20y &= 24.55 \\ -15x - 18y &= -22.77 \\ \hline 2y &= 1.78 \\ y &= 0.89 \end{aligned}$$

Substitute 0.89 for y in **(1)** and solve for x

$3x + 4(0.89) = 4.91$

$3x = 1.35$

$x = 0.45$

The cost of one doughnut is \$0.45.
The cost of one large coffee is \$0.89.

11. $x =$ speed of plane in still air
$y =$ speed of wind

$(x - y)\left(\dfrac{7}{6}\right) = 210$

$(x + y)\left(\dfrac{5}{6}\right) = 210$

$$\begin{aligned} x - y &= 180 \\ x + y &= 252 \\ \hline 2x &= 432 \end{aligned}$$

$x = 216$

$216 + y = 252$

$y = 36$

plane: 216 mph; wind:36 mph

13. $x =$ speed in still air
$y =$ speed of wind

$\dfrac{7}{2}(x - y) = 630$

$3(x + y) = 630$

(continued)

13. (continued)

$x-y=180$

$x+y=210$

$2x=390$

$x=195$

$195+y=210$

$y=15$

Speed of plane in still air = 195mph
Speed of wind = 15 mph

15. $x=$ number of free throws
$y=$ number of regular shots

$x+y=21,\ y=21-x$

$x+2y=32$

$x+2(21-x)=32$

$x+42-2x=32$

$-x=-10$

$x=10$

$y=21-10=11$

Number of free throws = 10
Number of regular shots = 11

17. $x=$ number of highway miles
$y=$ number of city miles

$x+y=432,\ y=432-x$

$\frac{x}{32}+\frac{y}{24}=16$

$\frac{x}{32}+\frac{432-x}{24}=16$

$24x+32(432-x)=12{,}288$

$24x+13{,}824-32x=12{,}288$

$-8x=-1536$

$x=192$

$y=432-x=240$

Carlos drove 192 highway miles and 240 city miles.

19. $x=$ cost of truck, $y=$ cost of car

$256x+183y=5{,}791{,}948.$

$y=\frac{5{,}791{,}948-256x}{183}$

$64x+107y=2{,}507{,}612$

$64x+107\left(\frac{5{,}791{,}948-256x}{183}\right)$

$=2{,}507{,}612$

$11{,}712x+619{,}738{,}436-27{,}392x$

$=458{,}892{,}996$

$-15{,}680x=-160{,}845{,}440$

$x=10{,}258$

$y=\frac{5{,}791{,}948-256(10{,}258)}{183}=17{,}300$

Cars cost \$10,258 and trucks cost \$17,300.

21. $x=$ number of 15-passenger vehicles
$y=$ number of 7-passenger vehicles
$z=$ number of 5-passenger vehicles

$15x+7y+5z=98$ **(1)**

$x+y+z=14$ **(2)**

$x+y=6$ **(3)**

Subtract **(3)** from **(2)**

$z=8$. Substitute $z=8$ into **(1)**

$15x+7y+5(8)=98$

$15x+7y=58$ **(4)**

Substitute $y=6-x$ from **(3)** into **(4)**

$15x+7(6-x)=58$

$15x+42-7x=58$

$8x=16$

$x=2,\ y=6-x=4$

Number of 15-passenger vehicles is 2, the number of 7-passenger vehicles is 4, and the number of 5-passenger vehicles is 8.

23. $x =$ number of adults
$y =$ number of high school students
$z =$ number of children
$5x+3y+2z=1010$ **(1)**
$7x+4y+3z=1390$ **(2)**
$x+y+z=300$ **(3)**
Multiply -7 times **(3)** and add to **(2)**
$-3y-4z=-710,\ 3y+4z=710$ **(4)**
Multiply -5 times **(3)** and add to **(1)**
$-2y-3z=-490,\ 2y+3z=490$ **(5)**
Add $2\cdot$**(4)** and $-3\cdot$**(5)**
$-z=-50$
$z=50$
Substitute $z=50$ into **(5)**
$2y+3(50)=490$
$2y=340$
$y=170$
Substitute $y=170, z=50$ into **(3)**
$x+170+50=300$
$x=80$

The number of adults attending was 80, the number of high school students was 170, and the number of children was 50.

25. $x =$ number of children under 12
$y =$ number of adults
$z =$ number of senior citizens
$x+y+z=12{,}000$ **(1)**
$0.25x+y+0.5z=10{,}700$ **(2)**
$0.35x+1.5y+0.5z=15{,}820$ **(3)**
Add -0.25 times **(1)** to **(2)**
$-0.25x-0.25y-0.25z=-3000$
$\underline{0.25x+y+0.5z=10{,}700}$
$0.75y+0.25z=7700$ **(4)**
Add -0.35 times **(1)** to **(3)**

(continued)

25. (continued)
$-0.35x-0.35y-0.35z=-4200$
$\underline{0.35x+1.5y+0.5z=15{,}820}$
$0.75y+0.25z=7700$ **(5)**
Add -1.15 times **(4)** to 0.75 times **(5)**
$-.08626y-0.2875z=-8855$
$\underline{0.8625y+0.1125z=8715}$
$-0.175z=-140$
$z=800$
Substitute 800 for z in **(1)**
$x+y=11{,}200$
Add -1 times **(3)** to **(2)**
$0.25x+y+0.5z=10{,}700$
$\underline{-0.35x-1.5y-0.5z=-15{,}820}$
$-0.1x-0.5y=-5120$ **(7)**
Add 0.5 times **(6)** to **(7)**
$0.5x+0.5y=5600$
$\underline{-0.1x-0.5y=-5120}$
$0.4x=480$
$x=1200$
Substitute 1200 for x in **(6)**
$1200+y=11{,}200$
$y=10{,}000$

children under 12 =1200
adults = 10,000
senior citizens = 800

27. $x =$ number of medium sandwiches
$y =$ number of large sandwiches
$z =$ number of extra large sandwiches
$x+y+z=24$ **(1)**
$6x+10y+14z=15.5(16)=248$ **(2)**
$2.5x+3y+4.5z=82$ **(3)**
Add -6 times (1) to (2) and -2.5 times (1) to (3)

(continued)

27. (continued)

$x+y+z=24$ **(1)**

$4y+8z=104$ **(2)**

$0.5y+2z=22$ **(3)**

Add $-\frac{1}{8}$ times **(2)** to **(3)**

$x+y+z=24$ **(1)**

$4y+8z=104$ **(2)**

$z=9$ **(3)**

Substitute $z=9$ back into **(2)**

$4y+8(9)=104$

$4y=32$

$y=8$

Substitute $y=8,\ z=9$ back into **(1)**

$x+8+9=24$

$x=7$

Nick sold 7 medium sandwiches, 8 large sandwiches, and 9 extra large sandwiches.

29. $x=$ number of Box A

$y=$ number of Box B

$z=$ number of Box C

$10x+5y+4z=51$ **(1)**

$3x+2y+z=16$ **(2)**

$3x+3y+2z=23$ **(3)**

Add $-\frac{3}{10}$ times **(1)** to **(2)** and **(3)**

$10x+5y+4z=51$ **(1)**

$0.5y-0.2z=0.7$ **(2)**

$1.5y+0.8z=7.7$ **(3)**

Add -3 times **(2)** to **(3)**

$10x+5y+4z=51$ **(1)**

$0.5y-0.2z=0.7$ **(2)**

$1.4z=5.6$ **(3)**

$z=4$

Substitute $z=4$ back into **(2)**

(continued)

29. (continued)

$0.5y-0.2(4)=0.7$ **(2)**

$0.5y=1.5$

$y=3$

Substitute $z=4,\ y=3$ back into **(1)**

$10x+5(3)+4(4)=51$

$10x=20$

$x=2$

The shipping manager can prepare 2 of the A Boxes, 3 of the B Boxes, and 4 of the C Boxes.

31. $x=$ measure of first angle

$y=$ measure of second angle

$z=$ measure of second angle

$x+y+z=180$

$x+y=\frac{1}{2}z$

$8x=y+z$

Clear fractions and write variables on LHS

$x+y+z=180$ **(1)**

$2x+2y-z=0$ **(2)**

$8x-y-z=0$ **(3)**

Add **(1)** and **(2)**

$3x+3y=180$

$x+y=60$ **(4)**

Add **(1)** and **(3)**

$9x=180$

$x=20$

Substitute $x=20$ into **(4)**

$20+y=60$

$y=40$

Substitute $x=20,\ y=40$ into **(1)**

$20+40+z=180,\ z=120$

The angles are $20°,\ 40°,\ 120°$.

33. x, y, z, q = number of A, B, C, and D packages respectively

$$\begin{array}{lllll} 42x & +20y & +0z & +10q & =134 \\ 20x & +10y & +20z & +0q & =150 \\ 34x & +0y & +10z & +20q & =178 \\ 50x & +35y & +30z & +40q & =405 \end{array}$$

Add $-\frac{20}{42}\times$ first eqn. to second eqn.
Add $-\frac{30}{42}\times$ first eqn. to third eqn.
Add $-\frac{50}{42}\times$ first eqn. to fourth eqn.

$$\begin{array}{lllll} 42x & +20y & +0z & +10q & =134 \\ 0x & +\frac{10}{21}y & +20z & -\frac{100}{21}q & =\frac{1810}{21} \\ 0x & -\frac{340}{21}y & +10z & +\frac{250}{21}q & =\frac{1460}{21} \\ 0x & +\frac{235}{21}y & +30z & +\frac{590}{21}q & =\frac{5155}{21} \end{array}$$

Add $34\times$ second eqn. to third eqn.
Add $-23.5\times$ second eqn. to fourth eqn.

$$\begin{array}{lllll} 42x & +20y & +0z & +10q & =134 \\ 0x & +\frac{10}{21}y & +20z & -\frac{100}{21}q & =\frac{1810}{21} \\ 0x & +0y & +690z & -150q & =3000 \\ 0x & +0y & -440z & +140q & =-1780 \end{array}$$

Add $\frac{44}{69}\times$ third eqn. to fourth eqn.

(continued)

33. (continued)

$$\begin{array}{lllll} 42x & +20y & +0z & +10q & =134 \\ 0x & +\frac{10}{21}y & +20z & -\frac{100}{21}q & =\frac{1810}{21} \\ 0x & +0y & +690z & -150q & =3000 \\ 0x & +0y & +0z & +\frac{1023}{23}q & =\frac{3060}{23} \end{array}$$

$\frac{1020}{23}q = \frac{3060}{23}$, $q = 3$

Substitute $q = 3$ into third equation

$690z - 150(3) = 3000$, $z = 5$

Substitute $z = 5$, $q = 3$ into second eqn

$\frac{10}{21}y + 20(5) - \frac{10}{21}(3) = \frac{1810}{21}$, $y = 1$

Substitute $y = 1$, $z = 5$, $q = 3$ into the first equation

$42x + 20(5) + 0(5) + 10(3) = 134$

$x = 2$

The scientist should use 2 A packages, 1 B package, 5 C packages, and 3 D packages.

Cumulative Review Problems

35. $0.06x + 0.15(0.5 - x) = 0.04$

$0.06x + 0.075 - 0.15x = 0.04$

$-0.09x = -0.035$

$x = \frac{7}{18}$

4.4 Exercises

1. $y \ge 2x - 1$

Test point: $(0, 0)$

$0 \ge 2(0) - 1$

$0 \ge -1$ True

(continued)

1. (continued)
   $x+y\le 6$
   Test point: (0,0)
   $0+0\le 6$
   $0\le 6$ True

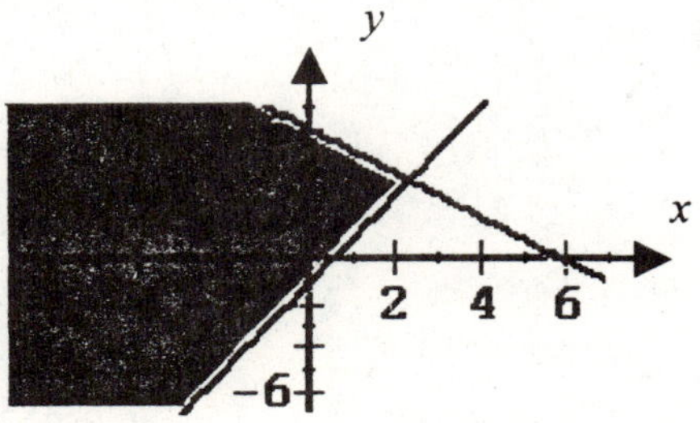

3. $y\ge -4x$
   Test Point: (0,2)
   $2\ge -4(0)$
   $2\ge 0$ True
   $y\ge 3x-2$
   Test point: (0,0)
   $0\ge 3(0)-2$
   $0\ge -2$ True

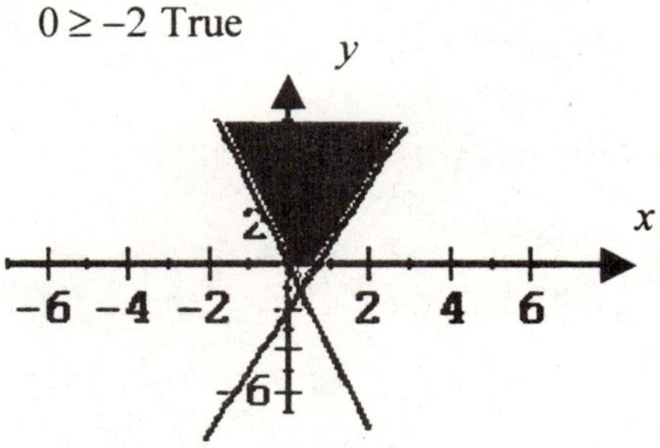

5. $y\ge 2x-3$
   Test point: (0,0)
   $0\ge 2(0)-3$
   $0\ge -3$ True
   $y\le \frac{2}{3}x$

(continued)

5. (continued)
   Test point: (0.2)
   $2\le \frac{2}{3}(0)$
   $2\le 0$ False

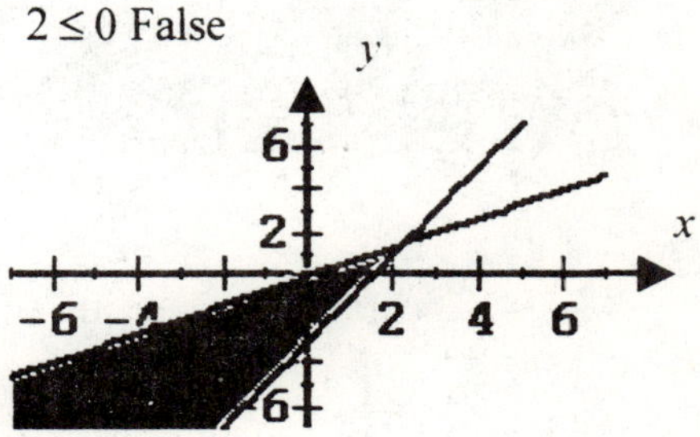

7. $x-y\ge 1$
   Test point: (0,0)
   $0-0\ge -1$
   $0\ge -1$ True
   $-3x-y\le 4$
   Test point: (0,0)
   $-3(0)-0\le 4$
   $0\le 4$ True

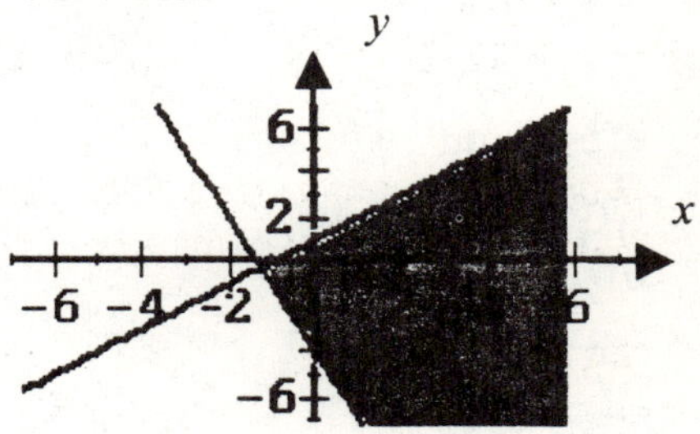

9. $x+2y<6$
   Test point: (0,0)
   $0+2(0)<6$
   $0<6$ True
   $y<3$
   Test point: (0,0)
   $0<3$ True

(continued)

9. (continued)

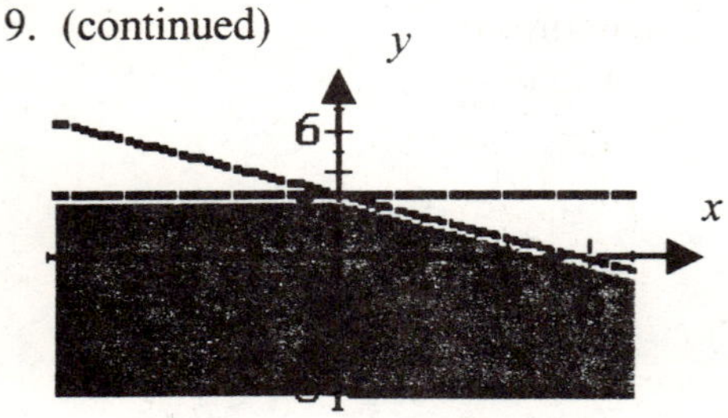

11. $y < 4,\ x > -2$

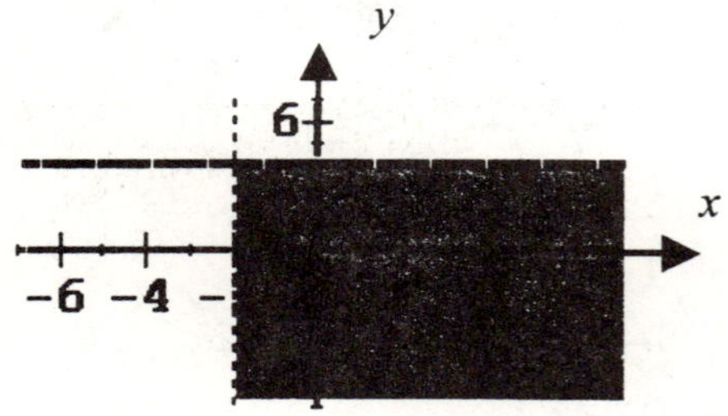

13. $3x + 2y < 6$

Test point: (0,0)

$3(0) + 2(0) < 6$

$0 < 6$ True

$3x + 2y > -6$

Test point: (0,0)

$3(0) + 2(0) > -6$

$0 > -6$ True

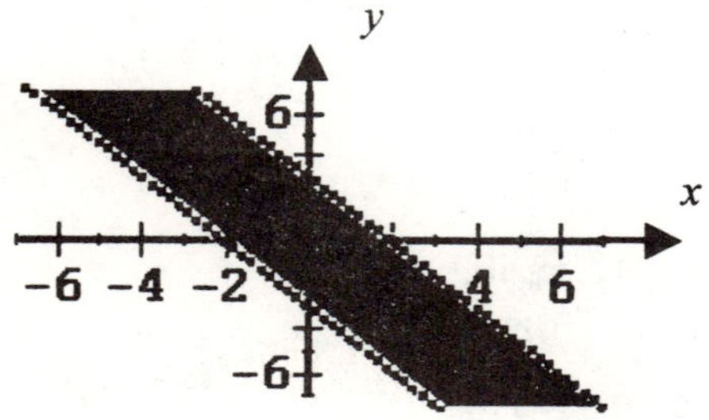

15. $x - 4y \geq -4$

Test point: (0,0)

$0 - 4(0) \geq -4$

$0 \geq -4$ True

$3x + y \leq 3$

Test point: (0,0)

$3(0) + 0 \leq 3$

$0 \leq 3$ True

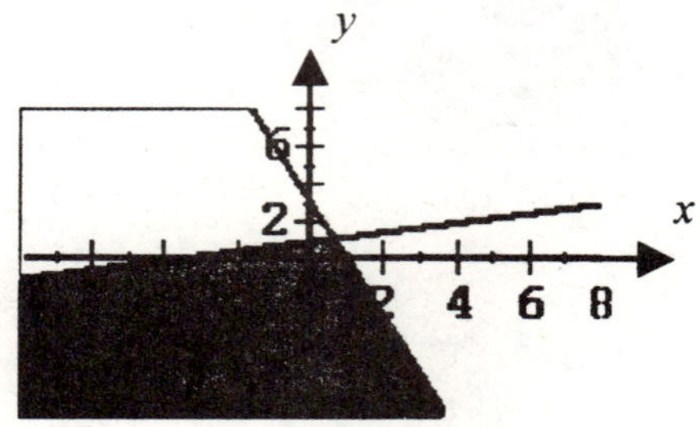

17. $x + y \leq 5$

Test point: (0,0)

$0 + 0 \leq 5$

$0 \leq 5$ True

$2x - y \geq 1$

Test point: (0,0)

$2(0) - 0 \geq 1$

$0 \geq 1$ False

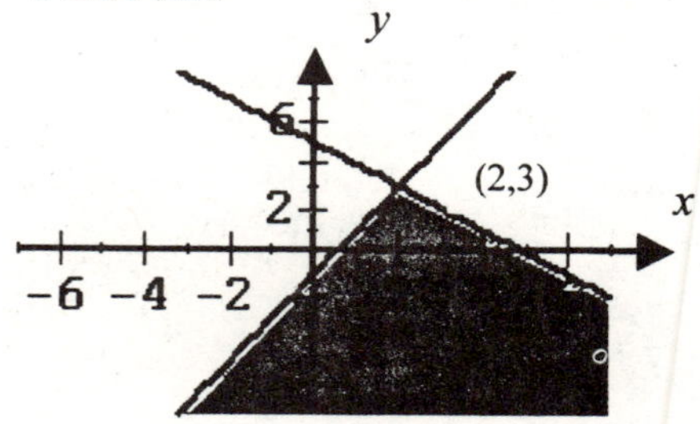

19. $x+3y\leq 12$
Test point: (0,0)
$0+3(0)\leq 12$
$0\leq 12$ True
$y<x$
Test point: (2,0)
$0<2$ True

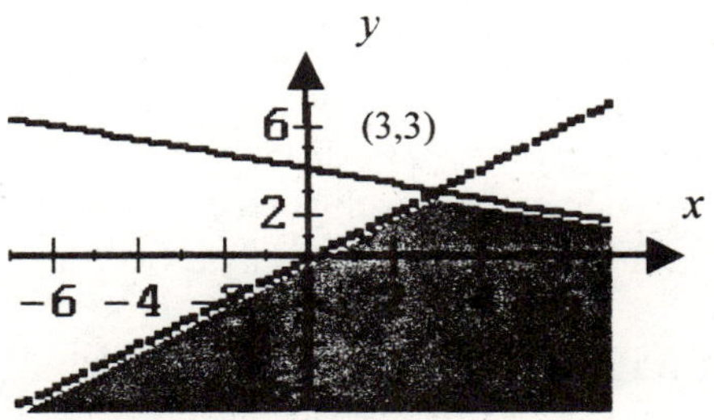

21. $x+y\geq 1$
Test point: (0,0)
$0+0\geq 1$
$0\geq 1$ False
$x-y\geq 1$
Test point: (0,0)
$0-0\geq 1$
$0\geq 1$ False
$x\geq 3$

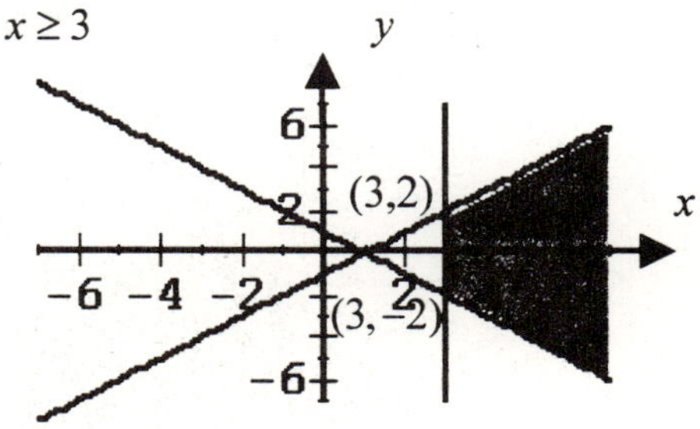

23. $y\leq 3x+6$
Test point: (0,0)
$0\leq 3(0)+6$
$0\leq 6$ True
(continued)

23. (continued)
$4x+3y\leq 3$
Test point: (0,0)
$4(0)+3(0)\leq 3$
$0\leq 3$ True
$x\geq -2,\ y\geq -3$

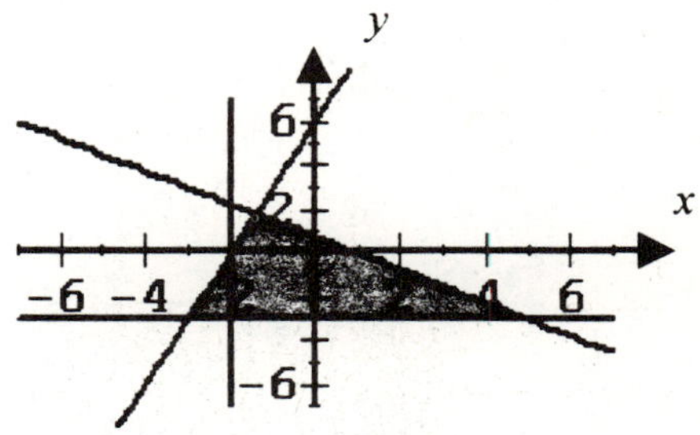

25. (a)
$N\leq 2D$
Test point: (2,2)
$2\leq 2(2)$
$2\leq 4$ True
$4N+3D\leq 20$
Test point: (2,2)
$4(2)+3(2)\leq 20$
$8+6\leq 20$
$14\leq 20,$ True
$N\geq 0,\ D\geq 0$

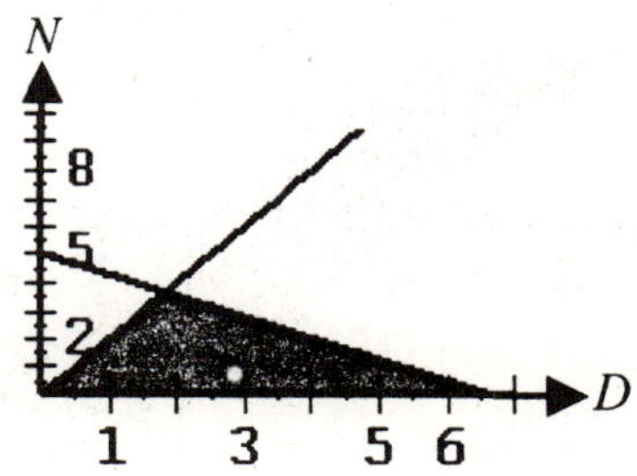

(b) Yes, (3,2) is in the shaded region.

(c) No, (1,4) is not in the shaded region.

Cumulative Review

27. $m = \dfrac{y_2 - y_1}{x_2 - x_1} = \dfrac{-4-(-2)}{3-(-1)} = -\dfrac{1}{2}$

29. $m = \dfrac{y_2 - y_1}{x_2 - x_1} = \dfrac{6-1}{2-(-2)} = \dfrac{5}{4}$

$$y - y_1 = m(x - x_1)$$
$$y - 6 = \frac{5}{4}(x-2)$$
$$y = \frac{5}{4}x - \frac{5}{2} + 6$$
$$y = \frac{5}{4}x + \frac{7}{2}$$

31. $x =$ money taken in or rainy day
$y =$ money taken in on sunny day

$$2x + 5y = 23,400 \Rightarrow y = \frac{23,400 - 2x}{5}$$
$$4x + 3y = 25,800$$
$$4x + 3\left(\frac{23,400 - 2x}{5}\right) = 25,800$$
$$20x + 70,200 - 6x = 129,000$$
$$14x = 58,800$$
$$x = 4200$$
$$y = \frac{23,400 - 2x}{5} = \frac{23,400 - 2(4200)}{5}$$
$$y = 3000$$
The Cape Cod Cinema takes in \$4200 on a rainy day and \$3000 on a sunny day.

33. $x =$ sales, $y =$ number of weeks
Hector: $0.05x + 200y = 7400$
$$y = \frac{7400 - 0.05x}{200}$$
Fernando: $0.08x + 100y = 9200$

(continued)

33. (continued)
$$0.08x + 100\frac{7400 - 0.05x}{200} = 9200$$
$$16x + 740,000 - 5x = 1,840,000$$
$$11x = 1,100,000$$
$$x = 100,000$$
$$y = \frac{7400 - 0.05x}{200}$$
$$y = \frac{7400 - 0.05(100,000)}{200} = 12$$

They had each worked 12 weeks and had \$100,000 in sales.

Chapter 4 Review Problems

1. $x + 2y = 8$
$x - y = 2$

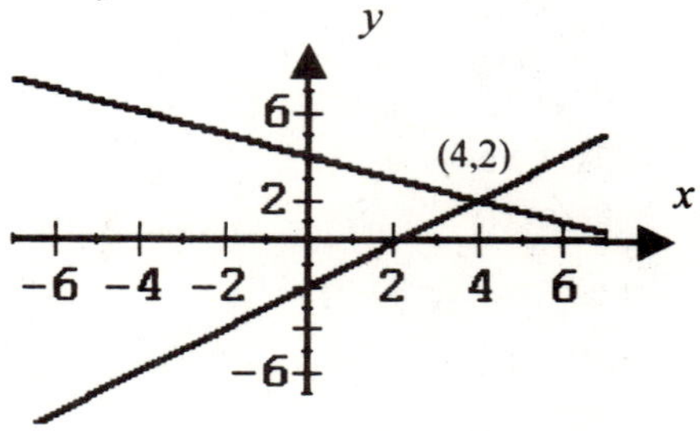

The solution is (4,2).

2. $x + y = 2$
$3x - y = 6$

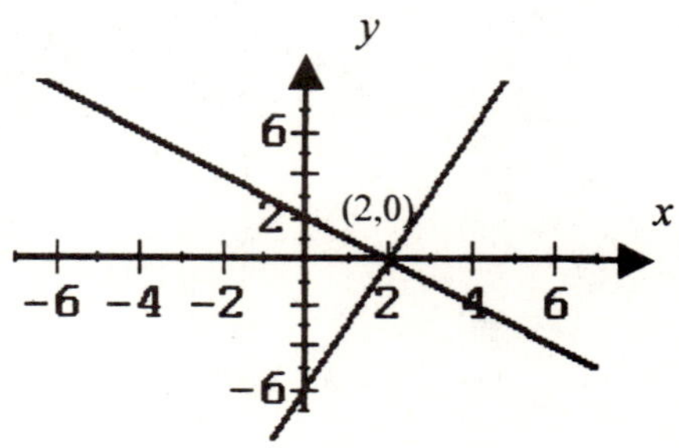

The solution is (2,0).

3. $2x+y=6$
   $3x+4y=4$

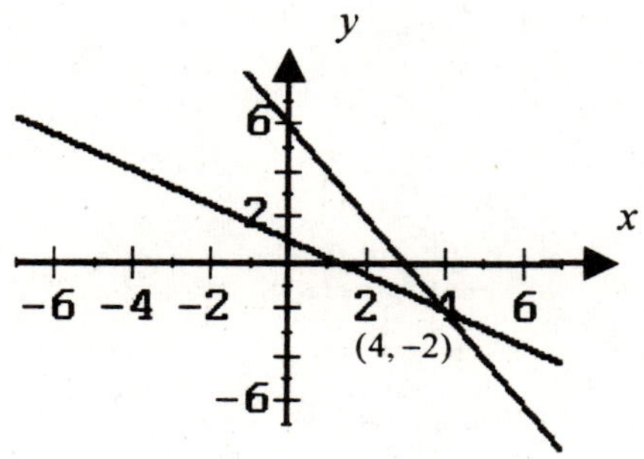

The solution is $(4,-2)$.

4. $3x-2y=-9,\ y=\dfrac{9+3x}{2}$

$$2x+y=1$$
$$2x+\frac{9+3x}{2}=1$$
$$4x+9+3x=2$$
$$7x=-7$$
$$x=-1$$
$$y=\frac{9+3x}{2}$$
$$y=\frac{9+3(-1)}{2}$$
$$y=3$$

5. $-6x-y=1, y=-6x-1$

$$3x-4y=31$$
$$3x-4(-6x-1)=31$$
$$3x+24x+4=31$$
$$27x=27$$
$$x=1$$
$$y=-6x-1$$
$$y=-6(1)-1$$
$$y=-7$$

6. $4x+3y=10,\ y=\dfrac{10-4x}{3}$

$$5x-y=3$$
$$5x-\frac{10-4x}{3}=3$$
$$15x-10+4x=9$$
$$19x=19,\ x=1$$
$$y=\frac{10-4x}{3}=\frac{10-4(1)}{3}=2$$

7. $-7x+y=-4,\ y=7x-4$

$$5x+2y=11$$
$$5x+2(7x-4)=11$$
$$5x+14x-8=11$$
$$19x=19$$
$$x=1$$
$$y=7x-4=7(1)-4=3$$

8. $-2x+5y=-12,\quad -6x+15y=-36$

$$3x+y=1,\qquad \underline{6x+2y=2}$$
$$17y=-34$$
$$y=-2$$
$$3x+y=1$$
$$3x-2=1$$
$$3x=3$$
$$x=1$$

9. $-4x+2y=-16,\quad -12x+6y=-48$

$$3x+5y=-1,\qquad \underline{12x+20y=-4}$$
$$26y=-52$$
$$y=-2$$
$$3x+5y=-1$$
$$3x+5(-2)=-1$$
$$3x=9$$
$$x=3$$

10. $7x-4y=2,$ $\quad -35x+20y=-10$
$6x-5y=-3,$ $\quad 24x-20y=-12$
$-11x=-22$
$x=2$

$6x-5y=-3$
$6(2)-5y=-3$
$-5y=-15$
$y=3$

11. $3x+4y=-6,$ $\quad 6x+8y=-12$
$-2x+7y=-25,$ $\quad -6x+21y=-75$
$29y=-87$
$y=-3$

$3x+4y=-6$
$3x+4(-3)=-6$
$3x=6$
$x=2$

12. $2x+4y=9,\ y=\dfrac{9-2x}{4}$
$3x+6y=8$
$3x+6\dfrac{9-2x}{4}=8$
$12x+54-12x=32$
$54=32$

No solution. Inconsistent system.

13. $x+5y=10,\ x=10-5y$
$y=2-\frac{1}{5}x=2-\frac{1}{5}(10-5y)$
$5y=10-10+5y$
$0=0$

Infinite number of solutions.
Dependent system of equations.

14. $7x+6y=-10$
$2x+y=0,\ y=-2x$
$7x+6(-2x)=-10$
$7x-12x=-10$
$-5x=-10$
$x=2$
$y=-2x=-2(2)=-4$

15. $3x+4y=1,\ y=\dfrac{1-3x}{4}$
$9x-2y=-4$
$9x-2\cdot\dfrac{1-3x}{4}=-4$
$36x-2+6x=-16$
$42x=-13$
$x=-\dfrac{1}{3}$
$y=\dfrac{1-3\left(-\frac{1}{3}\right)}{4}=\dfrac{1}{2}$

16. $x+\frac{1}{3}y=1,\ \ x=1-\frac{1}{3}y$
$\frac{1}{4}x-\frac{3}{4}y=-\frac{9}{4}$
$\frac{1}{4}\left(1-\frac{1}{3}y\right)-\frac{3}{4}y=-\frac{9}{4}$
$1-\frac{1}{3}y-3y=-9$
$-\frac{10}{3}y=-10$
$y=3$
$x=1-\frac{1}{3}y=1-\frac{1}{3}(3)=1-1$
$x=0$

17. $\frac{2}{3}x+y=\frac{14}{3}$

$\underline{\frac{2}{3}x-y=-\frac{22}{3}}$

$\frac{4}{3}x=-\frac{8}{3}$

$x=-2$

$\frac{2}{3}x+y=\frac{14}{3}$

$\frac{2}{3}(-2)+y=\frac{14}{3}$

$-4+3y=14$

$3y=18$

$y=6$

18. $9a+10b=7,\quad 36a+40b=28$

$6a-4b=10,\quad \underline{60a-40b=100}$

$96a=128$

$a=\frac{4}{3}$

$9a+10b=7$

$9\left(\frac{4}{3}\right)+10b=7$

$12+10b=7$

$10b=-5$

$b=-\frac{1}{2}$

19. $3a+5b=8,\quad 6a+10b=16$

$2a+4b=3,\quad \underline{-6a-12b=-9}$

$-2b=-7$

$b=-\frac{7}{2}$

(continued)

19. (continued)

$3a+5b=8$

$3a+5\left(-\frac{7}{2}\right)=8$

$6a-35=16$

$6a=51$

$a=\frac{17}{2}$

20. $x+3=3y+1,\ x=3y-2$

$1-2(x-2)=6y+1$

$1-2(3y-2-2)=6y+1$

$1-6y+8=6y+1$

$-12y=-8$

$y=\frac{2}{3}$

$x=3y-2=3\left(\frac{2}{3}\right)-2=2-2$

$x=0$

21. $10(x+1)-13=-8y,\ y=\frac{10x-3}{-8}$

$4(2-y)=5(x+1)$

$8-4y=5x+4$

$8-4\left(\frac{10x-3}{-8}\right)=5x+4$

$16+10x-3=10x+8$

$13=8$

No solution. Inconsistent system of equations.

22. $0.3x-0.2y=0.7,\quad 0.6x-0.4y=1.4$

$-0.6x+0.4y=0.3,\quad \underline{-0.6x+0.4y=0.3}$

$0=1.7$

No solution. Inconsistent system of equations.

23. $0.2x-0.1y=0.8,\quad 0.6x-0.3y=2.4$
$0.1x+0.3y=1.1,\quad \underline{0.1x+0.3y=1.1}$
$0.7x=3.5$
$x=5$

$0.2x-0.1y=0.8$
$0.2(5)-0.1y=0.8$
$-0.1y=-0.2$
$y=2$

24. $3x-2y-z=3 \quad (1)$
$2x+y+z=1 \quad (2)$
$-x-y+z=-4 \quad (3)$
Add (1) and (2)
$5x-y=4 \quad (4)$
Add (1) and (3)
$2x-3y=-1 \quad (5)$

Substitute y from (4) into (5)
$2x-3(5x-4)=-1$
$2x-15x+12=-1$
$-13x=-13$
$x=1$
Substitute into (4)
$5(1)-y=4$
$-y=-1$
$y=1$

Substitute $x=1,\ y=1$ into (3)
$-1-1+z=-4$
$z=-2$

$x=1,\ y=1,\ z=-2$ is the solution to the system of equations.

25. $-2x+y-z=-7 \quad (1)$
$x-2y-x=2 \quad (2)$
$6x+4y+2z=4 \quad (3)$
$\frac{1}{2}\times(1)+(2),$
$3\times(1)+(3),$
$-2x+y-z=-7 \quad (1)$
$0x-\frac{3}{2}y-\frac{3}{2}z=-\frac{3}{2} \quad (2)$
$0x+7y-z=-17 \quad (3)$
$\frac{14}{3}\times(2)+3,$
$-2x+y-z=-7 \quad (1)$
$0x-\frac{3}{2}y-\frac{3}{2}z=-\frac{3}{2} \quad (2)$
$0x+0y-8z=-24 \quad (3)$
$z=3$, from (2)
$0x-\frac{3}{2}y-\frac{3}{2}(3)=-\frac{3}{2} \quad (2)$
$3y+9=3$
$3y=-6$
$y=-2$, from (1)
$-2x+y-z=-7 \quad (1)$
$-2x+(-2)-3=-7$
$-2x-5=-7$
$-2x=-2$
$x=1$

$x=1,\ y=-2,\ z=3$ is the solution to the system of equations.

26. $2x+5y+3z=10$ **(1)**
$x+y+5z=42$ **(2)**
$2x+y=7$ **(3)**

Solve **(1)** for z
$z=3-2x-5y$ and substitute into **(2)**
$x+y+5(3-2x-5y)=42$
$x+y+15-10x-25y=42$
$-9x-24y=27$ **(4)**
Solve **(3)** for y, $y=7-2x$ and substitute into **(4)**
$-9x-24(7-2x)=27$
$-9x-168+48x=27$
$39x=195$
$x=5$
Substitute $x=5$ into $y=7-2x$
$y=7-2(5)=7-10=-3$
$y=-3$
Substitute $x=5, y=-3$ into
$z=3-2x-5y$
$z=3-2(5)-5(-3)=8$
$z=8$

$x=5,\ y=-3,\ z=8$ is the solution.

27. $x+2y+z=5$ **(1)**
$3x-8y=17$ **(2)**
$2y+z=-2$ **(3)**
Solve **(1)** for x, $x=5-2y-z$ and substitute into **(2)**
$3(5-2y-z)-8y=17$
$15-6y-3z-8y=17$
$3z=-14y-2$
$z=\dfrac{-14y-2}{3}.$

Substitute this into **(3)**
(continued)

27. (continued)
$2y+\dfrac{-14y-2}{3}=-2$
$6y-14y-2=-6$
$-8y=-4$
$y=\dfrac{1}{2}$

Substitute $y=\dfrac{1}{2}$ into **(3)**
$2\left(\dfrac{1}{2}\right)+z=-2$
$1+z=-2$
$z=-3$

Substitute $y=\dfrac{1}{2}$ into **(2)**
$3x-8\left(\dfrac{1}{2}\right)=17$
$3x-4=17$
$3x=21$
$x=7$
$x=7,\ y=\dfrac{1}{2},\ z=-3$ is the solution.

28. $2x-4y+3z=0$ **(1)**
$x-2y-5z=13$ **(2)**
$5x+3y-2z=19$ **(3)**

Multiply **(2)** by -2 and add to **(1)**
$13z=-26$
$z=-2$

Multiply **(2)** by -5 and add to **(3)**
$13y+23z=-46$
$13y+23(-2)=-46$
$13y=0$
$y=0$
(continued)

28. (continued)

Substitute $y=0,\ z=-2$ into **(2)**

$x-2(0)-5(-2)=13$

$x+10=13$

$x=3$

$x=3,\ y=0,\ z=-2$ is the solution.

29. $5x+2y+3z=10$ **(1)**

$6x-3y+4z=24$ **(2)**

$-2x+y+2z=2$ **(3)**

Add 3 times **(3)** to **(2)**

$10z=30$

$z=3$

Add -2 times **(3)** to **(1)**

$9x-z=6$

$9x-(3)=6$

$9x=9$

$x=1$

Substitute $x=1,\ z=3$ into **(3)**

$-2(1)+y+2(3)=2$

$-2+y+6=2$

$y=-2$

$x=1,\ y=-2,\ z=3$ is the solution.

30. $3x+2y=7$ **(1)**

$2x+7z=-26$ **(2)**

$5y+z=6$ **(3)**

Solve **(3)** for z and substitute into **(2)**

$2x+7(6-5y)=-26$

$2x+42-35y=-26$

$2x=35y-68$

$x=\dfrac{35y-68}{2}$

Substitute this into **(1)**

$3\left(\dfrac{35y-68}{2}\right)+2y=7$

(continued)

30. (continued)

$3\left(\dfrac{35y-68}{2}\right)+2y=7$

$105y-204+4y=14$

$109y=218$

$y=2$

Substitute $y=2$ into **(3)**

$5(2)+z=6$

$10+z=6$

$z=-4$

Substitute $y=2$ into **(1)**

$3x+2(2)=7$

$3x+4=7$

$3x=3$

$x=1$

$x=1,\ y=2,\ z=-4$ is the solution.

31. $x-y=2$ **(1)**

$5x+7y-5z=2$ **(2)**

$3x-5y+2z=-2$ **(3)**

Add 2 times **(2)** to 5 times **(3)**

$25x-11y=-6$ **(4)**

Solve **(1)** for x and substitute into **(4)**

$x=2+y$

$25(2+y)-11y=-6$

$50+25y-11y=-6$

$14y=-56$

$y=-4$

Substitute $y=-4$ into **(1)**

$x-(-4)=2$

$x+4=2$

$x=-2$

Substitute $x=-2,\ y=-4$ into **(3)**

(continued)

31. (continued)
$3(-2)-5(-4)+2z=-2$
$-6+20+2z=-2$
$2z=-16$
$z=-8$
$x=-2,\ y=-4,\ z=-8$ is the solution.

32. $v=$ speed of plane in still air
$w=$ speed of wind
$720=(v-w)\cdot 3 \qquad v-w=240$
$720=(v+w)(2.5) \qquad v+w=288$
$2v=528$
$v=264$
$w=288-v=288-264=24$
The speed of the plane in still air is 264 mph and the wind speed is 24 mph.

33. $x=$ number of new employees
$y=$ number of laid-off employees
$25x+10y=275$ **(1)**
$8x+3y=86$ **(2)**
Add 3 times **(1)** to -10 times **(2)**
$75x+30y=825$
$-80x-30y=-860$
$-5x=-35$
$x=7$
Substitute $x=7$ into **(1)**
$25(7)+10y=275$
$10y=100$
$y=10$
The company can train 7 new employees and 10 laid-off employees.

34. $x=$ number of laborers
$y=$ number of mechanics
$70x+90y=1950$ **(1)**

(continued)

34. (continued)
$80x+100y=2200$ **(2)**
Divide both equations by 10
$7x+9y=195$ **(1)**
$8x+10y=220$ **(2)**
Add -8 times **(1)** to 7 times **(2)**
$-2y=-20$
$y=10$
Substitute $y=10$ into **(2)**
$8x+10(10)=220$
$8x=120$
$x=15$
The circus hired 15 laborers and 10 mechanics.

35. $x=$ number of children's tickets
$y=$ number of adult tickets
$x+y=590$ **(1)**
$6x+11y=4790$ **(2)**
Solve **(1)** for y substitute into **(2)**
$6x+11(590-x)=4790$
$6x+6490-11x=4790$
$-5x=-1700$
$x=340$
$y=590-x=590-340=250$
There were 340 children's tickets sold and 250 adult tickets.

36. $x=$ cost of hat
$y=$ cost of shirt
$z=$ cost of pants
$2x+5y+4z=129$ **(1)**
$x+y+2z=42$ **(2)**
$2x+3y+z=63$ **(3)**
Add -2 times **(2)** and **(1)**
$3y=45$
$y=15$

(continued)

36. (continued)
Substitute $y=15$ into **(2)** and solve for *x*
$x+15+2z=42$
$x=27-2z$
Substitute this and $y=15$ into **(3)**
$2(27-2z)+3(15)+z=63$
$54-4z+45+z=63$
$-3z=-36$
$z=12$
Substitute $y=15$, $z=12$ into **(2)**
$x+15+2(12)=42$
$x+15+24=42$
$x=3$

The hats cost \$3, shirts \$15, and pants \$12.

37. $x=$ number of A packets
$y=$ number of B packets
$z=$ number of C packets
$2x+3y+4z=29$ **(1)**
$4x+y+3z=23$ **(2)**
$3x+y+2z=17$ **(3)**
Add 3 times **(1)** to -4 times **(2)**
$-10x+5y=-5$ **(4)**
Add 2 times **(2)** to -3 times **(3)**
$-x-y=-5$ **(5)**
Solve **(5)** for *y* and substitute into **(4)**
$-10x+5(-x+5)=-5$
$-10x-5x+25=-5$
$-15x=-30$
$x=2$
Substitute $x=2$ into **(5)**
$-2-y=-5$
$-y=-3$
$y=3$
(continued)

37. (continued)
Substitute $x=2$, $y=3$ into **(1)**
$2(2)+3(3)+4z=29$
$4+9+4z=29$
$4z=16$
$z=4$

She should use 2 of the A packets, 3 of the B packets, and 4 of the C packets.

38. $x=$ cost of jelly
$y=$ cost of peanut butter
$z=$ cost of honey
$4x+3y+5z=9.8$ **(1)**
$2x+2y+z=4.20$ **(2)**
$3x+4y+2z=7.70$ **(3)**
Add **(1)** and -5 times **(2)**
$-6x-7y=-11.2$ **(4)**
Add -2 times **(2)** to **(3)**
$-x=-0.7$
$x=0.7$
Substitute $x=0.7$ into **(4)**
$-6(0.7)-7y=-11.2$
$-7y=-7$
$y=1$
Substitute $x=0.7$, $y=1$ into **(2)**
$2(0.7)+2(1)+z=4.2$
$z=0.8$

The cost of a jar of jelly is \$0.70, the cost of a jar of peanut butter is \$1, and the cost of a jar of honey is \$0.80.

39. $x=$ number of buses
$y=$ number of station wagons
$z=$ number of sedans
$x+y+z=9$ **(1)**
$40x+8y+5z=127$ **(2)**
(continued)

39. (continued)

$8(3y)+5(2z)=126$

$24y+10z=126$ **(3)**

Add -40 times **(1)** to **(2)**

$-32y-35z=-233$ **(4)**

Add 32 times **(3)** to 24 times **(4)**

$-520z=-1560$

$z=3$

Substitute $z=3$ into **(3)**

$24y+10(3)=126$

$24y=96$

$y=4$

Substitute $y=4,\ z=3$ into **(1)**

$x+4+3=9$

$x=2$

They should use 2 buses, 4 station wagons, and 3 sedans.

40. $-x-5z=-5$ **(1)**

$13x+2z=2$ **(2)**

Solve **(1)** for x and substitute into **(2)**

$x=5-5z$

$13(5-5z)+2z=2$

$65-65z+2z=2$

$-63z=-63$

$z=1$

Substitute $z=1$ into **(1)**

$-x-5(1)=-5$

$-x=0$

$x=0$

The solution is $x=0,\ z=1$.

41. $x+y=10$ **(1)**

$6x+9y=70$ **(2)**

Solve **(1)** for y and substitute into **(2)**

(continued)

41. (continued)

$y=10-x$

$6x+9(10-x)=70$

$6x+90-9x=70$

$-3x=-20$

$x=\frac{20}{3}$

Substitute $x=\frac{20}{3}$ into **(1)**

$\frac{20}{3}+y=10$

$y=\frac{10}{3}$

The solution is $x=\frac{20}{3},\ y=\frac{10}{3}$.

42. $2x+5y=4$ **(1)**

$5x-7y=4$ **(2)**

Solve **(1)** for y and substitute into **(1)**

$y=\frac{4-2x}{5}$

$5x-7\frac{4-2x}{5}=-29$

$25x-28+14x=-145$

$39x=-117$

$x=-3$

Substitute $x=-3$ into **(1)**

$2(-3)+5y=4$

$-6+5y=4$

$5y=10$

$y=2$

The solution is $x=-3,\ y=2$.

43. $\frac{x}{2}-3y=-6$ **(1)**

(continued)

43. (continued)

$\frac{4}{3}x+2y=4$ **(2)**

Solve**(1)**for x and substitute into**(2)**

$x=6y-12$

$\frac{4}{3}(6y-12)+2y=4$

$24y-48+6y=12$

$30y=60$

$y=2$

$x=6y-12=6(2)-12=0$

The solution is $x=0,\ y=2$.

44. $\frac{3}{5}x-y=6$ **(1)**

$x+\frac{y}{3}=10$ **(2)**

Solve **(1)** for y and substitute into **(2)**

$y=\frac{3}{5}x-6$

$x+\frac{\frac{3}{5}x-6}{5}=10$

$5x+\frac{3}{5}x-6=50$

$25x+3x-30=250$

$28x=280$

$x=10$

Substitute $x=10$ into **(2)**

$10+\frac{y}{3}=10$

$y=0$

The solution is $x=10,\ y=0$.

45. $\frac{x+1}{5}=y+2$ **(1)**

(continued)

45. (continued)

$\frac{2y+7}{3}=x-y$ **(2)**

Solve**(1)** for y and substitute into**(2)**

$y=\frac{x+1}{5}-2$

$\frac{2\left(\frac{x+1}{5}-2\right)+7}{3}=x-\frac{x+1}{5}+2$

$2\left(\frac{x+1}{5}-2\right)+7=3x-\frac{3x+3}{5}+6$

$2(x+1-10)+35=15x-3x-3+30$

$2x-18+35=12x+27$

$10x=-10$

$x=-1$

$y=\frac{x+1}{5}-2=\frac{-1+1}{5}-2=-2$

The solution is $x=-1,\ y=-2$.

46. $3(2+x)=y+1$

$6+3x=y+1$

$y=3x+5$ **(1)**

$5(x-y)=-7-3y$

$5x-5y=-7-3y$

$5x-2y=-7$ **(2)**

Substitute y from **(1)** into **(2)**

$5x-2(3x+5)=-7$

$5x-6x-10=-7$

$-x=3$

$x=-3$

$y=3x+5=3(-3)+5=-9+5=-4$

The solution is $x=-3,\ y=-4$.

47. $7(x+3)=2y+25$

$7x+21=2y+25$

(continued)

47. (continued)

$7x-2y=4 \quad \textbf{(1)}$

$3(x-6)=-2(y+1)$

$3x-18=-2y-2$

$3x+2y=16 \quad \textbf{(2)}$

Add **(1)** and **(2)**

$10x=20$

$x=2$

Substitute $x=2$ into **(2)**

$3(2)+2y=16$

$2y=10$

$y=5$

The solution is $x=2,\ y=5$.

48. Multiply both equations by 10 to clear decimals.

$3x-4y=9 \quad \textbf{(1)}$

$2x-3y=4 \quad \textbf{(2)}$

Solve**(1)**for y and substitute into**(2)**

$y=\dfrac{3x-9}{4}$

$2x-3\dfrac{3x-9}{4}=4$

$8x-9x+27=16$

$-x=-11$

$x=11$

$y=\dfrac{3x-9}{4}=\dfrac{3(11)-9}{4}=6$

The solution is $x=11,\ y=6$.

49. Multiply both equations by 100 to clear decimals.

$5x+8y=-76 \quad \textbf{(1)}$

$4x-3y=5 \quad \textbf{(2)}$

Solve**(1)**for y and substitute into**(2)**

$y=\dfrac{-76-5x}{8}$

(continued)

49. (continued)

$4x-3\dfrac{-76-5x}{8}=5$

$32x+228+15x=40$

$47x=-188$

$x=-4$

$y=\dfrac{-76-5x}{8}=\dfrac{-76-5(-4)}{8}=-7$

The solution is $x=-4,\ y=-7$.

50. $x-\dfrac{y}{2}+\dfrac{1}{2}z=-1 \quad \textbf{(1)}$

$2x+\dfrac{5}{2}z=-1 \quad \textbf{(2)}$

$\dfrac{3}{2}y+2z=1 \quad \textbf{(3)}$

Solve **(2)** for x and **(3)** for y and substitute into **(1)**

$x=\dfrac{-\frac{5}{2}z-1}{2}$

$y=\dfrac{2(1-2z)}{3}=\dfrac{2-4z}{3}$

$\dfrac{-\frac{5}{2}z-1}{2}-\dfrac{\frac{2-4z}{3}}{2}+\dfrac{1}{2}z=-1$

$-\dfrac{5}{2}z-1-\dfrac{2-4z}{3}+z=-2$

$-15z-6-4+8z+6z=-12$

$-z=-2$

$z=2$

$x=\dfrac{-\frac{5}{2}z-1}{2}=\dfrac{-\frac{5}{2}(2)-1}{2}=-3$

$y=\dfrac{2-4z}{3}=\dfrac{2-4(2)}{3}=-2$

The solution is $x=-3,\ y=-2,\ z=2$.

51. $2x-3y+2z=0$ **(1)**

$x+2y-z=2$ **(2)**

$2x+y-3z=-1$ **(3)**

Add **(1)** and 2 times **(2)**

$4x-y=4$ **(4)**

Add -3 times**(2)** and **(3)**

$-x-5y=1$ **(5)**

Solve**(4)**for y and substitute into**(5)**

$y=4x-4$

$-x-5(4x-4)=1$

$-x-20x+20=1$

$-21x=-21$

$x=1$

$y=4x-4=4(1)-4=0$

Substitute $x=1$, $y=0$ into **(1)**

$2(1)-3(0)+2z=0$

$2+2z=0$

$2z=-2$

$z=-1$

The solution is $x=1$, $y=0$, $z=-1$.

52. $x-4y+4z=-1$ **(1)**

$2x-y+5z=-3$ **(2)**

$x-3y+z=4$ **(3)**

Add **(1)** and -4 times **(3)**

$-3x+8y=-17$ **(4)**

Add **(2)** and -5 times **(3)**

$-3x+14y=-23$ **(5)**

Subtract **(5)** from **(4)**

$-6y=6$

$y=-1$

Substitute $y=-1$ into **(4)**

$-3x+8(-1)=-17$

$-3x=-9$

$x=3$

(continued)

52. (continued)

Substitute $x=3$, $y=-1$ into **(3)**

$3-3(-1)+z=4$

$z=-2$

The solution is $x=3$, $y=-1$, $z=-2$

53. $x-2y+z=-5$ **(1)**

$2x+z=-10$ **(2)**

$y-z=15$ **(3)**

Solve **(3)** for y

$y=z+15$ **(4)**

Solve **(2)** for x

$x=\dfrac{-z-10}{2}$ **(5)**

Substitute **(4)** and **(5)** into **(1)**

$\dfrac{-z-10}{2}-2(z+15)+z=-5$

$-z-10-4z-60+2z=-10$

$=3z=60$

$z=-20$

Substitute $z=-20$ into **(4)** and **(5)**

$x=\dfrac{-(-20)-10}{2}=5$

$y=-20+15=-5$

The solution is $x=5$, $y=-5$, $z=-20$

54. $x-y\le 3$

Test point: (0,0)

$0-0\le 3$

$0\le 3$ True

$y\le -\frac{1}{4}x+2$

$0\le -\frac{1}{4}(0)+2$

$0\le 2$ True

(continued)

54. (continued)

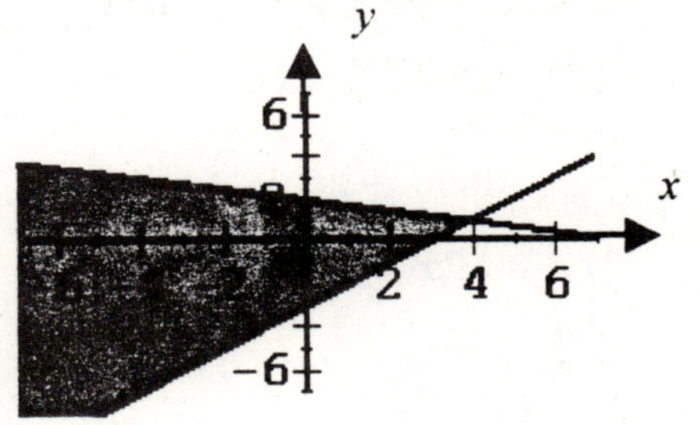

55. $-2x+3y<6$

Test point: (0,0)

$-2(0)+3(0)<6$

$0<6$ True

$y>-2$

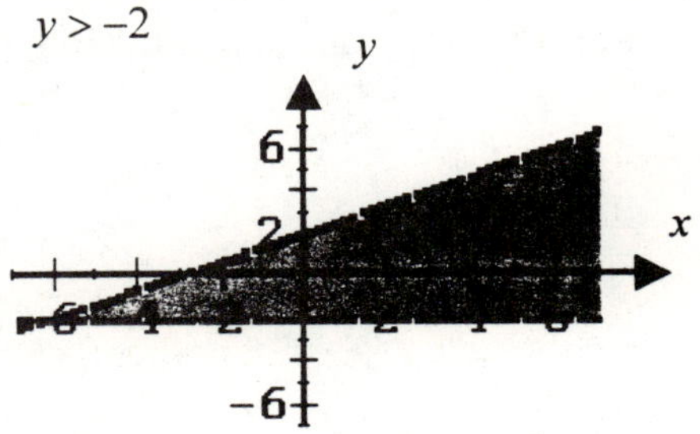

56. $x+y>1$

Test point: (0,0)

$0+0>1$

$0>1$ False

$2x-y<5$

Test point: (0,0)

$2(0)-0<5$

$0<5$ True

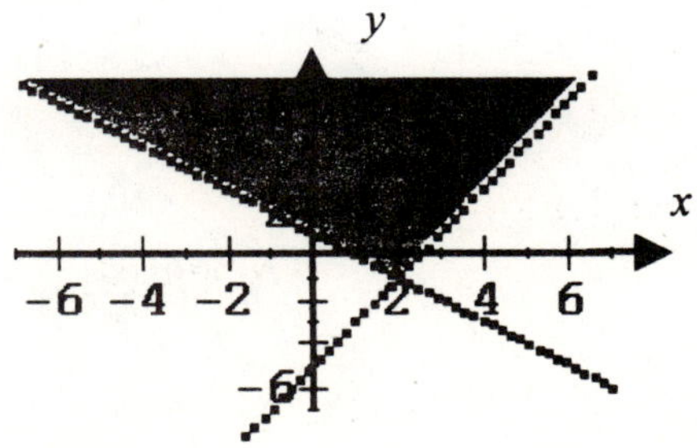

57. $x+y\ge 4$

Test point: (0,0)

$0+0\ge 4$

$0\ge 4$ False

$y\le x$

Test point: (2,0)

$0\le 2$ True

$x\le 6$

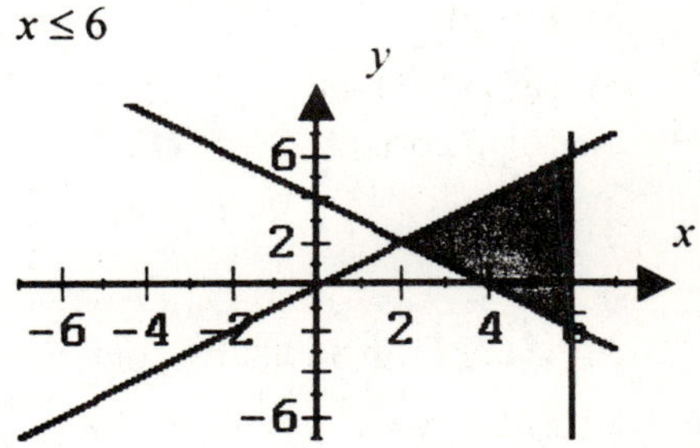

Chapter 4 Test

1. $3x-2y=-8$ **(1)**

$x+6y=4$ **(2)**

Solve**(2)** for x and substitute into **(1)**

$x=4-6y$

$3(4-6y)-2y=-8$

$12-18y-2y=-8$

$-20y=-20$

$y=1$

$x=4-6y=4-6(1)=-2$

The solution is $x=-2,\ y=1$.

2. $6x-2y=-2$ **(1)**

$3x+4y=14$ **(2)**

Add 2 times **(1)** to **(2)**

$15x=10$

$x=\frac{2}{3}$

(continued)

2. (continued)
$3x+4y=14$
$3\left(\frac{2}{3}\right)+4y=14$
$4y=12$
$y=3$
The solution is $x=\frac{2}{3},\ y=3$.

3. $\frac{1}{4}a-\frac{3}{4}b=-1$ **(1)**
$\frac{1}{3}a+b=\frac{5}{3}$ **(2)**
Solve **(2)** for b and substitute into **(1)**
$b=\frac{5}{3}-\frac{1}{3}a$
$\frac{1}{4}a-\frac{3}{4}\left(\frac{5}{3}-\frac{1}{3}a\right)=-1$
$a-5+a=-4$
$2a=1$
$a=\frac{1}{2}$
$b=\frac{5}{3}-\frac{1}{3}a=\frac{5}{3}-\frac{1}{3}\cdot\frac{1}{2}=\frac{3}{2}$
The solution is $a=\frac{1}{2},\ b=\frac{3}{2}$.

4. $7x-1=3(1+y)$ **(1)**
$7x-1=3+3y$
$7x-3y=4$ **(3)**
$1-6y=-14x-7$ **(2)**
$14x-6y=-8$
$7x-3y=-4$ **(4)**

Comparing **(3)** and **(4)** shows the system is inconsistent and has no solution.

5. $5x-3y=3$ **(1)**
$7x+y=25$ **(2)**
Add 3 times **(2)** to **(1)**
$26x=78$
$x=3$
$7x+y=25$
$7(3)+y=25$
$y=4$
The solution is $x=3,\ y=4$.

6. $\frac{1}{3}x+\frac{5}{6}y=2$ **(1)**
$\frac{3}{5}x-y=-\frac{7}{5}$ **(2)**
Solve **(2)** for y and substitute into **(1)**
$y=\frac{3}{5}x+\frac{7}{5}$
$\frac{1}{3}x+\frac{5}{6}\left(\frac{3}{5}x+\frac{7}{5}\right)=2$
$2x+3x+7=12$
$5x=5$
$x=1$
$y=\frac{3}{5}x+\frac{7}{5}$
$y=\frac{3}{5}(1)+\frac{7}{5}$
$y=2$
The solution is $x=1,\ y=2$.

7.
$$\begin{aligned}3x+5y-2z&=-5\\2x+3y-z&=-2\\2x+4y+6z&=18\end{aligned}$$

(continued)

7. (continued)

Multiply first equation by $-\frac{2}{3}$ and add to the second and third equation

$$\begin{aligned} 3x + 5y - 2z &= -5 \\ \frac{1}{3}y + \frac{1}{3}z &= \frac{4}{3} \\ \frac{2}{3}y + \frac{22}{3}z &= \frac{64}{3} \end{aligned}$$

Multiply second equation by 2 and add to the third equation

$$\begin{aligned} 3x + 5y - 2z &= -5 \\ \frac{1}{3}y + \frac{1}{3}z &= \frac{4}{3} \\ 8z &= 24 \end{aligned}$$

$8z = 24$

$z = 3$

From second equation

$-\frac{1}{3}y + \frac{1}{3}(3) = \frac{4}{3},\ y = -1$

From the first equation

$3x + 5(-1) - 2(3) = -5,\ x = 2$

The solution is $x = 2,\ y = -1,\ z = 3$.

8.

$$\begin{aligned} 3x + 2y &= 0 \\ 2x - y + 3z &= 8 \\ 5x + 3y + z &= 4 \end{aligned}$$

Multiply the first equation by $-\frac{2}{3}$ and add to the second equation and then multiply the first equation by $-\frac{5}{3}$ and add to the third equation

(continued)

8. (continued)

$$\begin{aligned} 3x + 2y &= 0 \\ -\frac{7}{3}y + 3z &= 8 \\ -\frac{1}{3}y + z &= 4 \end{aligned}$$

Multiply the second equation by $-\frac{1}{7}$ and add to the third equation

$$\begin{aligned} 3x + 2y &= 0 \\ -\frac{7}{3}y + 3z &= 8 \\ +\frac{4}{7}z &= \frac{20}{7} \end{aligned}$$

$\frac{4}{7}z = \frac{20}{7},\ z = 5$

From second equation

$-\frac{7}{3}y + 3(5) = 8,\ y = 3$

From the first equation

$3x + 2(3) = 0,\ x = -2$

The solution is $x = -2,\ y = 3,\ z = 5$.

9.

$$\begin{aligned} x + 5y + 4z &= -3 \\ x - y - 2z &= -3 \\ x + 2y + 3z &= -5 \end{aligned}$$

Multiply the first equation by -1 and add to the second and third equation

$$\begin{aligned} x + 5y + 4z &= -3 \\ -6y - 6z &= 0 \\ -3y - z &= -2 \end{aligned}$$

Multiply the second equation by $-\frac{1}{2}$ and add to the third equation

(continued)

9. (continued)

$$\begin{aligned} x + 5y + 4z &= -3 \\ -6y - 6z &= 0 \\ 2z &= -2 \end{aligned}$$

$2z = -2,\ z = -1$

From the second equation

$-6y - 6(-1) = 0,\ y = 1$

From the first equation

$x + 5(1) + 4(-1) = -3,\ x = -4$

The solution is $x = -4,\ y = 1,\ z = -1$.

10. v = speed of plane in still air
w = speed of wind

$1000 = (v + w)\cdot 2, \quad v + w = 500$

$1000 = (v - w)\cdot 2.5 \quad \underline{v - w = 400}$

$2v = 900$

$v = 450$

$450 + w = 500$

$w = 50$

The speed of the plane in still air is 450 mph. The speed of the wind is 50 mph.

11. x = number of station wagons
y = number of two-door sedans
z = number of four-door sedans

$$\begin{aligned} 5x + 4y + 3z &= 62 \\ 4x + 3y + 3z &= 52 \\ 3x + 2y + 2z &= 36 \end{aligned}$$

Multiply first equation by $-\frac{4}{5}$ and add to the second equation, then multiply the first equation by $-\frac{3}{5}$ and add to the third equation.

(continued)

11. (continued)

$$\begin{aligned} 5x + 4y + 3z &= 62 \\ -\frac{1}{5}y + \frac{3}{5}z &= \frac{12}{5} \\ -\frac{2}{5}y + \frac{1}{5}z &= -\frac{6}{5} \end{aligned}$$

Multiply the second equation by -2 and add to the third equation

$$\begin{aligned} 5x + 4y + 3z &= 62 \\ -\frac{1}{5}y + \frac{3}{5}z &= \frac{12}{5} \\ -z &= -6 \end{aligned}$$

$-z = -6,\ z = 6$

From the second equation

$-\frac{1}{5}y + \frac{3}{5}(6) = \frac{12}{5},\ y = 6$

From the first equation

$5x + 4(6) + 3(6) = 62,\ x = 4$

The line supervisor should send 4 station wagons, 6 two-door sedans, and 6 four-door sedans down the assembly line.

12. x = daily charge
y = mileage charge

$5x + 150y = 180 \quad$ **(1)**

$7x + 320y = 274 \quad$ **(2)**

Multiply **(1)** by $-\frac{7}{5}$ and add to **(2)**

$110y = 22$

$y = 0.2$

From first equation

$5x + 150(0.2) = 180$

$5x = 150$

$x = 30$

They charge $30 per day and $0.20 per mile.

13. $x+2y\le 6$

    Test point: (0,0)

    $0+2(0)\le 6$

    $0\le 6$ True

    $-2x+y\ge -2$

    Test point: (0,0)

    $-2(0)+0\ge -2$

    $0\ge -2$ True

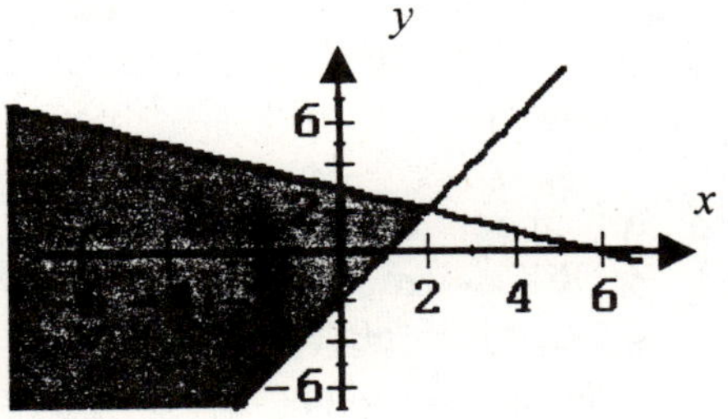

14. $3x+y\ge 8$

    Test point: (0,0)

    $3(0)+0\ge 8$

    $0\ge 8$ False

    $x-2y\ge 5$

    Test point: (0,0)

    $0-2(0)\ge 5$

    $0\ge 5$ False

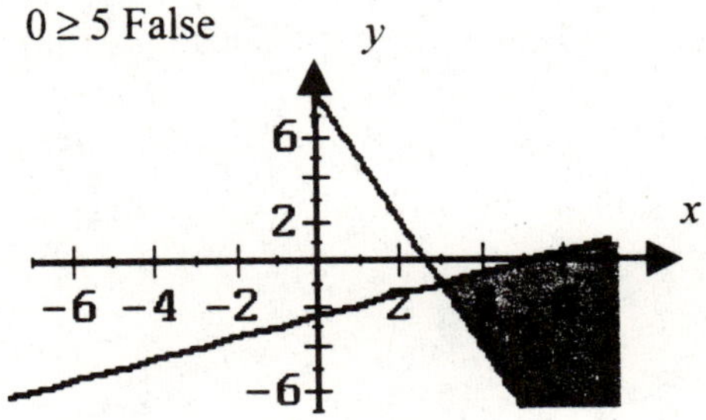

Cumulative Test for Chapters 1-4

1. $5\frac{1}{2}-3\frac{7}{8}=\frac{44}{8}-\frac{31}{8}=\frac{13}{8}=1\frac{5}{8}$

2. $(2-3)^3+20\div(-10)\left(\frac{1}{5}\right)$

   $=(-1)^3+(-2)\left(\frac{1}{5}\right)$

   $=-1-\frac{2}{5}$

   $=-1\frac{2}{5}$

3. $(-2)^2+(-3)^3=4+(-27)=-23$

4. $2x-4[x-3(2x+1)]$

   $=2x-4[x-6x-3]$

   $=2x-4[-5x-3]$

   $=2x+20x+12$

   $=22x+12$

5. $A=P(3+4rt)$

   $P=\frac{A}{3+4rt}$

6. $\frac{1}{4}x+5=\frac{1}{3}(x-2)$

   $3x+60=4x-8$

   $x=68$

7. $4x-8y=10$

x	y
-2	$-\frac{9}{4}$
0	$-\frac{5}{4}$
4	$\frac{3}{4}$

(continued)

7. (continued)

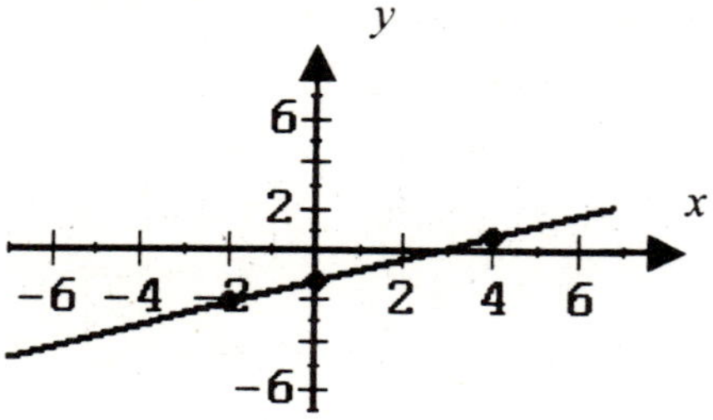

8. $m = \frac{y_2 - y_1}{x_2 - x_1} = \frac{-1-(-2)}{6-(-4)} = \frac{1}{10}$

9. $4x+3-13x-7 < 2(3-4x)$

$-9x-4 < 6-8x$

$-x < 10$

$x > -10$

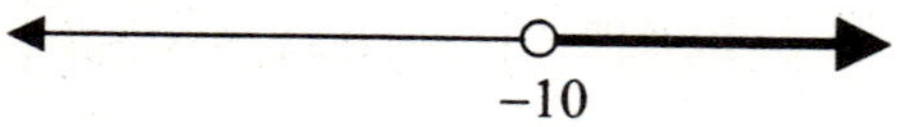

10. $\frac{2x-1}{3} \le 7$ and $2(x+1) \ge 12$

$2x-1 \le 21$ $2x+2 \ge 12$

$2x \le 22$ $2x \ge 10$

$x \le 11$ $x \ge 5$

$5 \le x \le 11$

11. $3x+6y=-2$

$6y=-3x-2$

$y=-\frac{1}{2}x-\frac{1}{3},\ m=-\frac{1}{2},\ m_\perp=2$

$y-y_1=m(x-x_1)$

$y-(-3)=2(x-2)$

$y+3=2x-4$

$y=2x-7$

12. $x =$ length of first side

$x+7 =$ length of second side

$2x-6 =$ length of third side

$x+x+7+2x-6=69$

$4x+1=69$

$4x=68$

$x=17$

$x+7=24$

$2x-6=28$

The first side is 17 meters, the second side is 24 meters, and the third side is 28 meters.

13. $x =$ amount invested at 7%

$6000-x =$ amount invested at 9%

$0.07x+0.09(6000-x)=510$

$0.07x+540-0.09x=510$

$-0.02x=-30$

$x=1500$

$6000-x=4500$

Victor invested \$1500 at 7% and \$4500 at 9%.

14. $5x+2y=2$ **(1)**

$4x+3y=-4$ **(2)**

Solve **(1)** for *y* and substitute into **(2)**

$2y=2-5x$

$y=\frac{2-5x}{2}$

$4x+3\frac{2-5x}{2}=-4$

$8x+6-15x=-8$

$-7x=-14$

$x=2$

$y=\frac{2-5x}{2}=\frac{2-5(2)}{2}=-4$

The solution is $x=2,\ y=-4$.

15.

$$\begin{aligned} 2x + y - z &= 4 \\ x + 2y - 2z &= 2 \\ x - 3y + z &= 4 \end{aligned}$$

Multiply the first equation by $-\frac{1}{2}$ and add to the second and third equations

$$\begin{aligned} 2x + y - z &= 4 \\ \frac{3}{2}y - \frac{3}{2}z &= 0 \\ -\frac{7}{2}y + \frac{3}{2}z &= 2 \end{aligned}$$

Multiply the second equation by $\frac{7}{3}$ and add to the third equation

$$\begin{aligned} 2x + y - z &= 4 \\ \frac{3}{2}y - \frac{3}{2}z &= 0 \\ -2z &= 2 \end{aligned}$$

$-2z = 2$

$z = -1$

From the second equation

$\frac{3}{2}y - \frac{3}{2}(-1) = 0$

$y = -1$

From the first equation

$2x + (-1) - (-1) = 4$

$2x = 4$

$x = 2$

The solution is $x = 2,\ y = -1,\ z = -1$.

16. $x =$ cost of shirt
$y =$ cost of slacks

$5x + 8y = 345$ **(1)**

$7x + 3y = 237$ **(2)**

Solve **(1)** for y and substitute into **(2)**

(continued)

16. (continued)

$8y = 345 - 5x$

$y = \frac{345 - 5x}{8}$

$7x + 3\frac{345 - 5x}{8} = 237$

$56x + 1035 - 15x = 1896$

$41x = 861$

$x = 21$

$y = \frac{345 - 5x}{8} = \frac{345 - 5(21)}{8} = 30$

The shirts cost \$21 and the slacks cost \$30.

17. $7x - 6y = 17$ **(1)**

$3x + y = 18$ **(2)**

Solve **(2)** for y and substitute into **(1)**

$y = 18 - 3x$

$7x - 6(18 - 3x) = 17$

$7x - 108 + 18x = 17$

$25x = 125$

$x = 5$

$y = 18 - 3x = 18 - 3(5) = 3$

The solution is $x = 5,\ y = 3$.

18.

$$\begin{aligned} x + 3y + z &= 5 \\ 2x - 3y - 2z &= 0 \\ x - 2y + 3z &= -9 \end{aligned}$$

Multiply first equation by -2 and add to the second equation, then multiply the first equation by -1 and add to the third equation.

$$\begin{aligned} x + 3y + z &= 5 \\ -9y - 4z &= -10 \\ -5y + 2z &= -14 \end{aligned}$$

(continued)

18. (continued)
Multiply the second equation by $-\frac{5}{9}$ and add to third equation.

$$\begin{aligned} x + 3y + z &= 5 \\ -9y - 4z &= -10 \\ +\frac{38}{9}z &= -\frac{76}{9} \end{aligned}$$

From third equation

$\frac{38}{9}z = -\frac{76}{9}$, $z = -2$

From second equation

$-9y - 4(-2) = -10$, $y = 2$

From first equation

$x + 3(2) + (-2) = 5$, $x = 1$

The solution is $x = 1$, $y = 2$, $z = -2$.

19. $-5x + 6y = 2$ **(1)**
$10x - 12y = -4$ **(2)**

Multiplying **(1)** by 2 and adding to **(2)** gives $0 = 0$. This is a dependent system of equations with an infinite number of solutions.

20. $x - y \geq -4$
Test point: (0,0)
$0 - 0 \geq -4$
$0 \geq -4$ True
$x + 2y \geq 2$
Test point: (0,0)
$0 + 2(0) \geq 2$
$0 \geq 2$ False

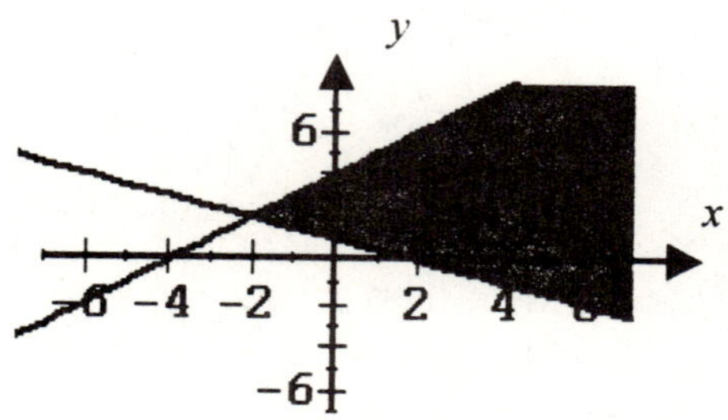

Chapter 5

Pretest Chapter 5

1. $\left(3^6\right)\left(3^{10}\right) = 3^{6+10} = 3^{16}$

2. $\left(-5x^3\right)\left(2x^4\right) = (-5 \cdot 2)\left(x^3 \cdot x^4\right)$
$= -10x^7$

3. $\left(-4ab^2\right)\left(2a^3b\right)(3ab)$
$= (-4)(2)(3)a^{1+3+1}b^{2+1+1}$
$= -24a^5b^4$

4. $\dfrac{x^{26}}{x^3} = x^{26-3} = x^{23}$

5. $\dfrac{12xy^2}{-6x^3y^4} = -\dfrac{2}{x^{3-1}y^{4-2}}$
$= -\dfrac{2}{x^2y^2}$

6. $\dfrac{-25a^0b^2c^3}{-15abc^2} = \dfrac{5b^{2-1}c^{3-2}}{3a^{1-0}}$
$= \dfrac{5bc}{3a}$

7. $\left(x^5\right)^{10} = x^{5(10)} = x^{50}$

8. $\left(-2x^2y\right)^3 = (-2)^3 x^{2\cdot 3}y^{1\cdot 3}$
$= -8x^6y^3$

9. $\left(\dfrac{4a^3b}{c^2}\right)^3 = \dfrac{4^3\left(a^3\right)^3b^3}{\left(c^2\right)^3}$
$= \dfrac{64a^9b^3}{c^6}$

10. $3x^2y^{-3}z^{-4} = \dfrac{3x^2}{y^3z^4}$

11. $\dfrac{-2a^{-3}}{b^{-2}c^4} = \dfrac{-2b^2}{a^3c^4}$

12. $(-3)^{-2} = \dfrac{1}{(-3)^2} = \dfrac{1}{9}$

13. $0.000638 = 6.38 \times 10^{-4}$

14. 1.894×10^{12}
$= 1.894000000000 \times 10^{12}$
$= 1{,}894{,}000{,}000{,}000$

15. $\left(3x^2 - 6x - 9\right) + \left(-5x^2 + 13x - 20\right)$
$= (3-5)x^2 + (-6+13)x - 9 - 20$
$= -2x^2 + 7x - 29$

16. $\left(2x^3 - 5x^2 + 7x\right) - \left(x^3 - 4x + 12\right)$
$= 2x^3 - 5x^2 + 7x - x^3 + 4x - 12$
$= x^3 - 5x^2 + 11x - 12$

17. $-3x(2-7x)=-6x+21x^2$

18. $2x^2\left(x^2+4x-1\right)$
$=2x^2\left(x^2\right)+2x^2(4x)-2x^2(1)$
$=2x^4+8x^3-2x^2$

19. $\left(x^3-3xy+y^2\right)\left(5xy^2\right)$
$=5x^4y^2-15x^2y^3+5xy^4$

20. $(x+5)(x+9)$
$=x^2+9x+5x+45$
$=x^2+14x+45$

21. $(3x-2y)(4x+3y)$
$=12x^2+9xy-8xy-6y^2$
$=12x^2+xy-6y^2$

22. $\left(2x^2-y^2\right)\left(x^2-3y^2\right)$
$=2x^4-6x^2y^2-x^2y^2+3y^4$
$=2x^4-7x^2y^2+3y^4$

23. $(8x-11y)(8x+11y)$
$=(8x)^2-(11y)^2$
$=64x^2-121y^2$

24. $(5x-3y)^2$
$=(5x)^2-2(5x)(3y)+(3y)^2$
$=25x^2-30xy+9y^2$

25. $(5ab-6)^2$
$=(5ab)^2+2(5ab)(-6)+(-6)^2$
$=25a^2b^2-60ab+36$

26. $\left(x^2-3x+2\right)(4x-1)$
$=4x^3-12x^2+8x-x^2+3x-2$
$=4x^3-13x^2+11x-2$

27. $(2x-3)(2x+3)\left(4x^2+9\right)$
$=\left(4x^2-9\right)\left(4x^2+9\right)$
$=16x^4-81$

28. $\left(42x^4-36x^3+66x^2\right)\div(6x)$
$=\dfrac{42x^4}{6x}-\dfrac{36x^3}{6x}+\dfrac{66x^2}{6x}$
$=7x^3-6x^2+11x$

29. $3x-2\,\overline{)\,15x^3-28x^2+18x-7}$ (quotient: $5x^2-6x+2$)

$$\begin{array}{r} 15x^3-10x^2 \\ \hline -18x^2+18x \\ -18x^2+12x \\ \hline 6x-7 \\ 6x-4 \\ \hline -3 \end{array}$$

$\left(15x^3-28x^2+18x-7\right)\div(3x-2)$
$=5x^2-6x+2-\dfrac{3}{3x-2}$

5.1 Exercises

1. When you multiply exponential expressions with the same base, keep the base the same and add the exponents.

3. $\frac{2^2}{2^3} \stackrel{?}{=} \frac{1}{2^{3-2}}$

$\frac{4}{8} \stackrel{?}{=} \frac{1}{2}$

$\frac{1}{2} = \frac{1}{2}$

5. $6x^{11}y$: Coefficient is 6, bases are x, y and exponents are 11, 1.

7. $2 \cdot 2 \cdot a \cdot a \cdot a \cdot b = 2^2 a^3 b$

9. $(-3)(a)(a)(b)(c)(b)(c)(c) = -3a^2b^2c^3$

11. $(3^8)(3^7) = 3^{8+7} = 3^{15}$

13. $(5^{10})(5^{16}) = 5^{10+16} = 5^{26}$

15. $(3^5)(8^2) = 3^5 \cdot 8^2$

17. $6x^2(-9x^3)$

$= -54x^{2+3}$

$= -54x^5$

19. $(-4x^2)(-3x^3)$

$= 12x^{2+3}$

$= 12x^5$

21. $(4x^8)(-3x^3)$

$= -12x^{8+3}$

$= -12x^{11}$

23. $\left(\frac{4}{5}x^5y\right)\left(\frac{15}{16}x^2y^4\right)$

$= \frac{3}{4}x^{5+2}y^{1+4}$

$= \frac{3}{4}x^7y^5$

25. $(2.3x^4w)(-3.5xy^4)$

$= -8.05x^{4+1}y^4w$

$= -8.05x^5y^4w$

27. $(8a)(2a^3b)(0) = 0$

29. $(-16x^2y^4)(-5xy^3)$

$= 80x^{2+1}y^{4+3}$

$= 80x^3y^7$

31. $(14a^5)(-2b^6) = -28a^5b^6$

33. $(-2x^3y^2)(0)(-3x^4y) = 0$

35. $(4x^3y)(0)(-12x^4y^2) = 0$

37. $(6w^5z^6)(-4xy) = -24w^5xyz^6$

39. $\frac{y^5}{y^8} = \frac{1}{y^{3-5}}$
$= \frac{1}{y^3}$

41. $\frac{y^{12}}{y^5} = y^{12-5}$
$= y^7$

43. $\frac{13^{20}}{13^{30}} = \frac{1}{13^{30-20}}$
$= \frac{1}{13^{10}}$

45. $\frac{-3^{18}}{-3^{14}} = 3^{18-14}$
$= 3^4$

47. $\frac{a^{13}}{4a^5} = \frac{a^{13-5}}{4}$
$= \frac{a^8}{4}$

49. $\frac{x^7}{y^9} = \frac{x^7}{y^9}$

51. $\frac{-12x^5y^3}{-24xy^3} = \frac{x^{5-1}y^{3-3}}{2}$
$= \frac{x^4}{2}$

53. $\frac{a^5b^6}{a^5b^6} = a^{5-5}b^{6-6}$
$= a^0b^0$
$= 1$

55. $\frac{48x^7y}{-24x^3y^6} = \frac{2x^{7-3}}{y^{6-1}}$
$= \frac{2x^4}{y^5}$

57. $\frac{3.1s^5t^3}{62s^8t} = \frac{t^{3-1}}{20s^{8-5}}$
$= \frac{t^2}{20s^3}$

59. $\frac{30x^5y^4}{5x^3y^4} = 6x^{5-3}y^{4-4}$
$= 6x^2$

61. $\frac{8^0x^2y^3}{16x^5y} = \frac{y^{3-1}}{16x^{5-2}}$
$= \frac{y^2}{16x^3}$

63. $\frac{18a^6b^3c^0}{24a^5b^3} = \frac{3}{4}a^{6-5}b^{3-3}c^0$
$= \frac{3}{4}a$

65. $\frac{25x^6}{35y^8} = \frac{5x^6}{7y^8}$

67. $\dfrac{(4x)(9x^4)}{12x^3} = \dfrac{36x^{1+4}}{12x^3}$
$= \dfrac{36x^5}{12x^3}$
$= 3x^{5-3}$
$= 3x^2$

69. $\dfrac{(9a^2b)(2a^3b^6)}{-27a^8b^7} = \dfrac{18a^5b^7}{-27a^8b^7}$
$= -\dfrac{2}{3a^3}$

71. $\dfrac{63a^5b^6}{-9a^4b} = -7ab^5$, so
$\dfrac{63a^5b^6}{-7ab^5} = -9a^4b$

73. Let $y = 2$: $\dfrac{y^8}{y^4} = \dfrac{2^8}{2^4} = \dfrac{256}{16} = 16$
$y^2 = 2^2 = 4$
$\dfrac{y^8}{y^4} \neq y^2$

75. $\dfrac{x^{5a}}{x^{2a}} = x^{5a-2a} = x^{3a}$

77. $\dfrac{(c^{3y})(c^4)}{c^{2y+2}}$
$= \dfrac{c^{3y+4}}{c^{2y+2}}$
$= c^{3y+4-2y-2} = c^{y+2}$

79. $(w^5)^8 = w^{5\cdot 8}$
$= w^{40}$

81. $(a^3b)^4 = a^{3\cdot 4}b^{1\cdot 4}$
$= a^{12}b^4$

83. $(m^3n^2p)^5 = m^{3(5)}n^{2(5)}p^5$
$= m^{15}n^{10}p^5$

85. $(3^2xy^2)^4 = 3^{2\cdot 4}x^{1\cdot 4}y^{2\cdot 4}$
$= 3^8x^4y^8$

87. $(-2a^5)^4 = (-2)^4a^{5\cdot 4}$
$= 16a^{20}$

89. $\left(\dfrac{12x}{y^2}\right)^5 = \dfrac{12^5x^{1\cdot 5}}{y^{2\cdot 5}}$
$= \dfrac{12^5x^5}{y^{10}}$

91. $\left(\dfrac{2a^4}{3b^3}\right)^4 = \dfrac{2^4a^{4(4)}}{3^4b^{3(4)}}$
$= \dfrac{16a^{16}}{81b^{12}}$

93. $(-2a^5b^2c)^5 = (-2)^5a^{5\cdot 5}b^{2\cdot 5}c^{1\cdot 5}$
$= -32a^{25}b^{10}c^5$

95. $(-4xy^0z^4)^3 = (-4)^3x^3y^{0\cdot 3}z^{4\cdot 3}$
$= -64x^3z^{12}$

97. $\dfrac{(4a^2b)^2}{(4ab^2)^3} = \dfrac{4^2a^4b^2}{4^3a^3b^6}$
$= \dfrac{a}{4b^4}$

99. $(-2a^2b^3)^3(ab^2)$
$= [(-2)^3a^6b^9](ab^2)$
$= -8a^7b^{11}$

101. $\left(\dfrac{4}{x^6}\right)^3 = \dfrac{4^3}{x^{6(3)}}$
$= \dfrac{64}{x^{18}}$

103. $\left(\dfrac{a^3b}{c^5d}\right)^5 = \dfrac{a^{15}b^5}{c^{25}d^5}$

105. $(4x^3y)^2(x^3y) = (4^2x^6y^2)(x^3y)$
$= 16x^9y^3$

107. $16x^{20}y^{16}z^{28} = (\pm 2x^5y^4z^7)^4$
Therefore, $\pm 2x^5y^4z^7$

Cumulative Review Problems

109. $-17 + (-32) + (-24) + 27$
$= -49 + (-24) + 27$
$= -73 + 27$
$= -46$

111. $\dfrac{-3}{4} \div \dfrac{-12}{80} = \dfrac{-3}{4} \cdot \dfrac{80}{-12} = \dfrac{1\cdot 20}{1\cdot 4} = 5$

113. $3x + y - z = 1$ (1)
$x - 2y + 3z = -9$ (2)
$-x + 5y + 6z = 21$ (3)
Add (2) and (3)
$3y + 9z = 12$ (4)
Add 3 times (3) to (1)
$3x + y - z = 1$
$-3x + 15y + 18z = 63$
$16y + 17z = 64$ (5)
Add 16 times (4) to -3 times (5)
$48y + 144z = 192$
$-48y - 51z = -192$
$93z = 0$
$z = 0$
Substitute 0 for z in (4)
$3y + 9(0) = 12$
$3y = 12$
$y = 4$
Substitute 4 for y and 0 for z in (1)
$3x + 4 - 0 = 1$
$3x = -3$
$x = -1$
The solution is $(-1, 4, 0)$.

5.2 Exercises

1. $3x^{-2} = \frac{3}{x^2}$

3. $\left(4x^2y\right)^{-2} = \frac{1}{\left(4x^2y\right)^2}$
$= \frac{1}{4^2x^{2(2)}y^2}$
$= \frac{1}{16x^4y^2}$

5. $\frac{3xy^{-2}}{z^{-3}} = \frac{3xz^3}{y^2}$

7. $\frac{(3x)^{-2}}{(3x)^{-3}} = \frac{(3x)^3}{(3x)^2}$
$= 3x$

9. $wx^{-5}y^3z^{-2} = \frac{wy^3}{x^5z^2}$

11. $\left(8^{-2}\right)\left(2^3\right) = \frac{1}{8^2} \cdot 2^3$
$= \frac{1}{64} \cdot 8$
$= \frac{1}{8}$

13. $\left(\frac{3xy^2}{z^4}\right)^{-2} = \left(\frac{z^4}{3xy^2}\right)^2$
$= \frac{z^8}{9x^2y^4}$

15. $\frac{x^{-2}y^{-3}}{x^4y^{-2}} = \frac{y^2}{x^4x^2y^3}$
$= \frac{1}{x^6y}$

17. $123,780 = 1.2378 \times 10^5$

19. $0.000742 = 7.42 \times 10^{-4}$

21. $7,652,000,000 = 7.652 \times 10^9$

23. $3.02 \times 10^5 = 302,000$

25. $3.3 \times 10^{-5} = 0.000033$

27. $9.83 \times 10^5 = 983,000$

29. $0.00000000000000000000000000167$
$= 1.67 \times 10^{-27}$ Kg

31. $7 \times 10^{-6} = 0.000007$ m

33. $\frac{(5,000,000)(16,000)}{8,000,000,000} = \frac{\left(5 \times 10^6\right)\left(1.6 \times 10^4\right)}{8 \times 10^9}$
$= \frac{8 \times 10^{10}}{8 \times 10^9}$
$= 1.0 \times 10^1$

35. $(0.0002)^5 = \left(2 \times 10^{-4}\right)^5$
$= 2^5 \times 10^{-20}$
$= 32 \times 10^{-20}$
$= 3.2 \times 10^{-19}$

37. $(150,000,000)(0.00005)(0.002)(30,000)$
$= (1.5\times10^8)(5\times10^{-5})(2\times10^{-3})(3\times10^4)$
$= 45\times10^4$
$= 4.5\times10^5$

39. $\dfrac{5.614\times10^{12}}{2.76\times10^8} = 2.03\times10^{12-8}$
$= \$2.03\times10^4$

41. $d = (276)(3.09\times10^{13})$
$= (2.76\times10^2)(3.09\times10^{13}) = 8.528\times10^{15}$
$r = 45,000 = 4.5\times10^4$
$d = rt$
$t = \dfrac{d}{r} = \dfrac{8.528\times10^{15}}{4.5\times10^4} = 1.90\times10^{11}$ hours

43. $d = (130)(3.09\times10^{13})$
$= (1.30\times10^2)(3.09\times10^{13})$
$= 4.017\times10^{15}$
$r = 35,000 = 3.5\times10^4$
$d = rt$
$t = \dfrac{d}{r} = \dfrac{4.017\times10^{15}}{3.5\times10^4} = 1.15\times10^{11}$ hours

45. $(3.9\times10^{26})(60)(60)(24)(7)(2)$
$= 4.717\times10^{32}$ joules

47. $\dfrac{5.28\times10^{11} - 3.61\times10^{11}}{3.61\times10^{11}} = \dfrac{1.67\times10^{11}}{3.61\times10^{11}}$
$= 0.463\times10^{11-11}$
$= 46.3\%$

Cumulative Review Problems

49. $-2.7-(-1.9) = -2.7+1.9 = -0.8$

51. $\dfrac{-3}{4}+\dfrac{5}{7} = \dfrac{-21}{28}+\dfrac{20}{28} = \dfrac{-21+20}{28} = -\dfrac{1}{28}$

53. x = Mario's salary
$x + 12,460$ = Gina's salary
$x - 8742$ = Alfonso's salary
$x + x + 12,460 + x - 8742 = 112,000$
$3x + 3718 = 112,000$
$3x = 108,282$
$x = 36,094$
$x + 12,460 = 48,554$
$x - 8742 = 27,352$
Mario earns \$36,094
Gina earns \$48,554
Alfonso earns \$27,352

55. $-x + 2y = 10 - x$
$2y = 10$
$y = 5$

Putting Your Skills to Work

1. $12.06 \times 10^8 - 8.92 \times 10^8 = 3.14 \times 10^8$

   3.14×10^8 acres

2. $19.24 \times 10^8 + 3.22 \times 10^8 = 22.46 \times 10^8$

   2.246×10^9 acres

3. $\dfrac{1.256 \times 10^9}{3.09 \times 10^8} = 0.4 \times 10^{9-8}$

   $= 0.4 \times 10^1$

   $= 4$ people

4. $(4.0 \times 10^7)(25) = 1 \times 10^9$

   $9.587 \times 10^9 + 1 \times 10^9 = 10.587 \times 10^9$

   10.587×10^9 acres of forest

5.3 Exercises

1. A polynomial in x is the sum of a finite number of terms of the form ax^n, where a is any real number and n is a whole number. An example is $3x^2 - 5x - 9$.

3. The degree of a polynomial in x is the largest exponent of x in any of the terms of the polynomial.

5. $6x^3y$, degree 4: monomial

7. $20x^5 + 6x^3 - 7x$, degree 5: trinomial

9. $5xy^2 - 3x^2y^3$, degree 5: binomial

11. $(-3x + 15) + (8x - 43)$

    $= (-3 + 8)x + 15 - 43$

    $= 5x - 28$

13. $(2x^2 - 8x + 7) + (-5x^2 - 2x + 3)$

    $= (2 - 5)x^2 + (-8 - 2)x + 7 + 3$

    $= -3x^2 - 10x + 10$

15. $\left(\dfrac{1}{2}x^2 + \dfrac{1}{3}x - 4\right) + \left(\dfrac{1}{3}x^2 + \dfrac{1}{6}x - 5\right)$

    $= \left(\dfrac{1}{2} + \dfrac{1}{3}\right)x^2 + \left(\dfrac{1}{3} + \dfrac{1}{6}\right)x - 4 - 5$

    $= \dfrac{5}{6}x^2 + \dfrac{1}{2}x - 9$

17. $(3.4x^3 - 5.6x^2 - 7.1x + 3.4)$

    $+(-1.7x^3 + 2.2x^2 - 6.1x - 8.8)$

    $= (3.4 - 1.7)x^3 + (-5.6 + 2.2)x^2$

    $+(-7.1 - 6.1)x + 3.4 - 8.8$

    $= 1.7x^3 - 3.4x^2 - 13.2x - 5.4$

19. $(2x - 19) - (-3x + 5)$

    $= (2x - 19) + (3x - 5)$

    $= (2 + 3)x - 19 - 5$

    $= 5x - 24$

21. $\left(\frac{1}{2}x^2+3x-\frac{1}{5}\right)-\left(\frac{2}{3}x^2-4x+\frac{1}{10}\right)$
$=\left(\frac{1}{2}x^2+3x-\frac{1}{5}\right)+\left(-\frac{2}{3}x^2+4x-\frac{1}{10}\right)$
$=\left(\frac{1}{2}-\frac{2}{3}\right)x^2+(3+4)x-\frac{1}{5}-\frac{1}{10}$
$=-\frac{1}{6}x^2+7x-\frac{3}{10}$

23. $(-3x^2+5x)-(2x^3-3x^2+10)$
$=(-3x^2+5x)+(-2x^3+3x^2-10)$
$=-2x^3+(-3+3)x^2+5x-10$
$=-2x^3+5x-10$

25. $(0.5x^4-0.7x^2+8.3)-(5.2x^4+1.6x+7.9)$
$=(0.5x^4-0.7x^2+8.3)+(-5.2x^4-1.6x-7.9)$
$=(0.5-5.2)x^4-0.7x^2-1.6x+8.3-7.9$
$=-4.7x^4-0.7x^2-1.6x+0.4$

27. $(7x-8)-(5x-6)+(8x+12)$
$=(7x-8)+(-5x+6)+(8x+12)$
$=(7-5+8)x-8+6+12$
$=10x+10$

29. $(5x^2y-6xy^2+2)+(-8x^2y+12xy^2-6)$
$=(5-8)x^2y+(-6+12)xy^2+2-6$
$=-3x^2y+6xy^2-4$

31. $(3x^4-4x^2-18)-(2x^4+3x^3+6)$
$=(3x^4-4x^2-18)+(-2x^4-3x^3-6)$
$=(3-2)x^4-3x^3-4x^2-18-6$
$=x^4-3x^3-4x^2-24$

33. $0.04x+5.2,\ x=1975-1970=5$
$0.04(5)+5.2=5.4$ mpg

35. $0.04x+5.2=7.2$
$0.04x=2$
$x=50$
year $=1970+50=2020$

37. $1.8x^2+22.2x+325,\ x=1990-1980=10$
$1.8(10)^2\ \_\ 22.2(10)+325=727$
727,000 prisoners

39. $1.8x^2+22.2x+325$
$2002-1980=22,\ 2006-1980=26$
$2002: 1.8(22)^2+22.2(22)+325$
$=1{,}684{,}600$
$2006: 1.8(26)^2+22.2(26)+325$
$=2{,}119{,}000$
Increase: $2{,}119{,}000-1{,}684{,}600$
$=434{,}400$ prisoners

41. $x^2+12x+2x^2=3x^2+12x$

Cumulative Review Problems

43. $3y - 8x = 2$
$-8x = -3y + 2$
$x = \dfrac{-3y + 2}{-8}$
$x = \dfrac{3y - 2}{8}$

45. x = amount spent in 1990
$x + 0.9x = 1.324$
$1.9x = 1.324$
$x = 0.6968$
\$696.8 billion was spent

47. $m = -4,\ (0, -3) \Rightarrow b = -3$
$y = -4x - 3$

5.4 Exercises

1. $-2x(6x^3 - x) = -12x^4 + 2x^2$

3. $-5x(3x^2 - 2x + 5) = -15x^3 + 10x^2 - 25x$

5. $2x^3(3x^4 - 2x^3 + 5x - 1)$
$= 6x^7 - 4x^6 + 10x^4 - 2x^3$

7. $\dfrac{1}{2}(2x + 3x^2 + 5x^3)$
$= x + \dfrac{3}{2}x^2 + \dfrac{5}{2}x^3$

9. $(5x^3 - 2x^2 + 6x)(-3xy^2)$
$= -15x^4y^2 + 6x^3y^2 - 18x^2y^2$

11. $(2b^2 + 3b - 4)(2b^2) = 4b^4 + 6b^3 - 8b^2$

13. $(x^3 - 3x^2 + 5x - 2)(3x)$
$= 3x^4 - 9x^3 + 15x^2 - 6x$

15. $(x^2y^2 - 6xy + 8)(-2xy)$
$= -2x^3y^3 + 12x^2y^2 - 16xy$

17. $(-7x^3 + 3x^2 + 2x - 1)(4x^2y)$
$= -28x^5y + 12x^4y + 8x^3y - 4x^2y$

19. $(3d^4 - 4d^2 + 6)(-2c^2d)$
$= -6c^2d^5 + 8c^2d^3 - 12c^2d$

21. $6x^3(2x^4 - x^2 + 3x + 9)$
$= 12x^7 - 6x^5 + 18x^4 + 54x^3$

23. $-4x^3(6x^4 - 3x^2 + 2)$
$= -24x^7 + 12x^5 - 8x^3$

25. $(x + 10)(x + 3) = x^2 + 13x + 30$

27. $(x + 6)(x + 2) = x^2 + 8x + 12$

29. $(x + 3)(x - 6) = x^2 - 3x - 18$

31. $(x-6)(x-5)=x^2-11x+30$

33. $(7x+1)(-2x-3)=-14x^2-23x-3$

35. $(x-9)(3x+4y)=3x^2+4xy-27x-36y$

37. $(3y+2)(4y-3)=12y^2-y-6$

39. $(5y-3)(4y-2)=20y^2-22y+6$

41. $-(3x-7)=-3x+7$
The last term should be +7.

43. $(5x-1)(5x+1)=25x^2-1$
It is the difference in two perfect squares.

45. $(3b^2-5c)(2b^2-7c)$
$=6b^4-21b^2c-10b^2c+35c^2$
$=6b^4-31b^2c+35c^2$

47. $(8x-2)^2=(8x-2)(8x-2)$
$=64x^2-32x+4$

49. $(5a^2-3b^2)^2=(5a^2-3b^2)(5a^2-3b^2)$
$=25a^4-15a^2b^2-15a^2b^2+9b^4$
$=25a^4-30a^2b^2+9b^4$

51. $(0.2x+3)(4x-0.3)$
$=0.8x^2-0.06x+12x-0.9$
$=0.8x^2+11.94x-0.9$

53. $(5x+8y)(6x-y)=30x^2+43xy-8y^2$

55. $\left(\frac{1}{3}x+\frac{1}{5}\right)\left(\frac{1}{3}x-\frac{1}{2}\right)=\frac{1}{9}x^2-\frac{1}{6}x+\frac{1}{15}x-\frac{1}{10}$
$=\frac{1}{9}x^2-\frac{1}{10}x-\frac{1}{10}$

57. $(5b-7c)(x-2b)$
$=5bx-10b^2-7cx+14bc$

59. $(2x-3)(5x+2)=10x^2+4x-15x-6$
$=10x^2-11x-6$

Cumulative Review Problems

61. $3(x-6)=-2(x+4)+6x$
$3x-18=-2x-8+6x$
$3x-18=4x-8$
$-10=x$

63. Let d = number of dimes
$d+3=$ number of quarters
$0.10d+0.25(d+3)=3.55$
$10d+25(d+3)=355$
$10d+25d+75=355$
$35d=280$
$d=8$
$d+3=11$
She has 8 dimes and 11 quarters.

65. $(6,-2)$ and $(2,0)$

$$m=\frac{0-(-2)}{2-6}=\frac{2}{-4}=-\frac{1}{2}$$

$$0=-\frac{1}{2}(2)+b\Rightarrow b=1$$

$$y=-\frac{1}{2}x+1$$

67. $1.5x+16,\ x=1994-1992=2$

$1.5(2)+16=19$

They spent 19 hours/week.

69. $1.5x+16,\ x=2003-1992=11$

$1.5(11)+16=32.5$

They will spent 32.5 hours/week.

5.5 Exercises

1. In the special case of $(a+b)(a-b)$, a binomial times a binomial is a <u>binomial</u>.

3. $(4x-7)^2=16x^2-56x+49$

The student left out the middle term which comes from the product of the two outer terms and the product of the two inner terms.

5. $(y-7)(y+7)=y^2-7^2=y^2-49$

7. $(x-9)(x+9)=x^2-(9)^2=x^2-81$

9. $(8x+3)(8x-3)=(8x)^2-(3)^2=64x^2-9$

11. $(2x-7)(2x+7)=(2x)^2-(7)^2=4x^2-49$

13. $(5x-3y)(5x+3y)=(5x)^2-(3y)^2$

$=25x^2-9y^2$

15. $(0.6x+3)(0.6x-3)=(0.6x)^2-3^2$

$=0.36x^2-9$

17. $(3y+1)^2=(3y)^2+2(3y)(1)+1^2$

$=9y^2+6y+1$

19. $(5x-4)^2=(5x)^2-2(5x)(4)+4^2$

$=25x^2-40x+16$

21. $(9x+5)^2=(9x)^2+2(9x)(5)+5^2$

$=81x^2+90x+25$

23. $(3x-7)^2=(3x)^2-2(3x)(7)+7^2$

$=9x^2-42x+49$

25. $\left(\frac{2}{3}x+\frac{1}{4}\right)^2$

$$=\left(\frac{2}{3}x\right)^2+2\left(\frac{2}{3}x\right)\left(\frac{1}{4}\right)+\left(\frac{1}{4}\right)^2$$

$$=\frac{4}{9}x^2+\frac{1}{3}x+\frac{1}{16}$$

27. $(6w+5z)^2=(6w)^2+2(6w)(5z)+(5z)^2$

$=36w^2+60wz+25z^2$

29. $(7x+3y)(7x-3y)=(7x)^2-(3y)^2$
$=49x^2-9y^2$

31. $(7c^3-6d)^2$
$=(7c^3)^2-2(7c^3)(6d)-(6d)^2$
$=49c^6-84c^3d-36d^2$

33. $(x^2+3x-2)(x-3)$
$=x(x^2+3x-2)-3(x^2+3x-2)$
$=x^3+3x^2-2x-3x^2-9x+6$
$=x^3-11x+6$

35. $(4x+1)(x^3-2x^2+x-1)$
$=4x(x^3-2x^2+x-1)+1(x^3-2x^2+x-1)$
$=4x^4-8x^3+4x^2-4x+x^3-2x^2+x-1$
$=4x^4-7x^3+2x^2-3x-1$

37. $(x+3)(x-1)(3x-8)$
$=(x^2+2x-3)(3x-8)$
$=3x^3+6x^2-9x-8x^2-16x+24$
$=3x^3-2x^2-25x+24$

39. $(3x+5)(x-2)(x-4)$
$=(3x+5)(x^2-6x+8)$
$=3x(x^2-6x+8)+5(x^2-6x+8)$
$=3x^3-18x^2+24x+5x^2-30x+40$
$=3x^3-13x^2-6x+40$

41. $(x+4)(2x-7)(x-4)$
$=(x+4)(x-4)(2x-7)$
$=(x^2)(2x-7)-16(2x-7)$
$=2x^3-7x^2-32x+112$

43. $(a^2-3a+2)(a^2+4a-3)$
$=a^2(a^2+4a-3)-3a(a^2+4a-3)$
$+2(a^2+4a-3)$
$=a^4+4a^3-3a^2-3a^3-12a^2+9a+2a^2$
$+8a-6$
$=a^4+a^3-13a^2+17a-6$

45. $V=(2x+1)(3x-2)(4x+3)$
$=(6x^2-x-2)(4x+3)$
$=24x^3-4x^2-8x+18x^2-3x-6$
$=24x^3+14x^2-11x-6$

Cumulative Review Problems

47. Let x = amount invested at 7%
$18,000-x$ = amount invested at 11%
$0.07x+0.11(18,000-x)=1540$
$7x+11(18,000-x)=154,000$
$7x+198,000-11x=154,000$
$-4x=-44,000$
$x=11,000$
$18,000-x=7000$
She invested \$11,000 at 7% and \$7000 at 11%.

49. $F = 1.8C + 32$
$F = 1.8(-33.4) + 32$
$F = -60.12 + 32$
$F = -28.12^0 F$

Putting Your Skills to Work

1. $(0.34)(4.5\times10^6) = 1.53\times10^6$ nanometers

2. $(1.53\times10^6 \text{ nanometers})\times\left(\frac{1\times10^{-9}\text{ meter}}{1\text{ nanometer}}\right)$
$\times\left(\frac{1\times\times10^3\text{ millimeters}}{1\text{ meter}}\right) = 1.53$ millimeters

3. $(0.34)(8.0\times10^9)$
$= 2.72\times10^9$ nanometers
$= \frac{2.72\times10^9}{1\times10^9} = 2.72$ meters

4. From Problem 2: E coli = 0.00153 meters
From Problem 3: human = 2.72 meters
$\frac{2.72}{0.00153} = 1778$ times larger

5.6 Exercises

1. $\frac{25x^4 - 15x^2 + 20x}{5x} = \frac{25x^4}{5x} - \frac{15x^2}{5x} + \frac{20x}{5x}$
$= 5x^3 - 3x + 4$

3. $\frac{8y^4 - 12y^3 - 4y^2}{4y^2} = \frac{8y^4}{4y^2} - \frac{12y^3}{4y^2} - \frac{4y^2}{4y^2}$
$= 2y^2 - 3y - 1$

5. $\frac{49x^6 - 21x^4 + 56x^2}{7x^2} = \frac{49x^6}{7x^2} - \frac{21x^4}{7x^2} + \frac{56x^2}{7x^2}$
$= 7x^4 - 3x^2 + 8$

7. $(48x^7 - 54x^4 + 36x^3) \div 6x^3$
$= \frac{48x^7}{6x^3} - \frac{54x^4}{6x^3} + \frac{36x^3}{6x^3}$
$= 8x^4 - 9x + 6$

9.
$$\begin{array}{r} 3x+5 \\ 2x+1\overline{)6x^2+13x+5} \\ \underline{6x^2+\ 3x} \\ 10x+5 \\ \underline{10x+5} \\ 0 \end{array}$$

$\frac{6x^2 + 13x + 5}{2x + 1} = 3x + 5$

Check: $(3x+5)(2x+1)$
$= 6x^2 + 3x + 10x + 5$
$= 6x^2 + 13x + 5$

11. $$\begin{array}{r} x-2 \\ x-7\overline{)x^2-9x-6} \\ \underline{x^2-7x} \\ -2x-6 \\ \underline{-2x+14} \\ -20 \end{array}$$

$$\frac{x^2-9x-6}{x-7} = x-2-\frac{20}{x-7}$$

Check: $(x-7)\left(x-2-\frac{20}{x-7}\right)$

$= x^2-2x-7x+14-20$

$= x^2-9x-6$

13. $$\begin{array}{r} 3x^2-4x+8 \\ x+1\overline{)3x^3-\ x^2+4x-2} \\ \underline{3x^3+3x^2} \\ -4x^2+4x-2 \\ \underline{-4x^2-4x} \\ 8x-2 \\ \underline{8x+8} \\ -10 \end{array}$$

$$\frac{3x^3-x^2+4x-2}{x+1} = 3x^2-4x+8-\frac{10}{x+1}$$

Check: $(x+1)\left(3x^2-4x+8-\frac{10}{x+1}\right)$

$= 3x^3-x^2+4x+8-10$

$= 3x^3-x^2+4x-2$

15. $$\begin{array}{r} 2x^2-3x-2 \\ 2x+5\overline{)4x^3+4x^2-19x-15} \\ \underline{4x^3+10x^2} \\ -6x^2-19x \\ \underline{-6x^2-15x} \\ -\ 4x-15 \\ \underline{-\ 4x-10} \\ -\ 5 \end{array}$$

$$\frac{4x^3+4x^2-19x-15}{2x+5} = 2x^2-3x-2-\frac{5}{2x+5}$$

Check:

$(2x+5)\left(2x^2-3x-2-\frac{5}{2x+5}\right)$

$= 4x^3+4x^2-19x-10-5$

$= 4x^3+4x^2-19x-15$

17. $$\begin{array}{r} 2x^2+x-2 \\ 3x-1\overline{)6x^3+\ x^2-7x+2} \\ \underline{6x^3-2x^2} \\ 3x^2-7x+2 \\ \underline{3x^2-\ x} \\ -6x+2 \\ \underline{-6x+2} \\ 0 \end{array}$$

$$\frac{6x^3+x^2-7x+2}{3x-1} = 2x^2+x-2$$

19.
$$\begin{array}{r} 6y^2+3y-8 \\ 2y-3\overline{)12y^3-12y^2-25y+31} \\ \underline{12y^3-18y^2} \\ 6y^2-25y \\ \underline{6y^2-9y} \\ -16y+31 \\ \underline{-16y+24} \\ 7 \end{array}$$

$$\frac{12y^3-12y^2-25y+31}{2y-3}$$
$$=6y^2+3y-8+\frac{7}{2y-3}$$

21.
$$\begin{array}{r} y^2-4y-1 \\ y+3\overline{)y^3-y^2-13y-12} \\ \underline{y^3+3y^2} \\ -4y^2-13y-12 \\ \underline{-4y^2-12y} \\ -y-12 \\ \underline{-y-3} \\ -9 \end{array}$$

$$\left(y^3-y^2-13y-12\right)\div(y+3)$$
$$=y^2-4y-1-\frac{9}{y+3}$$

23.
$$\begin{array}{r} y^3+2y^2-5y-10 \\ y-2\overline{)y^4+0y^3-9y^2+0y-5} \\ \underline{y^4-2y^3} \\ 2y^3-9y^2+0y-5 \\ \underline{2y^3-4y^2} \\ -5y^2+0y-5 \\ \underline{-5y^2+10y} \\ -10y-5 \\ \underline{-10y+20} \\ -25 \end{array}$$

$$\left(y^4-9y^2-5\right)\div(y-2)$$
$$=y^3+2y^2-5y-10-\frac{25}{y-2}$$

25.
$$\begin{array}{r} 2y^2+\tfrac{1}{2}y+\tfrac{7}{8} \\ 4y-1\overline{)8y^3+0y^2+3y-7} \\ \underline{8y^3-2y^2} \\ 2y^2+3y \\ \underline{2y^2-\tfrac{1}{2}y} \\ \tfrac{7}{2}y-7 \\ \underline{\tfrac{7}{2}y-\tfrac{7}{8}} \\ -\tfrac{49}{8} \end{array}$$

$$\frac{8y^3+3y-7}{4y-1}=2y^2+\frac{1}{2}y+\frac{7}{8}-\frac{49/8}{4y-1}$$

Cumulative Review Problems

27. $x-0.3x=77,000$
$0.7x=77,000$
$x=110,000$ gallons

29. x = number of cats in the study
$0.92x = 184$
$x = 200$
200 cats were in the study

Putting Your Skills to Work

1. $50.5 - 46.4 = 4.1$ million more

2. $54.4 - 53.7 = 0.7$ million
$= 700,000$ fewer

In Problems 3-6, use
$E = -0.00031x^3 - 0.024268x^2 + 0.975171x + 46.3761$

3. $x = 1998 - 1990 = 8$
$E = 52.5$ million

4. $x = 1994 - 1990 = 4$
$E = 49.9$ million

5. $x = 2006 - 1990 = 16$
$E = 54.5$ million

6. $x = 2009 - 1990 = 19$
$E = 54.0$ million

Chapter 5 Review Problems

1. $(-6a^2)(3a^5) = -18a^7$

2. $(5^{10})(5^{13}) = 5^{10+13} = 5^{23}$

3. $(3xy^2)(2x^3y^4) = 6x^4y^6$

4. $\frac{8^{20}}{8^3} = 8^{17}$

5. $\frac{7^{15}}{7^{27}} = \frac{1}{7^{27-15}} = \frac{1}{7^{12}}$

6. $\frac{x^{12}}{x^{17}} = \frac{1}{x^{17-12}} = \frac{1}{x^5}$

7. $\frac{y^{30}}{y^{16}} = y^{30-16} = y^{14}$

8. $\frac{3x^8y^0}{9x^4} = \frac{x^4}{3}$

9. $\frac{-15xy^2}{25x^6y^6} = -\frac{3}{5x^{6-1}y^{6-2}} = -\frac{3}{5x^5y^4}$

10. $\frac{-12a^3b^6}{18a^2b^{12}} = -\frac{2a}{3b^6}$

11. $(x^3)^8 = x^{3(8)} = x^{24}$

12. $(5xy^2)^3 = 5^3x^3(y^2)^3 = 125x^3y^6$

13. $(-3a^3b^2)^2 = (-3)^2a^{3(2)}b^{2(2)} = 9a^6b^4$

14. $\dfrac{2x^4}{3y^2} = \dfrac{2x^4}{3y^2}$

15. $\left(\dfrac{5ab^2}{c^3}\right)^2 = \dfrac{5^2a^2\left(b^2\right)^2}{\left(c^3\right)^2} = \dfrac{25a^2b^4}{c^6}$

16. $\left(\dfrac{x^0y^3}{4w^5z^2}\right)^3 = \dfrac{\left(y^3\right)^3}{4^3\left(w^5\right)^3\left(z^2\right)^3} = \dfrac{y^9}{64w^{15}z^6}$

17. $x^{-3} = \dfrac{1}{x^3}$

18. $x^{-5}y^{-11} = \dfrac{1}{x^5y^{11}}$

19. $\dfrac{2x^{-6}}{y^{-3}} = \dfrac{2y^3}{x^6}$

20. $2^{-1}x^5y^{-6} = \dfrac{x^5}{2y^6}$

21. $\left(2x^3\right)^{-2} = \dfrac{1}{\left(2x^3\right)^2} = \dfrac{1}{2^2x^{3(2)}} = \dfrac{1}{4x^6}$

22. $\dfrac{3x^{-3}}{y^{-2}} = \dfrac{3y^2}{x^3}$

23. $\dfrac{4x^{-5}y^{-6}}{w^{-2}z^8} = \dfrac{4w^2}{x^5y^6z^8}$

24. $\dfrac{3^{-3}a^{-2}b^5}{c^{-3}d^{-4}} = \dfrac{b^5c^3d^4}{3^3a^2} = \dfrac{b^5c^3d^4}{27a^2}$

25. $156{,}340{,}200{,}000 = 1.563402\times10^{11}$

26. $179{,}632 = 1.79632\times10^5$

27. $0.0078 = 7.8\times10^{-3}$

28. $0.00006173 = 6.173\times10^{-5}$

29. $1.2\times10^5 = 120{,}000$

30. $8.367\times10^{10} = 8.3670000000\times10^{10}$
$= 83{,}670{,}000{,}000$

31. $3\times10^6 = 3{,}000{,}000$

32. $2.5\times10^{-1} = 0.25$

33. $5.708\times10^{-8} = 0.00000005708$

34. $6\times10^{-9} = 000000006\times10^{-9}$
$= 0.000000006$

35. $\dfrac{(28{,}000{,}000)(5{,}000{,}000{,}000)}{7{,}000}$
$= \dfrac{\left(2.8\times10^7\right)\left(5\times10^9\right)}{7\times10^3}$
$= \dfrac{14\times10^{16}}{7\times10^3} = 2\times10^{13}$

36. $\left(3.12\times10^{5}\right)\left(2.0\times10^{6}\right)\left(1.5\times10^{8}\right)$
$=9.36\times10^{5+6+8}=9.36\times10^{19}$

37. $\left(1.6\times10^{-3}\right)\left(3.0\times10^{-5}\right)\left(2.0\times10^{-2}\right)$
$=9.6\times10^{-10}$

38. $\dfrac{(0.00078)(0.000005)(0.00004)}{0.002}$
$=\dfrac{\left(7.8\times10^{-4}\right)\left(5.0\times10^{-6}\right)\left(4.0\times10^{-5}\right)}{2.0\times10^{-3}}$

$=\dfrac{156\times10^{-15}}{2.0\times10^{-3}}=78\times10^{-12}=7.8\times10^{-11}$

39. $r=40{,}000=4\times10^{4}$
$t=365(24)=8760=8.76\times10^{3}$
$d=rt=\left(4\times10^{4}\right)\left(8.76\times10^{3}\right)$
$=35.04\times10^{7}$
$=3.504\times10^{8}$ km

40. Seconds in one day $=60\times60\times24=86{,}400$
$=8.64\times10^{4}$
Cycles of radiation in one second
$=9{,}192{,}631{,}770=9.19\times10^{9}$
Cycles of radiation in one day
$=\left(8.64\times10^{4}\right)\left(9.19\times10^{9}\right)$
$=79.4016\times10^{13}=7.94\times10^{14}$

41. $\dfrac{60}{1\times10^{-8}}=60\times10^{8}=6\times10^{9}$
In one minute the computer can perform 6×10^{9} operations.

42. $\left(2x^2-3x+5\right)+\left(-7x^2-8x-23\right)$
$=\left(2x^2-7x^2\right)+(-3x-8x)+(5-23)$
$=-5x^2-11x-18$

43. $\left(1.2x^2-3.4x+6\right)+\left(5.5x^2-7.6x-3\right)$
$=(1.2+5.5)x^2+(-3.4-7.6)x+6-3$
$=6.7x^2-11x+3$

44. $\left(x^3+x^2-6x+2\right)-\left(2x^3-x^2-5x-6\right)$
$=\left(x^3+x^2-6x+2\right)+\left(-2x^3x^2+5x+6\right)$
$=\left(x^3-2x^3\right)+\left(x^2+x^2\right)$
$+(-6x+5x)+(2+6)$
$=-x^3+2x^2-x+8$

45. $\left(4x^3-x^2-x+3\right)-\left(-3x^3+2x^2+5x-1\right)$
$=\left(4x^3-x^2-x+3\right)+\left(3x^3-2x^2-5x+1\right)$
$=(4+3)x^3+(-1-2)x^2+(-1-5)x+3+1$
$=7x^3-3x^2-6x+4$

46. $\left(9x^3y^3+3xy-4\right)-\left(4x^3y^3+2x^2y^2-7xy\right)$
$=\left(9x^3y^3+3xy-4\right)+\left(-4x^3y^3x^2y^2+7xy\right)$
$=\left(9x^3y^3-4x^3y^3\right)-2x^2y^2+(3xy+7xy)-4$
$=5x^3y^3-2x^2y^2+10xy-4$

47. $\frac{1}{2}x^2-\frac{3}{4}x+\frac{1}{5}-\left(\frac{1}{4}x^2-\frac{1}{2}x+\frac{1}{10}\right)$
$=\frac{1}{2}x^2-\frac{3}{4}x+\frac{1}{5}-\frac{1}{4}x^2+\frac{1}{2}x-\frac{1}{10}$
$=\left(\frac{1}{2}-\frac{1}{4}\right)x^2+\left(-\frac{3}{4}+\frac{1}{2}\right)x+\frac{1}{5}-\frac{1}{10}$
$=\frac{1}{4}x^2-\frac{1}{4}x+\frac{1}{10}$

48. $(5x^2+3x)+(-6x^2+2)-(5x-8)$
$=(5x^2+3x)+(-6x^2+2)+(-5x+8)$
$=(5x^2-6x^2)+(3x-5x)+(2+8)$
$=-x^2-2x+10$

49. $(2x^2-7)-(3x^2-4)+(-5x^2-6x)$
$=(2x^2-7)+(-3x^2+4)+(-5x^2-6x)$
$=(2-3-5)x^2-6x-7+4$
$=-6x^2-6x-3$

50. $(3x+1)(5x-1)=15x^2-3x+5x-1$
$=15x^2+2x-1$

51. $(7x-2)(4x-3)=28x^2-21x-8x+6$
$=28x^2-29x+16$

52. $(2x+3)(10x+9)$
$=20x^2+18x+30x+27$
$=20x^2+48x+27$

53. $5x(2x^2-6x+3)=10x^3-30x^2+15x$

54. $(3x^2y^2-5xy+6)(-2xy)$
$=3x^2y^2(-2xy)-5xy(-2xy)+6(-2xy)$
$=-6x^3y^3+10x^2y^2-12xy$

55. $(x^3-3x^2+5x-2)(4x)$
$=4x^4-12x^3+20x^2-8x$

56. $(5a+7b)(a-3b)=5a^2-15ab+7ab-21b^2$
$=5a^2-8ab-21b^2$

57. $(2x^2-3)(4x^2-5y)$
$=8x^4-10x^2y-12x^2+15y$

58. $-3x^2y(5x^4y+3x^2-2)$
$=-3x^2y(5x^4y)-3x^2y(3x^2)-3x^2y(-2)$
$=-15x^6y^2-9x^4y+6x^2y$

59. $(3x-2)^2$
$=9x^2+2(-2)(3)x+4$
$=9x^2-12x+4$

60. $(5x+3)(5x-3)=(5x)^2-3^2=25x^2-9$

61. $(7x+6y)(7x-6y)$
$=(7x)^2-(6y)^2$
$=49x^2-36y^2$

62. $(5a-2b)^2=(5a)^2-2(5a)(2b)+(2b)^2$
$=25a^2-20ab+4b^2$

63. $(8x+9y)^2 = 64x^2 + 2(8x)(9y) + 81y^2$
$= 64x^2 + 144xy + 81y^2$

64. $(x^2+7x+3)(4x-1)$
$= 4x^3 + 28x^2 + 12x - x^2 - 7x - 3$
$= 4x^3 + 27x^2 + 5x - 3$

65. $(x-6)(2x-3)(x+4)$
$= (2x^2 - 15x + 18)(x+4)$
$= 2x^3 - 15x^2 + 18x + 8x^2 - 60x + 72$
$= 2x^3 - 7x^2 - 42x + 72$

66. $(12y^3 + 18x^2 + 24y) \div (6y)$
$= \dfrac{12y^3}{6y} + \dfrac{18y^2}{6y} + \dfrac{24y}{6y}$
$= 2y^2 + 3y + 4$

67. $(30x^5 + 35x^4 - 90x^3) \div (5x^2)$
$= \dfrac{30x^5}{5x^2} + \dfrac{35x^4}{5x^2} - \dfrac{90x^3}{5x^2}$
$= 6x^3 + 7x^2 - 18x$

68. $(16x^3y^2 - 24x^2y + 32xy^2) \div (4xy)$
$= \dfrac{16x^3y^2}{4xy} - \dfrac{24x^2y}{4xy} + \dfrac{32xy^2}{4xy}$
$= 4x^2y - 6x + 8y$

69. $(106x^6 - 24x^5 + 38x^4 + 26x^3) \div (2x^3)$
$= \dfrac{106x^6}{2x^3} - \dfrac{24x^5}{2x^3} + \dfrac{38x^4}{2x^3} + \dfrac{26x^3}{2x^3}$
$= 53x^3 - 12x^2 + 19x + 13$

70.
$$\begin{array}{r} 4x+1 \\ 4x-3\overline{\big)16x^2-8x-3} \\ \underline{-16x^2+12x} \\ 4x-3 \\ \underline{4x-3} \end{array}$$

$(16x^2 - 8x - 3) \div (4x-3) = 4x+1$

71.
$$\begin{array}{r} 3x+7 \\ 5x+2\overline{\big)15x^2+41x+14} \\ \underline{15x^2+\ \ 6x} \\ 35x+14 \\ \underline{35x+14} \end{array}$$

72.
$$\begin{array}{r} 3x^2+2x+4 \\ 2x-1\overline{\big)6x^3+\ \ x^2+6x+5} \\ \underline{6x^3-3x^2\qquad\quad} \\ 4x^2+6x\quad \\ \underline{4x^2-2x\quad} \\ 8x+5 \\ \underline{8x-4} \\ 9 \end{array}$$

$(6x^3 + x^2 + 6x + 5) \div (2x-1)$
$= 3x^2 + 2x + 4 + \dfrac{9}{2x-1}$

73.
$$\begin{array}{r} 2x^2-5x+13-\frac{27}{x+2} \\ x+2\overline{)2x^3-x^2+3x-1} \\ \underline{2x^3+4x^2} \\ -5x^2+3x \\ \underline{-5x^2-10x} \\ 13x-1 \\ \underline{13x+26} \\ -27 \end{array}$$

74.
$$\begin{array}{r} 4x+1 \\ 3x+2\overline{)12x^2+11x+2} \\ \underline{12x^2+8x} \\ 3x+2 \\ \underline{3x+2} \end{array}$$

$$(12x^2+11x+2)\div(3x+2)=4x+1$$

75.
$$\begin{array}{r} 4x-5+\frac{11}{2x+1} \\ 2x+1\overline{)8x^2-6x+6} \\ \underline{8x^2+4x} \\ -10x+6 \\ \underline{-10x-5} \\ 11 \end{array}$$

76.
$$\begin{array}{r} x^2+3x+8 \\ x-3\overline{)x^3+0x^2-x-24} \\ \underline{x^3-3x^2} \\ 3x^2-x \\ \underline{3x^2-9x} \\ 8x-24 \\ \underline{8x-24} \end{array}$$

77.
$$\begin{array}{r} 2x^2+4x+5 \\ x-2\overline{)2x^3+0x^2-3x+1} \\ \underline{2x^3-4x^2} \\ 4x^2-3x \\ \underline{4x^2-8x} \\ 5x+1 \\ \underline{5x-10} \\ 11 \end{array}$$

$$(2x^3-3x+1)\div(x-2)$$
$$=2x^2+4x+5+\frac{11}{x-2}$$

78. $1.256\times10^9+1.74\times10^8$
$=1.256\times10^9+0.174\times10^9$
$=1.430\times10^9$ people

79. $1.018\times10^9+1.29\times10^8$
$=1.018\times10^9+0.129\times10^9$
$=1.147\times10^9$ people

80. $(9.11\times10^{-28})(30{,}000)$
$=(9.11\times10^{-28})(3\times10^4)$
$=27.33\times10^{-24}$
$=2.733\times10^{-23}$ grams

81. $(3.9\times10^{26})(60)(60)(24)$
$=3.9\times10^{26}(8.64\times10^4)$
$=33.696\times10^{30}$
$=3.3696\times10^{31}$ joules

82. $A = 2x(2y+1) - xy$
$= 4xy + 2x - xy$
$= 3xy + 2x$

83. $A = 2x(x) - 4y(y) = 2x^2 - 4y^2$

Chapter 5 Test

1. $(3^{10})(3^{24}) = 3^{10+24} = 3^{34}$

2. $\frac{3^{35}}{3^7} = 3^{35-7} = 3^{28}$

3. $(8^4)^6 = 8^{4\cdot 6} = 8^{24}$

4. $(-3xy^4)(-4x^3y^6) = 12x^{1+3}y^{4+6} = 12x^4y^{10}$

5. $\frac{-35x^8y^{10}}{25x^5y^{10}} = -\frac{7x^3}{5}$

6. $(-5xy^6)^3 = (-5)^3x^3(y^6)^3 = -125x^3y^{18}$

7. $\left(\frac{7a^7b^2}{3c^0}\right)^2 = \frac{7^2a^{7(2)}b^{2(2)}}{3^2c^{0(2)}} = \frac{49a^{14}b^4}{9}$

8. $\frac{4a^5b^6}{16a^{10}b^{12}} = \frac{1}{4a^{10-5}b^{12-6}} = \frac{1}{4a^5b^6}$

9. $4^{-3} = \frac{1}{4^3} = \frac{1}{64}$

10. $6a^{-4}b^{-3}c^5 = \frac{6c^5}{a^4b^3}$

11. $\frac{2x^{-3}y^{-4}}{w^{-6}z^8} = \frac{2w^6}{x^3y^4z^8}$

12. $0.0005482 = 5.482 \times 10^{-4}$

13. $5.82 \times 10^8 = 582{,}000{,}000$

14. $(4.0 \times 10^{-3})(3.0 \times 10^{-8})(2.0 \times 10^4)$
$= 24.0 \times 10^{-7} = 2.4 \times 10^{-6}$

15. $(2x^2 - 3x - 6) + (-4x^2 + 8x + 6)$
$= (2-4)x^2 + (-3+8)x + 6 - 6$
$= -2x^2 + 5x$

16. $(5x^2 + 6xy - 7y^2) - (2x^2 + 3xy + 6y)$
$= (5x^2 + 6xy - 7y^2) + (-2x^2 - 3xy - 6y)$
$= (5x^2 - 2x^2) + (6xy - 3xy) - 7y^2 - 6y$
$= 3x^2 + 3xy - 7y^2 - 6y$

17. $-7x^2(3x^3 - 4x^2 + 6x - 2)$
$= -21x^5 + 28x^4 - 42x^3 + 14x^2$

18. $(5x^2y^2 - 6xy + 2)(3x^2y)$
$= 5x^2y^2(3x^2y) - 6xy(3x^2y) + 2(3x^2y)$
$= 15x^4y^3 - 18x^3y^2 + 6x^2y$

19. $(5a-4b)(2a+3b)$
$=10a^2+15ab-8ab-12b^2$
$=10a^2+7ab-12b^2$

20. $(3x+2)(2x+1)(x-3)$
$=(6x^2+7x+2)(x-3)$
$=6x^3+7x^2+2x-18x^2-21x-6$
$=6x^3-11x^2-19x-6$

21. $(7x^2+2y^2)^2$
$=49x^4+2(7x^2)(2y^2)+4y^4$
$=49x^4+28x^2y^2+4y^4$

22. $(9x-2y)(9x+2y)=(9x)^2-(2y)^2$
$=81x^2-4y^2$

23. $(3x-2)(4x^3-2x^2+7x-5)$
$=3x(4x^3-2x^2+7x-5)$
$-2(4x^3-2x^2+7x-5)$
$=12x^4-6x^3+21x^2-15x-8x^3+4x^2$
$-14x+10$
$=12x^4-14x^3+25x^2-29x+10$

24. $(3x^2-5xy)(x^2+3xy)$
$=3x^4+9x^3y-5x^3y-15x^2y^2$
$=3x^4+4x^3y-15x^2y^2$

25. $\dfrac{15x^6-5x^4+25x^3}{5x^3}=\dfrac{15x^6}{5x^3}-\dfrac{5x^4}{5x^3}+\dfrac{25x^3}{5x^3}$
$=3x^3-x+5$

26.
$$\begin{array}{r} 2x^2-7x+4 \\ 4x+3\overline{)8x^3-22x^2-5x+12} \\ \underline{8x^3+6x^2} \qquad\qquad \\ -28x^2-5x \qquad \\ \underline{-28x^2-21x} \qquad \\ 16x+12 \\ \underline{16x+12} \end{array}$$

$\dfrac{8x^3-22x^2-5x+12}{4x+3}=2x^2-7x+4$

27.
$$\begin{array}{r} 2x^2+6x+12 \\ x-3\overline{)2x^3+0x^2-6x-36} \\ \underline{2x^3-6x^2} \qquad\qquad \\ 6x^2-6x \qquad \\ \underline{6x^2-18x} \qquad \\ 12x-36 \\ \underline{12x-36} \end{array}$$

$\dfrac{2x^3-6x-36}{x-3}=2x^2+6x+12$

28. $\dfrac{2.749\times10^8}{3.50\times10^6}\approx0.785\times10^2$
$\approx7.85\times10^1$ people/mi^2

29. $d=rt$
$=2.49\times10^4(7)(24)$
$=418\times10^4$
$=4.18\times10^6$ miles

Cumulative Test for Chapters 1-5

1. $7x(3x-4)-5x(2x-3)-(3x)^2$
$= 21x^2-28x-10x^2+15x-9x^2$
$= 2x^2-13x$

2. $2x^2-3xy+y^2 = 2(-2)^2-3(-2)(3)+3^2$
$= 2(4)+18+9$
$= 8+18+9 = 35$

3. $7x-3(4-2x) = 14x-(3-x)$
$7x-12+6x = 14x-3+x$
$13x-12 = 15x-3$
$-9 = 2x$
$-\frac{9}{2} = x$

4. $\frac{x-5}{3} = \frac{1}{4}x+2$
$12\left(\frac{x-5}{3}\right) = 12\left(\frac{1}{4}x+2\right)$
$4(x-5) = 3x+24$
$4x-20 = 3x+24$
$x-20 = 24$
$x = 44$

5. $4-7x < 11$
$-7x < 7$
$x > -1$

6. $B = \frac{1}{2}a(c+3f)$
$2B = a(c+3f)$
$\frac{2B}{a} = c+3f$
$\frac{2B}{a}-c = 3f$
$\frac{2B}{3a}-\frac{c}{3} = f$

7. $\frac{11{,}904}{x} = 0.96$
$11{,}904 = 0.96x$
$12{,}400 = x$
VBM has 12,400 employees.

8. $2y = 3x+44.$

x	y
−4	−4
0	2
2	5

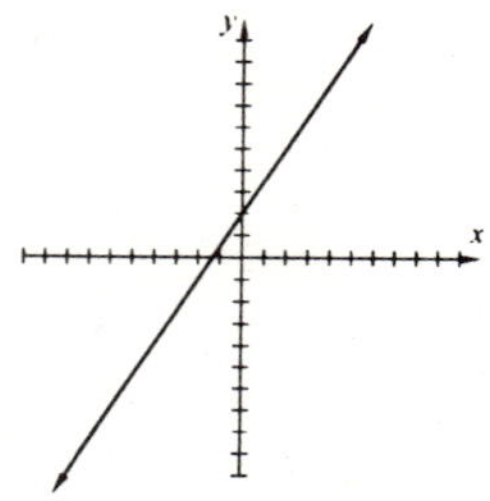

9. $(2,3)$ and $(-1,4)$
$m = \frac{4-3}{-1-2} = \frac{1}{-3} = -\frac{1}{3}$
$3 = -\frac{1}{3}(2)+b \Rightarrow \frac{11}{3}$
$y = -\frac{1}{3}x+\frac{11}{3}$

10. $y = x^2 + 2$

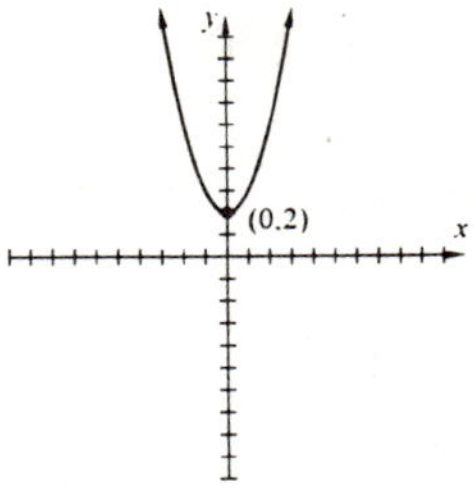

11. $f(x) = -x^2 - 5x + 7$
$f(-1) = -(-1)^2 - 5(-1) + 7$
$= -1 + 5 + 7$
$= 11$

12. $(3x-7)(5x-4)$
$= 15x^2 - 12x - 35x + 28$
$= 15x^2 - 47x + 28$

13. $(3x-5)^2 = (3x)^2 - 2(3x)(5) + 5^2$
$= 9x^2 - 30x + 25$

14. $(3x+2)(2x+1)(x-4)$
$= (3x+2)(2x^2 - 7x - 4)$
$= 6x^3 - 21x^2 - 12x + 4x^2 - 14x - 8$
$= 6x^3 - 17x^2 - 26x - 8$

15. $(-4x^2y^5)(5xy^3) = (-4)(5)x^{2+1}y^{5+3}$
$= -20x^3y^8$

16. $\dfrac{14x^8y^3}{-21x^5y^{12}} = -\dfrac{2x^3}{3y^9}$

17. $(-3xy^4z^2)^3 = (-3)^3x^3(y^4)^3(z^2)^3$
$= -27x^3y^{12}z^6$

18. $\dfrac{9x^{-3}y^{-4}}{w^2z^{-8}} = \dfrac{9z^8}{x^3y^4w^2}$

19. $1,360,000,000,000,000 = 1.36 \times 10^{15}$

20. $0.00056 = 5.6 \times 10^{-4}$

21. $\dfrac{(2.0\times10^{-12})(8.0\times10^{-20})}{4.0\times10^3} = \dfrac{16.0\times10^{-32}}{4.0\times10^3}$
4.0×10^{-35}

22. $(x^3 - 3x^2 - 5x + 20) - (-4x^3 - 10x^2 + x - 30)$
$= (x^3 - 3x^2 - 5x + 20) + (4x^3 + 10x^2 - x + 30)$
$= (x^3 + 4x^3) + (-3x^2 + 10x^2) + (-5x - x)$
$+ (20 + 30)$
$= 5x^3 + 7x^2 - 6x + 50$

23. $-6xy^2(6x^2 - 3xy + 8y^2)$
$= -36x^3y^2 + 18x^2y^3 - 48xy^4$

24. $(x^2 - 6x + 3)(2x^2 - 3x + 4)$
$= 2x^4 - 3x^3 + 4x^2 - 12x^3 + 18x^2 - 24x$
$+ 6x^2 - 9x + 12$
$= 2x^4 - 15x^3 + 28x^2 - 33x + 12$

25. $$\begin{array}{r|l} & x+5 \\ \hline x-3 & x^2+2x-12 \\ & \underline{x^2-3x} \\ & 5x-12 \\ & \underline{5x-12} \\ & 3 \end{array}$$

$$\frac{2x^3-6x-36}{x-3}=x+5+\frac{3}{x-3}$$

Chapter 6

Pretest Chapter 6

1. $2x^2 - 6xy + 12xy^2 = 2x(x - 3y + 6y^2)$

2. $3x(a-2b) + 4(a-2b)$
   $= (a-2b)(3x+4y)$

3. $36ab^2 - 18ab = 18ab(2b-1)$

4. $5a - 10b - 3ax + 6xb$
   $= 5(a-2b) - 3x(a-2b)$
   $= (a-2b)(5-3x)$

5. $3x^2 - 4y + 3xy - 4x$
   $= 3x^2 - 4x + 3xy - 4y$
   $= x(3x-4) + y(3x-4)$
   $= (3x-4)(x+y)$

6. $21x^2 - 14x - 9x + 6$
   $= 7x(3x-2) - 3(3x-2)$
   $= (3x-2)(7x-3)$

7. $x^2 - 22x - 48 = (x-24)(x+2)$

8. $x^2 - 8x + 15 = (x-5)(x-3)$

9. $x^2 + 9x + 8 = (x+8)(x+1)$

10. $2x^2 + 8x - 24 = 2(x^2 + 4x - 12)$
    $= 2(x+6)(x-2)$

11. $3x^2 - 6x - 189$
    $= 3(x^2 - 2x - 63)$
    $= 3(x-9)(x+7)$

12. $15x^2 - 16x + 4$
    $= 15x^2 - 6x - 10x + 4$
    $= 3x(5x-2)(5x-2)$
    $= (5x-2)(3x-2)$

13. $6y^2 + 5yz - 6z^2$
    $= (2y+3z)(3y-2z)$

14. $12x^2 + 44x + 40 = 4(3x^2 + 11x + 10)$
    $= 4(x^2 + 6x + 5x + 10)$
    $= 4[3x(x+2) + 5(x+2)]$
    $= 4(x+2)(3x+5)$

15. $81x^4 - 16$
    $= (9x^2)^2 - (4)^2$
    $= (9x^2+4)(9x^2-4)$
    $= (9x^2+4)(3x+2)(3x-2)$

16. $49x^2 - 28xy + 4y^2$
$= (7x)^2 - 2(7x)(2y) + (2y)^2$
$= (7x - 2y)^2$

17. $25x^2 + 80x + 64 = (5x + 8)^2$

18. $6x^3 + 15x^2 - 9x = 3x(2x^2 + 5x - 3)$
$= 3x(2x^2 + 6x - x - 3)$
$= 3x[2x(x + 3) - 1(x + 3)]$
$= 3x(x + 3)(2x - 1)$

19. $32x^2y^2 - 48xy^2 + 18y^2$
$= 2y^2(16x^2 - 24x + 9)$
$= 2y^2(4x - 3)^2$

20. $25x^2 + 81$ cannot be factored

21. $2x^2 + x - 3 = 0$
$(2x + 3)(x - 1) = 0$
$2x + 3 = 0 \quad x - 1 = 0$
$x = -\frac{3}{2} \quad x = 1$
Roots: $-\frac{3}{2}$, 1

22. $x^2 + 30 = 11x$
$x^2 - 11x + 30 = 0$
$(x - 5)(x - 6) = 0$
$x - 5 = 0$ or $x - 6 = 0$
$x = 5$ or $x = 6$

23. $\frac{3x^2 - 7x}{2} = 3$
$3x^2 - 7x = 6$
$3x^2 - 7x - 6 = 0$
$(3x + 2)(x - 3) = 0$

$3x + 2 = 0 \quad x - 3 = 0$
$x = -\frac{2}{3} \quad x = 3$
Roots: $-\frac{2}{3}$, 3

24. x = altitude
$2x - 1$ = base
$A = \frac{1}{2}ab$
$60 = \frac{1}{2}x(2x - 1)$
$120 = 2x^2 - x$
$0 = 2x^2 - x - 120$
$0 = (2x + 15)(x - 8)$
$2x + 15 = 0 \quad x - 8 = 0$
$2x = -15 \quad x = 0$
$x = -\frac{15}{2}$
Length cannot be negative so, altitude = 8 cm and length = $2(8) - 1 = 15$ cm.

6.1 Exercises

1. $3x^2$ and $5x^3$ are called factors.

3. The factoring is not complete because $6a^3 + 3a^2 - 9a$ contains a common factor of $3a$.

5. $2c^2 + 2c = 2c(c+1)$
   Check:
   $2c(c+1) = 2c^2 + 2c$

7. $18wz - 27w^2z = 9wz(2-3w)$
   Check:
   $9wz(2-3w) = 18wz - 27w^2z$

9. $8x^3 - 10x^2 - 14x = 2x(4x^2 - 5x - 7)$
   Check:
   $2x(4x^2 - 5x - 7)$
   $= 8x^3 - 10x^2 - 14x$

11. $12xy - 18yz - 36xz$
    $= 6(2xy - 3yz - 6xz)$
    Check:
    $6(2xy - 3yz - 6xz) = 12xy - 18yz - 36xz$

13. $2ab^3 + 3xb^2 - 5b^4 + 2b^2$
    $= b^2(2ab + 3x - b^2 + 2)$
    Check:
    $b^2(2ab + 3x - 5b^2 + 2)$
    $= 2ab^3 + 3xb^2 - 5b^4 + 2b^2$

15. $6x^9 - 8x^7 + 4x^5 = 2x^5(3x^4 - 4x^2 + 2)$
    Check:
    $2x^5(3x^4 - 4x^2 + 2) = 6x^9 - 8x^7 + 4x^5$

17. $9a^2b^2 - 36ab = 9ab(ab-4)$

19. $40a^2 - 16ab - 24a = 8a(5a - 2b - 3)$

21. $48xy - 24y^2 + 40y = 8y(6x - 3y + 5)$
    Check:
    $8y(6x - 3y + 5) = 48xy - 24y^2 + 40y$

23. $6(3a+b) - z(3a+b) = (6-z)(3a+6)$

25. $5x(x-7) + 3(x-7) = (x-7)(5x+3)$

27. $7x(3y+5z) - 6t(3y+5z)$
    $= (3y+5z)(7x-6t)$

29. $5a(bc-1) + b(bc-1) + c(bc-1)$
    $= (bc-1)(5a+b+c)$

31. $3c(bc-3a) - 2(bc-3a) - 6b(bc-3a)$
    $= (bc-3a)(3c-2-6b)$

33. $3x^2(x-2y) - (x-2y)$
    $= 3x^2(x-2y) - 1(x-2y)$
    $= (x-2y)(3x^2-1)$

35. $d(5x-3)-(5x-3)=(d-1)(5x-3)$

37. $C =$ Total cost
$C = 25.95a + 25.95b + 25.95c + 25.95d$
$= \$25.95(a+b+c+d)$

Cumulative Review Problems

39. Let $x =$ cost last year
$x + 0.04x = 27,040$
$1.04x = 27,040$
$x = 26,000$
$26,000

41. $\left(\frac{2a^2b^{-1}}{3a^3b^{-3}}\right)^2 = \left(\frac{2b^{-1-(-3)}}{3a^{3-2}}\right)^2 = \left(\frac{2b^2}{3a}\right)^2 = \frac{4b^4}{9a^2}$

43. 20% of $x = 130$
$0.20x = 130$
$x = 650$ people

6.2 Exercises

1. $ab - 3a + 4b - 12 = a(b-3) + 4(b-3)$
$= (a+4)(b-3)$
Check:
$(a+4)(b-3) = ab - 3a + 4b - 12$

3. $2ax + 6bx - ay - 3by$
$= 2x(a+3b) - y(a+3b)$
$= (a+3b)(2x-y)$
Check:
$(a+3b)(2x-y) = 2ax - ay + 6bx - 3by$
$= 2ax + 6bx - ay - 3by$

5. $x^3 - 4x^2 + 3x - 12$
$= x^2(x-4) + 3(x-4)$
$= (x-4)(x^2+3)$
Check:
$(x-4)(x^2+3) = x^3 + 3x - 4x^2 - 12$
$= x^3 - 4x^2 + 3x - 12$

7. $3ax + bx - 6a - 2b$
$= x(3a+b) - 2(3a+b)$
$= (3a+b)(x-2)$
Check:
$(3a+b)(x-2)$
$= 3ax - 6a + bx - 2b$
$= 3ax + bx - 6a - 2b$

9. $5a + 12bc + 10b + 6ac$
$= 5a + 10b + 6ac + 12bc$
$= 5(a+2b) + 6c(a+2b)$
$= (a+2b)(5+6c)$
Check:
$(a+2b)(5+6c)$
$= 5a + 6ac + 10b + 12bc$
$= 5a + 12bc + 10b + 6ac$

11. $5a - 5b - 2ax + 2xb$
$= 5(a-b) - 2x(a-b)$
$= (a-b)(5-2x)$
Check:
$(a-b)(5-2x)$
$= 5a - 2ax - 5b + 2bx$
$= 5a - 5b - 2ax + 2xb$

13. $y^2 - 2y - 3y + 6$
$= y(y-2) - 3(y-2)$
$= (y-2)(y-3)$

15. $14 - 7y + 2y - y^2$
$= 7(2-y) + y(2-y)$
$= (7+y)(2-y)$

17. $6ax - y + 2ay - 3x$
$= 6ax - 3x + 2ay - y$
$= 3x(2a-1) + y(2a-1)$
$= (2a-1)(3x+y)$

19. $2x^2 + 8x - 3x - 12$
$= 2x(x+4) - 3(x+4)$
$= (2x-3)(x+4)$

21. $28x^2 + 8xy^2 + 21xw + 6y^2w$
$= 4x(7x+2y^2) + 3w(7x+2y^2)$
$= (7x+2y^2)(4x+3w)$

23. Rearrange the terms so that factoring gives the same expression in parentheses.
$6a^2 - 12bd - 8ad + 9ab$
$= 6a^2 - 8ad + 9ab - 12bd$
$= 2a(3a-4d) + 3b(3a-4d)$
$= (3a-4d)(2a+3b)$

Cumulative Review Problems

25. $\frac{1}{3}(x-6) \geq \frac{1}{6}x - 3$
$2(x-6) \geq x - 18$
$2x - 12 \geq x - 18$
$x - 12 \geq -18$
$x \geq -6$

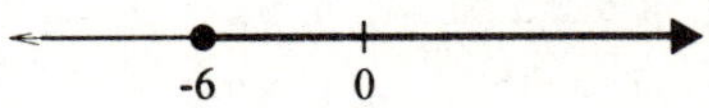

27. The dollar value in 1998 was
$126.9 + 15.1 = \$142$ million
Increase from 1998 to 2000 was
$0.20(142) = \$28.4$ million
The dollar value in 2000 was
$142 + 28.4 = \$170.4$ million

6.3 Exercises

1. Find two numbers whose product is 6 and whose sum is 5.

3. $x^2 + 2x + 1$; Product: 1, Sum: 2, + signs
$x^2 + 2x + 1 = (x+1)(x+1)$

5. $x^2+12x+35$; Product: 35, Sum: 12, + signs
$x^2+12x+35=(x+7)(x+5)$

7. x^2-4x+3; Product: 3, Sum: -4, $-$ signs
$x^2-4x+3=(x-1)(x-3)$

9. $x^2-11x+28$; Product: 28, Sum: -11, $-$ signs
$x^2-11x+28=(x-4)(x-7)$

11. x^2+x-12; Product: -12, Sum: 1,
Opposite signs with larger absolute value +.
$x^2+x-12=(x+4)(x-3)$

13. $x^2-13x-14$; Product: -14, Sum: -13,
Opposite signs with larger absolute value $-$.
$x^2-13x-14=(x-14)(x+1)$

15. $x^2+2x-35$; Product: -35, Sum: 2
Opposite signs with larger absolute value +.
$x^2+2x-35=(x+7)(x-5)$

17. $x^2-2x-24$; Product: -24, Sum: -2
Opposite signs with larger absolute value $-$.
$x^2-2x-24=(x-6)(x+4)$

19. $x^2+5x-14=(x+7)(x-2)$
Check: $(x+7)(x-2)=x^2-2x+7x-14$
$=x^2+5x-14$

21. $x^2-10x+24=(x-6)(x-4)$
Check: $(x-6)(x-4)=x^2-4x-6x+24$
$=x^2-10x+24$

23. $x^2+13x+30=(x+3)(x+10)$
Check: $(x+3)(x+10)=x^2+10x+3x+30$
$=x^2+13x+30$

25. $y^2-4y-5=(y-5)(y+1)$
Check: $(y-5)(y+1)=y^2+y-5y-5$
$=y^2-4y-5$

27. $a^2+6a-16=(a+8)(a-2)$
Check: $(a+8)(a-2)=a^2-2a+8a-16$
$=a^2+6a-16$

29. $x^2-12x+32=(x-4)(x-8)$
Check: $(x-4)(x-8)=x^2-8x-4x+32$
$=x^2-12x+32$

31. $x^2+4x-21=(x+7)(x-3)$
Check: $(x+7)(x-3)=x^2-3x+7x-21$
$=x^2+4x-21$

33. $x^2+13x+40=(x+5)(x+8)$
Check: $(x+5)(x+8)=x^2+8x+5x+40$
$=x^2+13x+40$

35. $x^2-21x-46=(x-23)(x+2)$
Check: $(x-23)(x+2)=x^2+2x-23x-46$
$=x^2-21x-46$

37. $x^2+9x-36=(x+12)(x-3)$

Check: $(x+12)(x-3)=x^2-3x+12x-36$

$=x^2+9x-36$

39. $x^2-2xy-15y^2=(x-5y)(x+3y)$

Check:

$(x-5y)(x+3y)=x^2+3xy-5xy-15y^2$

$=x^2-2xy-15y^2$

41. $x^2-16xy+63y^2=(x-9y)(x-7y)$

Check:

$(x-9y)(x-7y)=x^2-7xy-9xy+63y^2$

$=x^2-16xy+63y^2$

43. $2x^2-12x+16=2(x^2-6x+8)$

$=2(x-2)(x-4)$

45. $3x^2-6x-72=3(x^2-2x-24)$

$3(x-6)(x+4)$

47. $4x^2+24x+20=4(x^2+6x+5)$

$=4(x+1)(x+5)$

49. $7x^2+21x-70=7(x^2+3x-10)$

$=7(x+5)(x-2)$

51. $6x^2+18x+12=6(x^2+3x+2)$

$=6(x+2)(x+1)$

53. $3x^2-18x+15=3(x^2-6x+5)$

$=3(x-5)(x-1)$

55. $A=\pi(2x)^2-6^2$

$=4\pi x^2-36$

$=4(\pi x^2-9)$

Cumulative Review Problems

57. x = second angle

$2x+2$ = first angle

$3x-2$ = third angle

$x+2x+2+3x-2=180^\circ$

$6x=180^\circ$

$x=30^\circ$, $2x+2=62^\circ$, $3x-2=88^\circ$

second angle = 30°

first angle = 62°

third angle = 88°

59. $m=\frac{1}{2}$ and $(2,5)$

$5=\frac{1}{2}(2)+b\Rightarrow b=4$

$y=\frac{1}{2}x+4$

61. x = number of 30-min periods after midnight

$4(65)+85x=515$

$260+85x=515$

$85x=255$

$x=3$

The reception ended at 1:30 a.m.

63. $M = 3$
$T = 19 + 2M$
$T = 19 + 2(3)$
$T = 25$
25^0C

6.4 Exercises

1. $4x^2 + 13x + 3$; Grouping number: 12
$4x^2 + 13x + 3 = 4x^2 + 12x + x + 3$
$= 4x(x+3) + 1(x+3)$
$= (4x+1)(x+1)$
Check:
$(4x+1)(x+3) = 4x^2 + 12x + x + 3$
$= 4x^2 + 13x + 3$

3. $2x^2 - 5x + 2$; Grouping number: 4
$2x^2 - 5x + 2 = 2x^2 - 4x - x + 2$
$= 2x(x-2) - 1(x-2)$
$= (x-2)(2x-1)$
Check:
$(x-2)(2x-1) = 2x^2 - x - 4x + 2$
$= 2x^2 - 5x + 2$

5. $3x^2 - 4x - 7$; Grouping number: -21
$3x^2 - 4x - 7 = 3x^2 - 7x + 3x - 7$
$= x(3x-7) + (3x-7)$
$= (3x-7)(x+1)$
Check:
$= (3x-7)(x+1) = 3x^2 + 3x - 7x - 7$
$= 3x^2 - 4x - 7$

7. $2x^2 - 5x - 3$; Grouping number: -6
$2x^2 - 5x - 3 = 2x^2 - 6x + x - 3$
$= 2x(x-3) + (x-3)$
$= (x-3)(2x+1)$
Check:
$(x-3)(2x+1) = 2x^2 + x - 6x - 3$
$= 2x^2 - 5x - 3$

9. $5x^2 + 3x - 2$; Grouping number: -10
$5x^2 + 3x - 2 = 5x^2 + 5x - 2x - 2$
$= 5x(x+1) - 2(x+1)$
$= (x+1)(5x-2)$
Check:
$(x+1)(5x-2) = 5x^2 - 2x + 5x - 2$
$= 5x^2 + 3x - 2$

11. $15x^2 - 34x + 15$; Grouping number: 225
$15x^2 - 34x + 15 = 15x^2 - 25x - 9x + 15$
$= 5x(3x-5) - 3(3x-5)$
$= (5x-3)(3x-5)$
Check:
$(5x-3)(3x-5) = 15x^2 - 25x - 9x + 15$
$= 15x^2 - 34x + 15$

13. $2x^2 + 3x - 20$; Grouping number: -40
$2x^2 + 3x - 20 = 2x^2 + 8x - 5x - 20$
$= 2x(x+4) - 5(x+4)$
$= (x+4)(2x-5)$
Check:
$(x+4)(2x-5) = 2x^2 - 5x + 8x - 20$
$= 2x^2 + 3x - 20$

15. $9x^2+9x+2$; Grouping number: 18
$9x^2+9x+2=9x^2+3x+6x+2$
$=3x(3x+1)+2(3x+1)$
$=(3x+1)(3x+2)$
Check:
$(3x+1)(3x+2)=9x^2+6x+3x+2$
$=9x^2+9x+2$

17. $6x^2-5x-6$; Grouping number: -36
$6x^2-5x-6=6x^2-9x+4x-6$
$=3x(2x-3)+2(2x-3)$
$=(2x-3)(3x+2)$
Check:
$(2x-3)(3x+2)=6x^2+4x-9x-6$
$=6x^2-5x-6$

19. $6x^2-19x+10$; Grouping number: 60
$6x^2-19x+10=6x^2-4x-15x+10$
$=2x(3x-2)-5(3x-2)$
$=(3x-2)(2x-5)$
Check:
$(3x-2)(2x-5)=6x^2-15x-4x+10$
$=6x^2-19x+10$

21. $7x^2-5x-18=7x^2-14x+9x-18$
$=7x(x-2)+9(x-2)$
$=(7x+9)(x-2)$

23. $9y^2-13y+4=9y^2-9y-4y+4$
$=9y(y-1)-4(y-1)$
$=(y-1)(9y-4)$

25. $5a^2-13a-6=5a^2-15a+2a-6$
$=5a(a-3)+2(a-3)$
$=(a-3)(5a+2)$

27. $12x^2-20x+3=12x^2-18x-2x+3$
$=6x(2x-3)-(2x-3)$
$=(2x-3)(6x-1)$

29. $15x^2+4x-4=15x^2+10x-6x-4$
$=5x(3x+2)-2(3x+2)$
$=(3x+2)(5x-2)$

31. $12x^2+28x+15=12x^2+18x+10x+15$
$=6x(2x+3)+5(2x+3)$
$=(6x+5)(2x+3)$

33. $12x^2-16x-3=12x^2+2x-18x-3$
$=2x(6x+1)-3(6x+1)$
$=(6x+1)(2x-3)$

35. $2x^4+15x^2-8=2x^4+16x^2-x^2-8$
$=2x^2(x^2+8)-(x^2+8)$
$=(x^2+8)(2x^2-1)$

37. $4x^2+8xy-5y^2=4x^2+10xy-2xy-5y^2$
$=2x(2x+5y)-y(2x+5y)$
$=(2x+5y)(2x-y)$

39. $5x^2+16xy-16y^2=5x^2+20xy-4xy-16y^2$
$=5x(x+4y)-4y(x+4y)$
$=(x+4y)(5x-4y)$

41. $10x^2+22x+12=2(5x^2+11x+6)$
$=2(5x+6)(x+1)$

43. $12x^2-24x+9=3(4x^2-8x+3)$
$=3(2x-3)(2x-1)$

45. $10x^2-25x-15=5(2x^2-5x-3)$
$=5(2x^2-6x+x-3)$
$=5[2x(x-3)+1(x-3)]$
$=5(2x+1)(x-3)$

47. $6x^3-16x^2-6x=2x(3x^2-8x-3)$
$=2x(3x+1)(x-3)$

49. $12x^2+16x-35=(2x+5)(6x-7)$

51. $20x^2-27x+9=(4x-3)(5x-3)$

Cumulative Review Problems

53. $7x-3(4-2x)=2(x-3)-(5-x)$
$7x-12+6x=2x-6-5+x$
$13x-12=3x-11$
$10x=1$
$x=\frac{1}{10}$

55. Increase $=.18(31.1)=5.598$
$31.1+5.59\approx 36.7$
Approximately 36.7 million will be enrolled in 2002-2003.

6.5 Exercises

1. $81x^2-16=(9x)^2-(4)^2$
$=(9x+4)(9x-4)$

3. $16-9x^2=(4)^2-(3x)^2$
$=(4+3x)(4-3x)$

5. $9x^2-25=(3x)^2-(5)^2$
$=(3x+5)(3x-5)$

7. $4x^2-25=(2x)^2-(5)^2$
$=(2x+5)(2x-5)$

9. $36x^2-25=(6x)^2-(5)^2$
$=(6x+5)(6x-5)$

11. $1-49x^2=(1)^2-(7x)^2$
$=(1+7x)(1-7x)$

13. $16x^2-49y^2=(4x)^2-(7y)^2$
$=(4x+7y)(4x-7y)$

15. $25-121x^2=(5)^2-(11x)^2$
$=(5+11x)(5-11x)$

17. $81x^2-100y^2=(9x)^2-(10y)^2$
$=(9x+10y)(9x-10y)$

19. $25a^2-49=(5a)^2-(7)^2$
$=(5a+7)(5a-7)$

21. $9x^2+6x+1=(3x)^2+2(3x)(1)+(1)^2$
$=(3x+1)^2$

23. y^2-6y+9
$=y^2-2(y)(3)+3^2$
$=(y-3)^2$

25. $9x^2-24x+16$
$=(3x)^2-2(3x)(4)+4^2$
$=(3x-4)^2$

27. $49x^2+28x+4$
$=(7x)^2+2(7x)(2)+2^2$
$=(7x+2)^2$

29. $x^2+14x+49$
$=(x)^2+2(x)(7)+(7)^2$
$=(x+7)^2$

31. $25x^2-40x+16$
$=(5)^2-2(5x)(4)+4^2$
$=(5x-4)^2$

33. $81x^2+36xy+4y^2$
$=(9x)^2+2(9x)(2y)+(2y)^2$
$=(9x+2y)^2$

35. $25x^2-30xy+9y^2$
$=(5x)^2-2(5x)(3y)+(3y)^2$
$=(5x-3y)^2$

37. $16a^2+72ab+81b^2$
$=(4a)^2+2(4a)(9b)+(9b)^2$
$=(4a+9b)^2$

39. $9x^4-6x^2y+y^2$
$=(3x^2)^2-2(3x^2)(y)+y^2$
$=(3x^2-y)^2$

41. $49x^2 + 70x + 9 = (7x+1)(7x+9)$

43. $16x^4 - 1 = (4x^2)^2 - 1^2$
$= (4x^2+1)(4x^2-1)$
$= (4x^2+1)\left[(2x)^2 - (1)^2\right]$
$= (4x^2+1)(2x-1)(2x+1)$

45. $x^{10} - 36y^{10} = (x^5)^2 - (6y^5)^2$
$= (x^5+6y^5)(x^5-6y^5)$

47. $9x^{10} - 12x^5 + 4$
$= (3x^5)^2 - 2(3x^5)(2) + (2)^2$
$= (3x^5-2)^2$

49. You cannot factor $9x^2+1$ because there are no combinations of the product of two binomials that give $9x^2+1$.

51. $16 = 4^2$
$56 = (2)(4)(7)$
$c = 7^2$
$c = 49$
There is only one answer.

53. $16x^2 - 36 = 4(4x^2-9)$
$= 4(2x+3)(2x-3)$

55. $147x^2 - 3y^2 = 3(49x^2 - y^2)$
$= 3\left[(7x)^2 - y^2\right]$
$= 3(7x-y)(7x+y)$

57. $12x^2 - 36x + 27 = 3(4x^2 - 12x + 9)$
$= 3(2x-3)^2$

59. $98x^2 + 84x + 18 = 2(49x^2 + 42x + 9)$
$= 2(7x+3)^2$

61. $x^2 - 9x + 14 = (x-7)(x-2)$

63. $2x^2 + 5x - 3 = (2x-1)(x+3)$

65. $16x^2 - 121 = (4x+11)(4x-11)$

67. $9x^2 + 42x + 49 = (3x+7)^2$

69. $3x^2 + 6x - 45 = 3(x^2 + 2x - 15)$
$= 3(x+5)(x-3)$

71. $5x^2 - 80 = 5(x^2 - 16)$
$= 5(x+4)(x-4)$

73. $5x^2 + 20x + 20 = 5(x^2 + 4x + 4)$
$= 5(x+2)^2$

75. $2x^2 - 32x + 126 = 2(x^2 - 16x + 63)$
$= 2(x-9)(x-7)$

Cumulative Review Problems

77.
$$\begin{array}{r} x^2 + 3x + 4 \\ x-2 \overline{) x^3 + x^2 - 2x - 11} \\ \underline{x^3 - 2x^2} \\ 3x^2 - 2x \\ \underline{3x^2 - 6x} \\ 4x - 11 \\ \underline{4x - 8} \\ -3 \end{array}$$

$(x^3 + x^2 - 2x - 11) \div (x-2)$
$= x^2 + 3x + 4 - \dfrac{3}{x-2}$

79. Daily diet $= 0.2(150) = 3$ ounces
$.4(3) = 1.2$ ounces of greens
$.35(3) = 1.05$ ounces of bulk vegtable
$.25(3) = 0.75$ ounces of fruit

81. $A = 6288 - 700M$
$A = 6288 - 700(3.5)$
$A = 6288 - 2450$
$A = 3838$
3838 feet above sea level

6.6 Exercises

1. $6a^2 + 2ab - 3a = a(6a + 2b - 3)$
Check: $a(6a + 2b) = 6a^2 + 2ab - 3a$

3. $36x^2 - 9y^2 = 9(4x^2 - y^2)$
$= 9(2x + y)(2x - y)$
Check: $9(2x + y)(2x - y) = 9(4x^2 - y^2)$
$= 36x^2 - 9y^2$

5. $9x^2 - 12xy + 4y^2 = (3x - 2y)^2$
Check:
$(3x - 2y)^2 = 9x^2 + 2(3x)(-2y) + (-2y)^2$
$= 9x^2 - 12xy + 4y^2$

7. $x^2 + 8x + 15 = (x+5)(x+3)$
Check:
$(x+5)(x+3) = x^2 + 3x + 5x + 15$
$= x^2 + 8x + 15$

9. $15x^2 + 7x - 2 = (5x - 1)(3x + 2)$

11. $ax - 3ay - 6by + 2bx$
$= a(x - 3y) + 2b(x - 3y)$
$= (x - 3y)(a + 2b)$
Check:
$(x - 3y)(a + 2b) = ax + 2bx - 3ay - 6by$
$= ax - 3ay - 6by + 2bx$

13. $3x^4 - 12 = 3(x^4 - 4)$
$= 3(x^2 + 2)(x^2 - 2)$

15. $4x^2 - 12x + 9 = (2x - 3)(2x - 3)$
$= (2x - 3)^2$

17. $2x^2 - 11x + 12 = (2x - 3)(x - 4)$

19. $x^2 - 3xy - 70y^2 = (x - 10y)(x + 7y)$

21. $ax - 5a + 3x - 15 = a(x - 5) + 3(x - 5)$
$= (a + 3)(x - 5)$

23. $45x - 5x^3 = 5x(9 - x^2)$
$= 5x(3 + x)(3 - x)$

25. $5x^3y^3 - 10x^2y^3 + 5xy^3 = 5xy^3(x^2 - 2x + 1)$
$= 5xy^3(x - 1)^2$

27. $27xyz^2 - 12xy = 3xy(9z^2 - 4)$
$= 3xy(3z + 2)(3z - 2)$

29. $3x^2 + 6x - 105 = 3(x^2 + 2x - 35)$
$= 3(x + 7)(x - 5)$

31. $5x^2 - 30x + 40 = 5(x^2 - 6x + 8)$
$= 5(x - 4)(x - 2)$

33. $7x^2 - 2x^4 + 4 = -1(2x^4 - 7x^2 - 4)$
$= -1(2x^2 + 1)(x^2 - 4)$
$= -1(2x^2 + 1)(x + 2)(x - 2)$

35. $6x^2 - 3x + 2$ is prime.

37. $5x^2 + 10xy - 30y = 5(x^2 + 2xy - 6y)$

39. $30x^3 + 3x^2y - 6xy^2$
$= 3x(10x^2 + xy - 2y^2)$
$= 3x(2x + y)(5x - 2)$

41. $8x^2 + 28x - 16 = 4(2x^2 + 7x - 4)$
$= 4(2x - 1)(x + 4)$

43. A polynomial that cannot be factored by the methods of this chapter is called prime.

Cumulative Review Problems

45. Let x = salary in previous job
$x - 0.14x = 24,080$
$0.86x = 24,080$
$x = 28,000$
\$28,000

47. $\frac{1}{2}x \quad - y = 7 \qquad (1)$
$-3x + 2y = -22 \qquad (2)$
Multiply (1) by 2

$$x-2y=14$$
$$\underline{-3x+2y=-22}$$
$$-2x\quad=-8$$
$$x=4$$

Substitute 4 for x in (1)

$$\frac{1}{2}(4)-y=7$$
$$-y=5$$
$$y=-5$$
$$(4,-5)$$

49. x = number of magazines

$2x$ = number of softcover books

$2x-37$ = number of hardcover books

$$x+2x+2x-37=198$$
$$5x-37=198$$
$$5x=235$$
$$x=47$$

He has 47 magazines, 57 hardcover books, and 94 softcover books.

6.7 Exercises

1. $x^2-4x-21=0$

$$(x+3)(x-7)=0$$

$x+3=0 \qquad x-7=0$

$x=-3 \qquad x=7$

Check:

$(-3)^2-4(-3)-21\stackrel{?}{=}0 \qquad (7)^2-4(7)-21\stackrel{?}{=}$

$9+12-21\stackrel{?}{=}0 \qquad 49-28-21\stackrel{?}{=}0$

$0=0 \qquad 0=0$

3. $2x^2-5x-3=0$

$$2x^2-6x+x-3=0$$
$$2x(x-3)+1(x-3)=0$$
$$(2x+1)(x-3)=0$$

$2x+1=0 \qquad x-3=0$

$2x=-1 \qquad x=3$

$x=-\frac{1}{2}$

Check:

$$2\left(-\frac{1}{2}\right)^2-5\left(-\frac{1}{2}\right)-3\stackrel{?}{=}0$$
$$\frac{1}{2}+\frac{5}{2}-3\stackrel{?}{=}0$$
$$0=0$$

$$2(3)^2-5(3)-3\stackrel{?}{=}0$$
$$18-15-3\stackrel{?}{=}0$$
$$0=0$$

5. $2x^2-7x+6=0$

$$2x^2-3x-4x+6=0$$
$$x(2x-3)-2(2x-3)=0$$
$$(2x-3)(x-2)=0$$

$2x-3=0 \qquad x-2=0$

$x=\frac{3}{2} \qquad x=2$

Check:

$2\left(\frac{3}{2}\right)^2-7\left(\frac{3}{2}\right)+6\stackrel{?}{=}0 \qquad 2(2)^2-7(2)+6\stackrel{?}{=}0$

$\frac{9}{2}-\frac{21}{2}+\frac{12}{2}\stackrel{?}{=}0 \qquad 8-14+6\stackrel{?}{=}0$

$0=0 \qquad 0=0$

7. $6x^2 - 13x = -6$

$6x^2 - 13x + 6 = 0$

$(3x-2)(2x-3) = 0$

$3x - 2 = 0 \qquad 2x - 3 = 0$

$3x = 2 \qquad 2x = 3$

$x = \frac{2}{3} \qquad x = \frac{3}{2}$

Check:

$6\left(\frac{2}{3}\right)^2 - 13\left(\frac{2}{3}\right) \stackrel{?}{=} -6 \qquad 6\left(\frac{3}{2}\right)^2 - 13\left(\frac{3}{2}\right) \stackrel{?}{=} -6$

$\frac{8}{3} - \frac{26}{3} \stackrel{?}{=} -6 \qquad \frac{27}{2} - \frac{39}{2} \stackrel{?}{=} -6$

$-6 = -6 \qquad -6 = -6$

9. $x^2 + 13x = 0$

$x(x+13) = 0$

$x = 0 \qquad x + 13 = 0$

$x = -13$

Check:

$(-13)^2 + (13)(-13) \stackrel{?}{=} 0 \qquad 0^2 + 13(0) \stackrel{?}{=} 0$

$169 - 169 \stackrel{?}{=} 0 \qquad 0 \stackrel{?}{=} 0$

$0 = 0$

11. $8x^2 = 72$

$8x^2 - 72 = 0$

$8(x^2 - 9) = 0$

$8(x-3)(x+3) = 0$

$x - 3 = 0 \qquad x + 3 = 0$

$x = 3 \qquad x = -3$

Check:

$8(-3)^2 \stackrel{?}{=} 72 \qquad 8(3)^2 \stackrel{?}{=} 72$

$72 = 72 \qquad 72 = 72$

13. $5x^2 + 3x = 8x$

$5x^2 - 5x = 0$

$5x(x-1) = 0$

$5x = 0 \qquad x - 1 = 0$

$x = 0 \qquad x = 1$

Check:

$5(0)^2 + 3(0) \stackrel{?}{=} 8(0) \qquad 5(1)^2 + 3(1) \stackrel{?}{=} 8(1)$

$0 + 0 \stackrel{?}{=} 0 \qquad 5 + 3 \stackrel{?}{=} 8$

$0 = 0 \qquad 8 = 8$

15. $(x-5)(x+2) = -4(x+1)$

$x^2 - 3x - 10 = -4x - 4$

$x^2 + x - 6 = 0$

$(x+3)(x-2) = 0$

$x + 3 = 0 \qquad x - 2 = 0$

$x = -3 \qquad x = 2$

Check:

$(-3-5)(-3+2) \stackrel{?}{=} -4(-3+1)$

$-8(-1) \stackrel{?}{=} -4(-2)$

$8 = 8$

$(2-5)(2+2) \stackrel{?}{=} -4(2+1)$

$-3(4) \stackrel{?}{=} -4(3)$

$-12 = -12$

17.
$$4x^2 - 3x + 1 = -7x$$
$$4x^2 + 4x + 1 = 0$$
$$(2x+1)(2x+1) = 0$$
$$2x + 1 = 0$$
$$2x = -1$$
$$x = -\frac{1}{2}$$

Check:
$$4\left(-\frac{1}{2}\right)^2 - 3\left(-\frac{1}{2}\right) + 1 \stackrel{?}{=} -7\left(-\frac{1}{2}\right)$$
$$1 + \frac{3}{2} + 1 \stackrel{?}{=} \frac{7}{2}$$
$$\frac{7}{2} = \frac{7}{2}$$

19.
$$\frac{x^2}{2} - 8 + x = -8$$
$$x^2 - 16 + 2x = -16$$
$$x^2 + 2x = 0$$
$$x(x+2) = 0$$
$$x = 0 \qquad x + 2 = 0$$
$$x = -2$$

Check:
$$\frac{(0)^2}{2} - 8 + (0) \stackrel{?}{=} -8 \qquad \frac{(-2)^2}{2} - 8 + (-2) \stackrel{?}{=} -8$$
$$0 - 8 + 0 \stackrel{?}{=} -8 \qquad 2 - 8 - 2 \stackrel{?}{=} -8$$
$$-8 = -8 \qquad -8 = -8$$

21.
$$\frac{x^2 + 7x}{4} = -3$$
$$x^2 + 7x = -12$$
$$x^2 + 7x + 12 = 0$$
$$(x+4)(x+3) = 0$$
$$x + 4 = 0 \qquad x + 3 = 0$$
$$x = -4 \qquad x = -3$$

Check:
$$\frac{(-4)^2 + 7(-4)}{4} \stackrel{?}{=} -3 \qquad \frac{(-3)^2 + 7(-3)}{4} \stackrel{?}{=} -3$$
$$\frac{16 - 28}{4} \stackrel{?}{=} -3 \qquad \frac{9 - 21}{4} \stackrel{?}{=} -3$$
$$\frac{-12}{4} \stackrel{?}{=} -3 \qquad \frac{-12}{4} \stackrel{?}{=} -3$$
$$-3 = -3 \qquad -3 = -3$$

23.
$$\frac{9x^2 - 12x}{3} = 15$$
$$9x^2 12x = 45$$
$$9x^2 - 12x - 45 = 0$$
$$3(3x^2 - 4x - 15) = 0$$
$$3(3x+5)(x-3) = 0$$
$$3x + 5 = 0 \qquad x - 3 = 0$$
$$x = -\frac{5}{3} \qquad x = 3$$

Check:
$$\frac{9\left(-\frac{5}{3}\right)^2 - 12\left(-\frac{5}{3}\right)}{3} \stackrel{?}{=} 15 \qquad \frac{9(3)^2 12(3)}{3} \stackrel{?}{=} 15$$
$$\frac{25 + 20}{3} \stackrel{?}{=} 15 \qquad \frac{81 - 36}{3} \stackrel{?}{=} 15$$
$$15 = 15 \qquad 15 = 15$$

25. An equation in the form $ax^2 + bx = 0$ can always be solved by factoring out x.

27.
$$x = \text{length}$$
$$\frac{1}{2}x + 3 = \text{width}$$
$$x\left(\frac{1}{2}x + 3\right) = 140$$
$$\frac{x^2}{2} + 3x = 140$$
$$x^2 + 6x = 280$$
$$x^2 + 6x - 280 = 0$$
$$(x+20)(x-14) = 0$$
$$x + 20 = 0 \qquad x - 14 = 0$$
$$x = -20 \qquad x = 14$$
Width not negative $\quad \frac{1}{2}x + 3 = 10$

The length is 14 m and the width is 10 m.

29. $G = x^2 - x,\ x = 14$
$$G = (14)^2 - 14$$
$$= 182 \text{ games}$$

31.
$$G = x^2 - x,\ G = 210$$
$$x^2 - x = 210$$
$$x^2 - x - 210 = 0$$
$$(x+14)(x-15) = 0$$
$$x + 14 = 0 \qquad x - 15 = 0$$
$$x = -14 \qquad x = 15$$
A negative number of games is not allowed. There are 15 teams.

33.
$$s = -5t^2 + vt + h$$
$$v = 13 \qquad h = 6$$
$$s = -5t^2 + 13t + 6$$
When the ball hits the ground $s = 0$.
$$0 = -5t^2 + 13t + 6$$
$$0 = 5t^2 - 13t - 6$$
$$0 = (5t+2)(t-3)$$
$$5t + 2 = 0 \qquad t - 3 = 0$$
$$t = -\frac{2}{5} \qquad t = 3$$
Time can't be negative.

The ball will hit the ground at $t = 3$ sec.

$$s = -5(2)^2 + 13(2) + 6$$
$$= -20 + 26 + 6$$
$$= 12$$
After 2 seconds the ball is 12 m from the ground.

35.
$$C = 2x^2 - 7x$$
$$15 = 2x^2 - 7x$$
$$2x^2 - 7x - 15 = 0$$
$$(2x+3)(x-5) = 0$$
$$2x + 3 = 0 \qquad x - 5 = 0$$
$$x = -\frac{3}{2} \qquad x = 5$$
The number cannot be negative, so 5 additional helicopters were produced.

Use $T = 0.5\left(x^2 - x\right)$ in Exercise 37.

37. $x = 70$

$$T = 0.5\left[(70)^2 - 70\right]$$
$$= 0.5(4900 - 70)$$
$$= 2415 \text{ telephone calls}$$

39. $T = 153$

$$0.5\left(x^2 - x\right) = 306$$
$$x^2 - x - 306 = 0$$
$$(x+17)(x-18) = 0$$
$$x+17 = 0 \qquad x-18 = 0$$
$$x = -17 \qquad x = 18$$

39. The negative number of people is not allowed. 18 people worked that day.

Cumulative Review Problems

41. $\left(2x^2y^3\right)\left(-6xy^4\right) = -12x^3y^7$

43. $\dfrac{21a^5b^{10}}{-14ab^{12}} = -\dfrac{3a^4}{2b^2}$

Putting Your Skills to Work

1. $T = 0.5n(s+e)$

$$T = 0.5(20)\left[24,000 + (19 \cdot 1500 + 24,000)\right]$$
$$T = 0.5(20)(24,000 + 52,500)$$
$$T = 10(76,500)$$
$$T = 765,000$$
$$\$765,000$$

2. $T = 0.5n(s+e)$

$$T = 0.5(8)\left[150,000 + (7 \cdot 70,000 + 150,000)\right]$$
$$T = 0.5(8)(150,000 + 640,000)$$
$$T = 4(790,000)$$
$$T = 3,160,000$$
$$\$3,160,000$$

3. $56,000 - 40,000 = 16,000$

$$T = 0.5n(s+e)$$
$$1,000,000 =$$
$$0.5n\left[40,000 + (n-1)(16,000) + 40,000\right]$$
$$1,000,000 = 0.5n(64,000 + 16,000n)$$
$$1,000,000 = 32,000n + 8000n^2$$
$$0 = 8000n^2 + 32,000n - 1,000,000$$
$$0 = 8000\left(n^2 + 4n - 125\right)$$
If $n = 9,\ T = 936,000$

If $n = 10,\ T = 1,120,000$

10 years

4. $T = 0.5n(s+e)$

$$s = 780,000,000 - 10(15,000,000)$$
$$= 630,000,000$$
$$e = 630,000,000 - 15(15,000,000)$$
$$= 405,000,000$$
$$T = 0.5(15)(630,000,000) + 405,000,000$$
$$T = 7,762,500,000$$
$$\$7,762,500,000$$

Chapter 6 Review Problems

1. $15x^3y - 9x^2y^2 = 3x^2y(5x - 3y)$

2. $40x^2 - 32x = 8x(5x - 4)$

3. $7x^2y - 14xy^2 - 21x^3y^3$
$= 7xy(x - 2y - 3x^2y^2)$

4. $50a^4b^5 - 25a^4b^4 + 75a^5b^5$
$= 25a^4b^4(2b - 1 + 3ab)$

5. $27x^3 - 9x^2 = 9x^2(3x - 1)$

6. $2x - 4y + 6z + 12 = 2(x - 2y + 3z + 6)$

7. $2a(a + 3b) - 5(a + 3b) = (a + 3b)(2a - 5)$

8. $15x^3y + 6xy^2 + 3xy = 3xy(5x^2 + 2y + 1)$

9. $3ax - 7a - 6x + 14 = a(3x - 7) - 2(3x - 7)$
$= (3x - 7)(a - 2)$

10. $a^2 + 5ab - 4a - 20b = a(a + 5b) - (4a + 5b)$
$= (a + 5b)(a - 4)$

11. $x^2y + 3y - 2x^2 - 6 = y(x^2 + 3) - 2(x^2 + 3)$
$= (x^2 + 3)(y - 2)$

12. $4ad - 25c - 5d + 20ac = 4ad - 5d + 20ac - 25c$
$= d(4a - 5) + 5c(4a - 5)$
$= (4a - 5)(d + 5c)$

13. $15x^2 - 3x + 10x - 2 = 3x(5x - 1) + 2(5x - 1)$
$= (5x - 1)(3x + 2)$

14. $30w^2 - 18w + 5wz - 3z$
$= 6w(5w - 3) + z(5w - 3)$
$= (5w - 3)(6w + z)$

15. $x^2 - 2x - 35 = (x - 7)(x + 5)$

16. $x^2 - 10x + 24 = (x - 6)(x - 4)$

17. $x^2 + 14x + 48 = (x + 6)(x + 8)$

18. $x^2 + 8xy + 15y^2 = (x + 3y)(x + 5y)$

19. $x^4 + 13x^2 + 42 = (x^2 + 6)(x^2 + 7)$

20. $x^2 - 14x - 51 = (x - 17)(x + 3)$

21. $5x^2 + 20x + 15 = 5(x^2 + 4x + 3)$
$= 5(x + 3)(x + 1)$

22. $3x^2 + 39x + 36 = 3(x^2 + 13x + 12)$
$= 3(x + 12)(x + 1)$

23. $2x^2 - 28x + 96 = 2(x^2 - 14x + 48)$
$= 2(x - 8)(x - 6)$

24. $4x^2 - 44x + 120 = 4(x^2 - 11x + 30)$
$= 4(x - 6)(x - 5)$

25. $4x^2 + 7x - 15 = (4x - 5)(x + 3)$

26. $12x^2 + 11x - 5 = 12x^2 - 4x + 15x - 5$
$= 4x(3x-1) + 5(3x-1)$
$= (3x-2)(2x-3)$

27. $15x^2 + 7x - 4 = (3x-1)(5x+4)$

28. $6x^2 - 13x + 6 = 6x^2 - 4x - 9x + 6$
$= 2x(3x-2) - 3(3x-2)$
$= (3x-2)(2x-3)$

29. $2x^2 - x - 3 = (2x-3)(x+1)$

30. $3x^2 + 2x - 8 = (3x-4)(x+2)$

31. $20x^2 + 48x - 5 = (10x-1)(2x+5)$

32. $20x^2 + 21x - 5 = 20x^2 - 4x + 25x - 5$
$= 4x(5x-1) + 5(5x-1)$
$= (5x-1)(4x+5)$

33. $4a^2 - 11a - 3 = (4a+1)(a-3)$

34. $4a^2 - a - 3 = 4a^2 + 3a - 4a - 3$
$= a(4a+3) - 1(4a+3)$
$= (4a+3)(a-1)$

35. $6x^2 + 4x - 10 = 2(3x^2 + 2x - 5)$
$= 2(3x+5)(x-1)$

36. $6x^2 - 4x - 10 = 2(3x^2 - 2x - 5)$
$= 2(3x^2 - 5x + 3x - 5)$
$= 2[x(3x-5) + 1(3x-5)]$
$= 2(3x-5)(x+1)$

37. $4x^2 - 26x + 30 = 2(2x^2 - 13x + 15)$
$= 2(2x-3)(x-5)$

38. $4x^2 - 20x - 144 = 4(x^2 - 5x - 36)$
$= 4(x-9)(x+4)$

39. $12x^2 + 5x - 3 = (4x+3)(3x-1)$

40. $16x^2 + 14x - 30 = 2(8x^2 + 7x - 15)$
$= 2(8x^2 - 8x + 15x - 15)$
$= 2[8x(x-1) + 15(x-1)]$
$= 2(x-1)(8x+15)$

41. $6x^2 - 19xy + 10y^2 = (2x-5y)(3x-2y)$

42. $6x^2 - 32xy + 10y^2 = 2(3x^2 - 16xy + 5y^2)$
$= 2(3x^2 - 15xy - xy + 5y^2)$
$= 2[3x(x-5y) - y(x-5y)]$
$= 2(x-5y)(3x-y)$

43. $49x^2 - y^2 = (7x)^2 - (y)^2$
$= (7x+y)(7x-y)$

44. $16x^2 - 36y^2 = 4(4x^2 - 9y^2)$
$= 4[(2x)^2 - (3y)^2]$
$= 4(2x - 3y)(2x + 3y)$

45. $9x^2 - 12x + 4 = (3x - 2)^2$

46. $64x^2 - 1 = (8x)^2 - (1)^2 = (8x + 1)(8x - 1)$

47. $25x^2 - 36 = (5x)^2 - (6)^2$
$= (5x + 6)(5x - 6)$

48. $100x^2 - 9 = (10x)^2 - 3^2 = (10x - 3)(10x + 3)$

49. $1 - 49x^2 = (1)^2 - (7x)^2$
$= (1 + 7x)(1 - 7x)$

50. $4 - 49x^2 = 2^2 - (7x)^2 = (2 - 7x)(2 + 7x)$

51. $36x^2 + 12x + 1 = (6x + 1)^2$

52. $25x^2 - 20x + 4 = (5x)^2 - 2(5x)(2) + 2^2$
$= (5x - 2)^2$

53. $16x^2 - 24xy + 9y^2 = (4x - 3y)^2$

54. $49x^2 - 28xy + 4y^2$
$= (7x)^2 - 2(7x)(2y) + (2y)^2$
$= (7x - 2y)^2$

55. $2x^2 - 18 = 2(x^2 - 9)$
$= 2(x + 3)(x - 3)$

56. $3x^2 - 75 = 3(x^2 - 25) = 3(x^2 - 5^2)$
$= 3(x - 5)(x + 5)$

57. $8x^2 + 40x + 50 = 2(4x^2 + 20x + 25)$
$= 2(2x + 5)^2$

58. $50x^2 - 120x + 72 = 2(25x^2 - 60x + 36)$
$= 2(5x - 6)^2$

59. $4x^2 - 9y^2 = (2x)^2 - (3y)^2$
$= (2x + 3y)(2x - 3y)$

60. $x^2 + 6x + 9 = x^2 + 2(x)(3) + 3^2$
$= (x + 3)^2$

61. $x^2 - 9x + 18 = (x - 6)(x - 3)$

62. $x^2 + 13x - 30 = (x + 15)(x - 2)$

63. $6x^2 + x - 7 = (6x + 7)(x - 1)$

64. $10x^2 + x - 2 = 10x^2 - 4x + 5x - 2$
$= 2x(5x - 2) + 1(5x - 2)$
$= (5x - 2)(2x + 1)$

65. $12x+16=4(3x+4)$

66. $8x^2y^2-4xy=4xy(2xy-1)$

67. $50x^3y^2+30x^2y^2-10x^2y^2$
$=50x^3y^2+20x^2y^2$
$=10x^2y^2(5x+2)$

68. $26a^3b-13ab^3+52a^2b^4$
$=13ab(2a^2-b^2+4ab^3)$

69. $x^3-16x^2+64x=x(x^2-16x+64)$
$=x(x-8)^2$

70. $2x^2+40x+200=2(x^2+20x+100)$
$=2[x^2+2(x)(10)+10^2]$
$=2(x+10)^2$

71. $3x^2-18x+27=3(x^2-6x+9)$
$=3(x-3)^2$

72. $25x^3-60x^2+36x=x(25x^2-60x+36)$
$=x[(5x)^2-2(5x)(6)+6^2]$
$=x(5x-6)^2$

73. $7x^2-9x-10=(7x+5)(x-2)$

74. $4x^2-13x-12=(4x+3)(x-4)$

75. $9x^3y-4xy^3=xy(9x^2-4y^2)$
$=xy(3x+2y)(3x-2y)$

76. $3x^3a^3-11x^4a^2-20x^5a$
$=x^3a(3a^2-11xa-20x^2)$
$=x^3a(3a^2+4xa-15xa-20x^2)$
$=x^3a[a(3a+4x)-5x(3a+4x)]$
$=x^3a(3a+4x)(a-5x)$

77. $12a^2+14ab-10b^2=2(6a^2+7ab-5b^2)$
$=2(3a+5b)(2a-b)$

78. $16a^2-40ab+25b^2$
$=(4a)^2-2(4a)(5b)+(5b)^2$
$=(4a-5b)^2$

79. $7a-7-ab+b=7(a-1)-b(a-1)$
$=(a-1)(7-b)$

80. $3d-4-3cd+4c=3d-3cd-4+4c$
$=3d(1-c)-4(1-c)$
$=(1-c)(3d-4)$

81. $2x-1+2bx-b=(2x-1)+b(2x-1)$
$=(2x-1)(1+b)$

82. $5xb-35x+4by-28y$
$=5x(b-7)+4y(b-7)$
$=(b-7)(5x+4y)$

83. $2a^2x-15ax+7x=x(2a^2-15a+7)$
$=x(2a-1)(a-7)$

84. $x^5-17x^3+16x=x(x^4-17x^2+16)$
$=x(x^2-16)(x^2-1)$
$=x(x^2-4^2)(x^2-1^2)$
$=x(x-4)(x+4)(x-1)(x+1)$

85. $x^4-81y^{12}=(x^2)^2-(9y^6)^2$
$=(x^2+9y^6)(x^2-9y^6)$
$=(x^2+9y^6)(x+3y^3)(x-3y^3)$

86. $6x^4-x^2-15=6x^4-10x^2+9x^2-15$
$=2x^2(3x^2-5)+3(3x^2-5)$
$=(3x^2-5)(2x^2+3)$

87. $28yz-16xyz+x^2yz=yz(28-16x+x^2)$
$=yz(14-x)(2-x)$

88. $12x^3+17x^2+6x=x(12x^2+17x+6)$
$=x(12x^2+8x+9x+6)$
$=x[4x(3x+2)+3(3x+2)]$
$=x(3x+2)(4x+3)$

89. $16w^2-2w-5=(2w+1)(8w-5)$

90. $12w^2-12w+3=3(4w^2-4w+1)$
$=3[(2w)^2-2(2w)(1)+1^2]$
$=3(2w-1)^2$

91. $4y^3+10y^2-6y=2y(2y^2+5y-3)$
$=2y(2y-1)(y+3)$

92. $10y^2+33y-7=10y^2-2y+35y-7$
$=2y(5y-1)+7(5y-1)$
$=(5y-1)(2y+7)$

93. $8y^{10}-16y^8=8y^8(y^2-2)$

94. $49x^4-49=49(x^4-1)$
$=49[(x^2)^2-1^2]$
$=49(x^2-1)(x^2+1)$
$=49(x^2-1^2)(x^2+1)$
$=49(x-1)(x+1)(x^2+1)$

95. $x^2+13x+54$ is prime.

96. $8x^2-19x-6$ is prime.

97. $8y^5+4y^3-60y=4y(2y^4+y^2-15)$
$=4y(2y^2-5)(y^2+3)$

98. $9xy^2 + 3xy - 42x = 3x(3y^2 + y - 14)$
$= 3x(3y^2 + 7y - 6y - 14)$
$= 3x[y(3y+7) - 2(3y+7)]$
$= 3x(3y+7)(y-2)$

99. $16x^4y^2 - 56x^2y + 49 = (4x^2y - 7)^2$

100. $128x^3y - 2xy = 2xy(64x^2 - 1)$
$= 2xy[(8x)^2 - 1^2]$
$= 2xy(8x-1)(8x+1)$

101. $2ax + 5a - 10b - 4bx$
$= a(2x+5) - 2b(2x+5)$
$= (2x+5)(a-2b)$

102. $2x^3 - 9 + x^2 - 18x = 2x^3 + x^2 - 18x - 9$
$= x^2(2x+1) - 9(2x+1)$
$= (2x+1)(x^2 - 9)$
$= (2x+1)(x^2 - 3^2)$
$= (2x+1)(x-3)(x+3)$

103. $x^2 - 3x - 18 = 0$
$(x-6)(x+3) = 0$
$x - 6 = 0 \quad x + 3 = 0$
$x = 6 \quad x = -3$

104. $x^2 + 6x - 27 = 0$
$(x+9)(x-3) = 0$
$x + 9 = 0 \quad x - 3 = 0$
$x = -9 \quad x = 3$

105. $5x^2 = 2x - 7x^2$
$12x^2 - 2x = 0$
$2x(6x-1) = 0$
$2x = 0 \quad 6x - 1 = 0$
$x = 0 \quad x = \frac{1}{6}$

106. $8x^2 + 5x = 2x^2 - 6x$
$6x^2 + 11x = 0$
$x(6x+11) = 0$
$6x + 11 = 0 \quad x = 0$
$6x = -11 \quad x = 0$
$x = -\frac{11}{6}$

107. $2x^2 + 9x - 5 = 0$
$(2x-1)(x+5) = 0$
$2x - 1 = 0 \quad x + 5 = 0$
$x = \frac{1}{2} \quad x = -5$

108. $x^2 + 11x + 24 = 0$
$(x+8)(x+3) = 0$
$x + 8 = 0 \quad x + 3 = 0$
$x = -8 \quad x = -3$

109. $x^2 + 14x + 45 = 0$
$(x+9)(x+5) = 0$

$$x+9=0 \qquad x+5=0$$
$$x=-9 \qquad x=-5$$

110. $5x^2 = 7x+6$

$$5x^2-7x-6=0$$
$$(5x+3)(x-2)=0$$
$$5x+3=0 \qquad x-2=0$$
$$5x=-3 \qquad x=2$$
$$x=-\frac{3}{5} \qquad x=2$$

111. $3x^2+6x=2x^2-9$

$$x^2+6x+9=0$$
$$(x+3)^2=0$$
$$x+3=0$$
$$x=-3$$

112. $4x^2+9x-9=0$

$$(4x-3)(x+3)=0$$
$$4x-3=0 \qquad x+3=0$$
$$4x=3 \qquad x=-3$$
$$x=\frac{3}{4} \qquad x=-3$$

113. $5x^2-11x+2=0$

$$(5x-1)(x-2)=0$$
$$5x-1=0 \qquad x-2=0$$
$$x=\frac{1}{5} \qquad x=2$$

114. $x = \text{altitude}$

$x+3=\text{base}$

$$A=\frac{1}{2}ab$$
$$35=\frac{1}{2}x(x+3)$$
$$70=x^2+3x$$
$$0=x^2+3x-70$$
$$0=(x+10)(x-7)$$

$$x+10=0 \qquad x-7=0$$
$$x=-10 \qquad x=-7$$

Since length cannot be negative, $x=7$ and $x+3=7+3=10$.

altitude: 7 cm; base 10 cm

115. $x = \text{width}$

$2x+1=\text{length}$

$$A=\text{length}\times\text{width}$$
$$105=x(2x+1)$$
$$105=2x^2+x$$
$$0=2x^2+x-105$$
$$0=(2x+15)(x-7)$$

$$2x+15=0 \quad \text{or} \quad x-7=0$$
$$x=-\frac{15}{2} \qquad x=7$$
$$2x+1=15$$

Length cannot be negative so width = 7 feet and length = 15 feet.

116. $h = -16t^2 + 80t + 96$

$0 = -16t^2 + 80t + 96$

$0 = -16(t^2 - 5t - 6)$

$0 = -16(t-6)(t+1)$

$t - 6 = 0 \qquad t + 1 = 0$

$t = 6 \qquad t = -1$

Since the time must be positive, $t = 6$ seconds.

117. $480 = -5x^2 + 100x$

$5x^2 - 100x + 480 = 0$

$5(x^2 - 20x + 96) = 0$

$x - 12 = 0$ or $x - 8 = 0$

$x = 12 \qquad x = 8$

The current is 12 amperes or 8 amperes.

Putting Your Skills to Work

1. $\frac{6}{21} \div \frac{2}{21}$

$= \frac{6}{21}\left(\frac{21}{2}\right)$

= 3 times greater

2. $\frac{5}{21} \div \frac{1}{21}$

$= \frac{5}{21} \cdot \frac{21}{1}$

= 5 times greater

3. The denominator of the depreciation factors is $1 + 2 + 3 + 4 + 5 = 15$

Useable value $= 16,750 - 5,750 = 11,000$

Depreciation by year is;

1st year: $\frac{5}{15}(11,000) = \$3667$

2nd year: $\frac{4}{15}(11,000) = \$2933$

3rd year: $\frac{3}{15}(11,000) = \$2200$

4th year: $\frac{2}{15}(11,000) = \$1467$

5th year: $\frac{1}{15}(11,000) = \$733$

4. The denominator of the depreciation factors is $1 + 2 + 3 + 4 + 5 + 6 + 7 = 28$

Useable value $= 26,250 - 12,250 = 14,000$

Depreciation by year is;

1st year: $\frac{7}{28}(14,000) = \$3500$

2nd year: $\frac{6}{28}(14,000) = \$3000$

3rd year: $\frac{5}{28}(14,000) = \$2500$

4th year: $\frac{4}{28}(14,000) = \$2000$

5th year: $\frac{3}{28}(14,000) = \$1500$

6th year: $\frac{2}{28}(14,000) = \$1000$

7th year: $\frac{1}{28}(14,000) = \$500$

5. $V = 5700 - 30x$
$V = 30(190 - x),\ x = 80 - 20 = 60$
$V = 30(190 - 60)$
$= \$3900$

6. $V = 30(190 - x),\ V = 4500$
$4500 = 30(190 - x)$
$150 = 190 - x$
$x = 40$
$40 + 20 = 60$
It would have 60,000 miles on it.

Chapter 6 Test

1. $x^2 + 12x - 28 = (x + 14)(x - 2)$

2. $25x^2 - 49y^2 = (5x)^2 - (7y)^2$
$= (5x + 7y)(5x - 7y)$

3. $10x^2 + 27x + 5 = 10x^2 + 2x + 25x + 5$
$= 2x(5x + 1) + 5(5x + 1)$
$= (5x + 1)(2x + 5)$

4. $9a^2 - 30ab + 25b^2$
$= (3a)^2 - 2(3a)(5b) + (5b)^2$
$= (3a - 5b)^2$

5. $7x - 9x^2 + 14xy = x(7 - 9x + 14y)$

6. $3x^2 - 4wy - 2wx + 6xy$
$= 3x^2 - 2wx + 6xy - 4wy$
$= x(3x - 2w) + 2y(3x - 2w)$
$= (3x - 2w)(x + 2y)$

7. $6x^3 - 20x^2 + 16x = 2x(3x^2 - 10x + 8)$
$= 2x(3x^2 - 6x - 4x + 8)$
$= 2x[3x(x - 2) - 4(x - 2)]$
$= 2x(x - 2)(3x - 4)$

8. $5a^2c - 11abc + 2b^2c = c(5a^2 - 11ab + 2b^2)$
$= c(5a^2 - 10ab - ab + 2b^2)$
$= c[5a(a - 2b) - b(a - 2b)]$
$= c(a - 2b)(5a - b)$

9. $100x^4 - 16y^4 = 4(25x^4 - 4y^4)$
$= 4\left[(5x^2)^2 - (2y^2)^2\right]$
$= 4(5x^2 + 2y^2)(5x^2 - 2y^2)$

10. $9x^2 - 15xy + 4y^2 = 9x^2 - 12xy - 3xy + 4y^2$
$= 3x(3x - 4y) - y(3x - 4y)$
$= (3x - 4y)(3x - y)$

11. $7x^2 - 42x = 7x(x - 6)$

12. $36x^2 + 1 = (6x)^2 + (1)^2$
It is prime.

13. $3x^2+5x+1$ is prime.

14. $60xy^2-20x^2y-45y^3$
$$=-5y\left(-12xy+4x^2+9y^2\right)$$
$$=-5y\left(4x^2-12xy+9y^2\right)$$
$$=-5y\left[(2x)^2-2(2x)(3y)+(3y)^2\right]$$
$$=-5y(2x-3y)^2$$

15. $81x^2-1=(9x)^2-(1)^2$
$$=(9x+1)(9x-1)$$

16. $x^{16}-1=\left(x^8\right)^2-1^2=\left(x^8-1\right)\left(x^8+1\right)$
$$=\left[\left(x^4\right)^2-1^2\right]\left(x^8+1\right)$$
$$=\left(x^4-1\right)\left(x^4+1\right)\left(x^8+1\right)$$
$$=\left[\left(x^2\right)^2-1^2\right]\left(x^4+1\right)\left(x^8+1\right)$$
$$=\left(x^2-1\right)\left(x^2+1\right)\left(x^4+1\right)\left(x^8+1\right)$$
$$=\left(x^2-1^2\right)\left(x^2+1\right)\left(x^4+1\right)\left(x^8+1\right)$$
$$=(x-1)(x+1)\left(x^2+1\right)\left(x^4+1\right)\left(x^8+1\right)$$

17. $2ax+6a-5x-15=2a(x+3)-5(x+3)$
$$=(x+3)(2a-5)$$

18. $aw^2-8b+2bw^2-4a$
$$=aw^2-4a+2bw^2-8b$$
$$=a\left(w^2-4\right)+2b\left(w^2-4\right)$$
$$=\left(w^2-4\right)(a+2b)$$
$$=(w-2)(w+2)(a+2b)$$

19. $3x^2-3x-90=3\left(x^2-x-30\right)$
$$=3(x-6)(x+5)$$

20. $2x^3-x^2-15x=x\left(2x^2-x-15\right)$
$$=x\left(2x^2-6x+5x-15\right)$$
$$=x\left[2x(x-3)+5(x-3)\right]$$
$$=x(x-3)(2x+5)$$

21. $x^2+14x+45=0$
$$(x+9)(x+5)=0$$
$x+9=0$ $\quad$ $x+5=0$
$x=-9$ $\quad$ $x=-5$

22. $14+3x(x+2)=-7x$
$$14+3x^2+6x=-7x$$
$$3x^2+13x+14=0$$
$$(3x+7)(x+2)=0$$
$3x+7=0$ $\quad$ $x+2=0$
$3x=-7$ $\quad$ $x=-2$
$x=-\frac{7}{3}$ $\quad$ $x=-2$

23. $2x^2+x-10=0$
$$(2x+5)(x-2)=0$$
$2x+5=0$ $\quad$ $x-2=0$
$x=-\frac{5}{2}$ $\quad$ $x=2$

24.
$$\begin{aligned} x &= \text{width} \\ 2x-1 &= \text{length} \\ x(2x-1) &= 91 \\ 2x^2-x &= 91 \\ 2x^2-x-91 &= 0 \\ (2x+13)(x-7) &= 0 \end{aligned}$$

$$2x+13=0 \qquad x-7=0$$
$$2x=-13 \qquad x=7$$
$$x=-\frac{13}{2}$$

Since the width cannot be negative, $x = 7$ and $2x-1=2(7)-1=13$.
width = 7 miles, length = 13 miles.

Cumulative Test for Chapters 1 - 6

1. $-3.2-6.4+0.24-1.8+0.8$
$=-11.4+1.04$
$=-10.36$

2. $\left(-2x^3y^4\right)\left(-4xy^6\right)=(-2)(-4)x^{3+1}y^{4+6}$
$=8x^4y^{10}$

3. $(-3)^4=81$

4. $(9x-4)(3x+2)$
$=27x^2+18x-12x-8$
$=27x^2+6x-8$

5. $\left(2x^2-6x+1\right)(x-3)$
$=2x^3-6x^2+x-6x^2+18x-3$
$=2x^3-12x^2+19x-3$

6. $3x-4\ge 6x+5$
$-4\ge 3x+5$
$-9\ge 3x$
$-3\ge x$

7. $3x-(7-5x)=3(4x-5)$
$$\begin{aligned} 3x-7+5x &= 12x-15 \\ -7+8x &= 12x-15 \\ 8 &= 4x \\ 2 &= x \end{aligned}$$

8.
$$\begin{aligned} \frac{1}{2}x-3 &= \frac{1}{4}(3x+3) \\ 4\left(\frac{1}{2}x-3\right) &= 4\left[\frac{1}{4}(3x+3)\right] \\ 2x-12 &= 3x+3 \\ -12 &= x+3 \\ -15 &= x \end{aligned}$$

9.
$$\begin{aligned} s &= \frac{1}{2}(2a+3t) \\ 2s &= 2a+3t \\ 2s-2a &= 3t \\ \frac{2s-2a}{3} &= t \end{aligned}$$

10. $(-2,3)$ and $(0,5)$
$$m=\frac{5-3}{0-(-2)}=\frac{2}{2}=1$$

11. $h(x) = \dfrac{5}{x-7}$

$h(-3) = \dfrac{5}{-3-7} = \dfrac{5}{-10} = -\dfrac{1}{2}$

12. $2x + \dfrac{2}{3}y = -8 \qquad (1)$

$\dfrac{1}{5}x \quad + y = 2 \qquad (2)$

Multiply (1) by 3 and (2) by -30

$$\begin{array}{r} 6x + 2y = -24 \\ \underline{-6x - 30y = -60} \\ -28y = -84 \\ y = 3 \end{array}$$

Substitute 3 for y in (1)

$2x + \dfrac{2}{3}(3) = -8$

$2x = -10$

$x = -5$

$(-5, 3)$

13. $6x^2 - 5x + 1 = 6x^2 - 3x - 2x + 1$

$= 3x(2x-1) - 1(2x-1)$

$= (3x+2)(3x-1)$

14. $6x^2 + 5x - 4 = (3x+4)(2x-1)$

15. $9x^2 + 3x - 2 = 9x^2 + 6x - 3x - 2$

$= 3x(3x+2) - 1(3x+2)$

$= (3x+2)(3x-1)$

16. $121x^2 - 64y^2 = (11x)^2 - (8y)^2$

$= (11x + 8y)(11x - 8y)$

17. $4x + 120 - 80x^2 = -80x^2 + 4x + 120$

$= -4(20x^2 - x - 30)$

$= -4(20x^2 + 24x - 25x - 30)$

$= -4[4x(5x+6) - 5(5x+6)]$

$= -4(5x+6)(4x-5)$

18. $x^2 + 5x + 9$ is prime.

19. $16x^3 + 40x^2 + 25x = x(16x^2 + 40x + 25)$

$= x\left[(4x)^2 + 2(4x)(5) + 5^2\right]$

$= x(4x+5)^2$

20. $81x^4 - 16b^4 = (9x^2)^2 - (4b^2)^2$

$= (9x^2 + 4b^2)(9x^2 - 4b^2)$

$= (9x^2 + 4b^2)(3x + 2b)(3x - 2b)$

21. $2ax - 4bx + 3a - 6b$

$= 2x(a - 2b) + 3(a - 2b)$

$= (a - 2b)(2x + 3)$

22. $x^4 + 8x^2 + 15 = (x^2 + 5)(x^2 + 3)$

23. $x^2 + 5x - 24 = 0$

$(x+8)(x-3) = 0$

$x + 8 = 0 \qquad x - 3 = 0$

$x = -8 \qquad x = 3$

24. $3x^2 - 11x + 10 = 0$

$$(3x-5)(x-2) = 0$$

$$3x - 5 = 0 \qquad x - 2 = 0$$

$$x = \frac{5}{3} \qquad x = 2$$

25. $x = \text{base}$

$$2x + 7 = \text{altitude}$$

$$\frac{1}{2}x(2x+7) = 57$$

$$x(2x+7) = 114$$

$$2x^2 + 7x = 114$$

$$2x^2 + 7x - 114 = 0$$

$$(2x+19)(x-6) = 0$$

$$2x + 19 = 0 \qquad x - 6 = 0$$

$$2x = -19 \qquad x = 6$$

$$x = -\frac{19}{2}$$

Since length cannot be negative, $x = 6$ and $2x + 7 = 2(6) + 7 = 19$. The base is 6 miles and the altitude is 19 miles.

Chapter 7

Pretest Chapter 7

1. $$\frac{6a-4b}{2b-3a}=\frac{-2(-3a+2b)}{2b-3a}=\frac{-2(2b-3a)}{2b-3a}$$
$$=\frac{-2(2b-3a)}{2b-3a}=-2$$

2. $$\frac{x^2-x-6}{2x^2+7x+6}$$
$$=\frac{(x-3)(x+2)}{(2x+3)(x+2)}$$
$$=\frac{x-3}{2x+3}$$

3. $$\frac{a^2b+2ab^2}{2a^3+3a^2b-2ab^2}=\frac{ab(a+2b)}{a(2a^2+3ab-2b^2)}$$
$$=\frac{ab(a+2b)}{a(2a-b)(a+2b)}=\frac{ab(a+2b)}{a(2a-b)(a+2b)}$$
$$=\frac{b}{2a-b}$$

4. $$\frac{4a^2-b^2}{6a-6b}\cdot\frac{3a-3b}{6a+3b}$$
$$=\frac{(2a+b)(2a-b)}{6(a-b)}\cdot\frac{3(a-b)}{3(2a+b)}$$
$$=\frac{2a-b}{6}$$

5. $$\frac{x^2-6x+9}{x^2-x-6}\div\frac{x^2+2x-15}{x^2+2x}$$
$$=\frac{x^2-6x+9}{x^2-x-6}\cdot\frac{x^2+2x}{x^2+2x-15}$$
$$=\frac{(x-3)(x-3)}{(x-3)(x+2)}\cdot\frac{x(x+2)}{(x+5)(x-3)}$$
$$=\frac{(x-3)(x-3)}{(x-3)(x+2)}\cdot\frac{x(x+2)}{(x+5)(x-3)}$$
$$=\frac{x}{x+5}$$

6. $$\frac{x^2+5x+6}{3x^2+8x+4}\cdot\frac{6x^2-11x-10}{2x^2+x-15}$$
$$=\frac{(x+2)(x+3)}{(3x+2)(x+2)}\cdot\frac{(3x+2)(2x-5)}{(2x-5)(x+3)}$$
$$=1$$

7. $$\frac{xy+3y}{x^2-x}\div\frac{x+3}{x}=\frac{xy+3y}{x^2-x}\cdot\frac{x}{x+3}$$
$$=\frac{y(x+3)}{x(x-1)}\cdot\frac{x}{x+3}=\frac{y}{x-1}$$

8. $$\frac{3y+1}{y+2}+\frac{5}{y+2}$$
$$=\frac{3y+1+5}{y+2}$$
$$=\frac{3y+6}{y+2}$$
$$=\frac{3(y+2)}{y+2}=3$$

9. $\frac{2y-1}{2y^2+y-3}-\frac{2}{y-1}=\frac{2y-1}{(2y+3)(y-1)}-\frac{2}{y-1}$

$=\frac{2y-1}{(2y+3)(y-1)}-\frac{2(2y+3)}{(y-1)(2y+3)}$

$=\frac{2y-1-2(2y+3)}{(2y+3)(y-1)}=\frac{2y-1-4y-6}{(2y+3)(y-1)}$

$=\frac{-2y-7}{(2y+3)(y-1)}$

10. $\frac{4x+10}{x^2-25}-\frac{3}{x-5}$

$=\frac{4x+10}{(x+5)(x-5)}-\frac{3}{x-5}$

$=\frac{4x+10}{(x+5)(x-5)}-\frac{3}{x-5}\cdot\frac{x+5}{x+5}$

$=\frac{4x+10}{(x+5)(x-5)}-\frac{3x+15}{(x+5)(x-5)}$

$=\frac{4x+10-(3x+15)}{(x+5)(x-5)}$

$=\frac{4x+10-3x-15}{(x+5)(x-5)}$

$=\frac{x-5}{(x+5)(x-5)}$

$=\frac{1}{x+5}$

11. $\frac{x+3}{x^2-6x+9}+\frac{2x+3}{3x^2-9x}$

$=\frac{x+3}{(x-3)(x-3)}+\frac{2x+3}{3x(x-3)}$

$=\frac{3x(x+3)}{3x(x-3)(x-3)}+\frac{(2x+3)(x-3)}{3x(x-3)(x-3)}$

$=\frac{3x^2+9x+2x^2-3x-9}{3x(x-3)(x-3)}$

$=\frac{5x^2+6x-9}{3x(x-3)(x-3)}$

$=\frac{5x^2+6x-9}{3x(x-3)^2}$

12. $\frac{\frac{2}{a}-\frac{3}{a^2}}{5+\frac{1}{a}}$

$=\frac{\frac{2}{a}-\frac{3}{a^2}}{5+\frac{1}{a}}\cdot\frac{a^2}{a^2}$

$=\frac{2a-3}{5a^2+a}$

13. $\frac{\frac{a}{a+1}-\frac{2}{a}}{3a}=\frac{\frac{a(a)}{a(a+1)}-\frac{2(a+1)}{a(a+1)}}{3a}$

$=\frac{\frac{a^2-2(a+1)}{a(a+1)}}{3a}=\frac{\frac{a^2-2a-2}{a(a+1)}}{3a}$

$=\frac{a^2-2a-2}{a(a+1)}\cdot\frac{1}{3a}=\frac{a^2-2a-2}{3a^2(a+1)}$

14. $\frac{\frac{x-y}{x}-\frac{x+y}{y}}{\frac{x-y}{x}+\frac{x+y}{y}}$

$=\frac{\frac{x-y}{x}-\frac{x+y}{y}}{\frac{x-y}{x}+\frac{x+y}{y}}\cdot\frac{xy}{xy}$

$=\frac{y(x-y)-x(x+y)}{y(x-y)+x(x+y)}$

$=\frac{xy-y^2-x^2-xy}{xy-y^2+x^2+xy}$

$=\frac{-x^2-y^2}{x^2+2xy-y^2}$

15. $\frac{5}{2x}=2-\frac{2x}{x+1}$

$$(2x)(x+1)\left(\frac{5}{2x}\right)$$
$$=2x(x+1)(2)-(2x)(x+1)\left(\frac{2x}{x+1}\right)$$
$$(x+1)(5)=\left(2x^2+2x\right)(2)-(2x)(2x)$$
$$5x+5=4x^2+4x-4x^2$$
$$5x+5=4x$$
$$5=-x$$
$$-5=x$$

16. $\frac{24}{x^2-3x-10}=\frac{2}{x-5}+\frac{3}{x+2}$

$$\frac{24}{(x-5)(x+2)}=\frac{2}{x-5}+\frac{3}{x+2}$$
$$(x+2)(x-5)\left[\frac{24}{(x-5)(x+2)}\right]$$
$$=(x+2)(x-5)\left(\frac{2}{x-5}+\frac{3}{x+2}\right)$$
$$24=2(x+2)+3(x-5)$$
$$24=2x+4+3x-15$$
$$24=5x-11$$
$$35=5x$$
$$7=x$$

17. $\frac{5}{x}=\frac{7}{13}$

$$65=7x$$
$$\frac{65}{7}=x$$
$$x=9.3$$

18. $\frac{\text{U.S.}}{\text{British}}:\frac{3.10}{2}=\frac{x}{170}$

$$2x=527$$
$$x=263.50$$
$263.50

7.1 Exercises

1. $\frac{3a-9b}{a-3b}=\frac{3(a-3b)}{a-3b}=3$

3. $\frac{6x+18}{x^2+3x}=\frac{6(x+3)}{x(x+3)}=\frac{6}{x}$

5. $\frac{2x-8}{x^2-8x+16}=\frac{2(x-4)}{(x-4)(x-4)}=\frac{2}{x-4}$

7. $\frac{xy\left(x+y^2\right)}{x^2y^2}=\frac{x+y^2}{xy}$

9. $\frac{x^2+x-2}{x^2-x}=\frac{(x+2)(x-1)}{x(x-1)}=\frac{x+2}{x}$

11. $\frac{x^2-3x-10}{3x^2+5x-2}=\frac{(x-5)(x+2)}{(3x-1)(x+2)}$

$$=\frac{x-5}{3x-1}$$

13. $\frac{x^3-8x^2+16x}{x^2+2x-24}$

$$=\frac{x\left(x^2-8x+16\right)}{(x+6)(x-4)}$$

$$= \frac{x(x-4)(x-4)}{(x+6)(x-4)}$$

$$= \frac{x(x-4)}{x+6}$$

15. $\frac{3x^2+7x-6}{x^2+7x+12} = \frac{(3x-2)(x+3)}{(x+4)(x+3)} = \frac{3x-2}{x+4}$

17. $\frac{3x^2-8x+5}{4x^2-5x+1} = \frac{(3x-5)(x-1)}{(4x-1)(x-1)} = \frac{3x-5}{4x-1}$

19. $\frac{5x^2-27x+10}{5x^2+3x-2} = \frac{(5x-2)(x-5)}{(5x-2)(x+1)}$

$$= \frac{x-5}{x+1}$$

21. $\frac{6-3x}{2x-4} = \frac{-3x+6}{2x-4} = \frac{-3(x-2)}{2(x-2)} = \frac{-3}{2}$

23. $\frac{2x^2-7x-15}{25-x^2} = \frac{(2x+3)(x-5)}{(5-x)(5+x)}$

$$= \frac{-2x-3}{x+5} = \frac{-2x-3}{5+x}$$

25. $\frac{(4x+5)^2}{8x^2+6x-5} = \frac{(4x+5)^2}{(4x+5)(2x-1)}$

$$= \frac{(4x+5)(4x+5)}{(4x+5)(2x-1)} = \frac{4x+5}{2x-1}$$

27. $\frac{2x^2+9x-18}{30-x-x^2} = \frac{2x^2+9x-18}{-(x^2+x-30)}$

$$= \frac{(2x-3)(x+6)}{-(x+6)(x-5)}$$

$$= \frac{2x-3}{-x+5}$$

29. $\frac{a^2+2ab-3b^2}{2a^2+5ab-3b^2} = \frac{(a+3b)(a-b)}{(2a-b)(a+3b)} = \frac{a-b}{2a-b}$

31. $\frac{9x^2-4y^2}{9x^2+12xy+4y^2} = \frac{(3x-2y)(3x+2y)}{(3x+2y)(3x+2y)}$

$$= \frac{3x-2y}{3x+2y}$$

33. $\frac{6x^4-9x^3-6x^2}{12x^3+42x^2+18x} = \frac{3x^2(2x^2-3x-2)}{6x(2x^2+7x+3)}$

$$= \frac{3x^2(2x+1)(x-2)}{6x(2x+1)(x+3)}$$

$$= \frac{x(x-2)}{2(x+3)}$$

Cumulative Review Problems

35. $(3x-7)^2 = (3x)^2 - 2(3x)(7) + (7)^2$

$$= 9x^2 - 42x + 49$$

37. $\left(-2ab^0c^3\right)^3 = (-2)^3\, a^3b^0c^9 = -8a^3c^9$

39. x = number who speak Spanish

$x - 10,000,000 = 322,000,000$

$x = 332,000,000$

332,000,000 people speak Spanish

The number who speak Chinese

Mandarin $= 2(332,000,000) + 221,000,000$

$= 664,000,000 + 221,000,000$

$= 885,000,000$

7.2 Exercises

1. Factor the numerator and denominator completely and divide out any common factors.

3. $$\frac{2x-10}{x-4}\cdot\frac{x^2+5x+4}{x^2-4x-5} = \frac{2(x-5)}{x-4}\cdot\frac{(x+4)(x+1)}{(x-5)(x+1)} = \frac{2(x+4)}{x-4}$$

5. $$\frac{x^2+2x}{6x}\cdot\frac{3x^2}{x^2-4} = \frac{x(x+2)}{6x}\cdot\frac{3x^2}{(x+2)(x-2)} = \frac{x^2}{2(x-2)}$$

7. $$\frac{x^2+3x-10}{x^2+x-20}\cdot\frac{x^2-3x-4}{x^2+4x+3} = \frac{(x-2)(x+5)}{(x-4)(x+5)}\cdot\frac{(x-4)(x+1)}{(x+3)(x+1)} = \frac{x-2}{x+3}$$

9. $$(6x-5)\div\frac{36x^2-25}{6x^2+17x+10} = \frac{6x-5}{1}\cdot\frac{6x^2+17x+10}{36x^2-25} = \frac{6x-5}{1}\cdot\frac{(6x+5)(x+2)}{(6x+5)(6x-5)} = x+2$$

11. $$\frac{3x^2+12xy+12y^2}{x^2+4xy+3y^2}\div\frac{4x+8y}{x+y} = \frac{3(x+2y)(x+2y)}{(x+3y)(x+y)}\div\frac{4(x+2y)}{x+y} = \frac{3(x+2y)(x+2y)}{(x+3y)(x+y)}\cdot\frac{x+y}{4(x+2y)} = \frac{3(x+2y)}{4(x+3y)}$$

13. $$\frac{x^2+5x-14}{(x-5)}\div\frac{x^2+12x+35}{15-3x} = \frac{(x+7)(x-2)}{(x-5)}\div\frac{(x+7)(x+5)}{3(5-x)} = \frac{(x+7)(x-2)}{(x-5)}\cdot\frac{3(5-x)}{(x+7)(x+5)} = \frac{-3(x-2)}{x+5}$$

15. $$\frac{(x+5)^2}{3x^2-7x+2}\cdot\frac{x^2-4x+4}{(x+5)} = \frac{(x+5)(x+5)}{(3x-1)(x-2)}\cdot\frac{(x-2)(x-2)}{(x+5)} = \frac{(x+5)(x-2)}{3x-1}$$

17. $$\frac{y^2+4y-12}{y^2+2y-24}\cdot\frac{y^2-16}{y^2+2y-8} = \frac{(y+6)(y-2)}{(y+6)(y-4)}\cdot\frac{(y+4)(y-4)}{(y+4)(y-2)} = 1$$

19. By definition, the denominator of a rational expression cannot have the value zero. So the original expression cannot have a replacement value of 2 (or the first denominator would be zero). If we multiply the first fraction by the reciprocal of the second fraction, then the denominator $x+7$ cannot be zero. Thus x also cannot have a replacement value of -7.

Cumulative Review Problems

21. $$\begin{aligned} 5x^2 - 7x + 11 &= 5x^2 - x + 2 \\ -7x + 11 &= -x + 2 \\ 11 &= 6x + 2 \\ 9 &= 6x \\ \frac{9}{6} &= x \\ \frac{3}{2} &= x \end{aligned}$$

23. Area of road $= 8981(90)$
Area of sidewalk $= 8981(10.5)$
Difference in area $= 8981(90) - 8981(10.5)$
$= 8981(79.5)$
Difference in cost $= 8981(79.5)x$
$= \$713,989.5x$

25. Let x = the price of a gallon of milk
$4x = 2x + 4.76$
$2x = 4.76$
$x = 2.38$
\$2.38 for a gallon of milk

7.3 Exercises

1. The LCD would be a product that contains each factor. However, any repeated factor in any one denominator must be repeated the greatest number of times it occurs in any one denominator. So the LCD would be $(x+5)(x+3)^2$.

3. $$\frac{x}{x+5} + \frac{2x+1}{5+x} = \frac{x+2x+1}{x+5} = \frac{3x+1}{x+5}$$

5. $$\frac{x}{x-4} - \frac{x+1}{x-4} = \frac{x-(x+1)}{x-4} = \frac{x-x-1}{x-4} = -\frac{1}{x-4}$$

7. $$\frac{8x+3}{5x+7} - \frac{6x+10}{5x+7} = \frac{8x+3-(6x+10)}{5x+7}$$
$$\frac{8x+3-6x-10}{5x+7} = \frac{2x-7}{5x+7}$$

9. $5a^2, a^3$

   LCD $= 5a^3$

11. $16x^2y^3 = 2^4x^2y^3$

    $56xy^4 = 2^3 \cdot 7xy^4$

    $\text{LCD} = 2^4 \cdot 7x^2y^4 = 112x^2y^4$

13. $x^2 - 16 = (x+4)(x-4)$

    $x - 4 = x - 4$

    $\text{LCD} = (x+4)(x-4) = x^2 - 16$

15. $16x^2 - 8x + 1 = (4x-1)^2$

    $4x^2 + 7x - 2 = (4x-1)(x+2)$

    $\text{LCD} = (4x-1)^2(x+2)$

17. $\frac{5}{x^2-2x+1} + \frac{3}{x-1}$

    $= \frac{5}{(x-1)(x-1)} + \frac{3}{x-1} \cdot \frac{x-1}{x-1}$

    $= \frac{5+3x-3}{(x-1)(x-1)}$

    $= \frac{3x+2}{(x-1)(x-1)}$

19. $\frac{2}{y-1} + \frac{2}{y+1}$

    $= \frac{2}{y-1} \cdot \frac{y+1}{y+1} + \frac{2}{y+1} \cdot \frac{y-1}{y-1}$

    $= \frac{2y+2+2y-2}{(y-1)(y+1)} = \frac{4y}{(y-1)(y+1)}$

21. $\frac{5}{2ab} + \frac{1}{2a+b}$

    $= \frac{5}{2ab} \cdot \frac{2a+b}{2a+b} + \frac{1}{2a+b} \cdot \frac{2ab}{2ab}$

    $= \frac{10a+5b+2ab}{2ab(2a+b)}$

23. $\frac{x-3}{4x} + \frac{6}{x^2} = \frac{x-3}{4x} \cdot \frac{x}{x} + \frac{6}{x^2} \cdot \frac{4}{4}$

    $= \frac{x^2-3x+24}{4x^2}$

25. $\frac{3}{x^2-x-12} + \frac{2}{x^2+x-20}$

    $= \frac{3}{(x-4)(x+3)} \cdot \frac{x+5}{x+5} + \frac{2}{(x-4)(x+5)} \cdot \frac{x+3}{x+3}$

    $= \frac{3x+15+2x+6}{(x-4)(x+3)(x+5)}$

    $= \frac{5x+21}{(x-4)(x+3)(x+5)}$

27. $\frac{3x+5}{x^2+4x+3} + \frac{-x+5}{x^2+2x-3}$

    $= \frac{3x+5}{(x+3)(x+1)} \cdot \frac{x-1}{x-1} + \frac{-x+5}{(x+3)(x-1)} \cdot \frac{x+1}{x+1}$

    $= \frac{3x^2+2x-5-x^2+4x+5}{(x+3)(x+1)(x-1)}$

    $= \frac{2x^2+6x}{(x+3)(x+1)(x-1)} = \frac{2x(x+3)}{(x+3)(x+1)(x-1)}$

    $= \frac{2x}{(x+1)(x-1)}$

29. $\frac{a+1}{2}-\frac{a-1}{3}=\frac{a+1}{2}\cdot\frac{3}{3}-\frac{a-1}{3}\cdot\frac{2}{2}$
$=\frac{3a+3-2a+2}{2\cdot 3}$
$=\frac{a+5}{6}$

31. $\frac{6}{3x-4}-\frac{5}{4x-3}$
$=\frac{6}{3x-4}\cdot\frac{4x-3}{4x-3}-\frac{5}{4x-3}\cdot\frac{3x-4}{3x-4}$
$=\frac{24x-18-15x+20}{(3x-4)(4x-3)}$
$=\frac{9x+2}{(3x-4)(4x-3)}$

33. $\frac{1}{x^2-2x}-\frac{5}{x^2-4x+4}$
$=\frac{1}{x(x-2)}\cdot\frac{x-2}{x-2}-\frac{5}{(x-2)^2}\cdot\frac{x}{x}$
$=\frac{x-2-5x}{x(x-2)^2}=\frac{-4x-2}{x(x-2)^2}$

35. $\frac{2x}{x^2+5x+6}-\frac{x+1}{x^2+2x-3}$
$=\frac{2x}{(x+2)(x+3)}\cdot\frac{x-1}{x-1}-\frac{x+1}{(x+3)(x-1)}\cdot\frac{x+2}{x+2}$
$=\frac{2x^2-2x-(x^2+3x+2)}{(x+3)(x+2)(x-1)}$
$=\frac{x^2-5x-2}{(x+3)(x+2)(x-1)}$

37. $\frac{7y}{y-3}-\frac{2y}{3-y}$
$=\frac{7y}{y-3}-\frac{2y}{3-y}\cdot\frac{-1}{-1}$
$=\frac{7y-(-2y)}{y-3}=\frac{7y+2y}{y-3}$
$=\frac{9y}{y-3}$

39. $\frac{y-23}{y^2-y-20}+\frac{2}{y-5}$
$=\frac{y-23}{(y-5)(y+4)}+\frac{2}{y-5}\cdot\frac{y+4}{y+4}$
$=\frac{y-23+2y+8}{(y-5)(y+4)}=\frac{3y-15}{(y-5)(y+4)}$
$=\frac{3(y-5)}{(y-5)(y+4)}=\frac{3}{y+4}$

41. $\frac{x}{x-4}+\frac{5}{x+4}+\frac{10}{x^2-16}$
$=\frac{x}{x-4}\cdot\frac{x+4}{x+4}+\frac{5}{x+4}\cdot\frac{x-4}{x-4}+\frac{10}{(x+4)(x-4)}$
$=\frac{x^2+4x+5x-20+10}{(x+4)(x-4)}$
$=\frac{x^2+9x-10}{(x+4)(x-4)}$

43. $\frac{2y}{3y^2-8y-3}+\frac{1}{6y-2y^2}$
$=\frac{2y}{(3y+1)(y-3)}-\frac{1}{2y^2-6y}$
$=\frac{2y}{(3y+1)(y-3)}\cdot\frac{2y}{2y}-\frac{1}{2y(y-3)}\cdot\frac{3y+1}{3y+1}$
$=\frac{4y^2-3y-1}{2y(y-3)(3y+1)}=\frac{(4y+1)(y-1)}{2y(3y+1)(y-3)}$

45. $\dfrac{8}{15x+10}-\dfrac{3}{6x+4}$

$=\dfrac{8}{5(3x+2)}\cdot\dfrac{2}{2}-\dfrac{3}{2(3x+2)}\cdot\dfrac{5}{5}$

$=\dfrac{16-15}{2\cdot 5(3x+2)}$

$=\dfrac{1}{10(3x+2)}$

Cumulative Review Problems

47. $5ax = 2(ay-3bc)$

$5ax = 2ay-6bc$

$5ax+6bc = 2ay$

$\dfrac{5ax+6bc}{2a} = y$

49. $x+2y=-6$

$2y=-x-6$

$y=-\dfrac{1}{2}x-3$

$m=-\dfrac{1}{2},\ b=-3$

51. Change $= 14.6\% - 3.5\% = 11.1\%$

$0.111(5{,}100{,}000) = 566{,}100$

566,100 more people were unemployed

Putting Your Skills to Work

1. $x = 2006-1999 = 7$

$y = 124+10.845x$

$y = 124+108.45(7)$

$\approx \$883$ billion

2. $x = 2008-1999 = 9$

$y = 124+33.18x$

$y = 124+33.18(9)$

$\approx \$423$ billion

3. $x = 2004-1999 = 5$

Moderate growth - slow growth

$= 124+33.18x-(124-37.27x)$

$= 70.45x$

For $x = 5$, the difference is

$70.45(5) \approx \$352$ billion

4. $x = 2007-1999 = 8$

Strong growth - slow growth

$= 124+108.45x-(124-37.27x)$

$= 145.72x$

For $x = 8$, the difference is

$145.72(8) \approx \$1166$ billion or

$1.166 trillion

5. $I = \dfrac{5.6x+10.8}{x+3}$

For 2005, $x = 5$

$I = \dfrac{5.6(5)+10.8}{5+3} = 4.85$

For 2001, $x = 1$

$$I = \frac{5.6(1)+10.8}{1+3} = 4.1$$

Increase from 2001 to 2005

$= 4.85 - 4.1$

$= 0.75$ hundred billion

$= \$75$ billion

6. $I = \dfrac{5.6x+10.8}{x+3}$

For 2012, $x = 12$

$$I = \frac{5.6(12)+10.8}{12+3} = 5.2$$

For 2009, $x = 9$

$$I = \frac{5.6(19)+10.8}{9+3} = 5.1$$

Increase from 2009 to 2012

$= 5.2 - 5.1$

$= 0.1$ hundred billion

$= \$10$ billion

7.4 Exercises

1. $\dfrac{\frac{5}{xy}}{\frac{7}{y}} = \dfrac{5}{xy} \cdot \dfrac{y}{7} = \dfrac{5}{7x}$

3. $\dfrac{\frac{1}{x}+\frac{1}{y}}{\frac{1}{xy}} = \dfrac{\frac{1}{x}+\frac{1}{y}}{\frac{1}{xy}} \cdot \dfrac{(xy)}{(xy)}$

$$= \frac{\frac{xy}{x}+\frac{xy}{x}}{\frac{xy}{xy}} = \frac{y+x}{1} = y+x$$

5. $\dfrac{\frac{1}{x}+y}{\frac{1}{y}+x} = \dfrac{\frac{1}{x}+y}{\frac{1}{y}+x} \cdot \dfrac{(xy)}{(xy)}$

$$= \frac{\frac{xy}{x}+xy^2}{\frac{xy}{y}+x^2y} = \frac{y+xy^2}{x+y^2}$$

$$= \frac{y(1+xy)}{x(1+xy)} = \frac{y}{x}$$

7. $\dfrac{1-\frac{9}{x^2}}{\frac{3}{x}+1} = \dfrac{1-\frac{9}{x^2}}{\frac{3}{x}+1} \cdot \dfrac{x^2}{x^2}$

$$= \frac{x^2 - \frac{9x^2}{x^2}}{\frac{3x^2}{x}+x^2} = \frac{x^2-9}{3x+x^2}$$

$$= \frac{(x+3)(x-3)}{x(3+x)} = \frac{x-3}{x}$$

9. $\dfrac{\frac{2}{x+1}-2}{3} = \dfrac{\frac{2}{x+1}-2}{3} \cdot \dfrac{(x+1)}{(x+1)}$

$$= \frac{\frac{2(x+1)}{(x+1)} - 2(x+1)}{3(x+1)} = \frac{2-2x-2}{3(x+1)} = \frac{-2x}{3(x+1)}$$

11. $\dfrac{a+\frac{3}{a}}{\frac{a^2+2}{3a}} = \dfrac{a+\frac{3}{a}}{\frac{a^2+2}{3a}} \cdot \dfrac{3a}{3a}$

$$= \frac{3a^2+\frac{9a}{a}}{\frac{(a^2+2)(3a)}{3a}} = \frac{3a^2+9}{a^2+2}$$

13. $\dfrac{y-\frac{4}{y}}{1+\frac{3}{2y+1}} = \dfrac{y-\frac{4}{y}}{1+\frac{3}{2y+1}} \cdot \dfrac{y(2y+1)}{y(2y+1)}$

$$= \frac{y^2(2y+1)-4(2y+1)}{y(2y+1)+3y}$$

$$= \frac{(2y+1)(y^2-4)}{2y^2+y+3y}$$

$$= \frac{(2y+1)(y+2)(y-2)}{2y^2+4y}$$
$$= \frac{(2y+1)(y+2)(y-2)}{2y(y+2)}$$
$$= \frac{(2y+1)(y-2)}{2y}$$

15. $$\frac{\frac{x}{6}-\frac{1}{3}}{\frac{2}{3x}+\frac{5}{6}} = \frac{\frac{x}{6}(6x)-\frac{1}{3}(6x)}{\frac{2}{3x}(6x)+\frac{5}{6}(6x)}$$
$$= \frac{x^2-2x}{4+5x}$$

17. $$\frac{\frac{1}{x^2-9}+\frac{2}{x+3}}{\frac{3}{x-3}} = \frac{\frac{1}{(x+3)(x-3)}+\frac{2}{x+3}}{\frac{3}{x-3}}$$
$$= \frac{\frac{1}{(x+3)(x-3)}+\frac{2}{x+3}}{\frac{3}{x-3}} \cdot \frac{(x+3)(x-3)}{(x+3)(x-3)}$$
$$= \frac{(1)\frac{(x+3)(x-3)}{(x+3)(x-3)}+2\frac{(x+3)(x-3)}{x+3}}{\frac{3(x+3)(x-3)}{x-3}}$$
$$= \frac{1+2(x-3)}{3(x+3)} = \frac{1+2x-6}{3(x+3)} = \frac{2x-5}{3(x+3)}$$

19. $$\frac{\frac{2}{y-1}+2}{\frac{2}{y+1}-2} = \frac{\frac{2}{y-1}+2}{\frac{2}{y+1}-2} \cdot \frac{(y-1)(y+1)}{(y-1)(y+1)}$$
$$= \frac{\frac{2(y-1)(y+1)}{y-1}+2(y-1)(y+1)}{\frac{2(y-1)(y+1)}{(y+1)}-2(y-1)(y+1)}$$
$$= \frac{2(y+1)[1+(y-1)]}{2(y-1)[1-(y+1)]} = \frac{(y+1)(y)}{(y-1)(1-y-1)}$$
$$= \frac{y(y+1)}{(y-1)(-y)} = \frac{y+1}{-y+1}$$

21. No expression in any denominator can be allowed to be zero, since division by zero is undefined. So -3, 5, and 0 are not allowable replacements for the variable x.

Cumulative Review Problems

23. $$P = 2(\ell + w)$$
$$P = 2\ell + 2w$$
$$P - 2\ell = 2w$$
$$\frac{P-2\ell}{2} = w$$

25. $x - 2y \le 8$

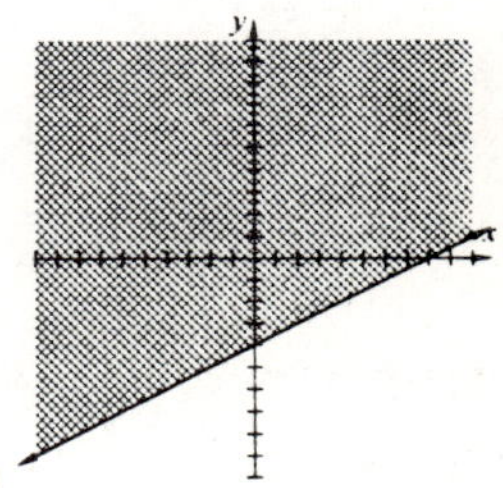

7.5 Exercises

1. $$\frac{x+1}{2x} = \frac{2}{3}$$
$$6x\left(\frac{x+1}{2x}\right) = 6x\left(\frac{2}{3}\right)$$
$$3(x+1) = 4x$$
$$3x + 3 = 4x$$
$$3 = x$$

Check: $$\frac{3+1}{2(3)} \stackrel{?}{=} \frac{2}{3}$$
$$\frac{4}{6} \stackrel{?}{=} \frac{2}{3}$$
$$\frac{2}{3} = \frac{2}{3}$$

3. $$\frac{1}{x+6}=\frac{4}{x}$$
$$x(x+6)\left(\frac{1}{x+6}\right)=x(x+6)\left(\frac{4}{x}\right)$$
$$x=4x+24$$
$$-24=3x$$
$$-8=x$$
Check: $$\frac{1}{-8+6}\overset{?}{=}\frac{4}{-8}$$
$$-\frac{1}{2}=-\frac{1}{2}$$

5. $$\frac{5}{3x-4}=\frac{3}{x-3}$$
$$(3x-4)(x-3)\frac{5}{3x-4}=\frac{3}{x-3}(3x-4)(x-3)$$
$$5(x-3)=3(3x-4)$$
$$5x-15=9x-12$$
$$-4x=3$$
$$x=-\frac{3}{4}$$
Check: $$\frac{5}{3(-\frac{3}{4})-4}\overset{?}{=}\frac{3}{(-\frac{3}{4})-3}$$
$$\frac{5}{\frac{9}{4}-\frac{16}{4}}\overset{?}{=}\frac{3}{-\frac{3}{4}-\frac{12}{4}}$$
$$5\cdot\frac{-4}{25}\overset{?}{=}3\cdot\frac{-4}{15}$$
$$-\frac{20}{25}\overset{?}{=}-\frac{12}{15}$$
$$-\frac{4}{5}=-\frac{4}{5}$$

7. $$\frac{2}{x}+\frac{x}{x+1}=1$$
$$x(x+1)\left(\frac{2}{x}\right)+x(x+1)\left(\frac{x}{x+1}\right)=x(x+1)(1)$$
$$(x+1)(2)+x^2=x^2+x$$
$$2x+2+x^2=x^2+x$$
$$2x+2=x$$
$$2=-x$$
$$-2=x$$
Check: $$\frac{2}{-2}+\frac{-2}{-2+1}\overset{?}{=}1$$
$$-1+2\overset{?}{=}1$$
$$1=1$$

9. $$\frac{x+1}{x}=1+\frac{x-2}{2x}$$
$$2x\left(\frac{x+1}{x}\right)=2x(1)+2x\left(\frac{x-2}{2x}\right)$$
$$2(x+1)=2x+x-2$$
$$2x+2=3x-2$$
$$2=x-2$$
$$4=x$$
Check: $$\frac{4+1}{4}\overset{?}{=}1+\frac{4-2}{2(4)}$$
$$\frac{5}{4}\overset{?}{=}\frac{10}{8}$$
$$\frac{5}{4}=\frac{5}{4}$$

11. $$\frac{x+1}{x}=\frac{1}{2}-\frac{4}{3x}$$
$$6x\left(\frac{x+1}{x}\right)=6x\left(\frac{1}{2}\right)-6x\left(\frac{4}{3x}\right)$$
$$6(x+1)=3x-8$$
$$6x+6=3x-8$$
$$3x+6=-8$$
$$3x=-14$$

$$x = \frac{-14}{3}$$

Check: $\frac{-\frac{14}{3}+1}{-\frac{14}{3}} \stackrel{?}{=} \frac{1}{2} - \frac{4}{3\left(-\frac{14}{3}\right)}$

$$\frac{-\frac{11}{3}}{-\frac{14}{3}} \stackrel{?}{=} \frac{1}{2} + \frac{4}{14}$$

$$\frac{11}{14} = \frac{11}{14}$$

13.
$$\frac{2}{x-6} - 5 = \frac{2(x-5)}{x-6}$$

$$(x-6)\left(\frac{2}{x-6}\right) - 5(x-6) = (x-6)\left[\frac{2(x-5)}{x-6}\right]$$

$$2 - 5x + 30 = 2x - 10$$

$$-5x + 32 = 2x - 10$$

$$42 = 7x$$

$$6 = x$$

Check:

$$\frac{2}{6-6} - 5 \stackrel{?}{=} \frac{2(6-5)}{6-6}$$

$$\frac{2}{0} - 5 \stackrel{?}{=} \frac{2}{0}$$

$x = 6$ makes the denominators zero so it is an extraneous solution. There is no solution

15. $\frac{2}{x+1} - \frac{1}{x-1} = \frac{2x}{x^2-1}$

$$\frac{2}{x+1} - \frac{1}{x-1} = \frac{2x}{(x+1)(x-1)}$$

$$(x+1)(x-1)\left(\frac{2}{x+1}\right) - (x+1)(x-1)\left(\frac{1}{x-1}\right)$$

$$= (x+1)(x-1)\left(\frac{2x}{(x+1)(x-1)}\right)$$

$$(x-1)(2) - (x+1) = 2x$$

$$2x - 2 - x - 1 = 2x$$

$$x - 3 = 2x$$

$$-3 = x$$

Check: $\frac{2}{-3+1} - \frac{1}{-3-1} \stackrel{?}{=} \frac{2(-3)}{(-3)^2 - 1}$

$$-1 + \frac{1}{4} \stackrel{?}{=} -\frac{6}{8}$$

$$-\frac{3}{4} = -\frac{3}{4}$$

17. $\frac{y+1}{y^2+2y-3} = \frac{1}{y+3} - \frac{1}{y-1}$

$$\frac{y+1}{(y+3)(y-1)} = \frac{1}{y+3} - \frac{1}{y-1}$$

$$(y+3)(y-1)\left(\frac{y+1}{(y+3)(y-1)}\right)$$

$$= (y+3)(y-1)\left(\frac{1}{y+3}\right) - (y+3)(y-1)\left(\frac{1}{y-1}\right)$$

$$y + 1 = y - 1 - y - 3$$

$$y + 1 = -4$$

$$y = -5$$

Check: $\frac{-5+1}{(-5)^2 + 2(-5) - 3} \stackrel{?}{=} \frac{1}{-5+3} - \frac{1}{-5-1}$

$$\frac{-4}{12} \stackrel{?}{=} -\frac{1}{2} + \frac{1}{6}$$

$$-\frac{4}{12} = -\frac{4}{12}$$

19.
$$\frac{79-x}{x} = 5 + \frac{7}{x}$$

$$x\left(\frac{79-x}{x}\right) = x(5) + x\left(\frac{7}{x}\right)$$

$$79 - x = 5x + 7$$

$$79 = 6x + 7$$

$$72 = 6x$$
$$12 = x$$

Check: $\dfrac{79-12}{12} \stackrel{?}{=} 5 + \dfrac{7}{12}$

$$\frac{67}{12} = \frac{67}{12}$$

21. $\dfrac{x+11}{x^2-5x+4} + \dfrac{3}{x-1} = \dfrac{5}{x-4}$

$$\frac{x+11}{(x-4)(x-1)} + \frac{3}{x-1} = \frac{5}{x-4}$$
$$(x-4)(x-1)\left(\frac{x+11}{(x-4)(x-1)}\right)$$
$$+(x-4)(x-1)\left(\frac{3}{x-1}\right) = (x-4)(x-1)\left(\frac{5}{x-4}\right)$$
$$x+11+(x-4)(3) = (x-1)(5)$$
$$x+11+3x-12 = 5x-5$$
$$4x-1 = 5x-5$$
$$-1 = x-5$$
$$4 = x \text{ Extraneous}$$

No solution.

23. $\dfrac{2x}{x+4} - \dfrac{8}{x-4} = \dfrac{2x^2+32}{x^2-16}$

$$\frac{2x}{x+4} - \frac{8}{x-4} = \frac{2x^2+32}{(x+4)(x-4)}$$
$$(x+4)(x-4)\left(\frac{2x}{x+4}\right) - (x+4)(x-4)\left(\frac{8}{x-4}\right)$$
$$= (x+4)(x-4)\left(\frac{2x^2+32}{(x+4)(x-4)}\right)$$
$$(x-4)(2x) - (x+4)(8) = 2x^2+32$$
$$2x^2-8x-8x-32 = 2x^2+32$$
$$2x^2-16x-32 = 2x^2+32$$
$$-16x-32 = 32$$
$$-16x = 64$$
$$x = -4 \text{ Extraneous}$$

No solution.

25. $\dfrac{6}{x-5} + \dfrac{3x+1}{x^2-2x-15} = \dfrac{5}{x+3}$

$$\frac{6}{x-5} + \frac{3x+1}{(x-5)(x+3)} = \frac{5}{x+3}$$
$$(x+3)(6)+(3x+1) = (x-5)(5)$$
$$6x+18+3x+1 = 5x-25$$
$$9x+19 = 5x-25$$
$$4x+19 = -25$$
$$4x = -44$$
$$x = -11$$

Check:

$$\frac{6}{-11-5} + \frac{3(-11)+1}{(-11)^2-2(-11)-15} \stackrel{?}{=} \frac{5}{-11+3}$$
$$\frac{3}{-8} - \frac{32}{128} \stackrel{?}{=} -\frac{5}{-8}$$
$$-\frac{3}{8} - \frac{2}{8} \stackrel{?}{=} -\frac{5}{8}$$
$$-\frac{5}{8} = -\frac{5}{8}$$

27. $x-2 \neq 0$, so $x \neq 2$

$x-4 \neq 0$, so $x \neq 4$

$$x^2-6x+8 \neq 0$$
$$(x-2)(x-4) \neq 0$$
$$x-2 \neq 0 \qquad x-4 \neq 0$$
$$x \neq 2 \qquad x \neq 4$$

Extraneous solutions $x = 2$, $x = 4$

Cumulative Review Problems

29. $x = \text{width}$

$$3x - 1 = \text{length}$$
$$P = 2L + 2W$$
$$54 = 2(3x-1) + 2x$$
$$54 = 6x - 2 + 2x$$
$$56 = 8x$$
$$7 = x$$
$$3x - 1 = 3(7) - 1 - 1 = 20$$

Width = 7 meters and length = 20 meters

31. $x = y^2 - 2$

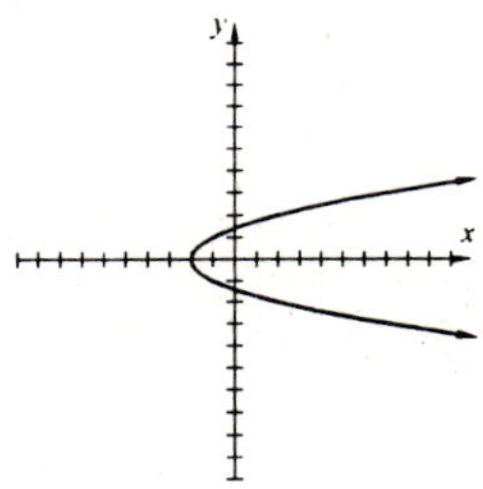

33. $4y^2 - 121z^2 = (2y)^2 - (11z)^2$
$$= (2y + 11z)(2y - 11z)$$

7.6 Exercises

1. $\frac{4}{9} = \frac{8}{x}$
$$4x = 72$$
$$x = 18$$

3. $\frac{x}{17} = \frac{12}{5}$
$$5x = 204$$
$$x = \frac{204}{5} \text{ or } 40\frac{4}{5}$$

5. $\frac{5}{3} = \frac{x}{8}$
$$40 = 3x$$
$$\frac{40}{3} = x \text{ or } x = 13\frac{1}{3}$$

7. $\frac{7}{x} = \frac{40}{130}$
$$40x = 910$$
$$x = \frac{910}{40} = 22.75$$

9. $d = \text{distance apart}$
$$\frac{d}{5.5} = \frac{15}{3/4}$$
$$\frac{3}{4}d = 82.5$$
$$d = 110 \text{ miles}$$

11. (a) x = the number of pounds he received
$$\frac{1}{1.53} = \frac{x}{800}$$
$$1.53x = 800$$
$$x = \frac{800}{1.53} = 522.88$$

He received 522.88 British pounds.

(b) d = the number of dollars he received

$$\frac{1}{1.53} = \frac{140}{d}$$
$$d = 140(1.53) = 214.2$$

He received \$214.20.

13. x = speed limit in miles per hour

$$\frac{x}{90} = \frac{62}{100}$$
$$100x = 5580$$
$$x = 55.8 \approx 56$$

The limit is 56 miles per hour.

15. x = miles from base of mountain

$$\frac{x}{\frac{3}{4}} = \frac{136}{3\frac{1}{2}}$$
$$\frac{7}{2}x = \frac{3}{4}(136)$$
$$\frac{7}{2}x = 102$$
$$x = 29.1$$

He is 29 miles away.

17. $x = 20,\ \ y = 29,\ \ m = 13$

$$\frac{n}{m} = \frac{y}{x}$$
$$\frac{n}{13} = \frac{29}{20}$$
$$20n = 377$$
$$n = \frac{377}{20} = 18\frac{17}{20} \text{ inches}$$

19. $\frac{d}{k} = \frac{a}{g}$

$$\frac{d}{12} = \frac{7}{9}$$
$$9d = 84$$
$$d = \frac{84}{9} = 9\frac{1}{3} \text{ in.}$$

21. $\frac{k}{d} = \frac{h}{b}$

$$\frac{k}{32} = \frac{24}{20}$$
$$k = \frac{768}{20} = 38\frac{2}{5} \text{ m}$$

23. x = length

$$\frac{30}{x} = \frac{5}{8}$$
$$5x = 8(30)$$
$$x = \frac{240}{5}$$
$$x = 48$$

48 inches long

25. x = height of kite on long line

$$\frac{x}{120} = \frac{5}{7}$$
$$7x = 600$$
$$x = 85.7 \approx 86$$

The kite is 86 meters high.

27. x = amount of accceleration in 11 seconds

$$\frac{x}{11} = \frac{3}{2}$$

$$2x = 33$$
$$x = 16.5$$
$$45 + 16.5 = 61.5$$

61.5 miles per hour

29.

	D	R	$T = \frac{D}{R}$
Helicopter	1050	s	$\frac{1050}{s}$
Airline	1250	$s+40$	$\frac{1250}{s+40}$

$$\frac{1250}{s+40} = \frac{1050}{s}$$
$$1050s = 1050s + 42,000$$
$$200s = 42,000$$
$$s = 210$$
$$s + 40 = 250$$

Helicopter flies at 210 km/hr

Airline flies at 250 km/hr

31. a. $\frac{75}{3.4} = \$22.06/\text{ounce}$

b. $\frac{42}{1.7} = \$24.71/\text{ounce}$

c. Larger $= \frac{10.2}{3.4}(75) = 225$

Smaller $= \frac{10.2}{1.7}(42) = 252$

Savings $= 252 - 225 = \$27$

33. x = time to mow together

$$\frac{1}{4} + \frac{1}{5} = \frac{1}{x}$$
$$5x + 4x = 20$$
$$9x = 20$$
$$x = \frac{20}{9} = 2\frac{2}{9} \text{ hours}$$
$$60\left(\frac{2}{9}\right) = 13.3$$

It will take $2\frac{2}{9}$ hours or 2 hours 13 minutes.

35. x = time to rake together

$$\frac{1}{6} + \frac{1}{8} = \frac{1}{x}$$
$$24x\left(\frac{1}{6}\right) + 24x\left(\frac{1}{8}\right) = 24x\left(\frac{1}{x}\right)$$
$$4x + 3x = 24$$
$$7x = 24$$
$$x = \frac{24}{7} = 3\frac{3}{7}$$

It will take $3\frac{3}{7}$ hours or 3 hours 26 minutes.

Cumulative Review Problems

37. $0.0000006316 = 6.316 \times 10^{-7}$

39. $\frac{x^{-3}y^{-2}}{z^4w^{-8}} = \frac{w^8}{x^3y^2z^4}$

Putting Your Skills to Work

1. $687 - 225 = 462$

   462 days more

2. $\frac{365(5)}{88} = 20.7$

   Approximately 21 orbits

3. $\frac{1}{S} = \frac{1}{a} - \frac{1}{b}$

   $\frac{1}{S} = \frac{1}{88} - \frac{1}{225}$

   $\frac{1}{S} = \frac{137}{19,800}$

   $S = 144.5$

   Approximately 145 days

4. $\frac{1}{S} = \frac{1}{a} - \frac{1}{b}$

   $\frac{1}{S} = \frac{1}{365} - \frac{1}{4333}$

   $\frac{1}{S} = \frac{3968}{1,581,545}$

   $S = 398.57$

   Approximately 399 days

Chapter 7 Review Problems

1. $\frac{4x-4y}{5y-5x} = \frac{4(x-y)}{-5(x-y)} = -\frac{4}{5}$

2. $\frac{bx}{bx-by} = \frac{bx}{b(x-y)}$

   $= \frac{x}{x-y}$

3. $\frac{2x^2+5x-3}{2x^2-9x+4} = \frac{(2x-1)(x+3)}{(2x-1)(x-4)} = \frac{x+3}{x-4}$

4. $\frac{3x^2+7x+2}{3x^2+13x+4}$

   $= \frac{(3x+1)(x+2)}{(3x+1)(x+4)}$

   $= \frac{x+2}{x+4}$

5. $\frac{x^2-9}{x^2-10x+21} = \frac{(x+3)(x-3)}{(x-7)(x-3)} = \frac{x+3}{x-7}$

6. $\frac{2x^2+18x+40}{3x+15}$

   $= \frac{2(x+5)(x+4)}{3(x+5)}$

   $= \frac{2(x+4)}{3}$

7. $\frac{4x^2+4x-3}{4x^2-2x} = \frac{(2x+3)(2x-1)}{2x(2x-1)} = \frac{2x+3}{2x}$

8. $\frac{x^3+3x^2}{x^3-2x^2-15x}$

   $= \frac{x^2(x+3)}{x(x-5)(x+3)}$

   $= \frac{x}{x-5}$

9. $\frac{2x^2-2xy-24y^2}{2x^2+5xy-3y^2}$

   $= \frac{2(x-4y)(x+3y)}{(2x-y)(x+3y)}$

   $= \frac{2(x-4y)}{2x-y}$

10. $\frac{4-y^2}{3y^2+5y-2}$

$=\frac{(2+y)(2-y)}{(3y-1)(y+2)}$

$=\frac{2-y}{3y-1}$

11. $\frac{5x^3-10x^2}{25x^4+5x^3-30x^2}$

$=\frac{5x^2(x-2)}{5x^2(5x+6)(x-1)}$

$=\frac{x-2}{(5x+6)(x-1)}$

12. $\frac{16x^2-4y^2}{4x-2y}$

$=\frac{4(2x+y)(2x-y)}{2(2x-y)}$

$=2(2x+y)$

$=4x+2y$

13. $\frac{3x^2-13x-10}{3x^2+2x}\cdot\frac{x^2-25x}{x^2-25}$

$=\frac{(3x+2)(x-5)}{x(3x+2)}\cdot\frac{x(x-25)}{(x+5)(x-5)}$

$=\frac{x-25}{x+5}$

14. $\frac{2x^2-18}{3y^2+3y}\div\frac{y^2+6y+9}{y^2+4y+3}$

$=\frac{2x^2-18}{3y^2+3y}\cdot\frac{y^2+4y+3}{y^2+6y+9}$

$=\frac{2(x+3)(x-3)}{3y(y+1)}\cdot\frac{(y+3)(y+1)}{(y+3)(y+3)}$

$=\frac{2(x+3)(x-3)}{3y(y+3)}$

15. $\frac{2y^2+3y-2}{2y^2+y-1}\div\frac{2y^2+y-1}{2y^2-3y-2}$

$=\frac{2y^2+3y-2}{2y^2+y-1}\cdot\frac{2y^2-3y-2}{2y^2+y-1}$

$=\frac{(2y-1)(y+2)}{(2y-1)(y+1)}\cdot\frac{(2y+1)(y-2)}{(2y-1)(y+1)}$

$=\frac{(y+2)(2y+1)(y-2)}{(y+1)^2(2y-1)}$

16. $\frac{6y^2+13y-5}{9y^2+3y}\div\frac{4y^2+20y+25}{12y^2}$

$=\frac{6y^2+13y-5}{9y^2+3y}\cdot\frac{12y^2}{4y^2+20y+25}$

$=\frac{(3y-1)(2y+5)}{3y(3y+1)}\cdot\frac{12y^2}{(2y+5)(2y+5)}$

$=\frac{4y(3y-1)}{(3y+1)(2y+5)}$

17. $\frac{3xy^2+12y^2}{2x^2-11x+5}\div\frac{2xy+8y}{8x^2+2x-3}$

$=\frac{3xy^2+12y^2}{2x^2-11x+5}\cdot\frac{8x^2+2x-3}{2xy+8y}$

$=\frac{3y^2(x+4)}{(2x-1)(x-5)}\cdot\frac{(2x-1)(4x+3)}{2y(x+4)}$

$=\frac{3y(4x+3)}{2(x-5)}$

18. $\frac{11}{x-2}\cdot\frac{2x^2-8}{44}=\frac{11}{x-2}\cdot\frac{2(x+2)(x-2)}{44}$
$=\frac{x+2}{2}$

19. $\frac{2x^2+10x+2}{8x-8}\cdot\frac{3x-3}{4x^2+20x+4}$
$=\frac{2(x^2+5x+1)}{8(x-1)}\cdot\frac{3(x-1)}{4(x^2+5x+1)}$
$=\frac{3}{16}$

20. $\frac{x^2-5xy-24y^2}{2x^2-2xy-24y^2}\cdot\frac{4x^2+4xy-24y^2}{x^2-10xy+16y^2}$
$=\frac{(x-8y)(x+3y)}{2(x-4y)(x+3y)}\cdot\frac{4(x+3y)(x-2y)}{(x-8y)(x-2y)}$
$=\frac{2(x+3y)}{x-4y}$

21. $\frac{7}{x+1}+\frac{4}{2x}=\frac{7}{x+1}\cdot\frac{2x}{2x}+\frac{4}{2x}\cdot\frac{x+1}{x+1}$
$=\frac{14x+4x+4}{2x(x+1)}$
$=\frac{18x+4}{2x(x+1)}$
$=\frac{9x+2}{x(x+1)}$

22. $5+\frac{1}{x}+\frac{1}{x+1}$
$=\frac{5}{1}\cdot\frac{x(x+1)}{x(x+1)}+\frac{1}{x}\cdot\frac{x+1}{x+1}+\frac{1}{x+1}\cdot\frac{x}{x}$
$=\frac{5x^2+5x}{x(x+1)}+\frac{x+1}{x(x+1)}+\frac{x}{x(x+1)}$
$=\frac{5x^2+5x+x+1+x}{x(x+1)}$
$=\frac{5x^2+7x+1}{x(x+1)}$

23. $\frac{2}{x^2-9}+\frac{x}{x+3}$
$=\frac{2}{(x+3)(x-3)}+\frac{x}{x+3}\cdot\frac{x-3}{x-3}$
$=\frac{2+x^2-3x}{(x+3)(x-3)}$
$=\frac{(x-2)(x-1)}{(x+3)(x-3)}$

24. $\frac{7}{x+2}+\frac{3}{x-4}$
$=\frac{7}{x+2}\cdot\frac{x-4}{x-4}+\frac{3}{x-4}\cdot\frac{x+2}{x+2}$
$=\frac{7x-28}{(x+2)(x-4)}+\frac{3x+6}{(x+2)(x-4)}$
$=\frac{7x-28+3x+6}{(x+2)(x-4)}$
$=\frac{10x-22}{(x+2)(x-4)}$

25. $\frac{x}{y}+\frac{3}{2y}+\frac{1}{y+2}$
$=\frac{x}{y}\cdot\frac{2(y+2)}{2(y+2)}+\frac{3}{2y}\cdot\frac{y+2}{y+2}+\frac{1}{y+2}\cdot\frac{2y}{2y}$
$=\frac{2x(y+2)+3y+6+2y}{2y(y+2)}$

$$= \frac{2xy+4x+5y+6}{2y(y+2)}$$

26. $\frac{4}{a}+\frac{2}{b}+\frac{3}{a+b}$

$$= \frac{4}{a}\cdot\frac{b(a+b)}{b(a+b)}+\frac{2}{b}\cdot\frac{a(a+b)}{a(a+b)}+\frac{3}{a+b}\cdot\frac{ab}{ab}$$

$$= \frac{4ab+4b^2}{ab(a+b)}+\frac{2a^2+2ab}{ab(a+b)}+\frac{3ab}{ab(a+b)}$$

$$= \frac{4ab+4b^2+2a^2+2ab+3ab}{ab(a+b)}$$

$$= \frac{2a^2+9ab+4b^2}{ab(a+b)}$$

$$= \frac{(2a+b)(a+4b)}{ab(a+b)}$$

27. $\frac{3x+1}{3x}-\frac{1}{x}=\frac{3x+1}{3x}-\frac{1}{x}\cdot\frac{3}{3}$

$$= \frac{3x+1-3}{3x}$$

$$= \frac{3x-2}{3x}$$

28. $\frac{x+4}{x+2}-\frac{1}{2x}$

$$= \frac{x+4}{x+2}\cdot\frac{2x}{2x}-\frac{1}{2x}\cdot\frac{x+2}{x+2}$$

$$= \frac{2x^2+8x}{2x(x+2)}-\frac{x+2}{2x(x+2)}$$

$$= \frac{2x^2+8x-(x+2)}{2x(x+2)}$$

$$= \frac{2x^2+8x-x-2}{2x(x+2)}$$

$$= \frac{2x^2+7x-2}{2x(x+2)}$$

29. $\frac{1}{x^2+7x+10}-\frac{x}{x+5}$

$$= \frac{1}{(x+2)(x+5)}-\frac{x}{x+5}\cdot\frac{x+2}{x+2}$$

$$= \frac{1-x^2-2x}{(x+2)(x+5)}$$

30. $\frac{27}{x^2-81}+\frac{3}{2(x+9)}$

$$= \frac{27}{(x+9)(x-9)}\cdot\frac{2}{2}+\frac{3}{2(x+9)}\cdot\frac{x-9}{x-9}$$

$$= \frac{54+3x-27}{2(x+9)(x-9)}$$

$$= \frac{3x+27}{2(x+9)(x-9)}$$

$$= \frac{3(x+9)}{2(x+9)(x-9)}$$

$$= \frac{3}{2(x-9)}$$

31. $\frac{\frac{3}{2y}-\frac{1}{y}}{\frac{4}{y}+\frac{3}{2y}}=\frac{\frac{3}{2y}-\frac{1}{y}}{\frac{4}{y}+\frac{3}{2y}}\cdot\frac{2y}{2y}$

$$= \frac{3-2}{8+3}=\frac{1}{11}$$

32. $\dfrac{\frac{2}{x}+\frac{1}{2x}}{x+\frac{x}{2}}$

$= \dfrac{\frac{2}{x}+\frac{1}{2x}}{x+\frac{x}{2}} \cdot \dfrac{2x}{2x}$

$= \dfrac{4+1}{2x^2+x^2} = \dfrac{5}{3x^2}$

33. $\dfrac{w-\frac{4}{w}}{1+\frac{2}{w}} = \dfrac{w-\frac{4}{w}}{1+\frac{2}{w}} \cdot \dfrac{w}{w}$

$= \dfrac{w^2-4}{w+2}$

$= \dfrac{(w+2)(w-2)}{w+2}$

$= w-2$

34. $\dfrac{1-\frac{w}{w-1}}{1+\frac{w}{1-w}}$

$= \dfrac{1-\frac{w}{x-1}}{1+\frac{w}{-(w-1)}}$

$= \dfrac{1-\frac{w}{w-1}}{1-\frac{w}{w-1}} = 1$

35. $\dfrac{1+\frac{1}{y^2-1}}{\frac{1}{y+1}-\frac{1}{y-1}} = \dfrac{1+\frac{1}{(y+1)(y-1)}}{\frac{1}{y+1}-\frac{1}{y-1}} \cdot \dfrac{(y+1)(y-1)}{(y+1)(y-1)}$

$= \dfrac{(y+1)(y-1)+1}{y-1-(y+1)}$

$= \dfrac{y^2}{-2} = -\dfrac{y^2}{2}$

36. $\dfrac{\frac{1}{y}+\frac{1}{x+y}}{1+\frac{2}{x+y}}$

$= \dfrac{\frac{1}{y}+\frac{1}{x+y}}{1+\frac{2}{x+y}} \cdot \dfrac{y(x+y)}{y(x+y)}$

$= \dfrac{x+y+y}{y(x+y)+2y}$

$= \dfrac{x+2y}{xy+y^2+2y}$

$= \dfrac{x+2y}{y(x+y+2)}$

37. $\dfrac{\frac{1}{a+b}-\frac{1}{a}}{b} = \dfrac{\frac{a-(a+b)}{a(a+b)}}{b}$

$= -\dfrac{1}{a(a+b)}$

38. $\dfrac{\frac{2}{a+b}-\frac{3}{b}}{\frac{1}{a+b}}$

$= \dfrac{\frac{2}{a+b}-\frac{3}{b}}{\frac{1}{a+b}} \cdot \dfrac{b(a+b)}{b(a+b)}$

$= \dfrac{2b-3(a+b)}{b}$

$= \dfrac{2b-3a-3b}{b} = \dfrac{-3a-b}{b}$

39. $\left(\dfrac{1}{x+2y}-\dfrac{1}{x-y}\right) \div \dfrac{2x-4y}{x^2-3xy+2y^2}$

$= \left(\dfrac{1}{x+2y}-\dfrac{1}{x-y}\right) \cdot \dfrac{x^2-3xy+2y^2}{2x-4y}$

$= \dfrac{x-y-(x+2y)}{(x+2y)(x-y)} \cdot \dfrac{(x-y)(x-2y)}{2(x-2y)}$

$= \dfrac{-3y}{2(x+2y)}$

40. $\frac{x+5y}{x-6y} \div \left(\frac{1}{5y} - \frac{1}{x+5y}\right)$

$$= \frac{x+5y}{x-6y} \div \left(\frac{1}{5y}\cdot\frac{x+5y}{x+5y} - \frac{1}{x+5y}\cdot\frac{5y}{5y}\right)$$

$$= \frac{x+5y}{x-6y} \div \left[\frac{x+5y}{5y(x+5y)} - \frac{5y}{5y(x+5y)}\right]$$

$$= \frac{x+5y}{x-6y} \div \frac{x+5y-5y}{5y(x+5y)}$$

$$= \frac{x+5y}{x-6y} \div \frac{x}{5y(x+5y)}$$

$$= \frac{x+5y}{x-6y} \cdot \frac{5y(x+5y)}{x}$$

$$= \frac{5y(x+5y)^2}{x(x-6y)}$$

41. $\frac{8}{a-3} = \frac{12}{a+3}$

$$(a-3)(a+3)\frac{8}{a-3} = (a-3)(a+3)\frac{12}{a+3}$$

$$8a+24 = 12a-36$$

$$60 = 4a$$

$$15 = a$$

42. $\frac{8a-1}{6a+8} = \frac{3}{4}$

$$\frac{8a-1}{2(3a+4)} = \frac{3}{4}$$

$$4(3a+4)\left[\frac{8a-1}{2(3a+4)}\right] = 4(3a+4)\left(\frac{3}{4}\right)$$

$$2(8a-1) = 3(3a+4)$$

$$16a-2 = 9a+12$$

$$7a-2 = 12$$

$$7a = 14$$

$$a = 2$$

43. $\frac{2x-1}{x} - \frac{1}{2} = -2$

$$2x\left(\frac{2x-1}{x}\right) - 2x\left(\frac{1}{2}\right) = -2(2x)$$

$$4x-2-x = -4x$$

$$7x = 2$$

$$x = \frac{2}{7}$$

44. $\frac{5-x}{x} - \frac{7}{x} = -\frac{3}{4}$

$$4x\left(\frac{5-x}{x}\right) - 4x\left(\frac{7}{x}\right) = 4x\left(-\frac{3}{4}\right)$$

$$4(5-x)-28 = -3x$$

$$20-4x-28 = -3x$$

$$-4x-8 = -3x$$

$$-8 = x$$

45. $\frac{5}{2} - \frac{2y+7}{y+6} = 3$

$$2(y+6)\left(\frac{5}{2}\right) - 2(y+6)\left(\frac{2y+7}{y+6}\right) = 2(y+6)(3)$$

$$5y+30-4y-14 = 6y+36$$

$$-20 = 5y$$

$$-4 = y$$

46. $\frac{5}{4} - \frac{1}{2x} = \frac{1}{x} + 2$

$$4x\left(\frac{5}{4}\right) - 4x\left(\frac{1}{2x}\right) = 4x\left(\frac{1}{x}\right) + 4x(2)$$

$$5x-2=4+8x$$
$$-2=4+3x$$
$$-6=3x$$
$$-2=x$$

47. $$\frac{7}{8x}-\frac{3}{4}=\frac{1}{4x}+\frac{1}{2}$$
$$8x\left(\frac{7}{8x}\right)-8x\left(\frac{3}{4}\right)=8x\left(\frac{1}{4x}\right)+8x\left(\frac{1}{2}\right)$$
$$7-6x=2+4x$$
$$7=2+10x$$
$$5=10x$$
$$\frac{1}{2}=x$$

48. $$\frac{1}{3x}+2=\frac{5}{6x}-\frac{1}{2}$$
$$6x\left(\frac{1}{3x}\right)+6x(2)=6x\left(\frac{5}{6x}\right)-6x\left(\frac{1}{2}\right)$$
$$2+12x=5-3x$$
$$15x=3$$
$$x=\frac{1}{5}$$

49. $$\frac{3}{y-3}=\frac{3}{2}+\frac{y}{y-3}$$
$$2(y-3)\left(\frac{3}{y-3}\right)$$
$$=2(y-3)\left(\frac{3}{2}\right)+2(y-3)\left(\frac{y}{y-3}\right)$$
$$6=3(y-3)+2y$$
$$6=3y-9+2y$$
$$6=5y-9$$
$$15=5y$$
$$3=y$$

Since $y=3$ causes a denominator in the original equation to equal 0, there is no solution.

50. $$\frac{x-8}{x-2}=\frac{2x}{x+2}-2$$
$$(x+2)(x-2)\left(\frac{x-8}{x-2}\right)$$
$$=(x+2)(x-2)\left(\frac{2x}{x+2}\right)-(x+2)(x-2)(2)$$
$$x^2-6x-16=2x^2-4x-2x^2+8$$
$$x^2-2x-24=0$$
$$(x-6)(x+4)=0$$

$x-6=0$ $\quad$ $x+4=0$

$x=6$ $\quad$ $x=-4$

51. $$\frac{3y-1}{3y}-\frac{6}{5y}=\frac{1}{y}-\frac{4}{15}$$
$$15y\left(\frac{3y-1}{3y}\right)-15y\left(\frac{6}{5y}\right)$$
$$=15y\left(\frac{1}{y}\right)-15y\left(\frac{4}{15}\right)$$
$$15y-5-18=15-4y$$
$$19y=38$$
$$y=2$$

52. $$\frac{9}{2}-\frac{7y-4}{y+2}=-\frac{1}{4}$$
$$4(y+2)\left(\frac{9}{2}\right)-4(y+2)\left(\frac{7y-4}{y+2}\right)=4(y+2)\left(-\frac{1}{4}\right)$$
$$18(y+2)-4(7y-4)=-(y+2)$$
$$18y+36-28y+16=-y-2$$
$$-10y+52=-y-2$$
$$52=9y-2$$
$$54=9y$$
$$6=y$$

53. $$\frac{y+18}{y^2-16}=\frac{y}{y+4}-\frac{y}{y-4}$$
$$\frac{y+18}{(y+4)(y-4)}=\frac{y}{y+4}-\frac{y}{y-4}$$
$$y+18=(y-4)(y)-(y+4)(y)$$
$$y+18=y^2-4y-y^2-4y$$
$$y+18=-8y$$
$$18=-9y$$
$$-2=y$$

54. $$\frac{4}{x^2-1}=\frac{2}{x-1}+\frac{2}{x+1}$$
$$4=(x+1)(2)+(x-1)(2)$$
$$4=2x+2+2x-2$$
$$4=4x$$
$$1=x$$

$x = 1$ is not an allowed solution.

There is no solution.

55. $$\frac{9y-3}{y^2+2y}-\frac{5}{y+2}=\frac{3}{y}$$
$$y(y+2)\left[\frac{9y-3}{y(y+2)}\right]-y(y+2)\left(\frac{5}{y+2}\right)=y(y+2)\left(\frac{3}{y}\right)$$
$$9y-3-5y=3y+6$$
$$y=9$$

56. $$\frac{2}{3-3y}+\frac{2}{2y-1}=\frac{4}{3y-3}$$
$$\frac{2}{-3(y-1)}+\frac{2}{2y-1}=\frac{4}{3(y-1)}$$
$$\frac{-2}{3(y-1)}+\frac{2}{2y-1}=\frac{4}{3(y-1)}$$
$$(2y-1)(-2)+3(y-1)(2)=(2y-1)(4)$$
$$-2(2y-1)+6(y-1)=4(2y-1)$$
$$-4y+2+6y-6=8y-4$$
$$2y-4=8y-4$$
$$-4=6y-4$$
$$0=6y$$
$$0=y$$

57. $$\frac{8}{5}=\frac{2}{x}$$
$$8x=10$$
$$x=\frac{5}{4}=1.25=1.3$$

58. $\frac{x}{4}=\frac{12}{17}$

$17x=(4)(12)$

$17x=48$

$\frac{17x}{17}=\frac{48}{17}$

$x=2.8$

59. $\frac{33}{10}=\frac{x}{8}$

$264=10x$

$26.4=x$

60. $\frac{5}{x}=\frac{22}{9}$

$22x=(5)(9)$

$22x=45$

$\frac{22x}{22}=\frac{45}{22}$

$x=2.0$

61. $\frac{13.5}{0.6}=\frac{360}{x}$

$13.5x=216$

$x=16$

62. $\frac{2\frac{1}{2}}{3\frac{1}{4}}=\frac{7}{x}$

$\frac{\frac{5}{2}}{\frac{13}{4}}=\frac{7}{x}$

$\frac{5}{2}x=\frac{91}{4}$

$x=9.1$

63. x = gallons to cover 400 square feet

$\frac{x}{400}=\frac{5}{240}$

$240x=2000$

$x=\frac{2000}{240}=8\frac{1}{3}$

8.3 gallons of paint are needed.

64. $\frac{3}{100}=\frac{5}{x}$

$3x=5(100)$

$3x=500$

$\frac{3x}{3}=\frac{500}{3}$

$x=\frac{500}{3}=166\frac{2}{3}$

167 cookies

65. x = gallons to get to Denver

$\frac{x}{1300}=\frac{7}{200}$

$200x=9100$

$x=\frac{91}{2}=45\frac{1}{2}$

He will use 46 gallons of gas.

66. $\frac{4}{122}=\frac{3}{x}$

$4x=(122)(3)$

$4x=366$

$\frac{4x}{4}=\frac{366}{4}$

$x=91.5$ miles

67.

	D	R	$T=\frac{D}{R}$
Train	180	$s+20$	$\frac{180}{s+20}$
Car	120	s	$\frac{120}{s}$

$$\frac{180}{s+20}=\frac{120}{s}$$
$$180s=120s+2400$$
$$60s=2400$$
$$s=40$$
$$s+20=60$$

Car's speed is 40 mph.

Trains speed is 60 mph.

68.

	Time for whole job	Part of job done in 1 hr
Professional	5	$1/5$
John	8	$1/8$
Together	x	$1/x$

$$\frac{1}{5}+\frac{1}{8}=\frac{1}{x}$$
$$40x\left(\frac{1}{5}\right)+40x\left(\frac{1}{8}\right)=40x\left(\frac{1}{x}\right)$$
$$8x+5x=40$$
$$13x=40$$
$$\frac{13x}{13}=\frac{40}{13}$$
$$x=3\frac{1}{13}$$
$$\frac{1}{13}\times\frac{60}{1}=5$$

3 hours 5 minutes

69. x = height of building

$$\frac{x}{450}=\frac{8}{3}$$
$$3x=3600$$
$$x=1200$$

Office building is 1200 feet tall.

70.
$$\frac{5.75}{3}=\frac{x}{95}$$
$$(5.75)(95)=3x$$
$$546.25=3x$$
$$\frac{546.25}{3}=\frac{3x}{3}$$
$$182.1=x$$

182 feet tall

71. x = time to paint together

$$\frac{1}{5}+\frac{1}{10}=\frac{1}{x}$$
$$10x\left(\frac{1}{5}\right)+10x\left(\frac{1}{10}\right)=10x\left(\frac{1}{x}\right)$$
$$2x+x=10$$
$$3x=10$$
$$x=3\frac{1}{3}$$

$3\frac{1}{3}$ hours or 3 hours, 20 minutes

72. x = time to plow together

$$\frac{1}{20}+\frac{1}{30}=\frac{1}{x}$$
$$60x\left(\frac{1}{20}\right)+60x\left(\frac{1}{30}\right)=60x\left(\frac{1}{x}\right)$$
$$3x+2x=60$$
$$5x=60$$
$$x=12\text{ hours}$$

Chapter 7 Test

1. $\frac{2ac+2ad}{3a^2c+3a^2d}=\frac{2a(c+d)}{3a^2(c+d)}=\frac{2}{3a}$

2. $\frac{8x^2-2x^2y^2}{y^2+4y+4}$
$=\frac{2x^2(2-y)(2+y)}{(y+2)^2}$
$=\frac{2x^2(2-y)}{y+2}$

3. $\frac{x^2+2x}{2x-1}\cdot\frac{10x^2-5x}{12x^3+24x^2}$
$=\frac{x(x+2)}{2x-1}\cdot\frac{5x(2x-1)}{12x^2(x+2)}$
$=\frac{5}{12}$

4. $\frac{x+2y}{12y^2}\cdot\frac{4y}{x^2+xy-2y^2}$
$=\frac{x+2y}{12y^2}\cdot\frac{4y}{(x+2y)(x-y)}$
$=\frac{1}{3y(x-y)}$

5. $\frac{2a^2-3a-2}{4a^2+a-14}\div\frac{2a^2+5a+2}{16a^2-49}$
$=\frac{2a^2-3a-2}{4a^2+a-14}\cdot\frac{16a^2-49}{2a^2+5a+2}$
$=\frac{(2a+1)(a-2)}{(4a-7)(a+2)}\cdot\frac{(4a+7)(4a-7)}{(2a+1)(a+2)}$
$=\frac{(a-2)(4a+7)}{(a+2)^2}$

6. $\frac{3}{x^2+x-6}+\frac{1}{x^2+3x-10}$
$=\frac{3}{(x+3)(x-2)}\cdot\frac{x+5}{x+5}+\frac{1}{(x+5)(x-2)}\cdot\frac{x+3}{x+3}$
$=\frac{3x+15+x+3}{(x+3)(x-2)(x+5)}$
$=\frac{4x+18}{(x+3)(x-2)(x+5)}$

7. $\frac{x-y}{xy}-\frac{a-y}{ay}=\frac{x-y}{xy}\cdot\frac{a}{a}-\frac{a-y}{ay}\cdot\frac{x}{x}$
$=\frac{ax-ay-ax+xy}{axy}$
$=\frac{xy-ay}{axy}$
$=\frac{y(x-a)}{axy}$
$=\frac{x-a}{ax}$

8. $\dfrac{3x}{x^2-3x-18}-\dfrac{x-4}{x-6}$

$=\dfrac{3x}{(x-6)(x+3)}-\dfrac{x-4}{x-6}$

$=\dfrac{3x}{(x-6)(x+3)}-\dfrac{x-4}{x-6}\cdot\dfrac{x+3}{x+3}$

$=\dfrac{3x}{(x-6)(x+3)}-\dfrac{x^2-x-12}{(x-6)(x+3)}$

$=\dfrac{3x-(x^2-x-12)}{(x-6)(x+3)}$

$=\dfrac{3x-x^2+x+12}{(x-6)(x+3)}$

$=\dfrac{-x^2+4x+12}{(x-6)(x+3)}$

$=\dfrac{-(x-6)(x+2)}{(x-6)(x+3)}$

$=-\dfrac{x+2}{x+3}$

9. $\dfrac{\frac{x}{3y}-\frac{1}{2}}{\frac{4}{3y}-\frac{2}{x}}=\dfrac{\frac{x}{3y}-\frac{1}{2}}{\frac{4}{3y}-\frac{2}{x}}\cdot\dfrac{6xy}{6xy}$

$=\dfrac{2x^2-3xy}{8x-12y}$

$=\dfrac{x(2x-3y)}{4(2x-3y)}$

$=\dfrac{x}{4}$

10. $\dfrac{\frac{2}{x+3}+\frac{1}{x}}{3x+9}$

$=\dfrac{\frac{2}{x+3}+\frac{1}{x}}{3(x+3)}$

$=\dfrac{\frac{2}{x+3}+\frac{1}{x}}{3(x+3)}\cdot\dfrac{x(x+3)}{x(x+3)}$

$=\dfrac{2x+x+3}{3x(x+3)^2}$

$=\dfrac{3x+3}{3x(x+3)^2}$

$=\dfrac{3(x+1)}{3x(x+3)^2}$

$=\dfrac{x+1}{x(x+3)^2}$

11. $\dfrac{2x^2+3xy-9y^2}{4x^2+13xy+3y^2}$

$=\dfrac{(2x-3y)(x+3y)}{(4x+y)(x+3y)}$

$=\dfrac{2x-3y}{4x+y}$

12. $\dfrac{1}{x+4}-\dfrac{2}{x^2+6x+8}$

$=\dfrac{1}{x+4}-\dfrac{2}{(x+4)(x+2)}$

$=\dfrac{1}{x+4}\cdot\dfrac{x+2}{x+2}-\dfrac{2}{(x+4)(x+2)}$

$=\dfrac{x+2}{(x+4)(x+2)}-\dfrac{2}{(x+4)(x+2)}$

$=\dfrac{x+2-2}{(x+4)(x+2)}$

$=\dfrac{x}{(x+4)(x+2)}$

13. $$\frac{15}{x}+\frac{9x-7}{x+2}=9$$
$$x(x+2)\left(\frac{15}{x}\right)+x(x+2)\left(\frac{9x-7}{x+2}\right)$$
$$=x(x+2)(9)$$
$$15x+30+9x^2-7x=9x^2+18x$$
$$30=10x$$
$$3=x$$

14. $$\frac{x-3}{x-2}=\frac{2x^2-15}{x^2+x-6}-\frac{x+1}{x+3}$$
$$\frac{x-3}{x-2}=\frac{2x^2-15}{(x+3)}-\frac{x+1}{x+3}$$
$$(x-2)(x+3)\left(\frac{x-3}{x-2}\right)$$
$$=(x-2)(x+3)\left[\frac{2x^2-15}{(x+3)(x-2)}\right]$$
$$-(x-2)(x+3)\left(\frac{x+1}{x+3}\right)$$
$$(x+3)(x-3)=2x^2-15-(x-2)(x+1)$$
$$x^2-9=2x^2-15-(x^2-x-2)$$
$$x^2-9=2x^2-15-x^2+x+2$$
$$x^2-9=x^2+x-13$$
$$-9=x-13$$
$$4=x$$

Check: $$\frac{4-3}{4-2}\stackrel{?}{=}\frac{2(4)^2-15}{4^2+4-6}-\frac{4+1}{4+3}$$
$$\frac{1}{2}\stackrel{?}{=}\frac{17}{14}-\frac{5}{7}$$
$$\frac{7}{14}\stackrel{?}{=}\frac{17}{14}-\frac{10}{14}$$
$$\frac{7}{14}=\frac{7}{14}$$

15. $$3-\frac{7}{x+3}=\frac{x-4}{x+3}$$
$$(x+3)(3)-(x+3)\left(\frac{7}{x+3}\right)=(x+3)\left(\frac{x-4}{x+3}\right)$$
$$3x+9-7=x-4$$
$$2x=-6$$
$$x=-3$$

$x=-3$ is not allowed.

There is no solution.

16. $$\frac{3}{3x-5}=\frac{7}{5x+4}$$
$$3(5x+4)=7(3x-5)$$
$$15x+12=21x-35$$
$$12=6x-35$$
$$47=6x$$
$$\frac{47}{6}=x$$

Check:
$$\frac{3}{3\left(\frac{47}{6}\right)-5}\stackrel{?}{=}\frac{7}{5\left(\frac{47}{6}\right)+4}$$
$$\frac{3}{\frac{111}{6}}\stackrel{?}{=}\frac{7}{\frac{259}{6}}$$
$$\frac{18}{111}\stackrel{?}{=}\frac{42}{259}$$
$$\frac{6}{37}=\frac{6}{37}$$

17. $$\frac{9}{x}=\frac{13}{5}$$
$$45=13x$$
$$\frac{45}{13}=x$$

18. $$\frac{9.3}{2.5}=\frac{x}{10}$$
$$\frac{10(9.3)}{2.5}=x$$
$$37.2=x$$

19. $$\frac{\text{time}}{\text{pages}}:\frac{55}{3}=\frac{x}{21}$$
$$3x=1155$$
$$x=385$$
385 minutes = 6 hours 25 minutes

20. x = Cost of wood for 92 days
$$\frac{x}{92}=\frac{100}{25}$$
$$25x=9200$$
$$x=368$$
It will cost $368.

21. $$\frac{\text{height}}{\text{shadow}}:\frac{6}{7}=\frac{87}{x}$$
$$6x=609$$
$$x=101.5$$
102 feet

Cumulative Test for Chapters 1 - 7

1. $$5(x-3)-2(4-2x)=7(x-1)-(x-2)$$
$$5x-15-8+4x=7x-7-x+2$$
$$9x-23=6x-5$$
$$3x-23=-5$$
$$3x=18$$
$$x=6$$

2. $$A=\pi r^2h$$
$$\frac{A}{\pi r^2}=\frac{\pi r^2h}{\pi r^2}$$
$$\frac{A}{\pi r^2}=h$$

3. $$4(2-x)<3$$
$$8-4x<3$$
$$-4x<-5$$
$$x>\frac{5}{4}$$

4. $$\frac{1}{4}x+\frac{3}{4}<\frac{2}{3}x-\frac{4}{3}$$
$$12\left(\frac{1}{4}x\right)+12\left(\frac{3}{4}\right)<12\left(\frac{2}{3}x\right)-12\left(\frac{4}{3}\right)$$
$$3x+9<8x-16$$
$$9<5x-16$$
$$25<5x$$
$$5<x$$
$$x>5$$

5. $y=2x-1$

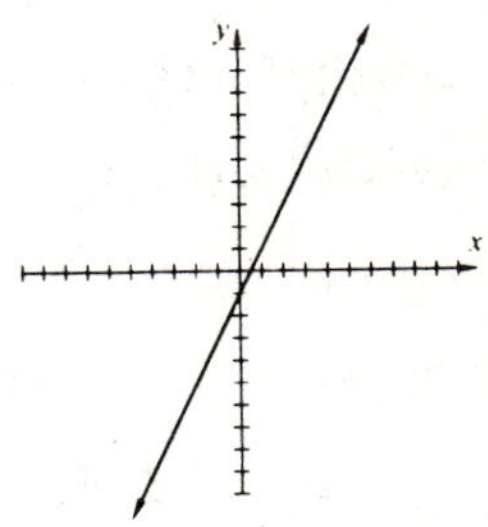

6. $y = x + 5 \Rightarrow m_1 = 1$

   Parallel to $y = x + 5 \Rightarrow m_2 = m_1 = 1$

   $y = mx + b, \ (-3, 0)$

   $0 = 1(-3) + b \Rightarrow b = 3$

   $y = x + 3$

7. $x - y - z = 8 \quad (1)$

   $2y - z = 1 \quad (2)$

   $3x + y + z = -4 \quad (3)$

   Add (1) and (3)

   $4x = 4$

   $x = 1$

   Add (3) and (2)

   $3x + 3y = -3 \quad (4)$

   Substitute 1 for x in (4)

   $3(1) + 3y = -3$

   $3y = -6$

   $y = -2$

   Substitute -2 for y in (2)

   $2(-2) - z = 1$

   $-z = 5$

   $z = -5$

   The solution is $(1, -2, -5)$.

8. $(-3x^2y)(4x^3y^6) = -3 \cdot 4x^{2+3}y^{1+6} = -12x^5y^7$

9. $(2x-7)^2 = (2x)^2 - 2(2x)(7) + (7)^2$

   $= 4x^2 - 28x + 49$

10. $3ax + 3bx - 2ay - 2by$

    $= 3x(a+b) - 2y(a+b)$

    $= (a+b)(3x-2y)$

11. $\dfrac{4x^2 - 25}{2x^2 + 9x - 35} = \dfrac{(2x+5)(2x-5)}{(2x-5)(x+7)}$

    $= \dfrac{2x+5}{x+7}$

12. $\dfrac{x^2 - 4}{x^2 - 25} \cdot \dfrac{3x^2 - 14x - 5}{3x^2 + 6x}$

    $= \dfrac{(x+2)(x-2)}{(x+5)(x-5)} \cdot \dfrac{(3x+1)(x-5)}{3x(x+2)}$

    $= \dfrac{(x-2)(3x+1)}{3x(x+5)}$

13. $\dfrac{2x^2 - 9x + 9}{8x - 12} \div \dfrac{x^2 - 3x}{2x}$

    $= \dfrac{2x^2 - 9x + 9}{8x - 12} \cdot \dfrac{2x}{x^2 - 3x}$

    $= \dfrac{(2x-3)(x-3)}{4(2x-3)} \cdot \dfrac{2x}{x(x-3)}$

    $= \dfrac{1}{2}$

14. $\dfrac{5}{2x+4} + \dfrac{3}{x-3}$

    $= \dfrac{5}{2(x+2)} \cdot \dfrac{x-3}{x-3} + \dfrac{3}{x-3} \cdot \dfrac{2(x+2)}{2(x+2)}$

    $= \dfrac{5x - 15 + 6x + 12}{2(x+2)(x-3)}$

    $= \dfrac{11x - 3}{2(x+2)(x-3)}$

15. $\dfrac{2x+1}{x^2+x-12}-\dfrac{x-1}{x^2-x-6}$

$=\dfrac{2x+1}{(x+4)(x-3)}-\dfrac{x-1}{(x-3)(x+2)}$

$=\dfrac{2x+1}{(x+4)(x-3)}\cdot\dfrac{x+2}{x+2}-\dfrac{x-1}{(x-3)(x+2)}\cdot\dfrac{x+4}{x+4}$

$=\dfrac{2x^2+5x+2}{(x+4)(x-3)(x+2)}-\dfrac{x^2+3x-4}{(x+4)(x-3)(x+2)}$

$=\dfrac{2x^2+5x+2-(x^2+3x-4)}{(x+4)(x-3)(x+2)}$

$=\dfrac{2x^2+5x+2-x^2-3x+4}{(x+4)(x-3)(x+2)}$

$=\dfrac{x^2+2x+6}{(x+4)(x-3)(x+2)}$

16. $\dfrac{3x-2}{3x+2}=2$

$(3x+2)\left(\dfrac{3x-2}{3x+2}\right)=(3x+2)(2)$

$3x-2=6x+4$

$-6=3x$

$-2=x$

17. $\dfrac{x-3}{x}=\dfrac{x+2}{x+3}$

$x(x+2)=(x-3)(x+3)$

$x^2+2x=x^2-9$

$2x=-9$

$x=-\dfrac{9}{2}$

18. $\dfrac{\frac{1}{x-3}+\frac{5}{x^2-9}}{\frac{6x}{x+3}}=\dfrac{\frac{1}{x-3}+\frac{5}{(x-3)(x-3)}}{\frac{6x}{x+3}}\cdot\dfrac{(x+3)(x-3)}{(x+3)(x-3)}$

$=\dfrac{x+3+5}{6x(x-3)}$

$=\dfrac{x+8}{6x(x-3)}$

19. $\dfrac{\frac{3}{a}+\frac{2}{b}}{\frac{5}{a^2}-\frac{2}{b^2}}=\dfrac{\frac{3}{a}+\frac{2}{b}}{\frac{5}{a^2}-\frac{2}{b^2}}\cdot\dfrac{a^2b^2}{a^2b^2}$

$=\dfrac{3ab^2+2a^2b}{5b^2-2a^2}$

20. $\dfrac{5}{7}=\dfrac{14}{x}$

$5x=98$

$x=\dfrac{98}{5}$

21. $\dfrac{\text{mi}}{\text{in.}}:\dfrac{130}{2\frac{1}{2}}=\dfrac{x}{4}$

$\left(2\dfrac{1}{2}\right)x=520$

$\dfrac{5}{2}x=520$

$x=208$

22. x = phone calls to make 110 sales

$\dfrac{x}{110}=\dfrac{22}{5}$

$5x=2420$

$x=484$

He must make 484 calls.

Chapter 8

Pretest Chapter 8

1. $(-3x^{\frac{1}{4}}y^{\frac{1}{2}})(-2x^{-\frac{1}{2}}y^{\frac{1}{3}}) = 6x^{\frac{1}{4}-\frac{1}{2}}y^{\frac{1}{2}+\frac{1}{3}}$
$$= 6x^{-\frac{1}{4}}y^{\frac{5}{6}}$$
$$= \frac{6y^{\frac{5}{6}}}{x^{\frac{1}{4}}}$$

2. $(-4x^{-\frac{1}{4}}y^{\frac{1}{3}})^3 = (-4)^3 x^{-\frac{3}{4}}y$
$$= -\frac{64y}{x^{\frac{3}{4}}}$$

3. $\dfrac{-18x^{-2}y^2}{-3x^{-5}y^{\frac{1}{3}}} = 6x^{-2+5}y^{2-\frac{1}{3}}$
$$= 6x^3y^{\frac{5}{3}}$$

4. $\left(\dfrac{27x^2y^{-5}}{x^{-4}y^4}\right)^{\frac{2}{3}} = \left(\dfrac{3^3x^6}{y^9}\right)^{\frac{2}{3}}$
$$= \left(\frac{3^2x^4}{y^6}\right)$$
$$= \frac{9x^4}{y^6}$$

5. $27^{-\frac{4}{3}} = \dfrac{1}{(3^3)^{\frac{4}{3}}}$
$$= \frac{1}{3^4}$$
$$= \frac{1}{81}$$

6. $\sqrt{169} + \sqrt[3]{-64} = 13 - 4 = 9$

7. $\sqrt[3]{27a^{12}b^6c^{15}} = \sqrt[3]{27}\sqrt[3]{a^{12}}\sqrt[3]{b^6}\sqrt[3]{c^{15}}$
$$= 3a^4b^2c^5$$

8. $\sqrt[4]{32x^8y^{15}} = \sqrt[4]{16\cdot 2x^8y^{12}y^3}$
$$= \sqrt[4]{16}\sqrt[4]{x^8}\sqrt[4]{y^{12}}\sqrt[4]{2y^3}$$
$$= 2x^2y^3\sqrt[4]{2y^3}$$

9. $3\sqrt{48y^3} - 2\sqrt[3]{16} + 3\sqrt[3]{54} - 5y\sqrt{12y}$
$$= 3\sqrt{16y^2\cdot 3y} - 2\sqrt[3]{8\cdot 2} + 3\sqrt[3]{27\cdot 2} - 5y\sqrt{4\cdot 3y}$$
$$= 12y\sqrt{3y} - 4\sqrt[3]{2} + 9\sqrt[3]{2} - 10y\sqrt{3y}$$
$$= 2y\sqrt{3y} + 5\sqrt[3]{2}$$

10. $(3\sqrt{3} - 5\sqrt{6})(\sqrt{12} - 3\sqrt{6})$
$$= 3\sqrt{36} - 9\sqrt{18} - 5\sqrt{72} + 15(6)$$
$$= 3(6) - 9\sqrt{9\cdot 2} - 5\sqrt{36\cdot 2} + 90$$
$$= 108 - 27\sqrt{2} - 30\sqrt{2}$$
$$= 108 - 57\sqrt{2}$$

11. $\dfrac{6}{\sqrt[3]{9x}}\cdot\dfrac{\sqrt[3]{3x^2}}{\sqrt[3]{3x^2}} = \dfrac{6\sqrt[3]{3x^2}}{\sqrt[3]{27x^3}}$
$$= \frac{6\sqrt[3]{3x^2}}{3x}$$
$$= \frac{2\sqrt[3]{3x^2}}{x}$$

12. $\dfrac{\sqrt{2}+\sqrt{3}}{\sqrt{2}-\sqrt{3}}\cdot\dfrac{\sqrt{2}+\sqrt{3}}{\sqrt{2}+\sqrt{3}} = \dfrac{2+2\sqrt{6}+3}{2-3}$
$$= -5 - 2\sqrt{6}$$

13. $\sqrt{3x+4}+2=x$

$\sqrt{3x+4}=x-2$

$3x+4=x^2-4x+4$

$x^2-7x=0$

$x(x-7)=0$

$x=0 \text{ or } x=7$

check:

$\sqrt{3(0)+4}+2\overset{?}{=}0$

$4\neq 0$

$\sqrt{3(7)+4}+2\overset{?}{=}7$

$7=7$

$x=7$ is the solution.

14. $\sqrt{2x+3}-\sqrt{x-2}=2$

$\sqrt{2x+3}=2+\sqrt{x-2}$

$2x+3=4+4\sqrt{x-2}+x-2$

$x+1=4\sqrt{x-2}$

$x^2+2x+1=16x-32$

$x^2-14x+33=0$

$(x-11)(x-3)=0$

$x=11 \text{ or } x=3$

check:

$\sqrt{2(11)+3}-\sqrt{11-2}\overset{?}{=}2$

$5-3=2$

$\sqrt{2(3)+3}-\sqrt{3-2}\overset{?}{=}2$

$3-1=2$

$x=3$, $x=11$ are solutions.

15. $(3-2i)-(-1+3i)=3-2i+1-3i$

$=4-5i$

16. $i^{15}+\sqrt{-25}=i^{4(3)+3}+i\sqrt{25}$

$=(i^4)^3\cdot i^3+5i$

$=(1)^3\cdot(-i)+5i$

$=4i$

17. $(3+5i)^2=9+30i+25i^2$

$=9+30i-25$

$=-16+30i$

18. $\dfrac{3+2i}{2+3i}\cdot\dfrac{2-3i}{2-3i}=\dfrac{6-5i-6i^2}{4+9}$

$=\dfrac{6-5i+6}{13}$

$=\dfrac{12-5i}{13}$

19. $y=kx^2$

$18=k(3)^2$

$k=2$

$y=2x^2$

$y=2(5)^2=50$

20. $y=\dfrac{k}{x}$

$12=\dfrac{k}{6}$

$k=72$

$y=\dfrac{72}{x^2}$

$y=\dfrac{72}{10^2}=7.2$

8.1 Exercises

1. $\left(\dfrac{4x^2y^{-3}}{x}\right)^2=\dfrac{4^2x^4y^{-6}}{x^2}=\dfrac{16x^2}{y^6}$

3. $\left(\frac{2a^{-1}b^3}{-3b^2}\right)^3 = \frac{8a^{-3}b^9}{-27b^6} = \frac{8b^3}{-27a^3}$

5. $(x^{\frac{3}{4}})^2 = x^{\frac{6}{4}} = x^{\frac{3}{2}}$

7. $(y^{12})^{\frac{2}{3}} = y^{12\cdot\frac{2}{3}} = y^8$

9. $\frac{x^{\frac{7}{12}}}{x^{\frac{1}{12}}} = x^{\frac{7}{12}-\frac{1}{12}} = x^{\frac{6}{12}} = x^{\frac{1}{2}}$

11. $\frac{x^3}{x^{\frac{1}{2}}} = x^{3-\frac{1}{2}} = x^{\frac{6}{2}-\frac{1}{2}} = x^{\frac{5}{2}}$

13. $x^{\frac{1}{7}}\cdot x^{\frac{3}{7}} = x^{\frac{1}{7}+\frac{3}{7}} = x^{\frac{4}{7}}$

15. $y^{\frac{3}{5}}\cdot y^{-\frac{1}{10}} = y^{\frac{6}{10}-\frac{1}{10}} = y^{\frac{1}{2}}$

17. $x^{-\frac{3}{4}} = \frac{1}{x^{\frac{3}{4}}}$

19. $a^{-\frac{5}{6}}b^{\frac{1}{3}} = \frac{b^{\frac{1}{3}}}{a^{\frac{5}{6}}}$

21. $6^{-\frac{1}{2}} = \frac{1}{6^{\frac{1}{2}}}$

23. $2a^{-\frac{1}{4}} = \frac{2}{a^{\frac{1}{4}}}$

25. $(x^{\frac{1}{2}}y^{\frac{1}{3}})(x^{\frac{1}{3}}y^{\frac{2}{3}}) = x^{\frac{3}{6}+\frac{2}{6}}y^{\frac{1}{3}+\frac{2}{3}} = x^{\frac{5}{6}}y$

27. $(7x^{\frac{1}{3}}y^{\frac{1}{4}})(-2x^{\frac{1}{4}}y^{-\frac{1}{6}}) = -14x^{\frac{7}{12}}y^{\frac{1}{12}}$

29. $6^2\cdot 6^{-\frac{2}{3}} = 6^{\frac{6}{3}-\frac{2}{3}} = 6^{\frac{4}{3}}$

31. $\frac{2x^{\frac{1}{5}}}{x^{-\frac{1}{2}}} = 2x^{\frac{2}{10}+\frac{5}{10}} = 2x^{\frac{7}{10}}$

33. $\frac{-20x^2y^{-\frac{1}{5}}}{5x^{-\frac{1}{2}}y} = -\frac{4x^{\frac{5}{2}}}{y^{\frac{6}{5}}}$

35. $\left(\frac{8a^2b^6}{a^{-1}b^3}\right)^{\frac{1}{3}} = \left(8a^3b^3\right)^{\frac{1}{3}} = 2ab$

37. $(-3x^{\frac{2}{5}}y^{\frac{3}{2}}z^{\frac{1}{3}})^2 = (-3)^2x^{\frac{4}{5}}y^3z^{\frac{2}{3}} = 9x^{\frac{4}{5}}y^3z^{\frac{2}{3}}$

39. $x^{\frac{2}{3}}(x^{\frac{4}{3}} - x^{\frac{1}{5}}) = x^{\frac{6}{3}} - x^{\frac{2}{3}+\frac{1}{5}} = x^2 - x^{\frac{13}{15}}$

41. $m^{\frac{7}{8}}(m^{-\frac{1}{2}} + 2m) = m^{\frac{7}{8}-\frac{4}{8}} + 2m^{\frac{7}{8}+\frac{8}{8}}$
$= m^{\frac{3}{8}} + 2m^{\frac{15}{8}}$

43. $\frac{(x^2\cdot x^{-\frac{3}{2}})^{\frac{1}{2}}}{x^{\frac{1}{2}}} = \frac{(x^{\frac{4}{2}-\frac{3}{2}})^{\frac{1}{2}}}{x^{\frac{1}{2}}} = \frac{(x^{\frac{1}{2}})^{\frac{1}{2}}}{x^{\frac{1}{2}}}$
$= \frac{x^{\frac{1}{4}}}{x^{\frac{1}{2}}} = \frac{1}{x^{\frac{1}{4}}}$

45. $(27)^{\frac{2}{3}} = (3^3)^{\frac{2}{3}} = 3^2 = 9$

47. $(4)^{\frac{3}{2}} = (2^2)^{\frac{3}{2}} = 2^3 = 8$

49. $(81)^{\frac{3}{4}} + (25)^{\frac{1}{2}} = (3^4)^{\frac{3}{4}} + (5^2)^{\frac{1}{2}}$
$= 3^3 + 5 = 27 + 5 = 32$

51. $2y^{\frac{1}{3}} + y^{-\frac{2}{3}} = 2y^{\frac{1}{3}} + \frac{1}{y^{\frac{2}{3}}}$

$$= \frac{2y^{\frac{1}{3}+\frac{2}{3}} + 1}{y^{\frac{2}{3}}}$$

$$= \frac{2y+1}{y^{\frac{2}{3}}}$$

53. $5^{-\frac{1}{4}} + x^{-\frac{1}{2}} = \frac{1}{5^{\frac{1}{4}}} + \frac{1}{x^{\frac{1}{2}}}$

$$= \frac{x^{\frac{1}{2}} + 5^{\frac{1}{4}}}{5^{\frac{1}{4}}x^{\frac{1}{2}}}$$

55. $6a^{\frac{4}{3}} - 8a^{\frac{3}{2}} = 2a(3a^{\frac{1}{3}} - 4a^{\frac{1}{2}})$

57. $x^b \div x^{\frac{1}{3}} = x^{\frac{1}{12}}$

$$x^{b-\frac{1}{3}} = x^{\frac{1}{12}}$$

$$b - \frac{1}{3} = \frac{1}{12}$$

$$b = \frac{1}{4}$$

59. $r = 0.62(V)^{\frac{1}{3}}$

$$r = 0.62(64)^{\frac{1}{3}}$$

$r = 0.62(4) = 2.48$ m

61. $r = \left(\frac{3V}{\pi h}\right)^{\frac{1}{2}}$

$$r = \left(\frac{3(3140)}{3.14(30)}\right)^{\frac{1}{2}} \approx 10 \text{ feet}$$

Cumulative Review Problems

63. $A = \frac{h}{2}(a+b)$

$$2A = ha + hb$$

$$hb = 2A - ha$$

$$b = \frac{2A - ha}{h} = \frac{2A}{h} - a$$

65. $y = \frac{ax}{a+12}$

$$75 = \frac{a \cdot 250}{a+12}$$

$$75a + 900 = 250a$$

$$175a = 900$$

$$a = \frac{900}{175} = \frac{36}{7} = 5\frac{1}{7}$$

To the nearest year, the child is 5 years old.

8.2 Exercises

1. A square root of a number is a value that when multiplied by itself is equal to the original number.

3. One answer is $\sqrt[3]{-8} = -2$ because $(-2)(-2)(-2) = -8$.

5. $\sqrt{64} = 8$

7. $\sqrt{25} + \sqrt{49} = 5 + 7 = 12$

9. $-\sqrt{\frac{1}{9}} = -\frac{1}{3}$

11. $\sqrt{36} - \sqrt{25} = 6 - 5 = 1$

13. $\sqrt{0.04} = 0.02$

15. $f(x) = \sqrt{10x+5}$
$f(0) = \sqrt{10(0)+5} = \sqrt{5} = 2.2$
$f(1) = \sqrt{10(1)+5} = \sqrt{15} = 3.9$
$f(2) = \sqrt{10(2)+5} = \sqrt{25} = 5$
$f(3) = \sqrt{10(3)+5} = \sqrt{35} = 5.9$
domain is $10x+5 \geq 0$
$10x \geq -5$
$x \geq -0.5$

17. $f(x) = \sqrt{0.5x-3}$
$f(6) = \sqrt{0.5(6)-3} = \sqrt{3-3} = \sqrt{0} = 0$
$f(8) = \sqrt{0.5(8)-3} = \sqrt{4-3} = \sqrt{1} = 1$
$f(14) = \sqrt{0.5(14)-3} = \sqrt{7-3} = \sqrt{4} = 2$
$f(16) = \sqrt{0.5(16)-3} = \sqrt{8-3} = \sqrt{5} = 2.2$
domain is $0.5x-3 \geq 0$
$0.5x \geq 3$
$x \geq 6$

19. $f(x) = \sqrt{x-3}$

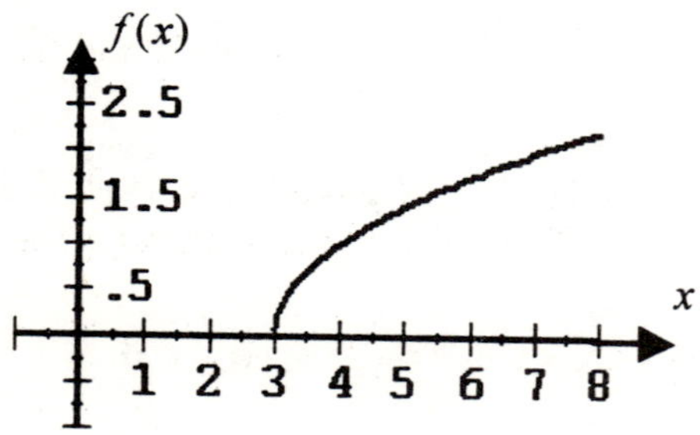

21. $f(x) = \sqrt{2x+4}$

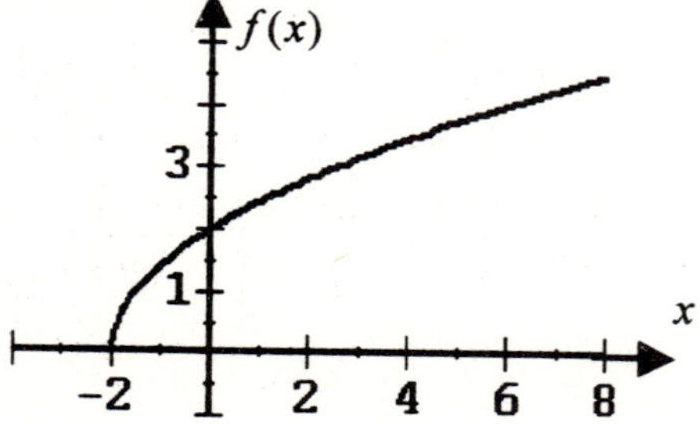

23. $\sqrt[3]{216} = 6$

25. $\sqrt[3]{64} = 4$

27. $\sqrt[3]{-8} = -2$

29. $\sqrt[4]{81} = 3$

31. $\sqrt[5]{(8)^5} = 8$

33. $\sqrt[8]{(5)^8} = 5$

35. $\sqrt[3]{-\frac{1}{64}} = -\frac{1}{4}$

37. $\sqrt[3]{\frac{27}{125}} = \frac{\sqrt[3]{27}}{\sqrt[3]{125}} = \frac{3}{5}$

39. $\sqrt{a} = a^{\frac{1}{2}}$

41. $\sqrt[4]{3y} = (3y)^{\frac{1}{4}}$

43. $\sqrt[9]{(a-b)^5} = (a-b)^{\frac{5}{9}}$

45. $\sqrt[5]{\sqrt{y}} = (y^{\frac{1}{2}})^{\frac{1}{5}} = y^{\frac{1}{10}}$

47. $(\sqrt[5]{2x})^3 = \left[(2x)^{\frac{1}{5}}\right]^3 = (2x)^{\frac{3}{5}}$

49. $\sqrt[5]{(-11)^5} = -11$

51. $\sqrt[4]{a^8b^4} = \sqrt[4]{(a^2)^4b^4} = a^2b$

53. $\sqrt{49x^2y^8} = \sqrt{7^2x^2(y^4)^2} = 7xy^4$

55. $\sqrt[4]{81a^{12}b^{20}} = \sqrt[4]{3^4(a^3)^4(b^5)^4} = 3a^3b^5$

57. $\sqrt[3]{8x^3y^9} = \sqrt[3]{2^3x^3(y^3)^3} = 2xy^3$

59. $x^{\frac{5}{6}} = (x^{\frac{1}{6}})^5 = (\sqrt[6]{x})^5$

61. $5^{-\frac{3}{5}} = \dfrac{1}{(5^{\frac{1}{5}})^3} = \dfrac{1}{(\sqrt[5]{5})^3} = \dfrac{1}{\sqrt[5]{125}}$

63. $(x+3y)^{\frac{4}{7}} = \left((x+3y)^{\frac{1}{7}}\right)^4 = \left(\sqrt[7]{x+3y}\right)^4$

65. $(-y)^{\frac{5}{7}} = \left((-y)^{\frac{1}{7}}\right)^5 = \left(\sqrt[7]{-y}\right)^5$

67. $(3ab)^{\frac{2}{7}} = \sqrt[7]{(3ab)^2} = \sqrt[7]{9a^2b^2}$

69. $27^{\frac{2}{3}} = (\sqrt[3]{27})^2 = 3^2 = 9$

71. $\left(\dfrac{4}{25}\right)^{\frac{1}{2}} = \sqrt{\dfrac{4}{25}} = \dfrac{2}{5}$

73. $(-125)^{\frac{2}{3}} = (\sqrt[3]{-125})^2 = (-5)^2 = 25$

75. $(36y^8)^{-\frac{1}{2}} = \dfrac{1}{(6^2(y^4)^2)^{\frac{1}{2}}} = \dfrac{1}{6y^4}$

77. $\sqrt{49x^8} = 7x^4$

79. $\sqrt{25a^{14}b^{18}} = 5a^7b^9$

81. $\sqrt{100x^{10}y^{12}z^2} = 10x^5y^6z$

83. $\sqrt[3]{-125a^6b^{15}c^{21}} = -5a^2b^5c^7$

85. $\sqrt{100x^2} = 10|x|$

87. $\sqrt[3]{-27x^9} = -3x^3$

89. $\sqrt[4]{x^{16}y^{40}} = x^4y^{10}$

91. $\sqrt[4]{a^4b^{20}} = \sqrt[4]{(ab^5)^4} = |ab^5|$

93. $\sqrt{49a^{12}b^4} = 7a^6b^2$

95. $C = 120\sqrt[3]{n} + 375$
$C = 120\sqrt[3]{216} + 375 = 120(6) + 375$
$C = 1095$
Cost is \$1095 per day.

Cumulative Review Problems

97. $\dfrac{1.24(0.30)}{1.52} = \dfrac{93}{380} = 0.2447368421...$

24.47% of the world's energy in 2020 will be produced in North America.

99. $2x^2 - x - 21 = (2x-7)(x+3)$

101. $\dfrac{6}{2x-4} + \dfrac{1}{12x+4}$
$= \dfrac{6\cdot 2(3x+1)}{2(x-2)\cdot 2(3x+1)} + \dfrac{1\cdot(x-2)}{4(3x+1)(x-2)}$
$= \dfrac{36x+12+x-2}{4(3x+1)(x-2)} = \dfrac{37x+10}{4(x+3)(x-2)}$

8.3 Exercises

1. $\sqrt{8} = \sqrt{4 \cdot 2} = \sqrt{4}\sqrt{2} = 2\sqrt{2}$

3. $\sqrt{18} = \sqrt{9 \cdot 2} = \sqrt{9}\sqrt{2} = 3\sqrt{2}$

5. $\sqrt{120} = \sqrt{4 \cdot 30} = \sqrt{4}\sqrt{30} = 2\sqrt{30}$

7. $\sqrt{44} = \sqrt{4 \cdot 11} = \sqrt{4}\sqrt{11} = 2\sqrt{11}$

9. $\sqrt{9x^3} = \sqrt{9}\sqrt{x^2}\sqrt{x} = 3x\sqrt{x}$

11. $\sqrt{60a^4b^5} = \sqrt{4 \cdot 15a^4b^4b}$
$= 2a^2b^2\sqrt{15b}$

13. $\sqrt{98x^5y^6z} = \sqrt{49 \cdot 2x^4xy^6z}$
$= 7x^2y^3\sqrt{2xz}$

15. $\sqrt[3]{8} = 2$

17. $\sqrt[3]{108} = \sqrt[3]{27 \cdot 4} = 3\sqrt[3]{4}$

19. $\sqrt[3]{56y} = \sqrt[3]{8 \cdot 7y} = 2\sqrt[3]{7y}$

21. $\sqrt[3]{8a^3b^8} = \sqrt[3]{2^3a^3b^6b^2} = 2ab^2\sqrt[3]{b^2}$

23. $\sqrt[3]{24x^6y^{11}} = \sqrt[3]{8 \cdot 3x^6y^9y^2}$
$= 2x^2y^3\sqrt[3]{3y^2}$

25. $\sqrt[4]{81kp^{23}} = \sqrt[4]{3^4kp^{20}p^3}$
$= 3p^5\sqrt[4]{kp^3}$

27. $\sqrt[5]{-32x^5y^6} = \sqrt[5]{(-2)^5x^5y^5y}$
$= -2xy\sqrt[5]{y}$

29. $\sqrt[4]{1792} = a\sqrt[4]{7}$
$\sqrt[4]{256 \cdot 7} = a\sqrt[4]{7}$
$\sqrt[4]{4^4 \cdot 7} = a\sqrt[4]{7}$
$4\sqrt[4]{7} = a\sqrt[4]{7}$
$a = 4$

31. $\sqrt{49} + \sqrt{100} = 7 + 10$
$= 17$

33. $\sqrt{3} + 7\sqrt{3} - 2\sqrt{3} = 6\sqrt{3}$

35. $3\sqrt{18} - \sqrt{2} = 3\sqrt{9 \cdot 2} - \sqrt{2}$
$= 9\sqrt{2} - \sqrt{2}$
$= 8\sqrt{2}$

37. $-2\sqrt{50} + \sqrt{32} - 3\sqrt{8}$
$= -2\sqrt{25 \cdot 2} + \sqrt{16 \cdot 2} - 3\sqrt{4 \cdot 2}$
$= -10\sqrt{2} + 4\sqrt{2} - 6\sqrt{2}$
$= -12\sqrt{2}$

39. $-5\sqrt{45} + 6\sqrt{20} + 3\sqrt{5}$
$= -5\sqrt{9 \cdot 5} + 6\sqrt{4 \cdot 5} + 3\sqrt{5}$
$= -15\sqrt{5} + 12\sqrt{5} + 3\sqrt{5}$
$= 0$

41. $\sqrt{44} - 3\sqrt{63x} + 4\sqrt{28x}$
$= \sqrt{4 \cdot 11} - 3\sqrt{9 \cdot 7x} + 4\sqrt{4 \cdot 7x}$
$= 2\sqrt{11} - 9\sqrt{7x} + 8\sqrt{7x}$
$= 2\sqrt{11} - \sqrt{7x}$

43. $\sqrt{200x^3} - x\sqrt{32x}$
$= \sqrt{100x^2 \cdot 2x} - x\sqrt{16 \cdot 2x}$
$= 10x\sqrt{2x} - 4x\sqrt{2x}$
$= 6x\sqrt{2x}$

45. $\sqrt[3]{16}+3\sqrt[3]{54}=\sqrt[3]{2^3\cdot 2}+3\sqrt[3]{3^3\cdot 2}$
$=2\sqrt[3]{2}+9\sqrt[3]{2}=11\sqrt[3]{2}$

47. $-2\sqrt[3]{125x^3y^4}+3y^2\sqrt[3]{8x^3}$
$=-2\sqrt[3]{5^3x^3y^3y}+3y^2\cdot 2x$
$=-10xy\sqrt[3]{y}+6xy^2$

49. $\sqrt{48}+\sqrt{27}+\sqrt{75}$
$=6.92820323$
$+5.196152423$
$+8.660254038=20.78460969$
$12\sqrt{3}=20.78460969$ which shows
$\sqrt{48}+\sqrt{27}+\sqrt{75}=12\sqrt{3}$

51. $I=\sqrt{\dfrac{P}{R}}=\sqrt{\dfrac{500}{10}}=\sqrt{50}\approx 7.071$
The current is approximately 7.071 amps.

53. $T=2\pi\sqrt{\dfrac{L}{32}}=2(3.14)\sqrt{\dfrac{8}{32}}\approx 3.14$
The period of the pendulum is approximately 3.14 seconds.

Cumulative Review Problems

55. $3y-2x=9$

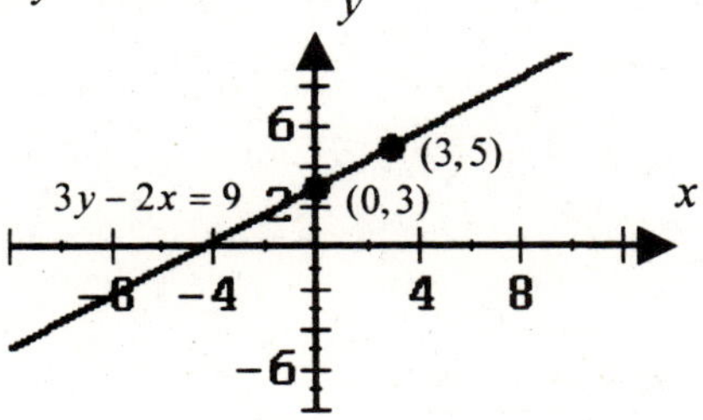

57. $\left(\dfrac{2x^2y}{3xy^3}\right)^{-2}=\left(\dfrac{2x}{3y^2}\right)^{-2}=\left(\dfrac{3y^2}{2x}\right)^{2}=\dfrac{9y^4}{4x^2}$

8.4 Exercises

1. $(2\sqrt{6})(-3\sqrt{2})=-6\sqrt{12}$
$=-6\sqrt{4\cdot 3}$
$=-12\sqrt{3}$

3. $(-3\sqrt{y})(\sqrt{5x})=-3\sqrt{5xy}$

5. $7\sqrt{x}(2\sqrt{3}-5\sqrt{x})=14\sqrt{3x}-35x$

7. $(3-\sqrt{2})(8+\sqrt{2})$
$=24-5\sqrt{2}-2$
$=22-5\sqrt{2}$

9. $(2\sqrt{3}+\sqrt{2})(2\sqrt{3}-4\sqrt{2})$
$=12-6\sqrt{6}-8$
$=4-6\sqrt{6}$

11. $(\sqrt{7}+4\sqrt{5x})(2\sqrt{7}+3\sqrt{5x})$
$=2\cdot 7+11\sqrt{35x}+12(5x)$
$=14+11\sqrt{35x}+60x$

13. $(\sqrt{3}+2\sqrt{2})(\sqrt{5}+\sqrt{3})$
$=\sqrt{15}+3+2\sqrt{10}+2\sqrt{6}$

15. $(\sqrt{x}-2\sqrt{3x})(\sqrt{x}+2\sqrt{3x})$
$=(\sqrt{x})^2-(2\sqrt{3x})^2$
$=x-12x$
$=-11x$

17. $(\sqrt{5}-2\sqrt{6})^2$
$=(\sqrt{5})^2-4\sqrt{5}\sqrt{6}+(2\sqrt{6})^2$
$=5-4\sqrt{30}+24$
$=29-4\sqrt{30}$

19. $(\sqrt{3x+4}+3)^2$
$= (\sqrt{3x+4})^2 + 6\sqrt{3x+4} + 9$
$= 3x + 4 + 6\sqrt{3x+4} + 9$
$= 3x + 13 + 6\sqrt{3x+4}$

21. $(6-5\sqrt{a})^2 = 36 - 60\sqrt{a} + 25a$

23. $(\sqrt[3]{x^2})(3\sqrt[3]{4x} - 4\sqrt[3]{x^5})$
$= 3\sqrt[3]{4x^3} - 4\sqrt[3]{x^7}$
$= 3x\sqrt[3]{4} - 4\sqrt[3]{x^6 \cdot x}$
$= 3x\sqrt[3]{4} - 4x^2\sqrt[3]{x}$

25. $\sqrt{\dfrac{49}{25}} = \dfrac{\sqrt{49}}{\sqrt{25}} = \dfrac{7}{5}$

27. $\sqrt{\dfrac{12x}{49y^6}} = \dfrac{\sqrt{4 \cdot 3x}}{\sqrt{49y^6}} = \dfrac{2\sqrt{3x}}{7y^3}$

29. $\sqrt[3]{\dfrac{8x^5y^6}{27}} = \dfrac{\sqrt[3]{8x^3y^6 \cdot x^2}}{\sqrt[3]{27}} = \dfrac{2xy^2\sqrt[3]{x^2}}{3}$

31. $\dfrac{\sqrt[3]{24x^3y^5}}{\sqrt[3]{3y^2}} = \sqrt[3]{\dfrac{24x^3y^5}{3y^2}} = \sqrt[3]{8x^3y^3} = 2xy$

33. $\dfrac{3}{\sqrt{2}} \cdot \dfrac{\sqrt{2}}{\sqrt{2}} = \dfrac{3\sqrt{2}}{2}$

35. $\sqrt{\dfrac{4}{3}} = \dfrac{\sqrt{4}}{\sqrt{3}} \cdot \dfrac{\sqrt{3}}{\sqrt{3}} = \dfrac{2\sqrt{3}}{3}$

37. $\dfrac{1}{\sqrt{5y}} \cdot \dfrac{\sqrt{5y}}{\sqrt{5y}} = \dfrac{\sqrt{5y}}{5y}$

39. $\dfrac{x}{\sqrt{5}-\sqrt{2}} \cdot \dfrac{\sqrt{5}+\sqrt{2}}{\sqrt{5}+\sqrt{2}} = \dfrac{x(\sqrt{5}+\sqrt{2})}{5-2}$
$= \dfrac{x(\sqrt{5}+\sqrt{2})}{3}$

41. $\dfrac{\sqrt{3}}{\sqrt{5}-2} \cdot \dfrac{\sqrt{5}+2}{\sqrt{5}+2} = \dfrac{\sqrt{15}+2\sqrt{3}}{5-4}$
$= \sqrt{15} + 2\sqrt{3}$

43. $\dfrac{\sqrt{x}}{\sqrt{3x}+\sqrt{2}} \cdot \dfrac{\sqrt{3x}-\sqrt{2}}{\sqrt{3x}-\sqrt{2}} = \dfrac{x\sqrt{3}-\sqrt{2x}}{3x-2}$

45. $\dfrac{\sqrt{5}+\sqrt{3}}{\sqrt{5}-\sqrt{3}} \cdot \dfrac{\sqrt{5}+\sqrt{3}}{\sqrt{5}+\sqrt{3}} = \dfrac{5+2\sqrt{15}+3}{5-3}$
$= \dfrac{8+2\sqrt{15}}{2}$
$= 4 + \sqrt{15}$

47. $\dfrac{\sqrt{3x}-2\sqrt{y}}{\sqrt{3x}+\sqrt{y}} \cdot \dfrac{\sqrt{3x}-\sqrt{y}}{\sqrt{3x}-\sqrt{y}}$
$= \dfrac{3x-3\sqrt{3xy}+2y}{3x-y}$

49. $\dfrac{x\sqrt{5}+1}{\sqrt{5}+2} \cdot \dfrac{\sqrt{5}-2}{\sqrt{5}-2} = \dfrac{5x-2x\sqrt{5}+\sqrt{5}-2}{5-4}$
$= 5x - 2x\sqrt{5} + \sqrt{5} - 2$

51. $\dfrac{5}{\sqrt{2}-\sqrt{3}} \cdot \dfrac{\sqrt{2}+\sqrt{3}}{\sqrt{2}+\sqrt{3}} = \dfrac{5(\sqrt{2}+\sqrt{3})}{2-3}$
$= -5(\sqrt{2}+\sqrt{3})$

53. $\dfrac{\sqrt[3]{x^2}}{\sqrt[3]{7x^2}} \cdot \dfrac{\sqrt[3]{7^2x}}{\sqrt[3]{7^2x}} = \dfrac{x\sqrt[3]{49}}{7x} = \dfrac{\sqrt[3]{49}}{7}$

55. $\dfrac{\sqrt{6}}{2\sqrt{3}-\sqrt{2}} = 1.194938299...$

$\dfrac{\sqrt{3}+3\sqrt{2}}{5} = 1.194938299...$

The decimal approximations are the same. The student worked correctly.

57. $\dfrac{\sqrt{3}+2\sqrt{7}}{8} \cdot \dfrac{\sqrt{3}-2\sqrt{7}}{\sqrt{3}-2\sqrt{7}}$

$= \dfrac{3-28}{8(\sqrt{3}+2\sqrt{7})}$

$= \dfrac{-25}{8(\sqrt{3}+2\sqrt{7})}$

59. $C = 0.18\dfrac{\sqrt{21}\sqrt{50}}{2}$

$C = 2.916333314...$

\$2.92 is the cost to fertilize the lawn.

61. $A = LW$

$A = (\sqrt{x}+5)(\sqrt{x}+3)$

$A = x + 8\sqrt{x} + 15 \text{ mm}^2$

Cumulative Review Problems

63. $2x+3y=13 \rightarrow 4x+6y=26$

$5x-2y=4 \rightarrow \underline{15x-6y=12}$

$19x = 38$

$x = 2$

$2(2)+3y=13$

$3y = 9$

$y = 3$

$x=2,\ y=3$ is the solution.

65. From the table, she reaches her goal on January 11th.

day	consumption
2nd	$280 \div 2 = 140$
5th	$140 \div 2 = 70$
8th	$70 \div 2 = 35$
11th	$35 \div 2 = 17.5$

8.5 Exercises

1. Isolate one of the radicals on one side of the equation.

3. $\sqrt{8x+1} = 5$

$8x+1 = 25$

$8x = 24$

$x = 3$

check: $\sqrt{8(3)+1} \stackrel{?}{=} 5$

$5 = 5$

5. $2x = \sqrt{x+3}$

$4x^2 = x+3$

$4x^2 - x - 3 = 0$

$(4x+3)(x-1) = 0$

$x = -\dfrac{3}{4}$ or $x = 1$

check: $2\left(-\dfrac{3}{4}\right) \stackrel{?}{=} \sqrt{-\dfrac{3}{4}+3}$

$-\dfrac{3}{2} \neq \dfrac{3}{2}$

$2(1) \stackrel{?}{=} \sqrt{1+3}$

$2 = 2$

$x = 1$ is the solution.

7. $y-\sqrt{y-3}=5$

$y-5=\sqrt{y-3}$

$y^2-10y+25=y-3$

$y^2-11y+28=0$

$(y-7)(y-4)=0$

$y=7$ or $y=4$

check:

$4-\sqrt{4-3}\stackrel{?}{=}5$

$3\neq 5$

$7-\sqrt{7-3}\stackrel{?}{=}5$

$5=5$

$y=7$ is the solution.

9. $\sqrt{y+1}-1=y$

$\sqrt{y+1}=y+1$

$y+1=y^2+2y+1$

$y^2+y=0$

$y(y+1)=0$

$y=0$ or $y=-1$

check: $\sqrt{0+1}-1\stackrel{?}{=}0$

$0=0$

$\sqrt{-1+1}-1\stackrel{?}{=}-1$

$-1=-1$

$y=0,\ y=-1$ is the solution.

11. $x-2\sqrt{x-3}=3$

$x-3=2\sqrt{x-3}$

$x^2-6x+9=4x-12$

$x^2-10x+21=0$

$(x-7)(x-3)=0$

$x=7$ or $x=3$

(continued)

11. (continued)

check:

$7-2\sqrt{7-3}\stackrel{?}{=}3$

$3=3$

$3-2\sqrt{3-3}\stackrel{?}{=}3$

$3=3$

$x=7,\ x=3$ is the solution.

13. $\sqrt{3x^2-x}=x$

$3x^2-x=x^2$

$2x^2-x=0$

$x(2x-1)=0$

$x=0$ or $x=\frac{1}{2}$

check:

$\sqrt{3(0)^2-0}\stackrel{?}{=}0$

$0=0$

$\sqrt{3\left(\frac{1}{2}\right)^2-\frac{1}{2}}\stackrel{?}{=}\frac{1}{2}$

$\frac{1}{2}=\frac{1}{2}$

$x=0,\ x=\frac{1}{2}$ is the solution.

15. $\sqrt[3]{2x+3}=2$

$2x+3=8$

$2x=5$

$x=\frac{5}{2}$

check: $\sqrt[3]{2\cdot\frac{5}{2}+3}\stackrel{?}{=}2$

$2=2$

$x=\frac{5}{2}$ is the solution.

17. $\sqrt[3]{4x-1}=3$

$4x-1=27$

$4x=28$

$x=7$

check:

$\sqrt[3]{4(7)-1}\stackrel{?}{=}3$

$3=3$

$x=7$ is the solution.

19. $\sqrt{x+4}=1+\sqrt{x-3}$

$x+4=1+2\sqrt{x-3}+x-3$

$3=\sqrt{x-3}$

$9=x-3$

$x=12$

check:

$\sqrt{12+4}\stackrel{?}{=}1+\sqrt{12-3}$

$4=4$

$x=12$ is the solution.

21. $\sqrt{x+6}=1+\sqrt{x+2}$

$x+6=1+2\sqrt{x+2}+x+2$

$9=4(x+2)$

$4x+8=9$

$4x=1$

$x=\frac{1}{4}$

check:

$\sqrt{\frac{1}{4}+6}\stackrel{?}{=}1+\sqrt{\frac{1}{4}+2}$

$\frac{5}{2}=\frac{5}{2}$

$x=\frac{1}{4}$ is the solution.

23. $\sqrt{6x+6}=1+\sqrt{4x+5}$

$6x+6=1+2\sqrt{4x+5}+4x+5$

$x^2=4x+5$

$x^2-4x-5=0$

$(x-5)(x+1)=0$

$x=5$ or $x=-1$

check:

$\sqrt{6(5)+6}\stackrel{?}{=}1+\sqrt{4(5)+5}$

$6=6$

$\sqrt{6(-1)+6}\stackrel{?}{=}1+\sqrt{4(-1)+5}$

$0\neq 2$

$x=5$ is the solution.

25. $\sqrt{2x+9}-\sqrt{x+1}=2$

$\sqrt{2x+9}=2+\sqrt{x+1}$

$2x+9=4+4\sqrt{x+1}+x+1$

$x+4=4\sqrt{x+1}$

$x^2+8x+16=16x+16$

$x^2-8x=0$

$x(x-8)=0$

$x=0$ or $x=8$

check:

$\sqrt{2(0)+9}-\sqrt{0+1}\stackrel{?}{=}2$

$2=2$

$\sqrt{2(8)+9}-\sqrt{8+1}\stackrel{?}{=}2$

$2=2$

$x=0$, $x=8$ is the solution.

27. $\sqrt{3x+4}+\sqrt{x+5}=\sqrt{7-2x}$

$3x+4+2\sqrt{3x+4}\sqrt{x+5}+x+5=7-2x$

$\sqrt{3x+4}\sqrt{x+5}=-1-3x$

$3x^2+19x+20=1+6x+9x^2$

$6x^2-13x-19=0$

$(6x-19)(x+1)=0$

$x=\frac{19}{6}$ or $x=-1$

check:

$$\sqrt{3\left(\frac{19}{6}\right)+4}+\sqrt{\frac{19}{6}+5}\stackrel{?}{=}\sqrt{7-2\left(\frac{19}{6}\right)}$$

$6.53197264\ldots \neq 0.8164965809\ldots$

$$\sqrt{3(-1)+4}+\sqrt{-1+5}\stackrel{?}{=}\sqrt{7-2(-1)}$$

$3=3$

$x=-1$ is the solution.

29. $2\sqrt{x}-\sqrt{x-5}=\sqrt{2x-2}$

$4x-4\sqrt{x}\sqrt{x-5}+x-5=2x-2$

$4\sqrt{x}\sqrt{x-5}=3x-3$

$16x^2-80x=9x^2-18x+9$

$7x^2-62x-9=0$

$(7x+1)(x-9)=0$

$x=-\frac{1}{7}$ or $x=9$

$x=-\frac{1}{7}$ does not check since it gives the square root of a negative.

$$2\sqrt{9}-\sqrt{9-5}\stackrel{?}{=}\sqrt{2(9)-2}$$

$4=4$

$x=9$ is the solution.

31. $x=\sqrt{5.326x-1.983}$

$x^2=5.326x-1.983$

$x^2-5.326x+1.983=0$

$x=0.4027856296$ or $x=4.92321437$

check:

$$0.4027856296\stackrel{?}{=}\sqrt{5.326(0.4027856296)-1.983}$$

$0.4027856296\stackrel{?}{=}0.4027856294$

$0.4028=0.4028$

$$4.92321437\stackrel{?}{=}\sqrt{5.326(4.92321437)-1.983}$$

$4.92321437\stackrel{?}{=}4.92321437$

$4.9232=4.9232$

$x=0.4028$, $x=4.9232$ is the solution.

33. (a) $V=2\sqrt{3S}$

$V^2=4(3S)$

$S=\frac{V^2}{12}$

(b) $S=\frac{18^2}{12}=27$ feet

35. $0.11y+1.25=\sqrt{3.7625+0.22x}$

$0.0121y^2+0.275y+1.5625=3.7625+0.22x$

$0.22x=0.0121y^2+0.275y-2.2$

$x=0.055y^2+1.25y-10$

37. $\sqrt{x^2-4x+c}=x-1$

$\sqrt{4^2-4(4)+c}=4-1$

$\sqrt{c}=3$

$c=9$

Cumulative Review Problems

39. $(4^3x^6)^{\frac{2}{3}} = 4^{3\cdot\frac{2}{3}}x^{6\cdot\frac{2}{3}} = 4^2x^4 = 16x^4$

41. $\sqrt[3]{-216x^6y^9} = -6x^2y^3$

43. $V = Ah = (4x^2+2x+9)(2x+3)$
$V = 8x^3+12x^2+4x^2+6x+18x+27$
$V = 8x^3+16x^2+24x+27 \text{ cm}^3$

45. $w =$ speed of current
$(12+w)\cdot 3 = (12-w)\cdot 5$
$36+3w = 60-5w$
$8w = 24$
$w = 3$

The current flows at 3 miles per hour.

8.6 Exercises

1. No. There is no real number that, when squared, will equal −9.

3. No. To be equal, the real number parts must be equal, and the imaginary number parts must be equal. $2 \neq 3$ and $3i \neq 2i$.

5. $\sqrt{-36} = \sqrt{36}\sqrt{-1} = 6i$

7. $\sqrt{-50} = \sqrt{25\cdot 2}\sqrt{-1} = 5i\sqrt{2}$

9. $\sqrt{-\frac{1}{4}} = \sqrt{\frac{1}{4}}\sqrt{-1} = \frac{1}{2}i$

11. $-\sqrt{-81} = -\sqrt{81}\sqrt{-1} = -9i$

13. $2+\sqrt{-3} = 2+\sqrt{3}\sqrt{-1} = 2+i\sqrt{3}$

15. $-3+\sqrt{-24} = -3+\sqrt{4\cdot 6}\sqrt{-1}$
$= -3+2i\sqrt{6}$

17. $x-3i = 5+yi$
$x = 5$
$y = -3$

19. $1.3-2.5yi = x-5i$
$x = 1.3$
$-2.5y = -5$
$y = 2$

21. $23+yi = 17-x+3i$
$23 = 17-x$
$x = -6$
$y = 3$

23. $\left(-\frac{3}{2}+\frac{1}{2}i\right)+\left(\frac{5}{2}-\frac{3}{2}i\right)$
$= -\frac{3}{2}+\frac{5}{2}+\left(\frac{1}{2}-\frac{3}{2}\right)i$
$= 1-i$

25. $(2.8-0.7i)-(1.6-2.8i)$
$= 2.8-1.6+(-0.7+2.8)i$
$= 1.2+2.1i$

27. $(2+3i)(2-i) = 4-2i+6i-3i^2$
$= 4+4i+3$
$= 7+4i$

29. $5i-2(-4+i) = 5i+8-2i = 8+3i$

31. $2i(5i-6) = 10i^2-12i = -10-12i$

33. $\left(\frac{1}{2}+i\right)^2 = \frac{1}{4}+i+i^2 = \frac{1}{4}+i-1 = -\frac{3}{4}+i$

35. $(i\sqrt{3})(i\sqrt{7}) = i^2\sqrt{21}$
$= -\sqrt{21}$

37. $(\sqrt{-3})(\sqrt{-2}) = (i\sqrt{3})(i\sqrt{2})$
$= i^2\sqrt{6}$
$= -\sqrt{6}$

39. $(\sqrt{-36})(\sqrt{-4}) = (i\sqrt{36})(i\sqrt{4})$
$= (6i)(2i)$
$= 12i^2$
$= -12$

41. $(3+\sqrt{-2})(4+\sqrt{-5})$
$= (3+i\sqrt{2})(4+i\sqrt{5})$
$= 12+3i\sqrt{5}+4i\sqrt{2}+i^2\sqrt{10}$
$= 12+3i\sqrt{5}+4i\sqrt{2}-\sqrt{10}$
$= 12-\sqrt{10}+(3\sqrt{5}+4\sqrt{2})i$

43. $i^{17} = (i^4)^4 \cdot i$
$= 1^4 \cdot i$
$= i$

45. $i^{24} = (i^4)^6$
$= 1^6$
$= 1$

47. $i^{46} = (i^4)^{11} \cdot i^2$
$= 1^{11}(-1)$
$= -1$

49. $i^{30} + i^{28} = (i^4)^7 \cdot i^2 + (i^4)^7$
$= 1^7(-1)+1^7$
$= -1+1$
$= 0$

51. $\frac{2+i}{3-i}\cdot\frac{3+i}{3+i} = \frac{6+5i+i^2}{9+1}$
$= \frac{6+5i-1}{10}$
$= \frac{5(1+i)}{10}$
$= \frac{1+i}{2}$

53. $\frac{3i}{4+2i}\cdot\frac{4-2i}{4-2i} = \frac{12i-6i^2}{16+4}$
$= \frac{6+12i}{20}$
$= \frac{3+6i}{10}$

55. $\frac{5-2i}{6i}\cdot\frac{-6i}{-6i} = \frac{-30i+12i^2}{36}$
$= \frac{-12-30i}{36}$
$= -\frac{2+5i}{6}$

57. $\frac{7}{5-6i}\cdot\frac{5+6i}{5+6i} = \frac{35+42i}{25+36}$
$= \frac{35+42i}{61}$

59. $\frac{5-2i}{3+2i}\cdot\frac{3-2i}{3-2i}$
$= \frac{15-10i-6i+4i^2}{9+4}$
$= \frac{15-16i-4}{13}$
$= \frac{11-16i}{13}$

61. $\frac{2-3i}{2+i}\cdot\frac{2-i}{2-i}$

$=\frac{4-2i-6i+3i^2}{4+1}$

$=\frac{4-8i-3}{5}$

$=\frac{1-8i}{5}$

63. $(29.3+56.2i)^2$

$=858.49+3293.32i+3158.44i^2$

$=858.49+3293.32i-3158.44$

$=-2299.95+3293.32i$

65. $Z=\frac{V}{I}$

$Z=\frac{3+2i}{3i}\cdot\frac{-3i}{-3i}$

$Z=\frac{-9i-6i^2}{9}$

$Z=\frac{6-9i}{9}$

$Z=\frac{2-3i}{3}$

Cumulative Review Problems

67. $x+3+2x-5+4x+2=105$

$7x=105$

$x=15$

$x+3=18$

$2x-5=25$

$4x+2=62$

18 hours producing juice in glass bottles, 25 hours producing juice in cans, and 62 hours producing juice in plastic bottles.

8.7 Exercises

1. Answers will vary. A person's weekly paycheck varies as the number of hours worked. $y=kx$ where y is the weekly salary, k is the hourly salary, and x is the number of hours worked.

3. $y=\frac{k}{x}$

5. $y=kx$, $15=k\cdot 40$, $k=\frac{3}{8}$

$y=\frac{3}{8}x$, $y=\frac{3}{8}\cdot 64=24$

7. $p=kd$, $21=k\cdot 50$, $k=\frac{21}{50}$

$p=\frac{21}{50}d$, $p=\frac{21}{50}\cdot 170=71.4$

The pressure would be 71.4 psi.

9. $d=kt^2$, $1=k\left(\frac{1}{4}\right)^2$, $k=16$

$d=16t^2=16(1)^2=16$

$d=16(2)^2=64$

It falls 16 feet in 1 second and 64 feet in 2 seconds.

11. $y=\frac{k}{x^2}$, $10=\frac{k}{2^2}$, $k=40$

$y=\frac{40}{x^2}$, $y=\frac{40}{0.5^2}=160$

13. $h=\frac{k}{t}$, $2000=\frac{k}{6}$, $k=12{,}000$

$h=\frac{12{,}000}{t}$, $h=\frac{12{,}000}{9}=1333.\bar{3}$

The heat loss is approximately 1333.3 Btu per hour.

15. $s=\frac{k}{t}$, $45=\frac{k}{6}$, $k=270$

$s=\frac{270}{t}$, $s=\frac{270}{9}=30$

The speed is 30 miles per hour.

17. $w=\frac{k}{l}$, $900=\frac{k}{8}$, $k=7200$

$w=\frac{7200}{l}$, $w=\frac{7200}{18}=400$

400 pounds can be safely supported.

19. $i=\frac{kf}{s}$, $4=\frac{k(700)}{200}$, $k=\frac{8}{7}$

$i=\frac{\frac{8}{7}f}{s}$, $i=\frac{\frac{8}{7}(500)}{250}\approx 2.3$

The intensity is approximately 2.3 oersteds.

21. $d=kav$, $222=k(37.8)(45)$, $k\approx 0.1305$

$d\approx 0.1305av$

$450\approx 0.1305(55)v$

$v\approx 62.7$

The Caravan must travel at approximately 62.7 miles per hour.

Cumulative Review Problems

23. $3x^2-8x+4=0$

$(3x-2)(x-2)=0$

$x=\frac{2}{3}$ or $x=2$

25. $488.75=1.0625p$

$p=460$

The original price was $460.

27. $x=$ number of gold leaf frames

$140x+95(110-x)=13.375$

$140x+10.450-95x=13.375$

$45x=2907$

$x=64.6$

$110-x=45.4$

He can buy 65 gold and 45 silver.

Putting Your Skills to Work

1. $D=\sqrt{\frac{3H}{2}}$

$D=\sqrt{\frac{3(10)}{2}}=3.9$ miles

$D=\sqrt{\frac{3(50)}{2}}=8.7$ miles

$D=\sqrt{\frac{3(100)}{2}}=1.22$ miles

2. $D=\sqrt{\frac{3(150)}{2}}=15$ miles

$D=\sqrt{\frac{3(250)}{2}}=19.4$ miles

H	D
10	3.9
50	8.7
100	12.2
150	15.0
250	19.4

(continued)

2. (continued)

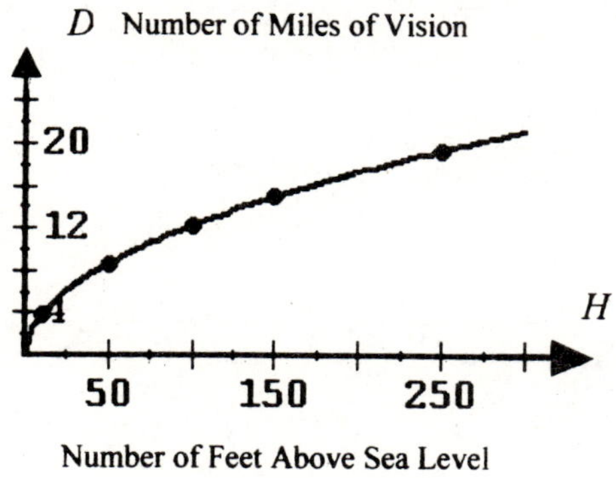

3. $D=\sqrt{\dfrac{3H}{2}}$

$D^3 = \dfrac{3H}{2}$

$H = \dfrac{2D^2}{3}$

$H = \dfrac{2(20)^2}{3} = 266.\overline{6}$

A sailor must be 266.7 feet above sea level to see a distance of 20 miles.

4. $\sqrt{\dfrac{3(210)}{2}} - \sqrt{\dfrac{3(150)}{2}} = 2.748239349...$

The viewing distance increased by approximately 2.7 miles.

Chapter 8 Review Problems

1. $(3xy^{\frac{1}{2}})(5x^2y^{-3}) = 15x^3y^{-\frac{5}{2}} = \dfrac{15x^3}{y^{\frac{5}{2}}}$

2. $\dfrac{3x^{\frac{2}{3}}}{6x^{\frac{1}{6}}} = \dfrac{x^{\frac{2}{3}-\frac{1}{6}}}{2} = \dfrac{x^{\frac{1}{2}}}{2}$

3. $(25a^3b^4)^{\frac{1}{2}} = 25^{\frac{1}{2}}a^{3\cdot\frac{1}{2}}b^{4\cdot\frac{1}{2}}$

$= 5a^{\frac{3}{2}}b^2$

4. $5^{\frac{1}{4}}\cdot 5^{\frac{1}{2}} = 5^{\frac{1}{4}+\frac{1}{2}} = 5^{\frac{3}{4}}$

5. $(2a^{\frac{1}{3}}b^{\frac{1}{4}})(-3a^{\frac{1}{2}}b^{\frac{1}{2}}) = -6a^{\frac{1}{3}+\frac{1}{2}}b^{\frac{1}{4}+\frac{1}{2}}$

$= -6a^{\frac{5}{6}}b^{\frac{3}{4}}$

6. $\dfrac{6x^{\frac{2}{3}}y^{\frac{1}{10}}}{12x^{\frac{1}{6}}y^{-\frac{1}{5}}} = \dfrac{x^{\frac{2}{3}-\frac{1}{6}}y^{\frac{1}{10}+\frac{1}{5}}}{2}$

$= \dfrac{x^{\frac{1}{2}}y^{\frac{3}{10}}}{2}$

7. $(2x^{-\frac{1}{5}}y^{\frac{1}{10}}z^{\frac{4}{5}})^{-5} = 2^{-5}x^{-\frac{1}{5}(-5)}y^{\frac{1}{10}(-5)}z^{\frac{4}{5}(-5)}$

$= \dfrac{xy^{-\frac{1}{2}}z^{-4}}{2^5}$

$= \dfrac{x}{32y^{\frac{1}{2}}z^4}$

8. $\left(\dfrac{49a^3b^6}{a^{-7}b^4}\right)^{\frac{1}{2}} = \left(49a^{10}b^2\right)^{\frac{1}{2}}$

$= 49^{\frac{1}{2}}a^{10(\frac{1}{2})}b^{2(\frac{1}{2})}$

$= 7a^5b$

9. $\dfrac{(x^{\frac{3}{4}}y^{\frac{2}{5}})^{\frac{1}{2}}}{x^{-\frac{1}{8}}} = x^{\frac{3}{4}\cdot\frac{1}{2}}x^{\frac{1}{8}}y^{\frac{2}{5}\cdot\frac{1}{2}}$

$= x^{\frac{3}{8}+\frac{1}{8}}y^{\frac{1}{5}}$

$= x^{\frac{1}{2}}y^{\frac{1}{5}}$

10. $\left(\dfrac{27x^{5n}}{x^{2n-3}}\right)^{\frac{1}{3}} = \left(27x^{5n-(2n-3)}\right)^{\frac{1}{3}}$
$= 27^{\frac{1}{3}}(x^{3n+3})^{\frac{1}{3}}$
$= 3x^{(3n+3)(\frac{1}{3})}$
$= 3x^{n+1}$

11. $(5^{\frac{6}{5}})^{\frac{10}{7}} = 5^{\frac{6}{5}\cdot\frac{10}{7}} = 5^{\frac{12}{7}}$

12. $2x^{\frac{1}{3}} + x^{-\frac{2}{3}} = 2x^{\frac{1}{3}} + \dfrac{1}{x^{\frac{2}{3}}}$
$= \dfrac{2x^{\frac{1}{3}+\frac{2}{3}} + 1}{x^{\frac{2}{3}}}$
$= \dfrac{2x+1}{x^{\frac{2}{3}}}$

13. $6x^{\frac{3}{2}} - 9x^{\frac{1}{2}} = 3x(2x^{\frac{1}{2}} - 3x^{-\frac{1}{2}})$

14. $\sqrt{\sqrt[5]{2x}} = \sqrt{(2x)^{\frac{1}{5}}} = ((2x)^{\frac{1}{5}})^{\frac{1}{2}} = (2x)^{\frac{1}{10}}$

15. $(2x+3y)^{\frac{4}{9}} = ((2x+3y)^{\frac{1}{9}})^4$
$= (\sqrt[9]{2x+3y})^4$

16. $\sqrt[3]{125} + \sqrt[4]{81} = 5 + 3 = 8$

17. $-\sqrt[6]{64} = -\sqrt[6]{2^6} = -2$
$\sqrt[6]{-64}$ is not a real number.

18. $27^{-\frac{4}{3}} = \dfrac{1}{(27^{\frac{1}{3}})^4} = \dfrac{1}{3^4} = \dfrac{1}{81}$

19. $\left(\dfrac{4}{9}\right)^{\frac{3}{2}} = \left(\left(\dfrac{4}{9}\right)^{\frac{1}{2}}\right)^3 = \left(\dfrac{2}{3}\right)^3 = \dfrac{8}{27}$

20. $\sqrt{99x^3y^6z^{10}} = \sqrt{9x^2y^6z^{10}\cdot 11x}$
$= 3xy^3z^5\sqrt{11x}$

21. $\sqrt[3]{-56a^8b^{10}c^{12}} = \sqrt[3]{-8a^6b^9c^{12}\cdot 7a^2b}$
$= -2a^2b^3c^4\sqrt[3]{7a^2b}$

22. $\sqrt{144x^{10}y^{12}z^0} = 12x^5y^6$

23. $\sqrt[3]{125a^9b^6c^{300}} = 5a^3b^2c^{100}$

24. $\sqrt[3]{y^3} = y$

25. $\sqrt{y^2} = |y|$

26. $\sqrt[4]{x^4y^4} = \sqrt[4]{(xy)^4} = |xy|$

27. $\sqrt[5]{x^{10}} = \sqrt[5]{(x^2)^5} = x^2$

28. $\sqrt[3]{x^{21}} = \sqrt[3]{(x^7)^3} = x^7$

29. $\sqrt{x^8} = \sqrt{(x^4)^2} = x^4$

30. $\sqrt{50} + 2\sqrt{32} - \sqrt{8} = 5\sqrt{2} + 8\sqrt{2} - 2\sqrt{2}$
$= 11\sqrt{2}$

31. $\sqrt{28} - 4\sqrt{7} + 5\sqrt{63}$
$= \sqrt{4\cdot 7} - 4\sqrt{7} + 5\sqrt{9\cdot 7}$
$= 2\sqrt{7} - 4\sqrt{7} + 15\sqrt{7}$
$= 13\sqrt{7}$

32. $\sqrt[3]{8}+3\sqrt[3]{16}-4\sqrt[3]{54}$
$=2+3\sqrt[3]{8\cdot 2}-4\sqrt[3]{27\cdot 2}$
$=2+6\sqrt[3]{2}-12\sqrt[3]{2}$
$=2-6\sqrt[3]{2}$

33. $2\sqrt{32x}-5x\sqrt{2}+\sqrt{18x}+2\sqrt{8x^2}$
$=2\sqrt{16\cdot 2x}-5x\sqrt{2}+\sqrt{9\cdot 2x}+2\sqrt{4x^2\cdot 2}$
$=8\sqrt{2x}-5x\sqrt{2}+3\sqrt{2x}+4x\sqrt{2}$
$=11\sqrt{2x}-x\sqrt{2}$

34. $(5\sqrt{12})(3\sqrt{6})=15\sqrt{72}$
$=15\sqrt{36\cdot 2}$
$=90\sqrt{2}$

35. $3\sqrt{x}(2\sqrt{8x}-3\sqrt{48}$
$=3\sqrt{x}(4\sqrt{2x}-12\sqrt{3})$
$=12x\sqrt{2}-36\sqrt{3x}$

36. $(5\sqrt{2}+\sqrt{3})(\sqrt{2}-2\sqrt{3})$
$=10-9\sqrt{6}-6=4-9\sqrt{6}$

37. $(5\sqrt{6}-2\sqrt{2})(\sqrt{6}-\sqrt{2})$
$=30-7\sqrt{12}+4$
$=34-7\sqrt{4\cdot 3}$
$=34-14\sqrt{3}$

38. $(2\sqrt{5}-3\sqrt{6})^2=20-12\sqrt{30}+54$
$=74-12\sqrt{30}$

39. $(\sqrt[3]{2x}+\sqrt[3]{6})(\sqrt[3]{4x^2}-\sqrt[3]{y})$
$=\sqrt[3]{8x^3}-\sqrt[3]{2xy}+\sqrt[3]{24x^2}-\sqrt[3]{6y}$
$=2x-\sqrt[3]{2xy}+2\sqrt[3]{3x^2}-\sqrt[3]{6y}$

40. $f(x)=\sqrt{5x+20}$
(a) $f(16)=\sqrt{5(16)+20}$
$=\sqrt{100}=10$
(b) domain: $5x+20\geq 0$
$5x\geq -20$
$x\geq -4$

41. $f(x)=\sqrt{36-4x}$
(a) $f(5)=\sqrt{36-4(5)}=\sqrt{16}=4$
(b) domain: $36-4x\geq 0$
$4x\leq 36$
$x\leq 9$

42. $f(x)=\sqrt{\frac{3}{4}x-\frac{1}{2}}$
(a) $f(1)=\sqrt{\frac{3}{4}\cdot 1-\frac{1}{2}}=\sqrt{\frac{1}{4}}=\frac{1}{2}$
(b) domain: $\frac{3}{4}x-\frac{1}{2}\geq 0$
$3x-2\geq 0$
$3x\geq 2$
$x\geq\frac{2}{3}$

43. $\sqrt{\frac{3x^2}{y}\cdot\frac{y}{y}}=\sqrt{\frac{3x^2y}{y^2}}=\frac{x\sqrt{3y}}{\sqrt{y^2}}$
$=\frac{x\sqrt{3y}}{y}$

44. $\frac{2}{\sqrt{3y}}\cdot\frac{\sqrt{3y}}{\sqrt{3y}}=\frac{2\sqrt{3y}}{3y}$

45. $\frac{3\sqrt{7x}}{\sqrt{21x}}=\frac{3\sqrt{7x}}{\sqrt{3}\cdot\sqrt{7x}}\cdot\frac{\sqrt{3}}{\sqrt{3}}=\frac{3\sqrt{3}}{3}=\sqrt{3}$

46. $\frac{2}{\sqrt{6}-\sqrt{5}}\cdot\frac{\sqrt{6}+\sqrt{5}}{\sqrt{6}+\sqrt{5}}$
$=\frac{2\sqrt{6}+2\sqrt{5}}{6-5}$
$=2\sqrt{6}+2\sqrt{5}$

47. $\frac{\sqrt{x}}{3\sqrt{x}+\sqrt{y}}\cdot\frac{3\sqrt{x}-\sqrt{y}}{3\sqrt{x}-\sqrt{y}}$
$=\frac{3x-\sqrt{xy}}{9x-y}$

48. $\frac{\sqrt{5}}{\sqrt{7}-3}\cdot\frac{\sqrt{7}+3}{\sqrt{7}+3}=\frac{\sqrt{35}+3\sqrt{5}}{7-9}$
$=\frac{-(\sqrt{35}+3\sqrt{5})}{2}$

49. $\frac{2\sqrt{3}+\sqrt{6}}{\sqrt{3}+2\sqrt{6}}\cdot\frac{\sqrt{3}-2\sqrt{6}}{\sqrt{3}-2\sqrt{6}}$
$=\frac{6-4\sqrt{18}+\sqrt{18}-12}{3-24}$
$=\frac{-6-3\sqrt{9\cdot 2}}{-21}$
$=\frac{-6-9\sqrt{2}}{-21}$
$=\frac{2+3\sqrt{2}}{7}$

50. $\frac{5\sqrt{2}-\sqrt{3}}{\sqrt{6}-\sqrt{3}}\cdot\frac{\sqrt{6}+\sqrt{3}}{\sqrt{6}+\sqrt{3}}$
$=\frac{5\sqrt{12}+5\sqrt{6}-\sqrt{18}-3}{6-3}$
$=\frac{5\sqrt{4\cdot 3}+5\sqrt{6}-\sqrt{9\cdot 2}-3}{3}$
$=\frac{10\sqrt{3}+5\sqrt{6}-3\sqrt{2}-3}{3}$

51. $\frac{3\sqrt{x}+\sqrt{y}}{\sqrt{x}-\sqrt{y}}\cdot\frac{\sqrt{x}+\sqrt{y}}{\sqrt{x}+\sqrt{y}}=\frac{3x+4\sqrt{xy}+y}{x-y}$

52. $\frac{2xy}{\sqrt[3]{16xy^5}}\cdot\frac{\sqrt[3]{4x^2y}}{\sqrt[3]{4x^2y}}=\frac{2xy\sqrt[3]{4x^2y}}{4xy^2}$
$=\frac{\sqrt[3]{4x^2y}}{2y}$

53. $\sqrt{-16}+\sqrt{-45}=\sqrt{16}\sqrt{-1}+\sqrt{45}\sqrt{-1}$
$=4i+i\sqrt{9\cdot 5}$
$=4i+3i\sqrt{5}$

54. $2x-3i+5=yi-2+\sqrt{6}$
$2x+5=-2+\sqrt{6}$
$2x=-7+\sqrt{6}$
$x=\frac{-7+\sqrt{6}}{2}$
$y=-3$

55. $(-12-6i)+(3-5i)=-12-6i+3-5i$
$=-9-11i$

56. $(2-i)-(12-3i)=2-i-12+3i$
$=-10+2i$

57. $(7+3i)(2-5i)=14-29i-15i^2$
$=14-29i+15$
$=29-29i$

58. $(8-4i)^2=64-64i+16i^2$
$=64-64i-16$
$=48-64i$

59. $2i(3+4i)=6i+8i^2$
$=-8+6i$

60. $3-4(2+i)=3-8-4i=-5-4i$

61. $i^{34}=(i^4)^8\cdot i^2=1^8(-1)=-1$

62. $i^{65}=(i^4)^{16}\cdot i=1^{16}\cdot i=i$

63. $\frac{7-2i}{3+4i}\cdot\frac{3-4i}{3-4i}=\frac{21-6i-28i+8i^2}{9+16}$
$=\frac{21-34i-8}{25}$
$=\frac{13-34i}{25}$

64. $\frac{5-2i}{1-3i}\cdot\frac{1+3i}{1+3i}=\frac{5-2i+15i-6i^2}{1+9}$
$=\frac{5+13i+6}{10}$
$=\frac{11+13i}{10}$

65. $\frac{4-3i}{5i}\cdot\frac{-5i}{-5i}=\frac{-20i+15i^2}{25}=\frac{-15-20i}{25}$
$=-\frac{3+4i}{5}$

66. $\frac{12}{3-5i}\cdot\frac{3+5i}{3+5i}=\frac{36+60i}{9+25}=\frac{36+60i}{34}$
$=\frac{18+30i}{17}$

67. $\frac{10-4i}{2+5i}\cdot\frac{2-5i}{2-5i}=\frac{20-58i+20i^2}{4+25}$
$=\frac{20-58i-20}{29}$
$=\frac{-58i}{29}$
$=-2i$

68. $2\sqrt{6x+1}=10$
$\sqrt{6x+1}=5$
$6x+1=25$
$6x=24$
$x=4$

check:

$2\sqrt{6(4)+1}\overset{?}{=}10$
$10=10$

69. $\sqrt[3]{3x-1}=\sqrt[3]{5x+1}$
$3x-1=5x+1$
$2x=-2$
$x=-1$

check:

$\sqrt[3]{3(-1)-1}\overset{?}{=}\sqrt[3]{5(-1)+1}$
$\sqrt[3]{-4}=\sqrt[3]{-4}$

70. $\sqrt{2x+1}=2x-5$
$2x+1=4x^2-20x+25$
$4x^2-22x+24=0$
$2x-11x+12=0$
$(x-4)(2x-3)=0$
$x=4$ or $x=-\frac{3}{2}$

check:

$\sqrt{2(4)+1}\overset{?}{=}2(4)-5$
$3=3$
$\sqrt{2\left(-\frac{3}{2}\right)+1}\overset{?}{=}2\left(-\frac{3}{2}\right)-5$
$\sqrt{-2}\neq-8$

$x=4$ is the solution.

71. $1+\sqrt{3x+1}=x$

$$\sqrt{3x+1}=x-1$$
$$3x+1=x^2-2x+1$$
$$x^2-5x=0$$
$$x(x-5)=0$$
$$x=0 \text{ or } x=5$$

check:

$$1+\sqrt{3(0)+1}\stackrel{?}{=}0$$
$$2\neq 0$$
$$1+\sqrt{3(5)+1}\stackrel{?}{=}5$$
$$5=5$$

$x=5$ is the solution.

72. $\sqrt{3x+1}-\sqrt{2x-1}=1$

$$\sqrt{3x+1}=\sqrt{2x-1}+1$$
$$3x+1=2x-1+2\sqrt{2x-1}+1$$
$$x+1=2\sqrt{2x-1}$$
$$x^2+2x+1=8x-4$$
$$x^2-6x+5=0$$
$$(x-5)(x-1)=0$$
$$x=5 \text{ or } x=1$$

check:

$$\sqrt{3(5)+1}-\sqrt{2(5)-1}\stackrel{?}{=}1$$
$$1=1$$
$$\sqrt{3(1)+1}-\sqrt{2(1)-1}\stackrel{?}{=}1$$
$$1=1$$

$x=5,\ x=1$ is the solution.

73. $\sqrt{7x+2}=\sqrt{x+3}+\sqrt{2x-1}$

$$7x+2=x+3+2\sqrt{x+3}\sqrt{2x-1}+2x-1$$
$$4x=2\sqrt{x+3}\sqrt{2x-1}$$
$$2x=\sqrt{x+3}\sqrt{2x-1}$$
$$4x^2=2x^2+5x-3$$
$$2x^2-5x+3=0$$
$$(x-1)(2x-3)=0$$
$$x=1 \text{ or } x=\frac{3}{2}$$

check:

$$\sqrt{7(1)+2}\stackrel{?}{=}\sqrt{1+3}+\sqrt{2(1)-1}$$
$$3=3$$
$$\sqrt{7\left(\frac{3}{2}\right)+2}\stackrel{?}{=}\sqrt{\left(\frac{3}{2}\right)+3}+\sqrt{2\left(\frac{3}{2}\right)-1}$$
$$\frac{5}{\sqrt{2}}=\frac{5}{\sqrt{2}}$$

$x=1,\ x=\frac{3}{2}$ is the solution.

74. $y=kx,\ 16=k\cdot 5,\ k=\frac{16}{5}$

$y=\frac{16}{5}x,\ y=\frac{16}{5}\cdot 3=\frac{48}{5}=9.6$

75. $y=kx,\ 5=k\cdot 20,\ k=\frac{1}{4}$

$y=\frac{1}{4}x,\ y=\frac{1}{4}\cdot 50=\frac{50}{4}=12.5$

76. $d=kv^2,\ 50=k\cdot 30^2,\ k=\frac{1}{18}$

$d=\frac{1}{18}v^2,\ d=\frac{1}{18}\cdot 55^2\approx 168.1$

The car stops in approximately 168.1 ft.

77. $t = k\sqrt{d}$, $2 = k\sqrt{64}$, $k = \dfrac{1}{4}$

$t = \dfrac{1}{4}\sqrt{d}$, $t = \dfrac{1}{4}\sqrt{196} = \dfrac{14}{4} = 3.5$

Object drops 196 ft. in 3.5 seconds.

78. $y = \dfrac{k}{x}$, $8 = \dfrac{k}{3}$, $k = 24$

$y = \dfrac{24}{x}$, $y = \dfrac{24}{48} = 0.5$

79. $V = \dfrac{k}{P}$, $70 = \dfrac{k}{24}$, $k = 1680$

$V = \dfrac{1680}{P}$, $100 = \dfrac{1680}{P}$, $P = 16.8$ lb/in^2

80. $y = k\dfrac{x}{z^2}$, $1 = k\cdot\dfrac{8}{4^2}$, $k = 2$

$y = 2\cdot\dfrac{x}{z^2}$, $y = 2\cdot\dfrac{6}{3^2} = \dfrac{4}{3}$

81. $V = khr^2$, $50 = k(5)(3)^2$, $k = \dfrac{10}{9}$

$V = \dfrac{10}{9}r^2h$, $V = \dfrac{10}{9}(4)^2(9) = 160$ cm^3

Chapter 8 Test

1. $(2x^{\frac{1}{2}}y^{\frac{1}{3}})(-3x^{\frac{1}{3}}y^{\frac{1}{6}}) = -6x^{\frac{1}{2}+\frac{1}{3}}y^{\frac{1}{3}+\frac{1}{6}}$

$= -6x^{\frac{5}{6}}y^{\frac{1}{2}}$

2. $\dfrac{7x^3}{4x^{\frac{3}{4}}} = \dfrac{7x^{\frac{9}{4}}}{4}$

3. $(8x^{\frac{1}{3}})^{\frac{3}{2}} = 8^{\frac{3}{2}}x^{\frac{1}{3}\cdot\frac{3}{2}} = 8^{\frac{3}{2}}x^{\frac{1}{2}}$

4. $6^{\frac{1}{5}}\cdot 6^{\frac{3}{5}} = 6^{\frac{1}{5}+\frac{3}{5}} = 6^{\frac{4}{5}}$

5. $8^{-\frac{2}{3}} = \dfrac{1}{\left(8^{\frac{1}{3}}\right)^2} = \dfrac{1}{2^2} = \dfrac{1}{4}$

6. $16^{\frac{5}{4}} = \left(16^{\frac{1}{4}}\right)^5 = 2^5 = 32$

7. $\sqrt{75a^4b^9} = \sqrt{25a^4b^8\cdot 3b}$

$= 5a^2b^4\sqrt{3b}$

8. $\sqrt{64x^6y^5} = \sqrt{64x^6y^4\cdot y}$

$= 8x^3y^2\sqrt{y}$

9. $\sqrt[3]{250x^4y^6} = \sqrt[3]{125x^3y^6\cdot 2x}$

$= 5xy^2\sqrt[3]{2x}$

10. $3\sqrt{48} - \sqrt[3]{54x^5} + 2\sqrt{27} + 2x\sqrt[3]{16x^2}$

$= 3\sqrt{16\cdot 3} - \sqrt[3]{27x^3\cdot 2x^2}$

$+ 2\sqrt{9\cdot 3} + 2x\sqrt[3]{8\cdot 2x^2}$

$= 12\sqrt{3} - 3x\sqrt[3]{2x^2} + 6\sqrt{3} + 4x\sqrt[3]{2x^2}$

$= 18\sqrt{3} + x\sqrt[3]{2x^2}$

11. $\sqrt{32} - 3\sqrt{8} + 2\sqrt{72}$

$= \sqrt{16\cdot 2} - 3\sqrt{4\cdot 2} + 2\sqrt{36\cdot 2}$

$= 4\sqrt{2} - 6\sqrt{2} + 12\sqrt{2}$

$= 10\sqrt{2}$

12. $2\sqrt{3}(3\sqrt{6} - 5\sqrt{2}) = 6\sqrt{18} - 10\sqrt{6}$

$= 6\sqrt{9\cdot 2} - 10\sqrt{6}$

$= 18\sqrt{2} - 10\sqrt{6}$

13. $(5\sqrt{3}-\sqrt{6})(2\sqrt{3}+3\sqrt{6})$
$=30+15\sqrt{18}-2\sqrt{18}-18$
$=12+13\sqrt{9\cdot 2}$
$=12+39\sqrt{2}$

14. $\frac{8}{\sqrt{20x}}\cdot\frac{\sqrt{20x}}{\sqrt{20x}}=\frac{8\sqrt{4\cdot 5x}}{20x}$
$=\frac{16\sqrt{5x}}{20x}$
$=\frac{4\sqrt{5x}}{5x}$

15. $\sqrt{\frac{xy}{3}}=\frac{\sqrt{xy}}{\sqrt{3}}\cdot\frac{\sqrt{3}}{\sqrt{3}}$
$=\frac{\sqrt{3xy}}{3}$

16. $\frac{5+2\sqrt{3}}{4-\sqrt{3}}\cdot\frac{4+\sqrt{3}}{4+\sqrt{3}}=\frac{20+13\sqrt{3}+6}{16-3}$
$=\frac{26+13\sqrt{3}}{13}$
$=2+\sqrt{3}$

17. $\sqrt{3x-2}=x$
$3x-2=x^2$
$x^2-3x+2=0$
$(x-2)(x-1)=0$
$x=2$ or $x-1$
check:
$\sqrt{3(2)-2}\stackrel{?}{=}2$
$2=2$
$\sqrt{3(1)-2}\stackrel{?}{=}1$
$1=1$
$x=2,\ x=1$ is the solution.

18. $5+\sqrt{x+15}=x$
$\sqrt{x+15}=x-5$
$x+15=x^2-10x+25$
$x^2-11x+10=0$
$(x-10)(x-11)=0$
$x=10$ or $x=11$
check:
$5+\sqrt{10+15}\stackrel{?}{=}10$
$10=10$
$5+\sqrt{11+15}\stackrel{?}{=}11$
$5+\sqrt{26}\neq 11$
$x=10$ is the solution.

19. $5-\sqrt{x-2}=\sqrt{x+3}$
$25-10\sqrt{x-2}+x-2=x+3$
$20=10\sqrt{x-2}$
$2=\sqrt{x-2}$
$4=x-2$
$x=6$
check:
$5-\sqrt{6-2}\stackrel{?}{=}\sqrt{6+3}$
$3=3$

$x=6$ is the solution.

20. $(8+2i)-3(2-4i)=8+2i-6+12i$
$=2+14i$

21. $i^{18}+\sqrt{-16}=(i^4)^4\cdot i^2+\sqrt{16}\sqrt{-1}$
$=1^4(-1)+4i$
$=-1+4i$

22. $(3-2i)(4+3i)=12+i-6i^2=12+i+6$
$=18+i$

23. $\frac{2+5i}{1-3i}\cdot\frac{1+3i}{1+3i}=\frac{2+11i+15i^2}{1+9}$
$=\frac{-13+11i}{10}$

24. $(6+3i)^2=36+36i+9i^2$
$=27+36i$

25. $i^{43}=(i^4)^{10}\cdot i^3=1^{10}\cdot(-i)=-i$

26. $y=\frac{k}{x},\ 9=\frac{k}{2},\ k=18$
$y=\frac{18}{x},\ y=\frac{18}{6}=3$

27. $y=k\cdot\frac{x}{z^2},\ 3=k\cdot\frac{8}{4^2},\ k=6$
$y=6\cdot\frac{x}{z^2},\ y=6\cdot\frac{5}{6^2}=\frac{5}{6}$

28. $d=kv^2,\ 30=k\cdot 30^2,\ k=\frac{1}{30}$
$d=\frac{1}{30}\cdot v^2,\ d=\frac{1}{30}\cdot 50^2\approx 83.3$ feet

Cumulative Test for Chapters 1-8

1. $\frac{7}{8}\cdot\frac{1}{2}+\frac{3}{8}\div\frac{1}{8}=\frac{7}{16}+3=3\frac{7}{16}$

2. $2a(3a^3-4)-3a^2(a-5)$
$=6a^4-8a-3a^3+15a^2$
$=6a^4-3a^3+15a^2-8a$

3. $7(12-14)^3-7+3\div(-3)$
$=7(-2)^3-7-1$
$=7(-8)-8$
$=-56-8=-64$

4. $y=-\frac{3}{4}x+2$
$4y=-3x+8$
$3x=-4y+8$
$x=\frac{-4y+8}{3}$

5. $3x-5y=15$

x	y
5	0
0	−3

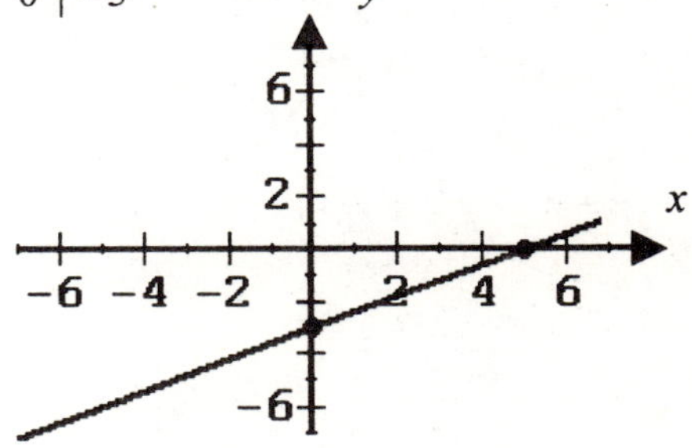

6. $16x^2+24x-16=8(2x^2+3x-2)$
$=8(2x-1)(x+2)$

7. Multiply first equation by -3 and add to second equation; then multiply first equation by -2 and add to third equation.

$$\begin{aligned} x+4y-z&=10\\ 3x+2y+z&=4\\ 2x-3y+2z&=-7\end{aligned}$$

$$\begin{aligned} x+4y-z&=10\\ -10y+4z&=-26\\ -11y+4z&=-27\end{aligned}$$

Multiply the second equation by $-\frac{11}{10}$ and add to the third equation.

(continued)

7. (continued)

$$\begin{array}{rrrcr} x & +4y & -z & = & 10 \\ & -10y & +4z & = & -26 \\ & & -\frac{2}{5}z & = & \frac{8}{5} \end{array}$$

$-\frac{2}{5}z = \frac{8}{5}$

$z = -4$, substitute in second equation

$-10y + 4(-4) = -26$

$y = 1$, substitute in first equation

$x + 4(1) - (-4) = 10$

$x = 2$

$x = 2,\ y = 1,\ z = -4$ is the solution.

8. $\frac{7x}{x^2 - 2x - 15} - \frac{2}{x-5}$

$= \frac{7x}{(x-5)(x+3)} - \frac{2(x+3)}{(x-5)(x+3)}$

$= \frac{7x - 2x - 6}{(x-5)(x+3)}$

$= \frac{5x-6}{(x-5)(x+3)}$

9. $P = 2L + 2W$

$48 = 2(2W+3) + 2W$

$24 = 2W + 3 + W$

$3W = 21$

$W = 7$

$2W + 3 = 17$

The length is 7 meters and the width is 17 meters.

10. $56x + 2 = 8b + 4x$

$4b = 26x + 1$

$b = \frac{26x+1}{4}$

11. $\frac{2x^{-3}y^{-4}}{4x^{-\frac{5}{2}}y^{\frac{7}{2}}} = \frac{1}{2}x^{-3-\left(-\frac{5}{2}\right)}y^{-4-\frac{7}{2}} = \frac{1}{2}x^{-\frac{1}{2}}y^{-\frac{15}{2}}$

$= \frac{1}{2x^{\frac{1}{2}}y^{\frac{15}{2}}}$

12. $(3x^{-\frac{1}{2}}y^2)^{-\frac{1}{3}} = 3^{-\frac{1}{3}}x^{-\frac{1}{2}\left(-\frac{1}{3}\right)}y^{2\left(-\frac{1}{3}\right)}$

$= \frac{x^{\frac{1}{6}}}{3^{\frac{1}{3}}y^{\frac{2}{3}}}$

13. $64^{-\frac{1}{3}} = \frac{1}{64^{\frac{1}{3}}} = \frac{1}{4}$

14. $\sqrt[3]{40x^5y^9} = \sqrt[3]{8\cdot 5x^3y^9x^2} = 2xy^3\sqrt[3]{5x^2}$

15. $\sqrt{80x} + 2\sqrt{45x} - 3\sqrt{20x}$

$= \sqrt{16\cdot 5x} + 2\sqrt{9\cdot 5x} - 3\sqrt{4\cdot 5x}$

$= 4\sqrt{5x} + 6\sqrt{5x} - 6\sqrt{5x}$

$= 4\sqrt{5x}$

16. $(2\sqrt{3} - 5\sqrt{2})(\sqrt{3} + 4\sqrt{2})$

$= 6 + 3\sqrt{6} - 40 = -34 + 3\sqrt{6}$

17. $\frac{\sqrt{3}+2}{2\sqrt{3}-5}\cdot\frac{2\sqrt{3}+5}{2\sqrt{3}+5} = \frac{6 + 9\sqrt{3} + 10}{12 - 25}$

$= -\frac{16 + 9\sqrt{3}}{13}$

18. $i^{21} + \sqrt{-16} + \sqrt{-49}$

$= (i^4)^5 \cdot i + \sqrt{16}\sqrt{-1} + \sqrt{49}\sqrt{-1}$

$= 1^5 \cdot i + 4i + 7i$

$= 12i$

19. $(3-4i)^2 = 3^2 - 2(3)(4i) + (4i)^2$
$= 9 - 24i + 16i^2 = 9 - 24i - 16$
$= -7 - 24i$

20. $\dfrac{1+4i}{1+3i} \cdot \dfrac{1-3i}{1-3i} = \dfrac{1+i-12i^2}{1^2+3^2} = \dfrac{13+i}{10}$

21.
$$\begin{aligned} x-3 &= \sqrt{3x+1} \\ x^2-6x+9 &= 3x+1 \\ x^2-9x+8 &= 0 \\ (x-1)(x-8) &= 0 \\ x = 1 &\text{ or } x = 8 \end{aligned}$$

check:

$1-3 \stackrel{?}{=} \sqrt{3(1)+1}$

$-2 \neq 2$

$8-3 \stackrel{?}{=} \sqrt{3(8)+1}$

$5 = 5$

$x = 8$ is the solution.

22.
$$\begin{aligned} 1+\sqrt{x+1} &= \sqrt{x+2} \\ 1+2\sqrt{x+1}+x+1 &= x+2 \\ \sqrt{x+1} &= 0 \\ x+1 &= 0 \\ x &= -1 \end{aligned}$$

check:

$1+\sqrt{-1+1} \stackrel{?}{=} \sqrt{-1+2}$

$1 = 1$

$x = 1$ is the solution.

23. $y = kx^2$, $12 = k \cdot 2^2$, $k = 3$
$y = 3x^2$, $y = 3 \cdot 5^2 = 75$

24. $I = \dfrac{k}{d^2}$. $120 = \dfrac{k}{10^2}$. $k = 12.000$
$I = \dfrac{12.000}{d^2}$. $I = \dfrac{12.000}{15^2} \approx 53.3$ lumens

Chapter 9

Pretest Chapter 9

1. $2x^2+3=39$
$$2x^2=36$$
$$x^2=18$$
$$x=\pm\sqrt{18}$$
$$x=\pm\sqrt{9\cdot 2}$$
$$x=\pm 3\sqrt{2}$$

2. $2x^2-4x-3=0$
$$(x-1)^2=\frac{5}{2}$$
$$x-1=\pm\sqrt{\frac{5}{2}\cdot\frac{2}{2}}$$
$$x-1=\pm\sqrt{\frac{10}{4}}$$
$$x-1=\pm\frac{\sqrt{10}}{2}$$
$$x=1\pm\frac{\sqrt{10}}{2}$$
$$x=\frac{2\pm\sqrt{10}}{2}$$

3. $8x^2-2x-7=0$
$$x=\frac{-b\pm\sqrt{b^2-4ac}}{2a}$$
$$x=\frac{-(-2)\pm\sqrt{(-2)^2-4(8)(-7)}}{2(8)}$$
$$x=\frac{2\pm\sqrt{228}}{16}=\frac{2\pm\sqrt{4\cdot 57}}{16}$$
$$x=\frac{2\pm 2\sqrt{57}}{16}=\frac{1\pm\sqrt{57}}{8}$$

4. $(x-1)(x+5)=2$
$$x^2+4x-5=2$$
$$x^2+4x-7=0$$
$$x=\frac{-b\pm\sqrt{b^2-4ac}}{2a}$$
$$x=\frac{-4\pm\sqrt{4^2-4(1)(-7)}}{2(1)}$$
$$x=\frac{-4\pm\sqrt{44}}{2}=\frac{-4\pm\sqrt{4\cdot 11}}{2}$$
$$x=\frac{-4\pm 2\sqrt{11}}{2}$$
$$x=-2\pm\sqrt{11}$$

5. $4x^2=-12x-17$
$$4x^2+12x+17=0$$
$$x=\frac{-b\pm\sqrt{b^2-4ac}}{2a}$$
$$x=\frac{-12\pm\sqrt{12^2-4\cdot 4\cdot 17}}{2\cdot 4}$$
$$x=\frac{-12\pm\sqrt{-128}}{8}$$
$$x=\frac{-12\pm\sqrt{64\cdot 2}\sqrt{-1}}{8}$$
$$x=\frac{-12\pm 8i\sqrt{2}}{8}$$
$$x=\frac{-3\pm 2i\sqrt{2}}{2}$$

6. $5x^2+4x-12=0$
$$(x+2)(5x-6)=0$$
$$x=-2,\ x=\frac{6}{5}$$

7. $7x^2+9x=14x^2-3x$

$7x^2-12x=0$

$x(7x-12)=0$

$x=0 \text{ or } 7x-12=0$

$x=0 \qquad x=\dfrac{12}{7}$

8. $\dfrac{18}{x}+\dfrac{12}{x+1}=9$

$18(x+1)+12x=9x(x+1)$

$18x+18+12x=9x^2+9x$

$9x^2-21x-18=0$

$3x^2-7x-6=0$

$(x-3)(3x+2)=0$

$x-3=0 \text{ or } 3x+2=0$

$x=3 \qquad x=-\dfrac{2}{3}$

9. $x^6-7x^3-8=0$

$(x^3)^2-7x^3-8=0$

$(x^3-8)(x^3+1)=0$

$x^3-8=0 \text{ or } x^3+1=0$

$x^3=8 \qquad x^3=-1$

$x=2 \qquad x=-1$

10. $w^{\frac{4}{3}}-6w^{\frac{2}{3}}+8=0$

$(w^{\frac{2}{3}})^2-6w^{\frac{2}{3}}+8=0$

$(w^{\frac{2}{3}}-4)(w^{\frac{2}{3}}-2)=0$

$w^{\frac{2}{3}}-4=0 \text{ or } w^{\frac{2}{3}}-2=0$

$w^2=64 \qquad w^2=8$

$w=\pm 8 \qquad w^2=8$

$w=\pm 2\sqrt{2}$

11. $3x^2+2wx+8w=0$

$x=\dfrac{-2w\pm\sqrt{(2w)^2-4(3)(8w)}}{2(3)}$

$x=\dfrac{-2w\pm\sqrt{4w^2-96w}}{6}$

$x=\dfrac{-2w\pm\sqrt{4(w^2-24w)}}{6}$

$x=\dfrac{-2w\pm 2\sqrt{w^2-24w}}{6}$

$x=\dfrac{-w\pm\sqrt{w^2-24w}}{3}$

12. $A=LW=(3W+1)W=52$

$3W^2+W-52=0$

$(3W+13)(W-4)=0$

$W-4=0 \text{ or } 3W+13=0$

$W=4 \qquad W=-\dfrac{13}{3}$

reject, $W>0$

$L=3W+1=13$

The width is 4 m and the length is 13 m.

13. $f(x)=3x^2+6x-9$

$x_{\text{vertex}}=\dfrac{-b}{2a}$

$x_{\text{vertex}}=\dfrac{-6}{2(3)}=-1$

$y_{\text{vertex}}=3(-1)^2+6(-1)-9$

$y_{\text{vertex}}=-12$

Vertex $(-1,-12)$

$f(0)=3(0)^2+6(0)-9=-9$

y-intecept $(0,-9)$

(continued)

13. (continued)

$3x^2+6x-9=0$

$x^2+2x-3=0$

$(x+3)(x-1)=0$

$x+3=0$ or $x-1=0$

$x=-3 \qquad x=1$

x-intercepts $(1,0)$, $(-3,0)$

14. $g(x)=-x^2+6x-5$

$x_{\text{vertex}}=\dfrac{-b}{2a}$

$x_{\text{vertex}}=\dfrac{-6}{2(-1)}=3$

$y_{\text{vertex}}=-(3)^2+6(3)-5$

$y_{\text{vertex}}=4$

Vertex $(3,4)$

$f(0)=-(0)^2+6(0)-5=-5$

y-intecept $(0,-5)$

$-x^2+6x-5=0$

$x^2-6x+5=0$

$(x-5)(x-1)=0$

$x-5=0$ or $x-1=0$

$x=5 \qquad x=1$

x-intercepts $(5,0)$, $(1,0)$

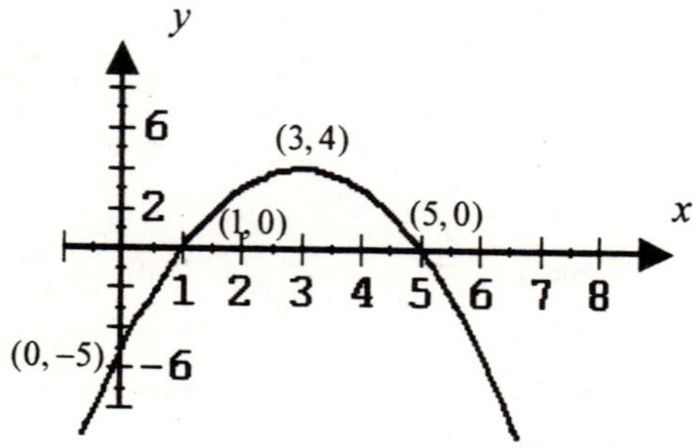

15. $x^2-x-6>0$

$(x-3)(x+2)=0$

$x-3=0$ or $x+2=0$

$x=3 \qquad x=-2$

Critical values: $x=-2, 3$

Test: x^2-x-6

Region	Test	Result
I	$x=-3$	$6>0$
II	$x=0$	$-6<0$
III	$x=4$	$6>0$

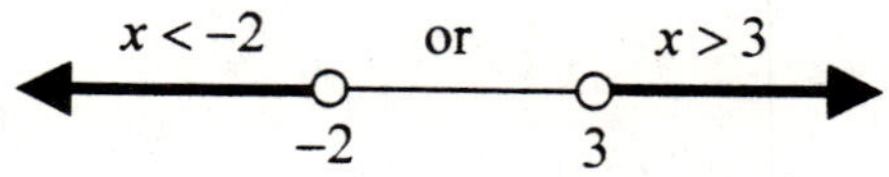

16. $2x^2+9x\le -9$

$2x^2+9x+9=0$

$(x+3)(2x+3)=0$

$x+3=0$ or $2x+3=0$

$x=-3 \qquad x=-\dfrac{3}{2}$

Critical values: $x=-3, -\dfrac{3}{2}$

Test: $2x^2+9x$

Region	Test	Result
I	$x=-4$	$-4>-9$
II	$x=-2$	$-10<-9$
III	$x=0$	$0>-9$

$-3\le x\le -\dfrac{3}{2}$

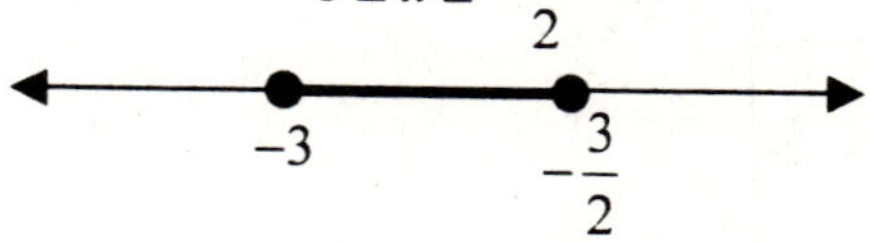

17. $4x^2 - 3x - 5 < 0$

$4x^2 - 3x - 5 = 0$

$$x = \frac{-(-3) \pm \sqrt{(-3)^2 - 4(4)(-5)}}{2(4)}$$

$x = -0.8, 1.6$

Critical values: $x = -3, -\frac{3}{2}$

Test: $4x^2 - 3x - 5$

Region	Test	Result
I	−1	$2 > 0$
II	0	$-5 < 0$
III	2	$5 > 0$

The solution is $-0.8 < x < 1.6$.

18. $|3x - 2| = 7$

$3x - 2 = 7$ or $3x - 2 = -7$

$3x = 9 \qquad 3x = -5$

$x = 3 \qquad x = -\frac{5}{3}$

19. $|8 - x| - 3 = 1 \Rightarrow |8 - x| = 4$

$8 - x = 4$ or $8 - x = -4$

$x = 4 \qquad x = 12$

20. $\left|\frac{2x+3}{4}\right| = 2$

$\frac{2x+3}{4} = 2$ or $\frac{2x+3}{4} = -2$

$2x + 3 = 8 \qquad 2x + 3 = -8$

$2x = 5 \qquad 2x = -11$

$x = \frac{5}{2} \qquad x = -\frac{11}{2}$

21. $|5x - 8| = |3x + 2|$

$5x - 8 = 3x + 2$ or $5x - 8 = -(3x + 2)$

$2x = 10 \qquad 5x - 8 = -3x - 2$

$x = 5 \qquad 8x = 6$

$x = \frac{3}{4}$

22. $|3x + 2| < 8 \Leftrightarrow -8 < 3x + 2 < 8$

$-10 < 3x < 6$

$-\frac{10}{3} < x < 2$

23. $\left|\frac{2}{3}x - \frac{1}{2}\right| \le 3 \Leftrightarrow -3 \le \frac{2}{3}x - \frac{1}{2} \le 3$

$-18 \le 4x - 3 \le 18$

$-15 \le 4x \le 21$

$-\frac{15}{4} \le x \le \frac{21}{4}$

24. $|2 - 5x - 4| > 13$

$2 - 5x - 4 < -13$ or $2 - 5x - 4 > 13$

$-5x - 2 < -13 \qquad -5x - 2 > 13$

$-5x < -11 \qquad -5x > 15$

$x > \frac{11}{5} \qquad x < -3$

25. $|2x - 7| \le 11 \Leftrightarrow -11 \le 2x - 7 \le 11$

$-4 \le 2x \le 18$

$-2 \le x \le 9$

9.1 Exercises

1. $x^2 = 100$

$x = \pm\sqrt{100}$

$x = \pm 10$

3. $x^2 + 81 = 0$
$x^2 = -81$
$x = \pm 9i$

5. $3x^2 - 45 = 0$
$3x^2 = 45$
$x^2 = 15$
$x = \pm\sqrt{15}$

7. $5x^2 - 10 = 0$
$5x^2 = 10$
$x^2 = 2$
$x = \pm\sqrt{2}$

9. $x^2 = -81$
$x = \pm 9i$

11. $6x^2 + 4 = 4x^2$
$2x^2 = -4$
$x^2 = -2$
$x = \pm i\sqrt{2}$

13. $(x+2)^2 = 18$
$x + 2 = \pm\sqrt{18} = \pm 3\sqrt{2}$
$x = -2 \pm 3\sqrt{2}$

15. $(3x+2)^2 = 5$
$3x + 2 = \pm\sqrt{5}$
$3x = -2 \pm \sqrt{5}$
$x = \frac{-2 \pm \sqrt{5}}{3}$

17. $(5x-2)^2 = 25$
$5x - 2 = \pm 5$
$5x = 2 \pm 5$
$x = \frac{2 \pm 5}{5} = \frac{7}{5}, -\frac{3}{5}$

19. $\left(\frac{x}{3} - 1\right)^2 = 45$
$\frac{x}{3} - 1 = \pm\sqrt{45}$
$\frac{x}{3} - 1 = \pm\sqrt{9 \cdot 5}$
$\frac{x}{3} - 1 = \pm 3\sqrt{5}$
$\frac{x}{3} = 1 \pm 3\sqrt{5}$
$x = 3 \pm 9\sqrt{5}$

21. $x^2 + 6x + 2 = 0$
$x^2 + 6x + 9 = -2 + 9$
$(x+3)^2 = 7$
$x + 3 = \pm\sqrt{7}$
$x = -3 \pm \sqrt{7}$

23. $x^2 - 12x = 4$
$x^2 - 12x + 36 = 4 + 36$
$(x-6)^2 = 40$
$x - 6 = \pm\sqrt{40}$
$x - 6 = \pm 2\sqrt{10}$
$x = 6 \pm 2\sqrt{10}$

25.
$$\frac{x^2}{3}-\frac{x}{3}=3$$
$$x^2-x=9$$
$$x^2-x+\frac{1}{4}=9+\frac{1}{4}$$
$$(x-\frac{1}{2})^2=\frac{37}{4}$$
$$x-\frac{1}{2}=\pm\frac{\sqrt{37}}{2}$$
$$x=\frac{1\pm\sqrt{37}}{2}$$

27. $7x^2+4x-5=0$
$$x^2+\frac{4}{7}x+\frac{4}{49}=\frac{5}{7}+\frac{4}{49}$$
$$\left(x+\frac{2}{7}\right)^2=\frac{39}{49}$$
$$x+\frac{2}{7}=\pm\frac{\sqrt{39}}{7}$$
$$x=\frac{-2\pm\sqrt{39}}{7}$$

29. $5x^2+4x-3=0$
$$x^2+\frac{4}{5}x+\frac{4}{25}=\frac{3}{5}+\frac{4}{25}$$
$$\left(x+\frac{2}{5}\right)^2=\frac{19}{25}$$
$$x+\frac{2}{5}=\pm\frac{\sqrt{19}}{5}$$
$$x=\frac{-2\pm\sqrt{19}}{5}$$

31.
$$2y^2-y=15$$
$$y^2-\frac{1}{2}y+\frac{1}{16}=\frac{15}{2}+\frac{1}{16}$$
$$\left(y-\frac{1}{4}\right)^2=\frac{121}{16}$$
$$y-\frac{1}{4}=\pm\frac{11}{4}$$
$$y=\frac{1\pm 11}{4}$$
$$y=3,\ y=-\frac{5}{2}$$

33.
$$2x^2+2=3x$$
$$x^2-\frac{3}{2}x+\frac{9}{16}=-1+\frac{9}{16}$$
$$\left(x-\frac{3}{4}\right)^2=-\frac{7}{16}$$
$$x-\frac{3}{4}=\pm\frac{i\sqrt{7}}{4}$$
$$x=\frac{3\pm i\sqrt{7}}{4}$$

35. $x^2+2x-5=0$
$$(-1+\sqrt{6})^2+2(-1+\sqrt{6})-5\stackrel{?}{=}0$$
$$1-2\sqrt{6}+6-2+2\sqrt{6}-5\stackrel{?}{=}0$$
$$0=0$$

37. $(x-7)^2(8)=648$
$$(x-7)^2=81$$
$$x-7=9$$
$$x=16$$

39. $L = 4t^2$

$t = \sqrt{\frac{L}{4}}$

$t = \sqrt{\frac{3.1}{4}}$

$t \approx 0.88$ second

41. $D = 16t^2$

$t = \sqrt{\frac{D}{16}}$

$t = \sqrt{\frac{3600}{16}}$

$t = 15$ seconds

Cumulative Review Problems

43. $\sqrt{b^2 - 4ac} = \sqrt{4^2 - 4(3)(-4)}$

$= \sqrt{64}$

$= 8$

45. $5x^2 - 6x + 8 = 5(-2)^2 - 6(-2) + 8$

$= 20 + 12 + 8$

$= 40$

9.2 Exercises

1. Place the quadratic in standard form. Find *a*, *b*, and *c*. Substitute these values into the quadratic formula.

3. If the discriminant in the quadratic formula is zero, then the quadratic equation will have one real solution.

5. $x^2 - x - 3 = 0$

$a = 1,\ b = -1,\ c = -3$

$x = \frac{-b \pm \sqrt{b^2 - 4ac}}{2a}$

$x = \frac{-(-1) \pm \sqrt{(-1)^2 - 4(1)(-3)}}{2(1)}$

$x = \frac{1 \pm \sqrt{13}}{2}$

7. $2x^2 + x - 4 = 0$

$a = 2,\ b = 1,\ c = -4$

$x = \frac{-b \pm \sqrt{b^2 - 4ac}}{2a}$

$x = \frac{-1 \pm \sqrt{1^2 - 4(2)(-4)}}{2(2)}$

$x = \frac{-1 \pm \sqrt{33}}{4}$

9. $x^2 = \frac{2}{3}x$

$x^2 - \frac{2}{3}x = 0$

$3x^2 - 2x = 0$

$a = 3,\ b = -2,\ c = 0$

$x = \frac{-b \pm \sqrt{b^2 - 4ac}}{2a}$

$x = \frac{-(-2) \pm \sqrt{(-2)^2 - 4(3)(0)}}{2(3)}$

$x = \frac{2 \pm \sqrt{4}}{6} = \frac{2 \pm 2}{6}$

$x = 0,\ x = \frac{2}{3}$

11. $6x^2 - x - 1 = 0$

$a = 6, b = -1, c = -1$

$$x = \frac{-b \pm \sqrt{b^2 - 4ac}}{2a}$$

$$x = \frac{-(-1) \pm \sqrt{(-1)^2 - 4(6)(-1)}}{2(6)}$$

$$x = \frac{1 \pm \sqrt{25}}{12}$$

$$x = \frac{1 \pm 5}{12}$$

$$x = \frac{1}{2}, x = -\frac{1}{3}$$

13. $4x^2 + 3x - 2 = 0$

$a = 4, b = 3, c = -2$

$$x = \frac{-b \pm \sqrt{b^2 - 4ac}}{2a}$$

$$x = \frac{-(3) \pm \sqrt{(3)^2 - 4(4)(-2)}}{2(4)}$$

$$x = \frac{-3 \pm \sqrt{41}}{8}$$

15. $3x^2 + 1 = 8$

$3x^2 - 7 = 0$

$a = 3, b = 0, c = -7$

$$x = \frac{-b \pm \sqrt{b^2 - 4ac}}{2a}$$

$$x = \frac{0 \pm \sqrt{0^2 - 4(3)(-7)}}{2(3)}$$

$$x = \frac{\pm\sqrt{84}}{6} = \frac{\pm 2\sqrt{21}}{6}$$

$$x = \frac{\pm\sqrt{21}}{3}$$

17. $2x(x + 3) - 3 = 4x - 2$

$2x^2 + 6x - 4x - 1 = 0$

$2x^2 + 2x - 1 = 0$

$a = 2, b = 2, c = -1$

$$x = \frac{-b \pm \sqrt{b^2 - 4ac}}{2a}$$

$$x = \frac{-(2) \pm \sqrt{(2)^2 - 4(2)(-1)}}{2(2)}$$

$$x = \frac{-2 \pm \sqrt{12}}{4} = \frac{-2 \pm 2\sqrt{3}}{4}$$

$$x = \frac{-1 \pm \sqrt{3}}{2}$$

19. $3x^2 + 5x + 1 = 5x + 4$

$3x^2 - 3 = 0$

$x^2 - 1 = 0$

$a = 1, b = 0, c = -1$

$$x = \frac{-b \pm \sqrt{b^2 - 4ac}}{2a}$$

$$x = \frac{-(0) \pm \sqrt{(0)^2 - 4(1)(-1)}}{2(1)}$$

$$x = \frac{\pm\sqrt{4}}{2}$$

$$x = \frac{\pm 2}{2}$$

$$x = \pm 1$$

21. $(x - 2)(x + 1) = \dfrac{2x + 3}{2}$

$x^2 - x - 2 = \dfrac{2x + 3}{2}$

$2x^2 - 2x - 4 = 2x + 3$

$2x^2 - 4x - 7 = 0$

(continued)

21. (continued)
$a = 2, b = -4, c = -7$

$$x = \frac{-b \pm \sqrt{b^2 - 4ac}}{2a}$$

$$x = \frac{-(-4) \pm \sqrt{(-4)^2 - 4(2)(-7)}}{2(2)}$$

$$x = \frac{4 \pm \sqrt{72}}{4}$$

$$x = \frac{4 \pm 6\sqrt{2}}{4}$$

$$x = \frac{2 \pm 3\sqrt{2}}{2}$$

23. $\frac{1}{x+2} + \frac{1}{x} = \frac{1}{3}$

$3x + 3(x+2) = x(x+2)$

$3x + 3x + 6 = x^2 + 2x$

$x^2 - 4x - 6 = 0$

$a = 1, b = -4, c = -6$

$$x = \frac{-b \pm \sqrt{b^2 - 4ac}}{2a}$$

$$x = \frac{-(-4) \pm \sqrt{(-4)^2 - 4(1)(-6)}}{2(1)}$$

$$x = \frac{4 \pm \sqrt{40}}{2}$$

$$x = \frac{4 \pm 2\sqrt{10}}{2}$$

$x = 2 \pm \sqrt{10}$

25. $\frac{1}{y} - y = \frac{5}{3}$

$3 - 3y^2 = 5y$

$3y^2 + 5y - 3 = 0$

(continued)

25. (continued)
$a = 3, b = 5, c = -3$

$$x = \frac{-b \pm \sqrt{b^2 - 4ac}}{2a}$$

$$x = \frac{-(5) \pm \sqrt{(5)^2 - 4(3)(-3)}}{2(3)}$$

$$x = \frac{-5 \pm \sqrt{61}}{6}$$

27. $\frac{1}{4} + \frac{6}{y+2} = \frac{6}{y}$

$y(y+2) + 6(4y) = 6(4(y+2))$

$y^2 + 2y + 24y = 24y + 48$

$y^2 + 2y - 48 = 0$

$a = 1, b = 2, c = -48$

$$y = \frac{-b \pm \sqrt{b^2 - 4ac}}{2a}$$

$$y = \frac{-(2) \pm \sqrt{(2)^2 - 4(1)(-48)}}{2(1)}$$

$$y = \frac{-2 \pm \sqrt{196}}{2} = \frac{-2 \pm 14}{2}$$

$y = 6,\ y = -8$

29. $x^2 - 2x + 4 = 0$

$a = 1, b = -2, c = 4$

$$x = \frac{-b \pm \sqrt{b^2 - 4ac}}{2a}$$

$$x = \frac{-(-2) \pm \sqrt{(-2)^2 - 4(1)(4)}}{2(1)}$$

$$x = \frac{2 \pm \sqrt{-12}}{2} = \frac{2 \pm \sqrt{4(3)}\sqrt{-1}}{2}$$

$$x = \frac{2 \pm 2i\sqrt{3}}{2}$$

$x = 1 \pm i\sqrt{3}$

31. $5x^2 = -3$
$5x^2 + 3 = 0$
$a = 5, b = 0, c = 3$
$$x = \frac{-b \pm \sqrt{b^2 - 4ac}}{2a}$$
$$x = \frac{-(0) \pm \sqrt{(0)^2 - 4(5)(3)}}{2(5)}$$
$$x = \frac{\pm\sqrt{-60}}{10} = \frac{\pm\sqrt{4(15)}\sqrt{-1}}{10}$$
$$x = \frac{\pm 2i\sqrt{15}}{10}$$
$$x = \frac{\pm i\sqrt{15}}{5}$$

33. $3x^2 + 4x = 2$
$3x^2 + 4x - 2 = 0$
$a = 3, b = 4, c = -2$
$b^2 - 4ac = 4^2 - 4(3)(-2) = 40$
2 irrational roots

35. $2x^2 + 10x + 8 = 0$
$a = 2, b = 10, c = 8$
$b^2 - 4ac = 10^2 - 4(2)(8) = 36$
2 rational roots

37. $9x^2 + 4 = 12x$
$9x^2 - 12x + 4 = 0$
$a = 9, b = -12, c = 4$
$b^2 - 4ac = (-12)^2 - 4(9)(4) = 0$
1 rational root

39. $13, -2$
$x = 13,\ x = -2$
$x - 13 = 0,\ x + 2 = 0$
$(x - 13)(x + 2) = 0$
$x^2 - 11x - 26 = 0$

41. $-5,\ -12$
$x = -5,\ x = -12$
$x + 5 = 0,\ x + 12 = 0$
$(x + 5)(x + 12) = 0$
$x^2 + 17x + 60 = 0$

43. $4i,\ -4i$
$x = 4i,\ x = -4i$
$x - 4i = 0,\ x + 4i = 0$
$(x - 4i)(x + 4i) = 0$
$x^2 + 16 = 0$

45. $3,\ -\frac{5}{2}$
$x = 3,\ x = -\frac{5}{2}$
$x - 3 = 0,\ 2x + 5 = 0$
$(x - 3)(2x + 5) = 0$
$2x^2 - x - 15 = 0$

47. $3x^2 + 5x - 9 = 0$
$$x = \frac{-5 \pm \sqrt{5^2 - 4(3)(-9)}}{2(3)}$$
$$x = \frac{-5 \pm \sqrt{133}}{6}$$
$x = 1.0888,\ x = -2.7554$

49. $20.6x^2 - 73.4x + 41.8 = 0$
$$x = \frac{-(-73.4) \pm \sqrt{(-73.4)^2 - 4(20.6)(41.8)}}{2(20.6)}$$
$$x = \frac{73.4 \pm \sqrt{1943.24}}{41.2}$$
$x = 2.8515,\ x = 0.7116$

51. $p = -100x^2 + 4200x - 39,476 = 0$

$$x = \frac{-4200 \pm \sqrt{4200^2 - 4(-100)(-39,476)}}{2(-100)}$$

$$x = \frac{-4200 \pm \sqrt{1,849,600}}{-200}$$

$x = 14.2,\ x = 27.8$

Fourteen or twenty-eight parachutes per day will produce a zero profit.

53. $p_{max} = -100(21)^2 + 4200(21) - 39,476$

$p_{max} = 4624$

The maximum profit is $4624 per day. 21 is the average of 14 and 28.

Cumulative Review Problems

55. $9x^2 - 6x + 3 - 4x - 12x^2 + 8$

$= -3x^2 - 10x + 11$

57. Current:

$2L + 2W = 50$

$L + W = 25 \Rightarrow L = 25 - W$

New:

$2(2L) + 2(3W) = 118$

$4L + 6W = 118$

$4(25 - W) + 6W = 118$

$100 - 4W + 6W = 118$

$2W = 18$

$W = 9$

$L = 25 - W = 16$

The width is 9 feet and the length is 16 feet.

9.3 Exercises

1. $x^4 - 9x^2 + 20 = 0$

$y = x^2$

$y^2 - 9y + 20 = 0$

$(y-5)(y-4) = 0$

$y = 5,\ y = 4$

$x^2 = 5,\ x^2 = 4$

$x = \pm\sqrt{5},\ x = \pm 2$

3. $x^4 + x^2 - 12 = 0$

$y = x^2$

$y^2 + y - 12 = 0$

$(y-3)(y+4) = 0$

$y = 3,\ y = -4$

$x^2 = 3,\ x^2 = -4$

$x = \pm\sqrt{3},\ x = \pm 2i$

5. $3x^4 = 10x^2 + 8$

$3x^4 - 10x^2 - 8 = 0$

$y = x^2$

$3y^2 - 10y - 8 = 0$

$(3y+2)(y-4) = 0$

$y = -\frac{2}{3},\ y = 4$

$x^2 = -\frac{2}{3},\ x^2 = 4$

$x = \pm\frac{i\sqrt{6}}{3},\ x = \pm 2$

7. $x^6 - 7x^3 - 8 = 0,\ y = x^3$

$y^2 - 7y - 8 = 0$

$(y-8)(y+1) = 0$

(continued)

7. (continued)
$y = 8,\ y = -1$
$x^3 = 8,\ x^3 = -1$
$x = 2,\ x = -1$

9. $x^6 - 5x^3 - 14 = 0$
$y = x^3$
$y^2 - 5y - 14 = 0$
$(y-7)(y+2) = 0$
$y = 7,\ y = -2$
$x^3 = 7,\ x^3 = -2$
$x = \sqrt[3]{7},\ x = \sqrt[3]{-2} = -\sqrt[3]{2}$

11. $x^8 = 3x^4 - 2$
$x^8 - 3x^4 + 2 = 0$
$y = x^4$
$y^2 - 3y + 2 = 0$
$(y-2)(y-1) = 0$
$y = 2,\ y = 1$
$x^4 = 2,\ x^4 = 1$
$x = \pm\sqrt[4]{2},\ x = \pm 1$

13. $3x^8 + 13x^4 = 10$
$3x^8 + 13x^4 - 10 = 0$
$y = x^4$
$3y^2 + 13y - 10 = 0$
$(3y-2)(y+5) = 0$
$y = \frac{2}{3},\ y = -5$
$x^4 = \frac{2}{3},\ x^4 = -5$, no real roots
$x = \pm\frac{\sqrt[4]{54}}{3}$

15. $x^{\frac{2}{3}} + 2x^{\frac{1}{3}} - 8 = 0$
$y = x^{\frac{1}{3}}$
$y^2 + 2y - 8 = 0$
$(y+4)(y-2) = 0$
$y = -4,\ y = 2$
$x^{\frac{1}{3}} = -4,\ x^{\frac{1}{3}} = 2$
$x = (-4)^3 = -64$
$x = 2^3 = 8$

17. $2x^{\frac{2}{3}} - 7x^{\frac{1}{3}} - 4 = 0$
$y = x^{\frac{1}{3}}$
$2y^2 - 7y - 4 = 0$
$(2y+1)(y-4) = 0$
$y = -\frac{1}{2},\ y = 4$
$x^{\frac{1}{3}} = -\frac{1}{2},\ x^{\frac{1}{3}} = 4$
$x = -\frac{1}{8},\ x = 64$

19. $3x^{\frac{1}{2}} - 14x^{\frac{1}{4}} - 5 = 0$
$y = x^{\frac{1}{4}}$
$3y^2 - 14y - 5 = 0$
$(3y+1)(y-5) = 0$
$y = -\frac{1}{3},\ y = 5$
$x^{\frac{1}{4}} = -\frac{1}{3},\ x = \frac{1}{81}$ (extraneous)
$x^{\frac{1}{4}} = 5,\ x = 5^4 = 625$

21. $2x^{\frac{1}{2}} - 14x^{\frac{1}{4}} - 5 = 0$

$y = x^{\frac{1}{4}}$

$2y^2 - 14y - 5 = 0$

$(y-2)(2y+3) = 0$

$y = 2, \ y = -\dfrac{3}{2}$

$x^{\frac{1}{4}} = 2, \ x^{\frac{1}{4}} = -\dfrac{3}{2}$

$x = \left(-\dfrac{3}{2}\right)^4 = \dfrac{81}{16}$ (extraneous)

$x = 2^4 = 16$

23. $x^{\frac{2}{5}} + x^{\frac{1}{5}} - 2 = 0$

$y = x^{\frac{1}{5}}$

$y^2 + y - 2 = 0$

$(y+2)(y-1) = 0$

$y = -2, \ y = 1$

$x^{\frac{1}{5}} = -2, \ x^{\frac{1}{5}} = 1$

$x = (-2)^5 = -32, \ x = 1^5 = 1$

25. $(x^2+x)^2 - 5(x^2+x) = -6$

$(x^2+x)^2 - 5(x^2+x) + 6 = 0$

$y = (x^2+x)$

$y^2 - 5y + 6 = 0$

$(y-2)(y-3) = 0$

$y = 2, \ y = 3$

$(x^2+x) = 2, \ (x^2+x) = 3$

$x^2 + x - 2 = 0, \ x^2 + x - 3 = 0$

$(x+2)(x-1) = 0, \ x = \dfrac{-1 \pm \sqrt{1^2 - 4(1)(-3)}}{2(1)}$

(continued)

25. (continued)

$x = -2, \ x = 1, \ x = \dfrac{-1 \pm \sqrt{13}}{2}$

27. $x - 5x^{\frac{1}{2}} + 6 = 0$

$y = x^{\frac{1}{2}}$

$y^2 - 5y + 6 = 0$

$(y-3)(y-2) = 0$

$y = 3, \ y = 2$

$x^{\frac{1}{2}} = 3, \ x^{\frac{1}{2}} = 2$

$x = 9, \quad x = 4$

29. $10x^{-2} + 7x^{-1} + 1 = 0$

$y = x^{-1}$

$10y^2 - 7y + 1 = 0$

$(5y+1)(2y+1) = 0$

$y = -\dfrac{1}{5}, \ y = -\dfrac{1}{2}$

$x^{-1} = -\dfrac{1}{5}, \ x^{-1} = -\dfrac{1}{2}$

$x = -5, \quad x = -2$

31. $15 - \dfrac{2x}{x-1} = \dfrac{x^2}{x^2 - 2x + 1}$

$15 - \dfrac{2x}{x-1} = \dfrac{x^2}{(x-1)(x-1)}$

$15(x-1)^2 - 2x(x-1) = x^2$

$15x^2 - 30x + 15 - 2x^2 + 2x = x^2$

$12x^2 - 28x + 15 = 0$

$(6x-5)(2x-3) = 0$

$x = \dfrac{5}{6}, \ x = \dfrac{3}{2}$

Cumulative Review Problems

33. $\sqrt{8x}+3\sqrt{2x}-4\sqrt{50x}$
$=\sqrt{4\cdot 2x}+3\sqrt{2x}-4\sqrt{25\cdot 2x}$
$=2\sqrt{2x}+3\sqrt{2x}-20\sqrt{2x}$
$=-15\sqrt{2x}$

35. $3\sqrt{2}(\sqrt{5}-2\sqrt{6})=3\sqrt{10}-6\sqrt{12}$
$=3\sqrt{10}-6\sqrt{4\cdot 3}$
$=3\sqrt{10}-12\sqrt{3}$

9.4 Exercises

1. $S=16t^2$
$t^2=\frac{S}{16}$
$t=\pm\sqrt{\frac{S}{16}}$
$t=\pm\frac{\sqrt{S}}{4}$

3. $A=\pi\left(\frac{d}{2}\right)^2=\pi\frac{d^2}{4}$
$d^2=\frac{4A}{\pi}$
$d=\pm\sqrt{\frac{4A}{\pi}}$

5. $3H=\frac{1}{2}ax^2$
$ax^2=6H$
$x^2=\frac{6H}{a}$
$x=\pm\sqrt{\frac{6H}{a}}$

7. $4(y^2+w)-5=7R$
$4(y^2+w)=5+7R$
$4y^2+4w=5+7R$
$y^2=\frac{7R-4w+5}{4}$
$y=\pm\frac{\sqrt{7R-4w+5}}{2}$

9. $Q=\frac{3mwM^2}{2c}$
$M^2=\frac{2Qc}{3mw}$
$M=\pm\sqrt{\frac{2Qc}{3mw}}$

11. $V=\pi(r^2+R^2)h$
$V=\pi r^2h+\pi R^2h$
$\pi r^2h=V-\pi R^2h$
$r^2=\frac{V-\pi R^2h}{\pi h}$
$r=\pm\sqrt{\frac{V-\pi R^2h}{\pi h}}$

13. $x^2+3bx-10b^2=0$
$(x+5b)(x-2b)=0$
$x=-5b,\ x=2b$

15. $P=EI-RI^2$
$RI^2-EI+P=0$
$a=R,\ b=-E,\ c=P$
$I=\frac{-(-E)\pm\sqrt{(-E)^2-4RP}}{2R}$
$I=\frac{E\pm\sqrt{E^2-4RP}}{2R}$

17. $10w^2 - 3qw - 4 = 0$

$a = 10,\ b = -3q,\ c = -4$

$$w = \frac{-(-3q) \pm \sqrt{(-3q)^2 - 4(10)(-4)}}{2(10)}$$

$$w = \frac{3q \pm \sqrt{9q^2 + 160}}{20}$$

19. $S = 2\pi rh + \pi r^2$

$\pi r^2 + 2\pi hr - S = 0$

$a = \pi,\ b = 2\pi h,\ c = -S$

$$r = \frac{-2\pi h \pm \sqrt{(2\pi h)^2 - 4\pi(-S)}}{2\pi}$$

$$r = \frac{-2\pi h \pm \sqrt{4\pi^2 h^2 + 4\pi S}}{2\pi}$$

$$r = \frac{-2\pi h \pm \sqrt{4\pi^2 h^2 + 4\pi S}}{2\pi}$$

$$r = \frac{-2\pi h \pm 2\sqrt{\pi^2 h^2 + \pi S}}{2\pi}$$

$$r = \frac{-\pi h \pm \sqrt{\pi^2 h^2 + \pi S}}{\pi}$$

21. $(a+1)x^2 + 5x + 2w = 0$

$$x = \frac{-5 \pm \sqrt{5^2 - 4(a+1)(2w)}}{2(a+1)}$$

$$x = \frac{-5 \pm \sqrt{25 - 8aw - 8w}}{2a+2}$$

23. $c^2 = a^2 + b^2$

$c^2 = 7^2 + \sqrt{3}^2$

$c^2 = 49 + 3 = 52$

$c = \sqrt{52}$

$c = \sqrt{4 \cdot 13}$

$c = 2\sqrt{13}$

25. $c^2 = a^2 + b^2$

$\sqrt{34}^2 = a^2 + \sqrt{19}^2$

$a^2 = 34 - 19$

$a^2 = 15$

$a = \sqrt{15}$

27. $c^2 = a^2 + b^2$

$144 = a^2 + 4a^2 = 5a^2$

$$a^2 = \frac{144}{5}$$

$$a = \sqrt{\frac{144}{5}} = \frac{12}{\sqrt{5}} = \frac{12\sqrt{5}}{5}$$

$$b = 2a = \frac{24\sqrt{5}}{5}$$

29. $c^2 = a^2 + b^2$

$c^2 = 12^2 + (c-6)^2$

$c^2 = 144 + c^2 - 12c + 36$

$12c = 180$

$c = 15$

$c - 6 = 9$

The hypotenuse is 15 miles long and the other leg is 9 miles long.

31. $c^2 = a^2 + b^2$

$11^2 = a^2 + (a+3)^2$

$121 = a^2 + a^2 + 6a + 9$

$2a^2 + 6a - 112 = 0$

$a = 2,\ b = 6,\ c = -112$

$$a = \frac{-6 + \sqrt{6^2 - 4(2)(-112)}}{2(2)} \approx 6.13$$

$a + 3 \approx 9.13$

The first leg was about 6.13 miles and the final leg was about 9.13 miles.

35. $b =$ base of triangle

$A = \frac{1}{2}bh = \frac{1}{2}b(2b+2) = 72$

$2b^2 + 2b = 144$

$b^2 + b - 72 = 0$

$(b-8)(b+9) = 0$

$b - 8 = 0, \quad b + 9 = 0$

$b = 8 \qquad b = -9, \text{ reject, } b > 0$

$2b + 2 = 18$

The base is 8 cm and the altitude is 18 cm.

37. $v =$ speed in rain

$s = vt, \quad t = \frac{s}{v}$

$\frac{225}{v} + \frac{150}{v+5} = 8$

$225(v+5) + 150v = 8v(v+5)$

$225v + 1125 + 150v = 8v^2 + 40v$

$8v^2 - 335v - 1125 = 0$

$(v-45)(8v+25) = 0$

$v = 45, \quad v = -\frac{25}{8} \text{ reject, } v > 0$

$v + 5 = 50$

The speed in the rain was 45 mph and the speed without rain was 50 mph.

39. $t =$ time from home to work

$1 \text{ hr } 16 \text{ min } = 1 + \frac{16}{60} = \frac{19}{15} \text{ hr}$

$\frac{19}{15} - t =$ time from work to home

(continued)

39. (continued)

The distance from home to work and the distance from work to home are the same.

$50t = 45\left(\frac{19}{15} - t\right)$

$50t = 57 - 45t$

$95t = 57$

$t = \frac{57}{95} = \frac{3}{5}$

$50 \cdot \frac{3}{5} = 30$

Bob lives 30 miles from his job.

41. $2003 - 1980 = 23$

$N = 1.11x^2 + 33.39x + 304.09$

$N = 1.11(23)^2 + 33.39(23) + 304.09$

$N = 1659.25 \text{ thousands}$

$N = 1,659,250$

To the nearest hundred, there will be 1,659,300 inmates in 2003.

43. $N = 1.11x^2 + 33.39x + 304.09$

$1,744,800 = 1744.8 \text{ thousands}$

$1.11x^2 + 33.39x + 304.09 = 1744.8$

$1.11x^2 + 33.39x - 1440.71 = 0$

$x = \frac{-33.39 + \sqrt{33.39^2 - 4(1.11)(-1440.71)}}{2(1.11)}$

$x = 23.99988462$

$1980 + 24 = 2004$

The number of inmates will reach 1,744,800 in the year 2004.

45. $w = \dfrac{12b^2}{\frac{5}{2}w + \frac{7}{2}b + \frac{21}{2}}$

$\frac{5}{2}w^2 + w\left(\frac{7}{2}b + \frac{21}{2}\right) - 12b^2 = 0$

$5w^2 + (7b + 21)w - 24b^2 = 0$

$w = \dfrac{-(7b+21) \pm \sqrt{(7b+21)^2 - 4(5)(-24b^2)}}{2(5)}$

$w = \dfrac{-7b - 21 \pm \sqrt{49b^2 + 294b + 441 + 480b^2}}{10}$

$w = \dfrac{-7b - 21 \pm \sqrt{529b^2 + 294b + 441}}{10}$

Cumulative Review Problems

47. $\dfrac{4}{\sqrt{3x}} \cdot \dfrac{\sqrt{3x}}{\sqrt{3x}}$

$= \dfrac{4\sqrt{3x}}{3x}$

49. $\dfrac{3}{\sqrt{x}+\sqrt{y}} \cdot \dfrac{\sqrt{x}-\sqrt{y}}{\sqrt{x}-\sqrt{y}}$

$= \dfrac{3(\sqrt{x}-\sqrt{y})}{x-y}$

51. $\dfrac{3ab}{\sqrt[3]{8ab^2}} \cdot \dfrac{\sqrt[3]{a^2b}}{\sqrt[3]{a^2b}}$

$= \dfrac{3ab\sqrt[3]{a^2b}}{2ab}$

$= \dfrac{3\sqrt[3]{a^2b}}{2}$

9.5 Exercises

1. $f(x) = x^2 - 2x - 8$

$x_{\text{vertex}} = \dfrac{-b}{2a}$

$x_{\text{vertex}} = \dfrac{-(-2)}{2(1)}$

$x_{\text{vertex}} = 1$

$f(x_{\text{vertex}}) = 1^2 - 2(1) - 8$

$y_{\text{vertex}} = -9$

$V(1, -9)$

$f(0) = 0^2 - 2(0) - 8$

$f(0) = -8$

$x^2 - 2x - 8 = 0$

$(x-4)(x+2) = 0$

$x = 4,\ \ x = -2$

$I(0, -8);\ \ (4, 0);\ \ (-2, 0)$

3. $f(x) = -x^2 - 4x + 12$

$x_{\text{vertex}} = \dfrac{-b}{2a}$

$x_{\text{vertex}} = \dfrac{-(-4)}{2(-1)}$

$x_{\text{vertex}} = -2$

$f(x_{\text{vertex}}) = -(-2)^2 - 4(-2) + 12$

$f(x_{\text{vertex}}) = 16$

$V(-2, 16)$

$f(0) = -0^2 - 4(0) + 12 = 12$

$-x^2 - 4x + 12 = 0$

$x^2 + 4x - 12 = 0$

$(x-2)(x+6) = 0$

$x = 2,\ \ x = -6$

$I(0, 12);\ \ (2, 0);\ \ (-6, 0)$

5. $p(x) = 3x^2 + 12x + 3$

$x_{\text{vertex}} = \frac{-b}{2a} = \frac{-(12)}{2(3)} = -2$

$f(x_{\text{vertex}}) = 3(-2)^2 + 12(-2) + 3 = -9$

$V(-2, -9)$

$f(0) = 3(0)^2 + 12(0) + 3 = 3$

$3x^2 + 12x + 3 = 0$

$x^2 + 4x + 1 = 0$

$a = 1, \quad b = 4, \quad c = 1$

$x = \frac{-4 \pm \sqrt{4^2 - 4(1)(1)}}{2(1)}$

$x = -0.3, \quad x = -3.7$

$I(0,3); \quad (-0.3,0); \quad (-3.7,0)$

7. $r(x) = x^2 - 6x + 8$

$x_{\text{vertex}} = \frac{-b}{2a}$

$x_{\text{vertex}} = \frac{-(-2)}{2(-3)}$

$x_{\text{vertex}} = -\frac{1}{3}$

$f(x_{\text{vertex}}) = -3\left(-\frac{1}{3}\right)^2 - 2\left(-\frac{1}{3}\right) - 6$

$f(x_{\text{vertex}}) = -\frac{17}{3}$

$V\left(-\frac{1}{3}, -\frac{17}{3}\right)$

$f(0) = -3(0)^2 - 2(0) - 6 = -6$

$-3x^2 - 2x - 6 = 0$

$a = -3, b = -2, c = -6$

$b^2 - 4ac = -68 < 0 \Rightarrow$ no x-intercepts

$I(0, -6)$

9. $f(x) = 2x^2 + 2x - 4$

$x_{\text{vertex}} = \frac{-b}{2a}$

$x_{\text{vertex}} = \frac{-(2)}{2(2)}$

$x_{\text{vertex}} = -\frac{1}{2}$

$f(x_{\text{vertex}}) = 2\left(-\frac{1}{2}\right)^2 + 2\left(-\frac{1}{2}\right) - 4$

$y_{\text{vertex}} = -\frac{9}{2}$

$V\left(-\frac{1}{2}, -\frac{9}{2}\right)$

$f(0) = 2(0)^2 + 2(0) - 4$

$f(0) = -4$

$2x^2 + 2x - 4 = 0$

$x^2 + x - 2 = 0$

$(x-1)(x+2) = 0$

$x = 1, \quad x = -2$

$I(0,-4); \quad (1,0); \quad (-2,0)$

11. $f(x) = x^2 - 6x + 8$

$\frac{-b}{2a} = \frac{-(-6)}{2(1)} = 3$

$f(3) = 3^2 - 6(3) + 8 = -1$

$V(3, -1)$

$f(0) = 0^2 - 6(0) + 8 = 8$

y-intercept: $(0,8)$

$x^2 - 6x + 8 = 0$

$(x-4)(x-2) = 0$

$x = 4, \quad x = 2$

x-intercepts: $(4,0), \ (2,0)$

(continued)

11. (continued)

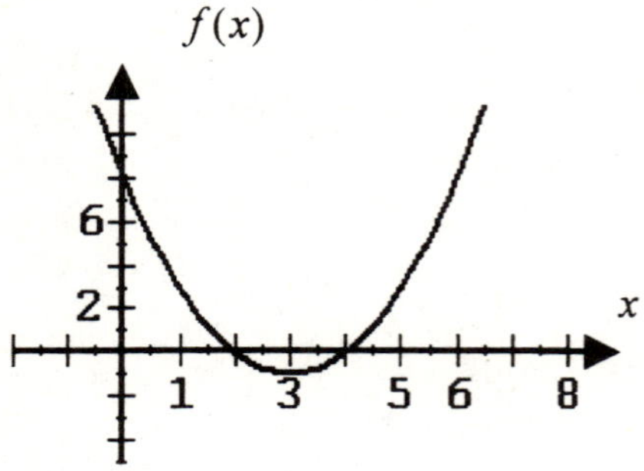

13. $g(x) = x^2 + 2x - 8$

$$\frac{-b}{2a} = \frac{-2}{2} = -1$$

$g(-1) = (-1)^2 + 2(-1) - 8 = -9$

$V(-1, -9)$

$g(0) = 0^2 + 2(0) - 8 = -8$

y-intercept: $(0, -8)$

$x^2 + 2x - 8 = 0$

$(x + 4)(x - 2) = 0$

$x = -4,\ \ x = 2$

x-intercepts: $(-4, 0),\ (2, 0)$

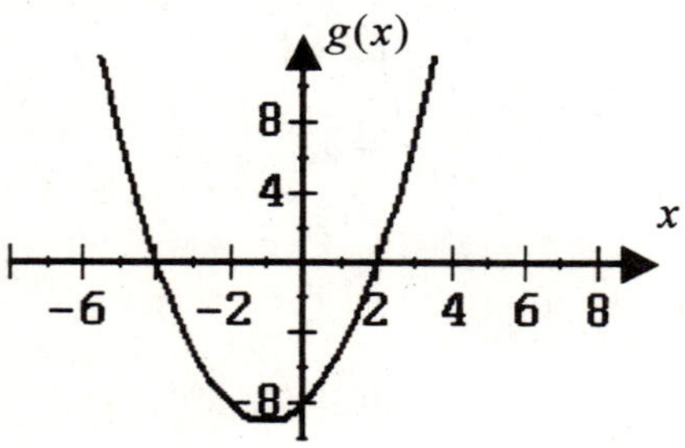

15. $p(x) = -x^2 + 4x - 3$

$$\frac{-b}{2a} = \frac{-4}{2(-1)} = 2$$

$p(2) = -2^2 + 4(2) - 3 = 1$

$V(2,1)$

(continued)

15. (continued)

$p(0) = -0^2 + 4(0) - 3 = -3$

y-intercept: $(0, -3)$

$-x^2 + 4x - 3 = 0$

$x^2 - 4x + 3 = 0$

$(x - 3)(x - 1) = 0$

$x = 3,\ \ x = 1$

x-intercepts: $(3, 0),\ (1, 0)$

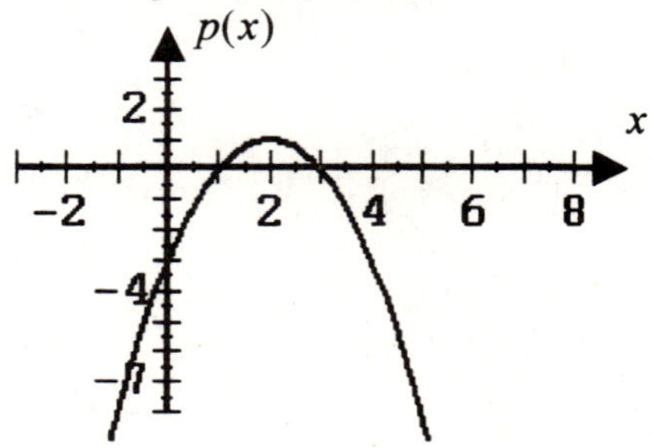

17. $r(x) = x^2 + 4x + 6$

$$\frac{-b}{2a} = \frac{-4}{2(1)} = -2$$

$r(-2) = (-2)^2 + 4(-2) + 6 = 2$

$V(-2, 2)$

$r(0) = 0^2 + 4(0) + 6 = 6$

y-intercept: $(0, 6)$

$x^2 + 4x + 6 = 0$

$a = 1,\ b = 4,\ c = 6$

$b^2 - 4ac = 4^2 - 4(1)(6) = -8 < 0$

x-intercepts: none

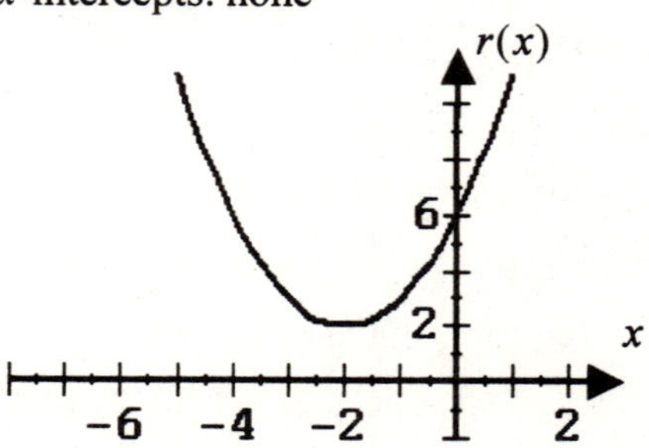

19. $f(x) = x^2 - 6x + 5$

$\dfrac{-b}{2a} = \dfrac{-(-6)}{2(1)} = 3$

$f(3) = 3^2 - 6(3) + 5 = -4$

$V(3,-4)$

$f(0) = 0^2 - 6(0) + 5 = 5$

y-intercept: $(0,5)$

$x^2 - 6x + 5 = 0$

$(x-5)(x-1) = 0$

$x = 5,\ \ x = 1$

x-intercept: $(5,0)$, $(1,0)$

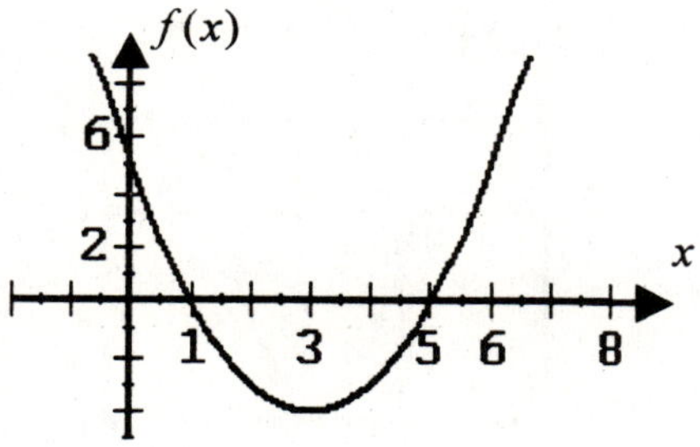

21. $g(x) = -x^2 + 6x - 9$

$\dfrac{-b}{2a} = \dfrac{-6}{2(-1)} = 3$

$g(3) = -3^2 + 6(3) - 9 = 0$

$V(3,0)$

$g(0) = -0^2 + 6(0) - 9 = -9$

y-intercept: $(0,-9)$

$-x^2 + 6x - 9 = 0$

$x^2 - 6x + 9 = 0$

$(x-3)(x-3) = 0$

$x = 3$

x-intercept: $(3,0)$

(continued)

21. (continued)

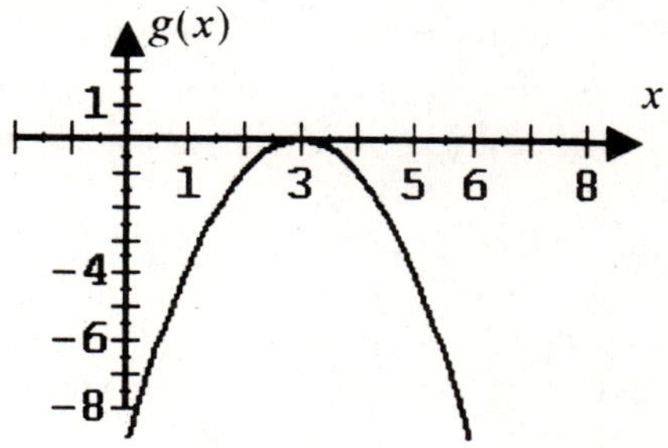

23. $N(x) = 0.18x^2 - 3.18x + 102.25$

$N(20) = 0.18(20)^2 - 3.18(20) + 102.25$
$= 110.65$ thousands
$= 110{,}650$

$N(40) = 0.18(40)^2 - 3.18(40) + 102.25$
$= 263.05$ thousands
$= 263{,}050$

$N(60) = 0.18(60)^2 - 3.18(60) + 102.25$
$= 559.45$ thousands
$= 559{,}450$

$N(80) = 0.18(80)^2 - 3.18(80) + 102.25$
$= 999.85$ thousands
$= 999{,}850$

$N(100) = 0.18(100)^2 - 3.18(100) + 102.25$
$= 1584.25$ thousands
$= 1{,}584{,}250$

25. From the graph, $N(70) \approx 750{,}000$ scuba divers with \$70,000 mean income.

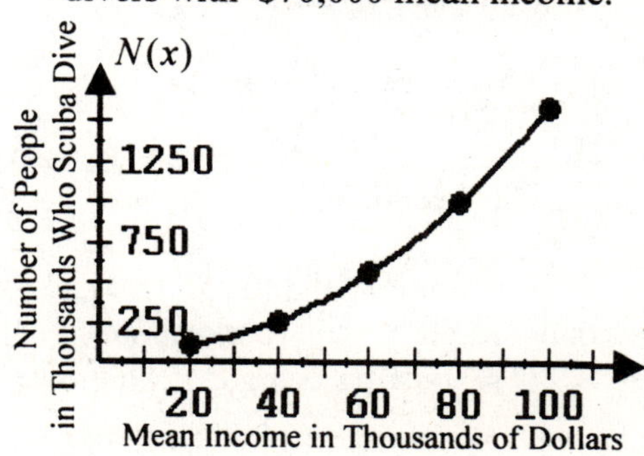

27. From the graph, $N(x) = 390$ for $x \approx 50$. This means that 390,000 people who scuba dive have a mean income of \$50,000.

29. $P(x) = -6x^2 + 312x - 3672$

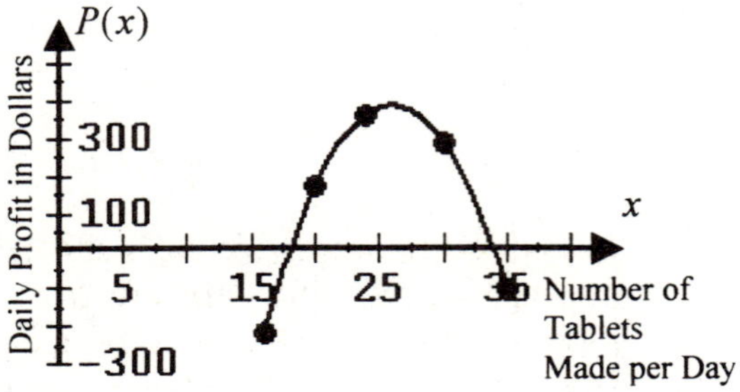

31. $P(x) = -6x^2 + 312x - 3672 = 360$

$6x^2 - 312x + 4032 = 0$

$x^2 - 52x + 672 = 0$

$(x-24)(x-28) = 0$

$x = 24, \quad x = 28$

24 or 28 tablets per day will give \$360 profit because the parabola has vertical axis of symmetry.

33. $d(t) = -16t^2 + 32t + 40$

$\dfrac{-b}{2a} = \dfrac{-32}{2(-16)} = 1$

$d(1) = -16(1^2) + 32(1) + 40 = 56$

$-16t^2 + 32t + 40 = 0$

$t = \dfrac{-32 - \sqrt{32^2 - 4(-16)(40)}}{2(-16)}$

$t = 2.870828693...$

The maximum height is 56 feet and the ball will reach the ground about 2.9 seconds after being thrown upward.

35. $y = 2.3x^2 - 4.4x + 7.59$

$\dfrac{-b}{2a} = \dfrac{-(-4.4)}{2(1)} = 2.2$

$2.2^2 - 4.4(2.2) + 7.59 = 2.75$

$V(2.2, 2.8)$

$0^2 - 4.4(0) + 7.59$

y-intercept: $(0, 7.59)$

$a = 1,\ b = -4.4,\ c = 7.59$

$b^2 - 4ac = (-4.4)^2 - 4(2)(7.59)$

$b^2 - 4ac = -41.36 < 0$

x-intercepts: none

37. $y = 2.3x^2 - 5.4x - 1.6$

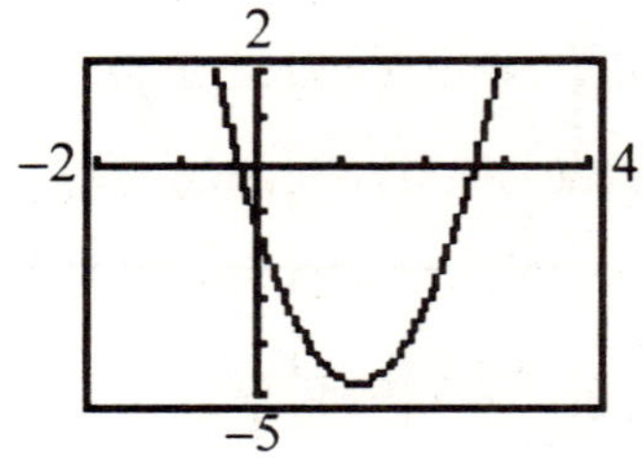

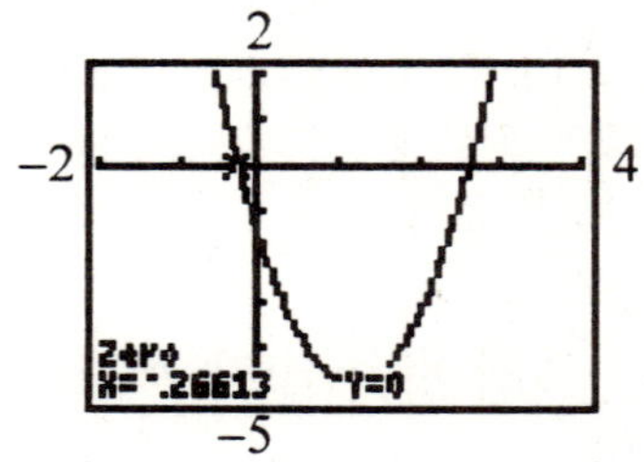

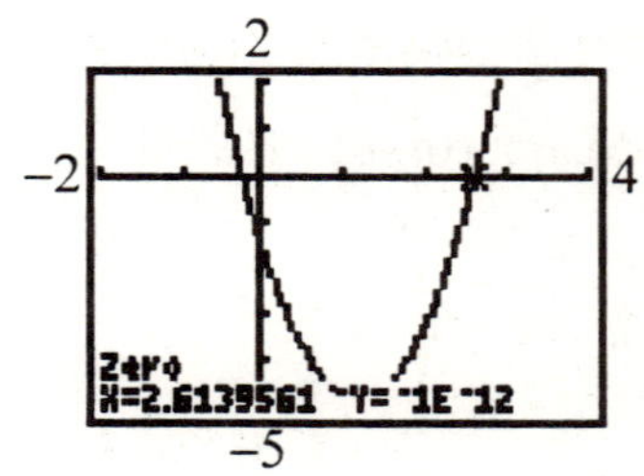

x-intercepts: $(-0.3, 0)$, $(2.6, 0)$

Cumulative Review Problems

39. $9x+5y=6$

$2x-5y=-17$

$11x \quad =-11$

$x=-1$

$9(-1)+5y=6$

$5y=15$

$y=3$

$(-1,3)$

41. $3x-y+2z=12$

$2x-3y+z=5$

$x+3y+8z=22$

Multiply the first equation by $-\frac{2}{3}$ and add to the second equation, then multiply the first equation by $-\frac{1}{3}$ and add to the third equation.

$3x-y+2z=12$

$-\frac{7}{3}y-\frac{1}{3}z=-3$

$\frac{10}{3}y+\frac{22}{3}z=18$

Multiply the second equation by $\frac{10}{7}$ and add to the third equation

$3x-y+2z=12$

$-\frac{7}{3}y-\frac{1}{3}z=-3$

$\frac{48}{7}z=\frac{96}{7}, \quad z=2$

(continued)

41. (continued)

Substitute $z=2$ into second equation.

$-\frac{7}{3}y-\frac{1}{3}(2)=-3, \quad y=1$

Substitute $y=1, z=2$ into first equation.

$3x-1+2(2)=12, \quad x=3$

$(3,1,2)$

9.6 Exercises

1. The critical points divide the number line into regions. All values of x in a given region produce results that are greater than zero, or else all the values of x in a given region produce results that are less than zero.

3. $x^2+x-12<0$

$x^2+x-12=0$

$(x+4)(x-3)=0$

$x=-4, \quad x=3$

Critical points: $-4, 3$

Test: x^2+x-12

Region	Test	Result
I	−5	$8>0$
II	0	$-12<0$
III	4	$8>0$

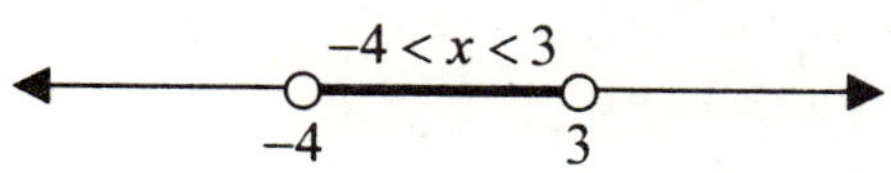

5. $2x^2+x-3<0$

$2x^2+x-3=0$

$(2x+3)(x-1)=0$

$x=-\frac{3}{2}, \quad x=1$

(continued)

5. (continued)

Critical points: $-\frac{3}{2}, 1$

Test: $2x^2 + x - 3$

Region	Test	Result
I	-2	$3 > 0$
II	0	$-3 < 0$
III	2	$7 > 0$

$-\frac{3}{2} < x < 1$

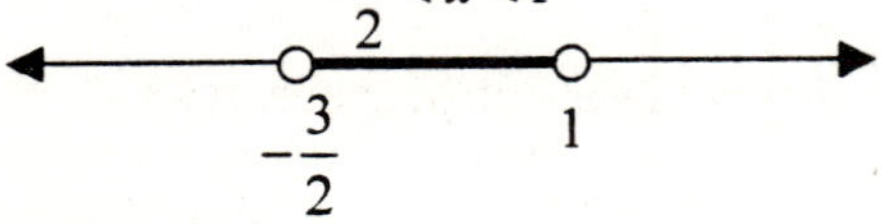

7. $x^2 \geq 4$

$x^2 - 4 \geq 0$

$x^2 - 4 = 0$

$(x+2)(x-2) = 0$

$x = -2, \ x = 2$

Critical points: $-2, 2$

Test: $x^2 - 4$

Region	Test	Result
I	-3	$5 > 0$
II	0	$-4 < 0$
III	3	$5 > 0$

$-2 \leq x$ or $x \geq 2$

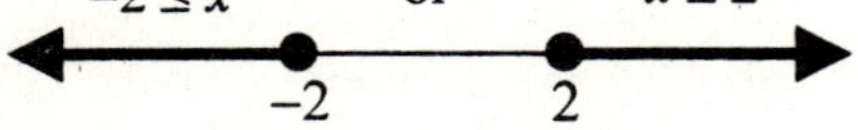

9. $5x^2 \leq 4x + 1$

$5x^2 - 4x - 1 \leq 0$

$5x^2 - 4x - 1 = 0$

$(5x+1)(x-1) = 0$

$x = -\frac{1}{5}, \ x = 1$

(continued)

9. (continued)

Critical points: $-\frac{1}{5}, 1$

Test: $5x^2 - 4x - 1$

Region	Test	Result
I	-1	$8 > 0$
II	0	$-1 < 0$
III	2	$11 > 0$

$-\frac{1}{5} \leq x \leq 1$

11. $20 - x - x^2 > 0$

$x^2 + x - 20 < 0$

$x^2 + x - 20 = 0$

$(x+5)(x-4) = 0$

$x = -5, \ x = 4$

Critical points: $-5, 4$

Test: $20 - x - x^2$

Region	Test	Result
I	-6	$-10 < 0$
II	0	$20 > 0$
III	5	$-10 < 0$

$-5 < x < 4$

13. $6x^2 - 5x > 6$

$6x^2 - 5x - 6 > 0$

$6x^2 - 5x - 6 = 0$

$(2x-3)(3x+2) = 0$

$x = \frac{3}{2}, \ x = -\frac{2}{3}$

Critical points: $\frac{3}{2}, -\frac{2}{3}$

Test: $6x^2 - 5x - 6$

(continued)

13. (continued)

Region	Test	Result
I	-1	$5>0$
II	0	$-6<0$
III	2	$8>0$

$x<-\frac{2}{3}$ or $x>\frac{3}{2}$

15. $-2x+30\ge x(x+5)$

$-2x+30\ge x^2+5x$

$0\ge x^2+7x-30$

$x^2+7x-30\le 0$

$x^2+7x-30=0$

$(x+10)(x-3)=0$

$x=-10,\ x=3$

Critical points: $-10, 3$

Test: $x^2+7x-30$

Region	Test	Result
I	-12	$30>0$
II	0	$-30<0$
III	5	$30>0$

$-10\le x\le 3$

17. $x^2-4x\le -4$

$x^2-4x+4\le 0$

$x^2-4x+4=0$

$(x-2)(x-2)=0$

Critical point: 2

Test: x^2-4x+4

Region	Test	Result
I	0	$4>0$
II	3	$1>0$

$x=2$

19. $x^2-2x>4$

$x^2-2x-4>0$

$x^2-2x-4=0$

$a=1,\ b=-2,\ c=-4$

$x=\frac{-(-2)\pm\sqrt{(-2)^2-4(1)(-4)}}{2(1)}$

$x=\frac{2\pm\sqrt{20}}{2}=1\pm\sqrt{5}$

$x\approx 3.2, -1.2$

Approximate critical points: $3.2, -1.2$

Test: x^2-2x-4

Region	Test	Result
I	-2	$4>0$
II	0	$-4<0$
III	4	$4>0$

$x<1-\sqrt{5}\approx -1.2$ or $x>1+\sqrt{5}\approx 3.2$

21. $x^2-6x<-7$

$x^2-6x+7<0$

$x^2-6x+7=0$

$a=1,\ b=-6,\ c=7$

$x=\frac{-(-6)\pm\sqrt{(-6)^2-4(1)(7)}}{2(1)}$

$x=\frac{6\pm\sqrt{8}}{2}$

$x\approx 4.4,\ 1.6$

Approximate critical points: 4.4, 1.6

Test: x^2-6x+7

Region	Test	Result
I	0	$7>0$
II	1	$-2<0$
III	3	$2>0$

$1.6\approx 3-\sqrt{2}<x<3+\sqrt{2}\approx 4.4$

23. $2x^2 \geq x^2 - 4$
$x^2 \geq -4$
All real numbers satisfy this inequality.

25. $s = -16t^2 + 640t$
$-16t^2 + 640t > 6000$
$t^2 - 40t + 375 < 0$
$t^2 - 40t + 375 = 0$
$(t-15)(t-25) = 0$
$t = 15,\ t = 25$
Critical points: 15, 25
Test: $t^2 - 40t + 375$

Region	Test	Result
I	0	$375 > 0$
II	20	$-25 < 0$
III	30	$75 > 0$

$15 < t < 25$
The height will be greater than 6000 feet for times between 15 and 25 seconds.

27. Profit $= -20(x^2 - 220x + 2400)$
$-20(x^2 - 220x + 2400) > 0$
$x^2 - 220x + 2400 < 0$
$x^2 - 220x + 2400 = 0$
$x = \dfrac{220 \pm \sqrt{38,800}}{2} \approx 208.5,\ 11.5$
Approximate critical points: 208.4, 11.5
Test: $x^2 - 220x + 2400$

Region	Test	Result
I	0	$2400 > 0$
II	20	$-1600 < 0$
III	300	$26,400 > 0$

$\approx 11.5 < x < 208.5$

(continued)

27. (continued)
(a) The profit will be greater than zero for production levels between approximately 11.5 and 208.5 units.
(b) $-20(50^2 - 220(50) + 2400)$
$= 122,000$
The daily profit is \$122,000 when 50 units are manufactured.
(c) $-20(60^2 - 220(60) + 2400)$
$= 144,000$
The daily profit is \$144,000 when 60 units are manufactured.

Cumulative Review Problems

29. $\dfrac{0+81+92+80+\text{ E5 } + \text{ E6}}{6} \geq 70$
$\text{E5} + \text{E6} \geq 420 - 253$
$\text{E5} + \text{E6} \geq 167$
She must score a combined total of 167 points on the two remaining tests. Any two scores that total 167 will be sufficient to participate in synchronized swimming.

31. $\frac{1}{2}y + \frac{1}{3}x = -1$

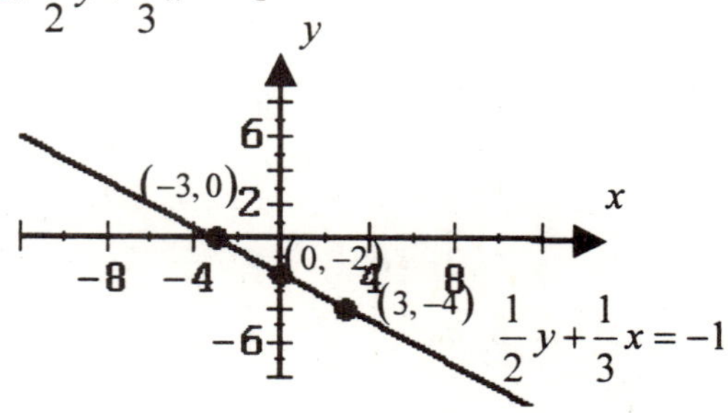

33. $16y^2 - 1 = (4y+1)(4y-1)$

Putting Your Skills to Work

1. $V = 4 \cdot \pi r^2 h = 4(3.14)(28)^2(4000)$
$V = 39,388,160 \text{ ft}^3$

2. $V = 4\pi(r_{\text{outer}}^2 - r_{\text{inner}}^2)h$

   $V = 4(3.14)(28^2 - 25^2)(4000)$

   $V = 7{,}988{,}160 \text{ ft}^3$

3. $y = 0.00205x^2$, $y = 726$, $x = 660$

   $y = 0$

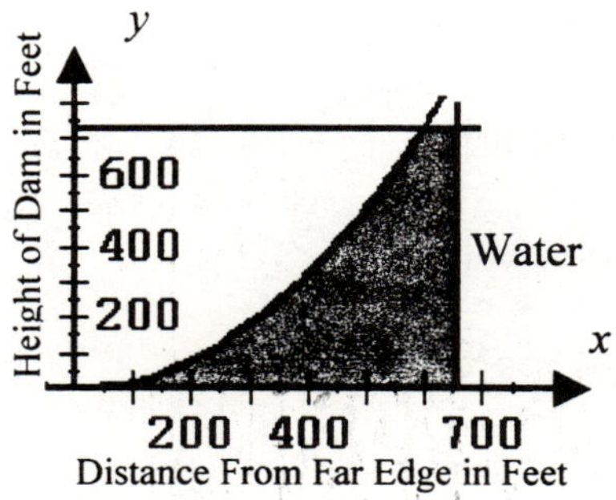

4. $t =$ thickness

   $400 = 0.00205x^2$, $x \approx 441.7$

   $t = 660 - x \approx 218.3$

   From the equations it is approximately 218 ft thick or wide when the dam is 400 ft high which is in agreement with an estimate from the graph.

9.7 Exercises

1. $|x| = 30$

   $x = 30$ or $x = -30$

   check: $|30| \stackrel{?}{=} 30 \qquad |-30| \stackrel{?}{=} 30$

   $30 = 30 \qquad 30 = 30$

3. $|2x - 5| = 13$

   $2x - 5 = 13$ or $2x - 5 = -13$

   $2x = 18 \qquad 2x = -8$

   $x = 9 \qquad x = -4$

   (continued)

3. (continued)

   check: $|2 \cdot 9 - 5| \stackrel{?}{=} 13 \qquad |2(-4) - 5| \stackrel{?}{=} 13$

   $|18 - 5| \stackrel{?}{=} 13 \qquad |-8 - 5| \stackrel{?}{=} 13$

   $|13| \stackrel{?}{=} 13 \qquad |-13| \stackrel{?}{=} 13$

   $13 = 13 \qquad 13 = 13$

5. $|5 - 4x| = 11$

   $5 - 4x = 11$ or $5 - 4x = -11$

   $-4x = 6 \qquad -4x = -16$

   $x = -\dfrac{3}{2} \qquad x = 4$

   check: $\left|5 - 4\left(-\dfrac{3}{2}\right)\right| \stackrel{?}{=} 11 \qquad |5 - 4(4)| \stackrel{?}{=} 11$

   $|5 + 6| \stackrel{?}{=} 11 \qquad |5 - 16| \stackrel{?}{=} 11$

   $|11| \stackrel{?}{=} 11 \qquad |-11| \stackrel{?}{=} 11$

   $11 = 11 \qquad 11 = 11$

7. $\left|\dfrac{1}{2}x - 3\right| = 2$

   $\dfrac{1}{2}x - 3 = 2$ or $\dfrac{1}{2}x - 3 = -2$

   $x - 6 = 4 \qquad x - 6 = -4$

   $x = 10 \qquad x = 2$

   check: $\left|\dfrac{1}{2} \cdot 10 - 3\right| \stackrel{?}{=} 2 \qquad \left|\dfrac{1}{2} \cdot 2 - 3\right| \stackrel{?}{=} 2$

   $|5 - 3| \stackrel{?}{=} 2 \qquad |1 - 3| \stackrel{?}{=} 2$

   $|-2| \stackrel{?}{=} 2 \qquad |1 - 3| \stackrel{?}{=} 2$

   $2 = 2 \qquad 2 = 2$

9. $|2.3-0.3x|=1$

$2.3-0.3x=1$ or $2.3-0.3x=-1$

$23-3x=10$ $\quad 23-3x=-10$

$-3x=-13$ $\quad -3x=-33$

$x=\frac{13}{3}$ $\quad x=11$

check: $\left|2.3-0.3\left(\frac{13}{3}\right)\right|\stackrel{?}{=}1$ $\quad |2.3-0.3(11)|\stackrel{?}{=}1$

$|2.3-1.3|\stackrel{?}{=}1$ $\quad |2.3-3.3|\stackrel{?}{=}1$

$|1|\stackrel{?}{=}1$ $\quad |-1|\stackrel{?}{=}1$

$1=1$ $\quad 1=1$

11. $|x+2|-1=7$

$|x+2|=8$

$x+2=8$ or $x+2=-8$

$x=6$ $\quad x=-10$

check: $|6+2|-1\stackrel{?}{=}7$ $\quad |-10+2|-1\stackrel{?}{=}7$

$|8|-1\stackrel{?}{=}7$ $\quad |-8|-1\stackrel{?}{=}7$

$8-1\stackrel{?}{=}7$ $\quad 8-1\stackrel{?}{=}7$

$7=7$ $\quad 7=7$

13. $\left|\frac{x-1}{3}\right|-2=1,$ $\quad \left|\frac{x-1}{3}\right|=3$

$\frac{x-1}{3}=3$ or $\frac{x-1}{3}=-3$

$x-1=9$ $\quad x-1=-9$

$x=10$ $\quad x=-8$

(continued)

13. (continued)

check: $\left|\frac{10-1}{3}\right|-2\stackrel{?}{=}1$ $\quad \left|\frac{-8-1}{3}\right|-2\stackrel{?}{=}1$

$\left|\frac{9}{3}\right|-2\stackrel{?}{=}1$ $\quad \left|\frac{-9}{3}\right|-2\stackrel{?}{=}1$

$|3|-2\stackrel{?}{=}1$ $\quad |-3|-2\stackrel{?}{=}1$

$3-2\stackrel{?}{=}1$ $\quad 3-2\stackrel{?}{=}1$

$1=1$ $\quad 1=1$

15. $\left|1-\frac{3}{4}x\right|+4=7$

$\left|1-\frac{3}{4}x\right|=3$

$1-\frac{3}{4}x=3$ or $1-\frac{3}{4}x=-3$

$4-3x=12$ $\quad 4-3x=-12$

$-3x=8$ $\quad -3x=-16$

$x=-\frac{8}{3}$ $\quad x=\frac{16}{3}$

check: $\left|1-\frac{3}{4}\cdot\frac{-8}{3}\right|+4\stackrel{?}{=}7$ $\quad \left|1-\frac{3}{4}\cdot\frac{16}{3}\right|+4\stackrel{?}{=}7$

$|1+2|+4\stackrel{?}{=}7$ $\quad |1-4|+4\stackrel{?}{=}7$

$|3|+4\stackrel{?}{=}7$ $\quad |-3|+4\stackrel{?}{=}7$

$3+4\stackrel{?}{=}7$ $\quad 3+4\stackrel{?}{=}7$

$7=7$ $\quad 7=7$

17. $|x+6|=|2x-3|$

$x+6=2x-3$ or $x+6=-2x+3$

$-x=9$ $\quad 3x=-3$

$x=9$ $\quad x=-1$

(continued)

17. (continued)

check: $|9+6|\stackrel{?}{=}|2\cdot 9-3|$ $\quad |-1+6|\stackrel{?}{=}|2(-1)-3|$

$|15|\stackrel{?}{=}|18-3|$ $\quad |-5|\stackrel{?}{=}|-2-3|$

$15\stackrel{?}{=}|15|$ $\quad 5\stackrel{?}{=}|-5|$

$15=15$ $\quad 5=5$

19. $\left|\frac{x-1}{2}\right|=|2x+3|$

$\frac{x-1}{2}=2x+3$ or $\frac{x-1}{2}=-2x-3$

$x-1=4x+6$ $\quad x-1=-4x-6$

$-3x=7$ $\quad 5x=-5$

$x=-\frac{7}{3}$ $\quad x=-1$

21. $|1.5x-2|=|x-0.5|$

$1.5x-2=x-0.5$ or $1.5x-2=-x+0.5$

$15x-20=10x-5$ $\quad 15x-20=-10x+5$

$5x=15$ $\quad 25x=25$

$x=3$ $\quad x=1$

23. $|3-x|=\left|\frac{x}{2}+3\right|$

$3-x=\frac{x}{2}+3$ or $3-x=-\frac{x}{2}-3$

$6-2x=x+6$ $\quad 6-2x=-x-6$

$-3x=0$ $\quad -x=-12$

$x=0$ $\quad x=12$

25. $|-0.74x-8.26|=5.36$

$-0.74x-8.26=5.36$ or $-0.74x-8.26=-5.36$

$-0.74x=13.62$ $\quad -0.74x=2.9$

$x=-18.41$ $\quad x=-3.92$

27. $|4(x-2)|+1=19$

$|4x-8|=18$

$4x-8=18$ or $4x-8=-18$

$4x=26$ $\quad 4x=10$

$x=6.5$ $\quad x=2.5$

check: $|4(6.5-2)|+1\stackrel{?}{=}19$ $\quad |4(2.5-2)|+1\stackrel{?}{=}19$

$|18|+1\stackrel{?}{=}19$ $\quad |-18|+1\stackrel{?}{=}19$

$18+1\stackrel{?}{=}19$ $\quad 18+1\stackrel{?}{=}19$

$19=19$ $\quad 19=19$

29. $\left|\frac{3}{4}x+9\right|=0$

$\frac{3}{4}x+9=0$ $\quad$ check: $\left|\frac{3}{4}(-12)+9\right|\stackrel{?}{=}0$

$3x+36=0$ $\quad |-9+9|\stackrel{?}{=}0$

$3x=-36$ $\quad |0|\stackrel{?}{=}0$

$x=-12$ $\quad 0=0$

31. $\left|\frac{3}{4}x-\frac{2}{3}\right|=-8.$ No solution.

33. $\left|\frac{3x+1}{2}\right|=\frac{3}{4}$

$\frac{3x+1}{2}=\frac{3}{4}$ or $\frac{3x+1}{2}=-\frac{3}{4}$

$6x+2=3$ $\quad 6x+2=-3$

$6x=1$ $\quad x=-5$

$x=\frac{1}{6}$ $\quad x=-\frac{6}{5}$

(continued)

33. (continued)

check: $\left|\dfrac{3\cdot\frac{1}{6}+1}{2}\right| \stackrel{?}{=} \dfrac{3}{4}$ $\qquad \left|\dfrac{3\left(-\frac{5}{6}\right)+1}{2}\right| \stackrel{?}{=} \dfrac{3}{4}$

$\left|\dfrac{\frac{3}{2}}{2}\right| \stackrel{?}{=} \dfrac{3}{4} \qquad \left|\dfrac{-\frac{3}{2}}{2}\right| \stackrel{?}{=} \dfrac{3}{4}$

$\left|\dfrac{\frac{3}{2}}{2}\right| \stackrel{?}{=} \dfrac{3}{4} \qquad \left|\dfrac{-\frac{3}{2}}{2}\right| \stackrel{?}{=} \dfrac{3}{4}$

$\left|\dfrac{3}{4}\right| \stackrel{?}{=} \dfrac{3}{4} \qquad \left|-\dfrac{3}{4}\right| \stackrel{?}{=} \dfrac{3}{4}$

$\dfrac{3}{4} = \dfrac{3}{4} \qquad \dfrac{3}{4} = \dfrac{3}{4}$

35. $|6x-0.3| = |5.6x+11.9|$

$6x-0.3 = 5.6x+11.9$ or $6x-0.3 = -5.6x = 11.9$

$0.4x = 12.2 \qquad 11.6x = -11.6$

$x = 30.5 \qquad x = -1$

check: $|6(30.5)-0.3| \stackrel{?}{=} |5.6(30.5)+11.9|$

$|182.7| \stackrel{?}{=} |182.7|$

$182.7 = 182.7$

$|6(-1)-0.3| \stackrel{?}{=} |5.6(-1)+11.9|$

$|-6.3| \stackrel{?}{=} |6.3|$

$6.3 = 6.3$

37. $\dfrac{|x-3|}{-4} = -2, \quad |x-3| = 8$

$x-3 = 8$ or $x-3 = -8$

$x = 11 \qquad x = -5$

(continued)

37. (continued)

check: $\dfrac{|11-3|}{-4} \stackrel{?}{=} -2 \qquad \dfrac{|-5-3|}{-4} \stackrel{?}{=} -2$

$\dfrac{|8|}{-4} \stackrel{?}{=} -2 \qquad \dfrac{|-8|}{-4} \stackrel{?}{=} -2$

$\dfrac{8}{-4} \stackrel{?}{=} -2 \qquad \dfrac{8}{-4} \stackrel{?}{=} -2$

$-2 = -2 \qquad -2 = -2$

39. $\left(\dfrac{2x^{-2}y}{z^{-1}}\right)^3 = \dfrac{2^3x^{-6}y^3}{z^{-3}} = \dfrac{8y^3z^3}{x^6}$

41. $\dfrac{t}{4} + \dfrac{t}{6} = 1$

$3t + 2t = 12$

$5t = 12$

$t = 2.4$ hours together

9.8 Exercises

1. $|x| \le 8$

$-8 \le x \le 8$

3. $|x| > 5$

$x < -5$ *or* $x > 5$

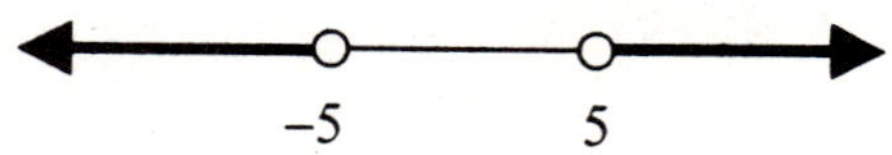

5. $|x+4.5| < 5$

$-5 < x + 4.5 < 5$

$-9.5 < x < 0.5$

7. $|x-3|\le 5$

$-5\le x-3\le 5$

$-2\le x\le 8$

9. $|2x-5|\le 7$

$-7\le 2x-5\le 7$

$-2\le 2x\le 12$

$-1\le x\le 6$

11. $|0.2x-0.7|\le 0.3$

$-0.3\le 0.2x-0.7\le 0.3$

$-3\le 2x-7\le 3$

$4\le 2x\le 10$

$2\le x\le 5$

13. $\left|x-\frac{3}{2}\right|<\frac{1}{2}$

$-\frac{1}{2}<x-\frac{3}{2}<\frac{1}{2}$

$-1<2x-3<1$

$2<2x<4$

$1<x<2$

15. $\left|\frac{1}{4}x+2\right|<6$

$-6<\frac{1}{4}x+2<6$

$-24<x+8<24$

$-32<x<16$

17. $\left|\frac{3}{4}(x-1)\right|<6$

$-6<\frac{3}{4}(x-1)<6$

$-24<3x-3<24$

$-21<3x<27$

$-7<x<9$

19. $\left|\frac{3x-2}{4}\right|<3$

$-3<\frac{3x-2}{4}<3$

$-12<3x-2<12$

$-10<3x<14$

$-\frac{10}{3}<x<\frac{14}{3}$

21. $|x+2|>5$

$x+2<-5$ *or* $x+2>5$

$x<-7$ $\quad$ $x>3$

23. $|x-1|\ge 2$

$x-1\le -2$ *or* $x-1\ge 2$

$x\le -1$ $\quad$ $x\ge 3$

25. $|3x-8|\ge 7$

$3x-8\le -7$ *or* $3x-8\ge 7$

$3x\le 1$ $\quad$ $3x\ge 15$

$x\le \frac{1}{3}$ $\quad$ $x\ge 5$

27. $\left|3-\frac{3}{4}x\right|>9$

$3-\frac{3}{4}x<-9$ *or* $3-\frac{3}{4}x>9$

$12-3x<-36$ $\quad$ $12-3x>36$

$-3x<-48$ $\quad$ $-3x>24$

$x>16$ $\quad$ $x<-8$

29. $\left|\frac{1}{5}x-\frac{1}{10}\right|>2$

$\frac{1}{5}x-\frac{1}{10}<-2$ *or* $\frac{1}{5}x-\frac{1}{10}>2$

$2x-1<-20 \qquad 2x-1>20$

$2x<-19 \qquad 2x>21$

$x<-\frac{19}{2} \qquad x>\frac{21}{2}$

31. $\left|\frac{1}{3}(x-2)\right|<5$

$-5<\frac{1}{3}(x-2)<5$

$-15<x-2<5$

$-13<x<7$

33. $|m-s|\le 0.12$

$|m-18.65|\le 0.12$

$-0.12\le m-18.65\le 0.12$

$18.53\le m\le 18.77$

35. $|n-p|\le 0.05$

$|n-9.68|\le 0.05$

$-0.05\le n-9.68\le 0.05$

$9.63\le n\le 9.73$

37. $12<4x-8<-12$ implies $12<-12$ which is a false statement.

Cumulative Review Problems

39. $(6-4)^3\div(-4)+2^2$

$=2^3\div(-4)+2^2$

$=8\div(-4)+4$

$=-2+4$

$=2$

41. $\frac{-2}{x^2-25}+\frac{3}{2x+10}$

$=\frac{-2(2)}{(x+5)(x-5)(2)}+\frac{3(x-5)}{2(x+5)(x-5)}$

$=\frac{-4+3x-15}{2(x+5)(x-5)}=\frac{3x-19}{2(x+5)(x-5)}$

Chapter 9 Review Problems

1. $5x^2=100$

$x^2=20$

$x=\pm\sqrt{20}=\pm\sqrt{4\cdot 5}$

$x=\pm 2\sqrt{5}$

2. $(x+8)^2=81$

$x+8=\pm 9$

$x=-8\pm 9$

$x=1,\ x=-17$

3. $x^2+8x+13=0$

$x^2+8x+16=-13+16$

$(x+4)^2=3$

$x+4=\pm\sqrt{3}$

$x=-4\pm\sqrt{3}$

4. $4x^2-8x+1=0$

$x^2-2x+1=-\frac{1}{4}+1$

$(x-1)^2=\frac{3}{4}$

$x-1=\pm\frac{\sqrt{3}}{2}$

$x=1\pm\frac{\sqrt{3}}{2}$

5. $3x^2-10x+6=0$
$a=3,\ b=-10,\ c=6$
$$x=\frac{-(-10)\pm\sqrt{(-10)^2-4(3)(6)}}{2(3)}$$
$$x=\frac{10\pm\sqrt{28}}{6}$$
$$x=\frac{10\pm\sqrt{4\cdot 7}}{6}$$
$$x=\frac{5\pm\sqrt{7}}{3}$$

6. $x^2-6x-4=0$
$a=1,\ b=-6,\ c=-4$
$$x=\frac{-(-6)\pm\sqrt{(-6)^2-4(1)(-4)}}{2(1)}$$
$$x=\frac{6\pm\sqrt{52}}{2}=\frac{6\pm\sqrt{4\cdot 13}}{2}$$
$$x=3\pm\sqrt{13}$$

7. $4x^2-12x+9=0$
$$(2x-3)^2=0$$
$$2x-3=0$$
$$x=\frac{3}{2}$$

8. $3x^2-8x+6=0$
$a=3,\ b=-8,\ c=6$
$$x=\frac{-(-8)\pm\sqrt{(-8)^2-4(3)(6)}}{2(3)}$$
$$x=\frac{8\pm\sqrt{-8}}{6}=\frac{8\pm\sqrt{4\cdot 2}\sqrt{-1}}{6}=\frac{8\pm 2i\sqrt{2}}{6}$$
$$x=\frac{4\pm i\sqrt{2}}{3}$$

9. $6x^2-23x=4x$
$$6x^2-27x=0$$
$$3x(2x-9)=0$$
$$3x=0,\quad 2x-9=0$$
$$x=0,\ x=\frac{9}{2}$$

10. $12x^2-29x+15=0$
$$(3x-5)(4x-3)=0$$
$$3x-5=0,\quad 4x-3=0$$
$$x=\frac{5}{3},\ x=\frac{3}{4}$$

11. $x^2-3x-23=5$
$$x^2-3x-28=0$$
$$(x-7)(x+4)=0$$
$$x-7=0,\quad x+4=0$$
$$x=7,\ x=-4$$

12. $3x^2+7x+13=13$
$$3x^2+7x=0$$
$$x(3x+7)=0$$
$$x=0,\quad 3x+7=0$$
$$x=0,\quad x=-\frac{7}{3}$$

13. $3x^2-2x=15x-10$
$$3x^2-17x+10=0$$
$$(3x-2)(x-5)=0$$
$$3x-2=0,\quad x-5=0$$
$$3x-2=0,\quad x-5=0$$
$$x=\frac{2}{3},\ x=5$$

14. $6x^2+12x-24=0$

$x^2+2x-4=0$

$a=1,\ b=2,\ c=-4$

$x=\dfrac{-2\pm\sqrt{2^2-4(1)(-4)}}{2(1)}$

$x=\dfrac{-2\pm\sqrt{20}}{2}=\dfrac{-2\pm\sqrt{4\cdot 5}}{2}$

$x=-1\pm\sqrt{5}$

15. $4x^2-3x+2=0$

$a=4,\ b=-3,\ c=2$

$x=\dfrac{-(-3)\pm\sqrt{(-3)^2-4(4)(2)}}{2(4)}$

$x=\dfrac{3\pm\sqrt{-23}}{8}$

$x=\dfrac{3\pm\sqrt{23}\sqrt{-1}}{8}$

$x=\dfrac{3\pm i\sqrt{23}}{8}$

16. $3x^2+5x+1=0$

$a=3,\ b=5,\ c=1$

$x=\dfrac{-5\pm\sqrt{5^2-4(3)(1)}}{2(3)}$

$x=\dfrac{-5\pm\sqrt{13}}{6}$

17. $3x(3x+2)-2=3x$

$9x^2+6x-2-3x=0$

$9x^2+3x-2=0$

$(3x+2)(3x-1)=0$

$3x+2=0,\ \ 3x-1=0$

$x=-\dfrac{2}{3},\ x=\dfrac{1}{3}$

18. $10x(x-2)+10=2x$

$10x^2-20x+10-2x=0$

$10x^2-22x+10=0$

$5x^2-11x+5=0$

$a=5,\ b=-11,\ c=5$

$x=\dfrac{-(-11)\pm\sqrt{(-11)^2-4(5)(5)}}{2(5)}$

$x=\dfrac{11\pm\sqrt{21}}{10}$

19. $\dfrac{5}{6}x^2-x+\dfrac{1}{3}=0$

$5x^2-6x+2=0$

$a=5,\ b=-6,\ c=2$

$x=\dfrac{-(-6)\pm\sqrt{(-6)^2-4(5)(2)}}{2(5)}$

$x=\dfrac{6\pm\sqrt{-4}}{10}$

$x=\dfrac{6\pm\sqrt{4}\sqrt{-1}}{10}$

$x=\dfrac{6\pm 2i}{10}$

$x=\dfrac{3\pm i}{5}$

20. $\dfrac{4}{5}x^2+x+\dfrac{1}{5}=0$

$4x^2+5x+1=0$

$(4x+1)(x+1)=0$

$4x+1=0,\ x+1=0$

$x=-\dfrac{1}{4},\ x=-1$

21. $y+\frac{5}{3y}+\frac{17}{6}=0$

$6y^2+10+17y=0$

$6y^2+17y+10=0$

$(6y+5)(y+2)=0$

$6y+5=0,\ y+2=0$

$y=-\frac{5}{6},\ y=-2$

22. $\frac{19}{y}-\frac{15}{y^2}+10=0$

$19y-15+10y^2=0$

$10y^2+19y-15=0$

$(2y+5)(5y-3)=0$

$2y+5=0,\ 5y-3=0$

$y=-\frac{5}{2},\ y=\frac{3}{5}$

23. $\frac{15}{y^2}-\frac{2}{y}=1$

$15-2y=y^2$

$y^2+2y-15=0$

$(y+5)(y-3)=0$

$y+5=0,\ y-3=0$

$y=-5,\ y=3$

24. $y-18+\frac{81}{y}=0$

$y^2-18y+81=0$

$(y-9)^2=0$

$y-9=0$

$y=9$

25. $(3y+2)(y-1)=7(-y+1)$

$3y^2-y-2=-7y+7$

$3y^2+6y-9=0$

$y^2+2y-3=0$

$(y+3)(y-1)=0$

$y+3=0,\ y-1=0$

$y=-3,\ y=1$

26. $y(y+1)+(y+2)^2=4$

$y^2+y+y^2+4y+4=4$

$2y^2+5y=0$

$y(2y+5)=0$

$y=0,\ 2y+5=0$

$y=0,\ y=-\frac{5}{2}$

27. $\frac{2x}{x+3}+\frac{3x-1}{x+1}=3$

$2x(x+1)+(3x-1)(x+3)=3(x+3)(x+1)$

$2x^2+2x+3x^2+8x-3=3x^2+12x+9$

$2x^2-2x-12=0$

$x^2-x-6=0$

$(x-3)(x+2)=0$

$x-3=0,\ x+2=0$

$x=3,\ x=-2$

28. $\frac{4x+1}{2x+5}+\frac{3x}{x+4}=2$

$(4x+1)(x+4)+3x(2x+5)=2(2x+5)(x+4)$

$4x^2+17x+4+6x^2+15x=4x^2+26x+40$

$6x^2+6x-36=0$

$x^2+x-6=0$

$(x+3)(x-2)=0$

$x+3=0,\ x-2=0$

$x=-3,\ x=2$

29. $2x^2+5x-3=0$
    $a=2,\ b=5,\ c=-3$
    $b^2-4ac=5^2-4(2)(-3)=49=7^2$
    Two rational solutions.

30. $3x^2-7x-12=0$
    $a=3, b=-7, c=-12$
    $b^2-4ac=(-7)^2-4(3)(-12)=193$
    Two irrational solutions.

31. $4x^2-6x+5=0$
    $a=4,\ b=-6,\ c=5$
    $b^2-4ac=(-6)^2-4(4)(5)=-44$
    Two complex solutions.

32. $25x^2-20x+4=0$
    $a=25, b=-20, c=4$
    $b^2-4ac=(-20)^2-4(25)(4)=0$
    One rational solution.

33. $5,\ -5$
    $x=5,\ x=-5$
    $x-5=0,\ x+5=0$
    $(x-5)(x+5)=0$
    $x^2-25=0$

34. $3i,\ -3i$
    $x=3i,\ x=-3i$
    $x-3i=0,\ x+3i=0$
    $(x-3i)(x+3i)=0$
    $x^2-9i^2=0$
    $x^2+9=0$

35. $4\sqrt{2},\ -4\sqrt{2}$
    $x=4\sqrt{2},\ x=-4\sqrt{2}$
    $x-4\sqrt{2}=0,\ x+4\sqrt{2}=0$
    $(x-4\sqrt{2})(x+4\sqrt{2})=0$
    $x^2-(4\sqrt{2})(4\sqrt{2})=0$
    $x^2-32=0$

36. $-\frac{3}{4},\ -\frac{1}{2}$
    $x=-\frac{3}{4},\ x=-\frac{1}{2}$
    $4x=-3,\ 2x=-1$
    $4x+3=0,\ 2x+1=0$
    $(4x+3)(2x+1)=0$
    $8x^2+10x+3=0$

37. $x^4-6x^2+8=0$
    $y=x^2$
    $y^2-6y+8=0$
    $(y-4)(y-2)=0$
    $y-4=0, y-2=0$
    $y=4,\ y=2,$
    $x^2=4,\ x^2=2$
    $x=\pm 2,\ x=\pm\sqrt{2}$

38. $2x^6-5x^3-3=0$
    $x^3=-\frac{1}{2},\ x^2=3$
    $x=-\sqrt[3]{\frac{1}{2}}=-\frac{\sqrt[3]{4}}{2},\ x=\sqrt[3]{3}$

(continued)

38. (continued)

$y = x^3$

$2y^2 - 5y - 3 = 0$

$(2y+1)(y-3) = 0$

$2y+1 = 0, y-3 = 0$

$y = -\frac{1}{2}, y = 3$

39. $x^{\frac{2}{3}} + 9x^{\frac{1}{3}} = -8$

$y = x^{\frac{1}{3}}$

$y^2 + 9y + 8 = 0$

$(y+8)(y+1) = 0$

$y+8 = 0, y+1 = 0$

$y = -8, y = -1$

$x^{\frac{1}{3}} = -8, x^{\frac{1}{3}} = -1$

$x = -512, x = -1$

40. $3x^{\frac{1}{2}} - 11x^{\frac{1}{4}} = 4$

$y = x^{\frac{1}{4}}$

$3y^2 - 11y - 4 = 0$

$(3y+1)(y-4) = 0$

$3y+1 = 0, y-4 = 0$

$y = -\frac{1}{3}, y = 4$

$x^{\frac{1}{4}} = -\frac{1}{3}$ (not a valid real root), $x^{\frac{1}{4}} = 4$

$x = 4^4 = 256$

41. $(2x-5)^2 + 4(2x-5) + 3 = 0$

$y = 2x-5$

$y^2 + 4y + 3 = 0$

$(y+3)(y+1) = 0$

$y+3 = 0, y+1 = 0$

$y = -3, y = -1$ (continued)

41. (continued)

$2x-5 = -3, 2x-5 = -1$

$2x = 2, 2x = 4$

$x = 1, x = 2$

42. $1 + 4x^{-8} = 5x^{-4}$

$x^8 + 4 = 5x^4$

$x^8 - 5x^4 + 4 = 0$

$y = x^4$

$y^2 - 5y + 4 = 0$

$(y-4)(y-1) = 0$

$y-4 = 0, y-1 = 0$

$y = 4, y = 1$

$x^4 = 4, x^4 = 1$

$x = \pm 4^{\frac{1}{4}} = \pm(2^2)^{\frac{1}{4}}, x = \pm 1$

$x = \pm\sqrt{2}, x = \pm 1$

43. $A = \frac{2B^2C}{3H}$

$3AH = 2B^2C$

$B^2 = \frac{3AH}{2C}$

$B = \pm\sqrt{\frac{3AH}{2C}}$

44. $2H = 3g(a^2 + b^2)$

$b^2 = \frac{2H}{3g} - a^2$

$b = \pm\sqrt{\frac{2H}{3g} - a^2}$

45. $20d^2 - xd - x^2 = 0$

$(4d - x)(5d + x) = 0$

$4d - x = 0, 5d + x = 0$

$d = \frac{x}{4}, d = -\frac{x}{5}$

46. $yx^2 - 3x - 7 = 0$
$a = y,\ b = -3,\ c = -7$
$$x = \frac{-(-3) \pm \sqrt{(-3)^2 - 4(y)(-7)}}{2y}$$
$$x = \frac{3 \pm \sqrt{9 + 28y}}{2y}$$

47. $3y^2 - 4ay + 2a = 0$
$$y = \frac{-(-4a) \pm \sqrt{(-4a)^2 - 4(3)(2a)}}{2(3)}$$
$$y = \frac{4a \pm \sqrt{16a^2 - 24a}}{6}$$
$$y = \frac{4a \pm \sqrt{4(4a^2 - 6a)}}{6}$$
$$y = \frac{4a \pm 2\sqrt{4a^2 - 6a}}{6}$$
$$y = \frac{2a \pm \sqrt{4a^2 - 6a}}{3}$$

48. $PV = 5x^2 + 3y^2 + 2x$
$5x^2 + 2x + 3y^2 - PV = 0$
$a = 5,\ b = 2,\ c = 3y^2 - PV$
$$x = \frac{-2 \pm \sqrt{2^2 - 4(5)(3y^2 - PV)}}{2(5)}$$
$$x = \frac{-2 \pm \sqrt{4(1 - 15y^2 + 5PV)}}{10}$$
$$x = \frac{-2 \pm 2\sqrt{1 - 15y^2 + 5PV}}{10}$$
$$x = \frac{-1 \pm \sqrt{1 - 15y^2 + 5PV}}{5}$$

49. $c^2 = a^2 + b^2$
$16^2 = a^2 + 4^2$
$a^2 = 256 - 16 = 240 = 16(15)$
$a = 4\sqrt{15}$

50. $c^2 = a^2 + b^2 = (3\sqrt{2})^2 + 2^2$
$c^2 = 18 + 4 = 22$
$c = \sqrt{22}$

51. $c^2 = a^2 + b^2$
$6^2 = 5^2 + b^2$
$b^2 = 36 - 25 = 11$
$b = \sqrt{11} \approx 3.3$
The car is approximately 3.3 miles from the observer.

52. $A = LW = 203$
$(4W + 1)W = 203$
$4W^2 + W - 203 = 0$
$(W - 7)(4W + 29) = 0$
$W - 7 = 0,\ 4W + 29 = 0$
$W = 7,\ W = -\frac{29}{4}$ reject, $W > 0$
$L = 4W + 1 = 29$

The width is 7 m and the length is 29 m.

53. $A = \frac{1}{2}bh = 70$
$b(2b + 6) = 140$
$2b^2 + 6b - 140 = 0$
$b^2 + 3b - 70 = 0$
$(b - 7)(b + 10) = 0$
$b - 7 = 0,\ b + 10 = 0$
$b = 7,\ b = -10$ reject, $b > 0$
$h = 2b + 6 = 20$

The base is 7 cm and the altitude is 20 cm.

54. $v =$ speed with no rain

$$200 = vt_1 \Rightarrow t_1 = \frac{200}{v}$$

$$90 = (v-5)t_2 \Rightarrow t_2 = \frac{90}{v-5}$$

$$t_1 + t_2 = \frac{200}{v} + \frac{90}{v-5} = 6$$

$$200(v-5) + 90v = 6v(v-5)$$

$$200v - 1000 + 90v = 6v^2 - 30v$$

$$6v^2 - 320v + 1000 = 0$$

$$(v-50)(6v-20) = 0$$

$$v - 50 = 0,\ 6v - 20 = 0$$

$$v = 50,\ v = \frac{10}{3} \text{ reject, } v > 5$$

$$v - 5 = 45$$

The speed before the rain was 50 mph and 45 mph in the rain.

55. $v =$ cruising speed

$$60 = vt_1 \Rightarrow t_1 = \frac{60}{v}$$

$$5 = (v-15)t_2 \Rightarrow t_2 = \frac{5}{v-15}$$

$$t_1 + t_2 = \frac{60}{v} + \frac{5}{v-15} = 4$$

$$60(v-15) + 5v = 4v(v-15)$$

$$60v - 900 + 5v = 4v^2 - 60v$$

$$4v^2 - 125v + 900 = 0$$

$$(v-20)(4v-45) = 0$$

$$v - 20 = 0,\ 4v - 45 = 0$$

$$v = 20,\ v = \frac{45}{4} = 11.25 \text{ reject, } v > 15$$

$$v - 15 = 5$$

The cruising speed is 20 mph and the trolling speed is 5 mph.

56. $(10+2x)(6+2x) - 10(6) = 100$

$$60 + 32x + 4x^2 - 60 = 100$$

$$4x^2 + 32x - 100 = 0$$

$$x^2 + 8x - 25 = 0$$

$$x = \frac{-8 + \sqrt{8^2 - 4(1)(-25)}}{2} \approx 2.4$$

The walkway should be approximately 2.4 feet wide.

57. $(40+2x)(30+2x) - 40(3) = 296$

$$1200 + 140x + 4x^2 - 1200 = 296$$

$$4x^2 + 140x - 296 = 0$$

$$x^2 + 35x - 74 = 0$$

$$(x-2)(x+37) = 0$$

$$x = 2,\ x = -37 \text{ reject, } x > 0$$

The walkway should be 2 feet wide.

58. $g(x) = -x^2 + 6x - 11$

$$\frac{-b}{2a} = \frac{-6}{2(-1)} = 3$$

$$g(3) = -3^2 + 6(3) - 11 = -2$$

$$V(3, -2)$$

$$g(0) = -0^2 + 6(0) - 11 = -11$$

y-intercept: $(0, -11)$

$$b^2 - 4ac = 6^2 - 4(-1)(-11) = -8 < 0$$

x-intercepts: none

59. $f(x) = x^2 + 10x + 25$

$\frac{-b}{2a} = \frac{-10}{2(1)} = -5$

$f(-5) = (-5)^2 + 10(-5) + 25 = 0$

$V(-5, 0)$

$f(0) = 25$

y-intercept: $(0, 25)$

$x^2 + 10x + 25 = 0$

$(x+5)^2 = 0$

$x = -5$

x-intercept: $(-5, 0)$

60. $f(x) = x^2 + 4x + 3$

$\frac{-b}{2a} = \frac{-4}{2(1)} = -2$

$f(-2) = (-2)^2 + 4(-2) + 3 = -1$

$V(-2, -1)$

$f(0) = 3$

y-intercept: $(0, 3)$

$x^2 + 4x + 3 = 0$

$(x+3)(x+1) = 0$

$x = -3$, $x = -1$

x-intercepts: $(-3, 0)$, $(-1, 0)$

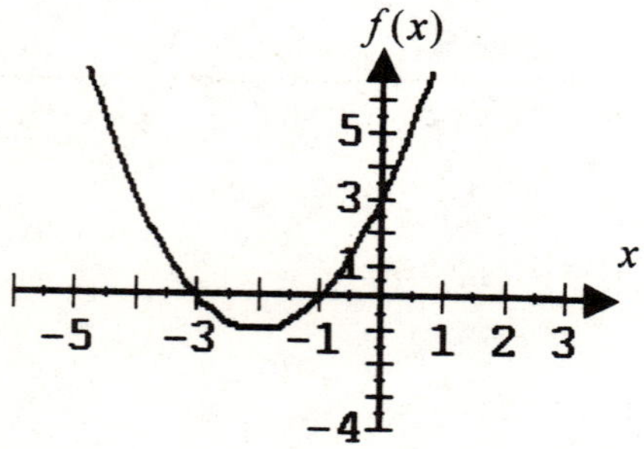

61. $f(x) = x^2 + 6x + 5$

$\frac{-b}{2a} = \frac{-6}{2(1)} = -3$

$f(-3) = (-3)^2 + 6(-3) + 5 = -4$

$V(-3, -4)$

$f(0) = 5$

y-intercept: $(0, 5)$

$x^2 + 6x + 5 = 0$

$(x+5)(x+1) = 0$

$x = -5$, $x = -1$

x-intercepts: $(-5, 0)$, $(-1, 0)$

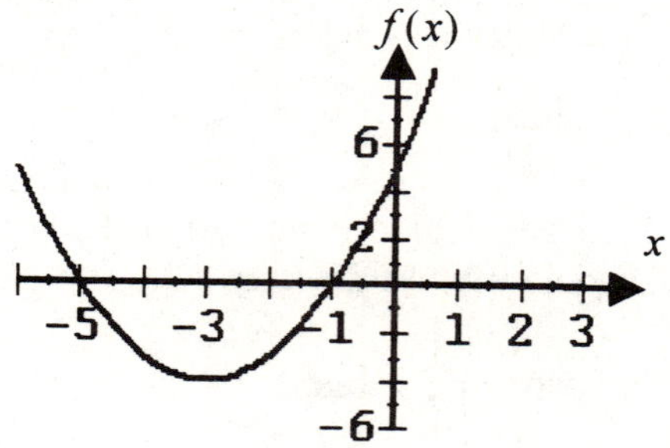

62. $f(x) = -x^2 + 6x - 5$

$\frac{-b}{2a} = \frac{-6}{2(-1)} = 3$

$f(3) = -3^2 + 6(3) - 5 = 4$

$V(3, 4)$

(continued)

62. (continued)

$f(0) = -5$

y-intercept: $(0, -5)$

$-x^2 + 6x - 5 = 0$

$x^2 - 6x + 5 = 0$

$(x-1)(x-5) = 0$

$x = 1,\ x = 5$

x-intercepts: $(1,0),\ (5,0)$

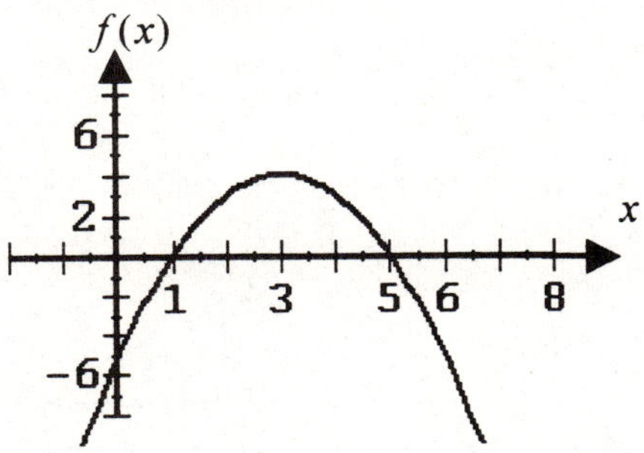

63. $h(t) = -16t^2 + 400t + 40$

$\dfrac{-b}{2a} = \dfrac{-400}{2(-16)} = 12.5$

$h(1.25) = -16(1.25)^2 + 400(12.5) + 40$

$h(12.5) = 2540$

$-16t^2 + 400t + 40 = 0$

$t = \dfrac{-400 - \sqrt{400^2 - 4(-16)(40)}}{2(-16)}$

$t = 25.09960317...$

The maximum height is 2540 feet. The amount of time for the complete flight is 25.1 seconds.

64. $R(x) = x(1200 - x) = -x^2 + 1200x$

$\dfrac{-b}{2a} = \dfrac{-1200}{2(-1)} = 600$

The maximum revenue will occur if the price is \$600 for each unit.

65. $x^2 + 4x - 18 < 0$

$x^2 + 7x - 18 = 0$

$(x+9)(x-2) = 0$

$x = -9,\ x = 2$

Critical points: $-9,\ 2$

Test: $x^2 + 7x - 18$

Region	Test	Results
I	−10	$12 > 0$
II	0	$-18 < 0$
III	3	$12 > 0$

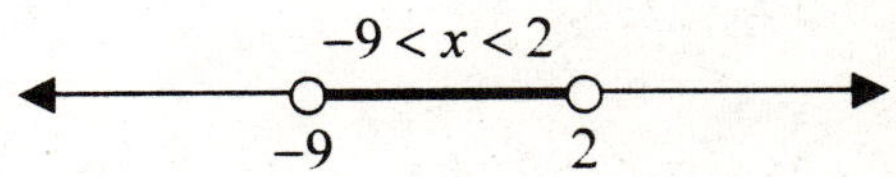

66. $x^2 + 4x - 21 < 0$

$x^2 + 4x - 21 = 0$

$(x+7)(x-3) = 0$

$x = -7,\ x = 3$

Critical points: $-7,\ 3$

Test: $x^2 + 4x - 21$

Region	Test	Results
I	−8	$11 > 0$
II	0	$-21 < 0$
III	4	$11 > 0$

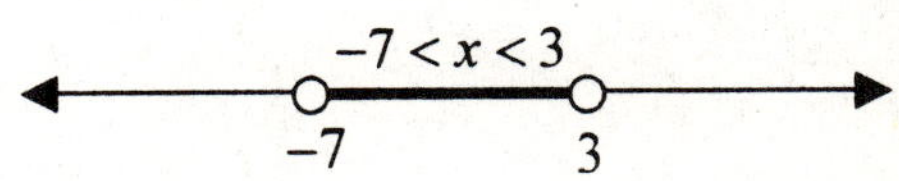

67. $x^2 - 9x + 20 > 0$

$x^2 - 9x + 20 = 0$

$(x-5)(x-4) = 0$

$x = 5,\ x = 4$

Critical points: 4, 5

Test: $x^2 - 9x + 20$

(continued)

67. (continued)

Region	Test	Results
I	0	$20 > 0$
II	4.5	$-0.25 < 0$
III	6	$2 > 0$

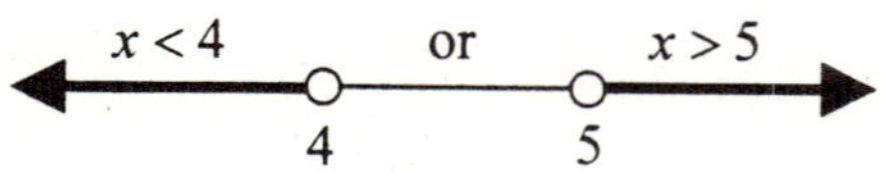

68. $x^2 - 11x + 28 > 0$

$x^2 - 11x + 28 = 0$

$(x-7)(x-4) = 0$

$x = 7,\ x = 4$

Critical points: 7, 4

Test: $x^2 - 11x + 28$

Region	Test	Results
I	0	$28 > 0$
II	6	$-2 < 0$
III	8	$4 > 0$

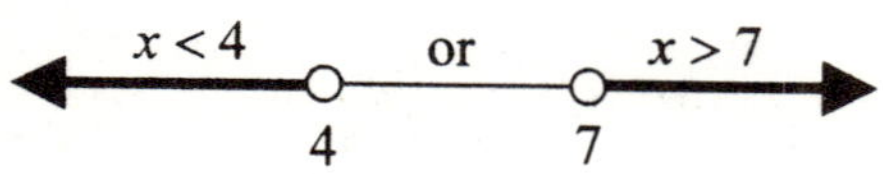

69. $2x^2 - 5x - 3 \le 0$

$2x^2 - 5x - 3 = 0$

$(2x+1)(x-3) = 0$

$x = -\frac{1}{2},\ x = 3$

Critical points: $-\frac{1}{2}$, 3

Test: $2x^2 - 5x - 3$

Region	Test	Results
I	-1	$4 > 0$
II	0	$-3 < 0$
III	4	$9 > 0$

(continued)

69. (continued)

$-\frac{1}{2} \le x \le 3$

70. $3x^2 - 5x - 2 \le 0$

$3x^2 - 5x - 2 = 0$

$(3x+1)(x-2) = 0$

$x = -\frac{1}{3},\ x = 2$

Critical points: $-\frac{1}{3}$, 2

Test: $3x^2 - 5x - 2$

Region	Test	Results
I	-1	$6 > 0$
II	0	$-2 < 0$
III	3	$10 > 0$

$-\frac{1}{3} \le x \le 2$

71. $16x^2 - 25 > 0$

$16x^2 - 25 = 0$

$x^2 = \frac{25}{16}$

$x = \pm\frac{5}{4}$

Critical points: $-\frac{5}{4}$, $\frac{5}{4}$

Test: $16x^2 - 25$

Region	Test	Results
I	-2	$39 > 0$
II	0	$-25 < 0$
III	2	$39 > 0$

$x < -\frac{5}{4}$ or $x > \frac{5}{4}$

72. $9x^2 - 4 > 0$

$9x^2 - 4 = 0$

$x^2 = \frac{4}{9}$

$x = \pm\frac{2}{3}$

Critical points: $-\frac{2}{3}, \frac{2}{3}$

Test: $9x^2 - 4$

Region	Test	Results
I	−2	$32 > 0$
II	0	$-4 < 0$
III	2	$32 > 0$

$x < -\frac{2}{3}$ or $x > \frac{2}{3}$

73. $x^2 - 9x > 4 - 7x$

$x^2 - 2x - 4 = 0$

$x = \frac{-(-2) \pm \sqrt{(-2)^2 - 4(1)(-4)}}{2}$

$x = 1 \pm \sqrt{5}$

Critical points: $1-\sqrt{5},\ 1+\sqrt{5}$

Test: $x^2 - 2x - 4$

Region	Test	Results
I	−2	$4 > 0$
II	0	$-4 < 0$
III	4	$4 > 0$

$x < 1-\sqrt{5} \approx -1.2$ or $x > 1+\sqrt{5} \approx 3.2$

74. $4x^2 - 8x \le 12 + 5x^2$

$-x^2 - 8x - 12 \le 0$

$x^2 + 8x + 12 \ge 0$

$x^2 + 8x + 12 = 0$

(continued)

74. (continued)

$(x+6)(x+2) = 0$

$x = -6,\ x = -2$

Critical points: $-6,\ -2$

Test: $x^2 + 8x + 12$

Region	Test	Results
I	−8	$12 > 0$
II	−4	$-4 < 0$
III	0	$12 > 0$

$x \le -6$ or $x \ge -2$

75. $x^2 + 13x > 16 + 7x$

$x^2 + 6x - 16 > 0$

$x^2 + 6x - 16 = 0$

$(x+8)(x-2) = 0$

$x+8 = 0,\ x-2 = 0$

$x = -8,\ x = 2$

Critical points: $-8,\ 2$

Test: $x^2 + 6x - 16$

Region	Test	Results
I	−10	$24 > 0$
II	0	$-16 < 0$
III	4	$24 > 0$

$x < -8$ or $x > 2$

76. $3x^2 - 12x > -11$

$3x^2 - 12x + 11 > 0$

$3x^2 - 12x + 11 = 0$

$x = \frac{12 \pm \sqrt{(-12)^2 - 4(3)(11)}}{2(3)} = 1.4,\ 2.6$

Critical points: 1.4, 2.6

Test: $3x^2 - 12x + 11$

(continued)

76. (continued)

Region	Test	Results
I	0	$11 > 0$
II	2	$-1 < 0$
III	3	$2 > 0$

$x < 1.4$ or $x > 2.6$

77. $-2x^2 + 7x + 12 \le -3x^2 + x$
$x^2 + 6x + 12 \le 0$
$a = 1,\ b = 6,\ c = 12$
$b^2 - 4ac = 6^2 - 4(1)(12) = -12 < 0$
No real solution.

78. $4x^2 + 12x + 9 < 0$
$(2x + 3)^2 < 0$
No real solution.

79. $|x + 1| = 8$
$x + 1 = 8$ or $x + 1 = -8$
$x = 7 \qquad x = -9$

80. $|4x - 5| = 7$
$4x - 5 = 7$ or $4x - 5 = -7$
$4x = 12 \qquad 4x = -2$
$x = 3 \qquad x = -\dfrac{1}{2}$

81. $|3x + 2| = 20$
$3x + 2 = 20$ or $3x + 2 = -20$
$3x = 18 \qquad 3x = -22$
$x = 6 \qquad x = -\dfrac{22}{3}$

82. $|3 - x| = |5 - 2x|$
$3 - x = 5 - 2x$ or $3 - x = -(5 - 2x)$
$2x - x = 5 - 3 \qquad 3 - x = -5 + 2x$
$x = 2 \qquad 3x = 8$
$x = \dfrac{8}{3}$

83. $\left|\dfrac{1}{4}x - 3\right| = 8$
$\dfrac{1}{4}x - 3 = 8$ or $\dfrac{1}{4}x - 3 = -8$
$x - 12 = 32 \qquad x - 12 = -32$
$x = 44 \qquad x = -20$

84. $|2x - 8| + 7 = 12$
$|2x - 8| = 5$ or $2x - 8 = -5$
$2x - 8 = 5 \qquad 2x = 3$
$2x = 13 \qquad x = \dfrac{3}{2}$
$x = \dfrac{13}{2}$

85. $|x + 7| < 15$
$-15 < x + 7 < 15$
$-22 < x < 8$

86. $|x + 9| < 18$
$-18 < x + 9 < 18$
$-27 < x < 9$

87. $\left|\frac{1}{2}x+2\right|<\frac{7}{4}$

$$-\frac{7}{4}<\frac{1}{2}x+2<\frac{7}{4}$$
$$-7<2x+8<7$$
$$-15<2x<-1$$
$$-\frac{15}{2}<x<-\frac{1}{2}$$

88. $|2x-1|\geq 9$

$2x-1\leq -9$ or $2x-1\geq 9$

$2x\leq -8$ $\quad$ $2x\geq 10$

$x\leq -4$ $\quad$ $x\geq 5$

89. $|3x-1|\geq 2$

$3x-1\leq -2$ or $3x-1\geq 2$

$3x\leq -1$ $\quad$ $3x\geq 3$

$x\leq -\frac{1}{3}$ $\quad$ $x\geq 1$

90. $|3(x-1)|\geq 5$

$3(x-1)\leq -5$ or $3(x-1)\geq 5$

$3x-3\leq -5$ $\quad$ $3x-3\geq 5$

$3x\leq -2$ $\quad$ $3x\geq 8$

$x\leq -\frac{2}{3}$ $\quad$ $x\geq \frac{8}{3}$

91. $(x+4)(x-2)(3-x)>0$

$(x+4)(x-2)(3-x)=0$

$x=-4,\ x=2,\ x=3$

Critical points: $-4,\ 2,\ 3$

Test: $(x+4)(x-2)(3-x)$

(continued)

91. (continued)

Region	Test	Result
I	–5	$56>0$
II	0	$-24<0$
III	2.5	$\frac{13}{8}>0$
IV	4	$-16<0$

$x<-4$ or $2<x<3$

92. $(x+1)(x+4)(2-x)<0$

$(x+1)(x+4)(2-x)=0$

$x=-1,\ x=-4,\ x=2$

Critical points: $-1,\ -4,\ 2$

Test: $(x+1)(x+4)(2-x)$

Region	Test	Result
I	–5	$28>0$
II	–2	$-8<0$
III	0	$8>0$
IV	3	$-28<0$

$-4<x<-1$ or $x>2$

Chapter 9 Test

1. $8x^2+9x=0$

$x(8x+9)=0$

$x=0,\ 8x+9=0$

$x=0,\ x=-\frac{9}{8}$

2. $8x^2+10x=3$

$8x^2+10x-3=0$

$(4x-1)(2x+3)=0$

$4x-1=0, 2x+3=0$

$x=\frac{1}{4}, x=-\frac{3}{2}$

3. $$\frac{3x}{2}-\frac{8}{3}=\frac{2}{3x}$$
$$9x^2-16x-4=0$$
$$(9x+2)(x-2)=0$$
$$9x+2=0,\ x-2=0$$
$$x=-\frac{2}{9},\ x=2$$

4. $x(x-3)-30=5(x-2)$
$$x^2-3x-30=5x-10$$
$$x^2-8x-20=0$$
$$(x-10)(x+2)=0$$
$$x-10=0,\ x+2=0$$
$$x=10,\ x=-2$$

5. $7x^2-4=52$
$$7x^2=56$$
$$x^2=8$$
$$x=\pm\sqrt{8}=\pm\sqrt{4\cdot 2}$$
$$x=\pm 2\sqrt{2}$$

6. $$\frac{2x}{2x+1}-\frac{6}{4x^2-1}=\frac{x+1}{2x-1}$$
$$\frac{2x}{2x+1}-\frac{6}{(2x+1)(2x-1)}=\frac{x+1}{2x-1}$$
$$2x(2x-1)-6=(x+1)(2x+1)$$
$$4x^2-2x-6=2x^2+3x+1$$
$$2x^2-5x-7=0$$
$$(2x-7)(x+1)=0$$
$$2x-7=0,\ x+1=0$$
$$x=\frac{7}{2},\ x=-1$$

7. $2x^2-6x+5=0$
$$a=2,\ b=-6,\ c=5$$
$$x=\frac{-(-6)\pm\sqrt{(-6)^2-4(2)(5)}}{2(2)}$$
$$x=\frac{6\pm\sqrt{-4}}{4}$$
$$x=\frac{6\pm\sqrt{4}\sqrt{-1}}{4}$$
$$x=\frac{6\pm 2i}{4}$$
$$x=\frac{3\pm i}{2}$$

8. $2x(x-3)=-3$
$$2x^2-6x+3=0$$
$$a=2,\ b=-6,\ c=3$$
$$x=\frac{-(-6)\pm\sqrt{(-6)^2-4(2)(3)}}{2(2)}$$
$$x=\frac{6\pm\sqrt{12}}{4}=\frac{6\pm\sqrt{4\cdot 3}}{4}$$
$$x=\frac{3\pm\sqrt{3}}{2}$$

9. $x^4-9x^2+14=0$
$$y=x^2$$
$$y^2-9y+14=0$$
$$(y-7)(y-2)=0$$
$$y-7=0,\ y-2=0$$
$$y=7,\ y=2$$
$$x^2=7,\ x^2=2$$
$$x=\pm\sqrt{4},\ x=\pm\sqrt{2}$$

10. $3x^{-2}-11x^{-1}-20=0$

$3-11x-20x^2=0$

$20x^2+11x-3=0$

$(5x-1)(4x+3)=0$

$5x-1=0,\ 4x+3=0$

$x=\frac{1}{5},\ x=-\frac{3}{4}$

11. $x^{\frac{2}{3}}-2x^{\frac{1}{3}}-12=0$

$y=x^{\frac{1}{3}}$

$y^2-2y-12=0$

$y=\frac{2\pm\sqrt{(-2)^2-4(1)(-12)}}{2}=\frac{2\pm\sqrt{52}}{2}$

$y=\frac{2\pm\sqrt{4\cdot13}}{2}=\frac{2\pm2\sqrt{13}}{2}=1\pm\sqrt{13}$

$x^{\frac{1}{3}}=1\pm\sqrt{13}$

$x=(1\pm\sqrt{13})^3$

12. $B=\frac{xyw}{z^2}$

$z^2=\frac{xyw}{B}$

$z=\pm\sqrt{\frac{xyw}{B}}$

13. $5y^2+2by+6w=0$

$y=\frac{-2b\pm\sqrt{(2b)^2-4(5)(6w)}}{2(5)}$

$y=\frac{-2b\pm\sqrt{4(b^2-30w)}}{10}$

$y=\frac{-b\pm\sqrt{b^2-30w}}{5}$

14. $A=LW=(3W+1)W=80$

$3W^2+W-80=0$

$(W-5)(3W+16)=0$

$W=5,\ W=-\frac{16}{3}$ reject. $W>0$

$L=3W+1=16$

The length is 16 miles and the width is 5 miles.

15. $c^2=a^2+b^2$

$c^2=6^2+(2\sqrt{3})^2$

$c^2=36+12$

$c^2=48$

$c=\sqrt{48}$

$c=\sqrt{16\cdot3}$

$c=4\sqrt{3}$

16. $6=vt_1\Rightarrow t_1=\frac{6}{v}$

$3=(v+1)t_2\Rightarrow t_2=\frac{3}{v+1}$

$t_1+t_2=4$

$\frac{6}{v}+\frac{3}{v+1}=4$

$6(v+1)+3v=4v(v+1)$

$6v+6+3v=4v^2+4v$

$4v^2-5v-6=0$

$(v-2)(4v+3)=0$

$v-2=0,\ 4v+3=0$

$v=2,\ v=-\frac{3}{4}$ reject, $v>0$

$v+1=3$

They paddled 2 mph for the first part and 3 mph after lunch.

17. $f(x)=-x^2-6x-5$

$\frac{-b}{2a}=\frac{-(-6)}{2(-1)}=-3$

$f(-3)=-(-3)^2-6(-3)-5$

$f(-3=4$

$V(-3,4)$

$f(0)=-5$

y-intercept: $(0,-5)$

$x^2+6x+5=0$

$(x+1)(x+5)=0$

$x+1=0,\ x+5=0$

$x=-1,\ x=-5$

x-intercepts: $(-1,0),\ (-5,0)$

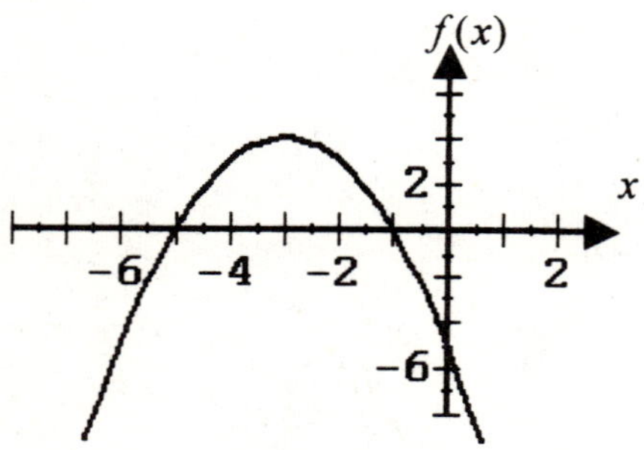

18. $2x^2+3x\geq 27$

$2x^2+3x-27\geq 0$

$2x^2+3x-27=0$

$(2x+9)(x-3)=0$

$x=-\frac{9}{2},\ x=3$

Critical Points: $-\frac{9}{2},\ 3$

Test: $2x^2+3x-27$

(continued)

18. (continued)

Region	Test	Result
I	−5	$8>0$
II	0	$-27<0$
III	4	$17>0$

$x\leq -\frac{9}{2}$ or $x\geq 3$

19. $-3x^2+10x+8\geq 0$

$-3x^2+10x+8\geq 0$

$3x^2-10x-8\leq 0$

$3x^2-10x-8=0$

$(3x+2)(x-4)=0$

$3x+2=0,\ x-4=0$

$x=-\frac{2}{3},\ x=4$

Critical points: $-\frac{2}{3},\ 4$

Test: $3x^2-10x-8$

Region	Test	Result
I	−1	$5>0$
II	0	$-8<0$
III	5	$17>0$

$-\frac{2}{3}\leq x\leq 4$

20. $x^2+3x-7>0$

$x=\frac{-3\pm\sqrt{3^2-4(1)(-7)}}{2}$

$x=-4.5,\ x=1.5$

Critical points: $-4.5,\ 1.5$

Test: x^2+3x-7

(continued)

20. (continued)

Region	Test	Result
I	−5	$3>0$
II	0	$-7<0$
III	2	$3>0$

$x<-4.5$ or $x>1.5$

21. $|5x-2|=37$

$5x-2=37$ or $5x-2=-37$

$5x=39 \qquad 5x=-35$

$x=\frac{39}{5} \qquad x=-7$

22. $\left|\frac{1}{2}x+3\right|-2=4$

$\left|\frac{1}{2}x+3\right|=6$

$\frac{1}{2}x+3=6$ or $\frac{1}{2}x+3=-6$

$x+6=12 \qquad x+6=-12$

$x=6 \qquad x=-18$

23. $|7x-3|\le 18$

$-18\le 7x-3\le 18$

$-15\le 7x\le 21$

$-\frac{15}{7}\le x\le 3$

24. $|3x+1|>7$

$3x+1<-7$ or $3x+1>7$

$3x<-8 \qquad 3x>6$

$x<-\frac{8}{3} \qquad x>2$

Cumulative Test for Chapters 1-9

1. $(-3x^{-2}y^3)^4=(-3)^4x^{-2\cdot 4}y^{3\cdot 4}$

$=\frac{81y^{12}}{x^8}$

2. $\frac{1}{2}a^3-2a^2+3a-\frac{1}{4}a^3-6a+a^2$

$=\frac{1}{4}a^3-a^2-3a$

3. $a(2y+b)=3ay-4$

$2ay+ab=3ay-4$

$ay=ab+4$

$y=\frac{ab+4}{a}$

4. $6x-3y=-12$

x	y
0	4
−2	0

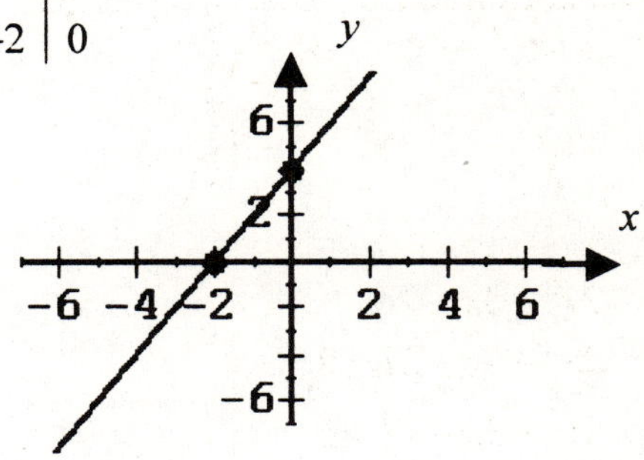

5. $2y+x=8$

$y=-\frac{1}{2}x+4 \Rightarrow m=-\frac{1}{2} \Rightarrow m_{\parallel}=-\frac{1}{2}$

$y-y_1=m_{\parallel}(x-x_1)$

$y-(-1)=-\frac{1}{2}(x-6)$

$2y+2=-x+6$

$x+2y=4$

6. $V = \frac{4}{3}\pi r^3$

$V = \frac{4}{3}\pi(2)^3$

$V = \frac{32\pi}{3} \text{ in}^3$

7. $125x^3 - 27y^2$

$= (5x-3y)(25x^2+15xy+9y^2)$

8. $\sqrt{72x^3y^6} = \sqrt{36x^2y^6 \cdot 2x}$

$= 6xy^3\sqrt{2x}$

9. $(3+\sqrt{2})(\sqrt{6}+\sqrt{3})$

$= 3\sqrt{6}+3\sqrt{3}+\sqrt{12}+\sqrt{6}$

$= 4\sqrt{6}+3\sqrt{3}+\sqrt{4\cdot 3}$

$= 4\sqrt{6}+3\sqrt{3}+2\sqrt{3}$

$= 4\sqrt{6}+5\sqrt{3}$

10. $\frac{3}{\sqrt{11}} \cdot \frac{\sqrt{11}}{\sqrt{11}} = \frac{3\sqrt{11}}{11}$

11. $3x^2+12x = 26x$

$3x^2 - 14x = 0$

$x(3x-14) = 0$

$x = 0, \ x = \frac{14}{3}$

12. $12x^2 = 11x - 2$

$12x^2 - 11x + 2 = 0$

$(4x-1)(3x-2) = 0$

$4x-1 = 0, \ 3x-2 = 0$

$x = \frac{1}{4}, \ x = \frac{2}{3}$

13. $44 = 3(2x-3)^2 + 8$

$3(2x-3)^2 = 36$

$(2x-3)^2 = 12$

$2x-3 = \pm\sqrt{12}$

$2x-3 = \pm\sqrt{4\cdot 3}$

$2x-3 = \pm 2\sqrt{3}$

$2x = 3 \pm 2\sqrt{3}$

$x = \frac{3\pm 2\sqrt{3}}{2}$

14. $3 - \frac{4}{x} + \frac{5}{x^2} = 0$

$3x^2 - 4x + 5 = 0$

$x = \frac{-(-4)\pm\sqrt{(-4)^2-4(3)(5)}}{2(3)}$

$x = \frac{4\pm\sqrt{-44}}{6}$

$x = \frac{4\pm\sqrt{4\cdot 11}\sqrt{-1}}{6}$

$x = \frac{4\pm 2i\sqrt{11}}{6}$

$x = \frac{2\pm i\sqrt{11}}{3}$

15. $\sqrt{x-12} = \sqrt{x} - 2$

$x - 12 = x - 4\sqrt{x} + 4$

$4\sqrt{x} = 16$

$\sqrt{x} = 4$

$x = 16$

check:

$\sqrt{16-12} \stackrel{?}{=} \sqrt{16} - 2$

$2 = 2$

$x = 16$

16. $x^{\frac{2}{3}} + 9x^{\frac{1}{3}} + 18 = 0$

$y = x^{\frac{1}{3}}$

$y^2 + 9y + 18 = 0$

$(y+6)(y+3) = 0$

$y = -6,\ y = -3$

$x^{\frac{1}{3}} = -6,\ x^{\frac{1}{3}} = -3$

$x = -216,\ x = -27$

check:

$$(-216)^{\frac{2}{3}} + 9(-216)^{\frac{1}{3}} + 18 \stackrel{?}{=} 0$$
$$0 = 0$$

$$(-27)^{\frac{2}{3}} + 9(-27)^{\frac{1}{3}} + 18 \stackrel{?}{=} 0$$
$$0 = 0$$

$x = -216,\ x = -27$

17. $2y^2 + 5wy - 7z = 0$

$$y = \frac{-5w \pm \sqrt{(5w)^2 - 4(2)(-7z)}}{2(2)}$$

$$y = \frac{-5w \pm \sqrt{25w^2 + 56z}}{4}$$

18. $3y^2 + 16z^2 = 5w$

$$3y^2 = 5w - 16z^2$$

$$y^2 = \frac{5w - 16z^2}{3}$$

$$y = \pm\sqrt{\frac{5w - 16z^2}{3} \cdot \frac{3}{3}}$$

$$y = \pm\sqrt{\frac{15w - 48z^2}{9}}$$

$$y = \pm\frac{\sqrt{15w - 48z^2}}{3}$$

19. $c^2 = a^2 + b^2$

$\left(\sqrt{31}\right)^2 = 4^2 + b^2$

$31 = 16 + b^2,\ b^2 = 15$

$b = \sqrt{15}$

20. $A = \frac{1}{2}bh$

$A = \frac{1}{2}b(3b+3)$

$45 = \frac{1}{2}b(3b+3)$

$90 = 3b^2 + 3b$

$b^2 + b - 30 = 0$

$(b-5)(b+6) = 0$

$b - 5 = 0,\ b + 6 = 0$

$b = 5,\ b = -6$ reject, $b > 0$

$h = 3b + 3$

$h = 18$

The base of the triangle is 5 meters and the altitude of the triangle is 18 meters.

21. $f(x) = -x^2 + 8x - 12$

$a = -1,\ b = 8,\ c = -12$

$x_{\text{vertex}} = \frac{-b}{2a} = \frac{-8}{2(-1)} = 4$

$y_{\text{vertex}} = f(x_{\text{vertex}})$

$y_{\text{vertex}} = f(4) = -4^2 + 8(4) - 12 = 4$

$V(4,4)$

$f(0) = -12$, y-intercept: $(0,-12)$

$-x^2 + 8x - 12 = 0,\ x^2 - 8x + 12 = 0$

$(x-6)(x-2) = 0$

$x - 6 = 0,\ x - 2 = 0$

$x = 6,\ x = 2$

x-intercepts: $(6,0)$, $(2,)$

22.

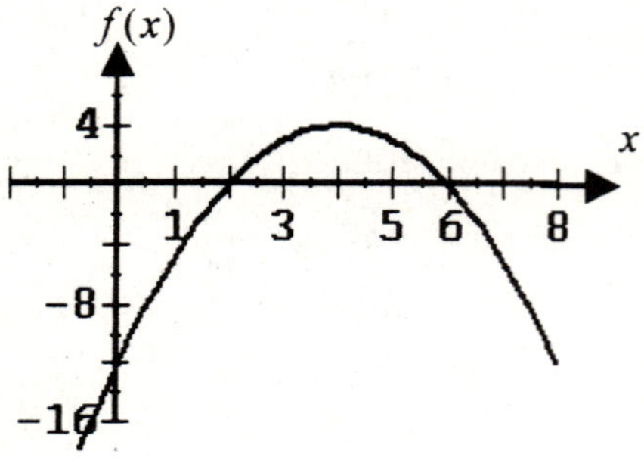

23. $6x^2 - x \le 2$

$6x^2 - x - 2 \le 0$

$6x^2 - x - 2 = 0$

$(3x-2)(2x+1) = 0$

$3x - 2 = 0,\ 2x + 1 = 0$

$x = \frac{2}{3},\ x = -\frac{1}{2}$

Critical points: $\frac{2}{3},\ -\frac{1}{2}$

Test: $6x^2 - x - 2$

Region	Test	Results
I	−1	$5 > 0$
II	0	$-2 < 0$
III	1	$3 > 0$

$-\frac{1}{2} \le x \le \frac{2}{3}$

24. $x^2 > -2x + 15$

$x^2 + 2x - 15 > 0$

$x^2 + 2x - 15 = 0$

$(x+5)(x-3) = 0$

$x + 5 = 0,\ x - 3 = 0$

$x = -5,\ x = 3$

(continued)

24. (continued)

Critical points: $-5,\ 3$

Test: $x^2 + 2x - 15$

Region	Test	Results
I	−6	$9 > 0$
II	0	$-15 < 0$
III	5	$20 > 0$

$x < -5$ or $x > 3$

25. $|3x+1| = 16$

$3x + 1 = 16$ or $3x + 1 = -16$

$3x = 15 \qquad 3x = -17$

$x = 5 \qquad x = -\frac{17}{3}$

26. $\left|\frac{1}{2}x + 2\right| \le 8$

$-8 \le \frac{1}{2}x + 2 \le 8$

$-16 \le x + 4 \le 16$

$-20 \le x \le 12$

27. $|3x-4| > 11$

$3x - 4 < -11$ or $3x - 4 > 11$

$3x < -7 \qquad 3x > 15$

$x < -\frac{7}{3} \qquad x > 5$

Chapter 10

Pretest Chapter 10

1. $(x-h)^2+(y-k)^2=r^2$

$(x-8)^2+(y-(-2))^2=\sqrt{7}^2$

$(x-8)^2+(y+2)^2=7$

2. $(x_1,y_1)=(-6,-2),\ (x_2,y_2)=(-3,4)$

$d=\sqrt{(x_2-x_1)^2+(y_2-y_1)^2}$

$d=\sqrt{(-3-(-6))^2+(4-(-2))^2}$

$d=\sqrt{45}=\sqrt{9(5)}$

$d=3\sqrt{5}$

3. $x^2+y^2-2x-4y+1=0$

$x^2-2x+1+y^2-4y+4=-1+1+4$

$(x-1)^2+(y-2)^2=4=2^2$

Center $=(1,2)$

radius $=2$

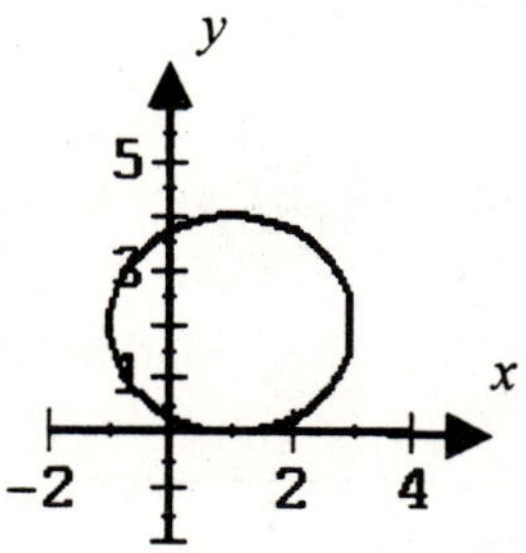

4. $x=(y+1)^2+2$

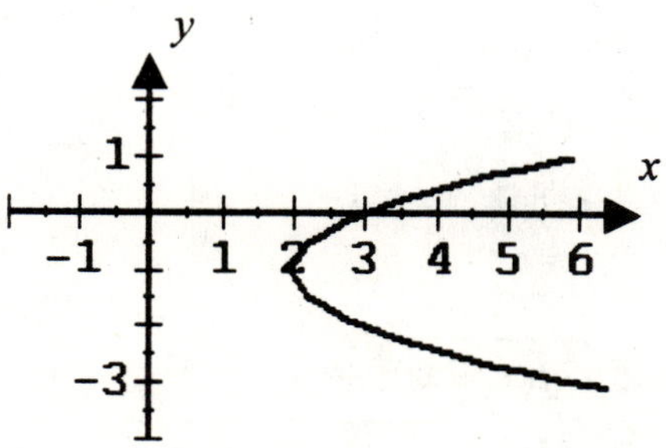

5. $x^2=y-4x-1$

$y=x^2+4x+1$

$y=x^2+4x+4-3$

$y=(x+2)^2-2$

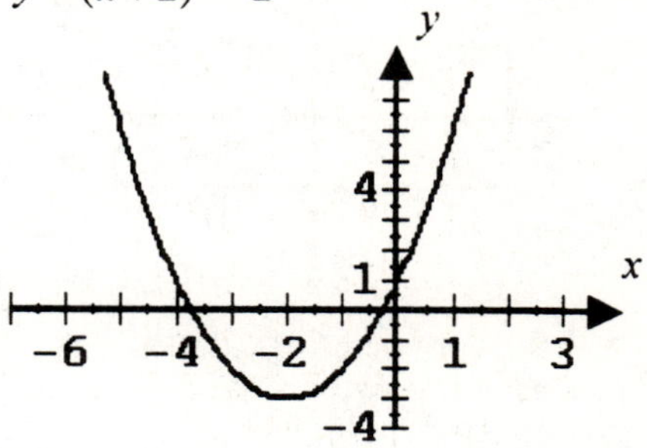

6. $4x^2+y^2-36=0$

$4x^2+y^2=36$

$\frac{x^2}{9}+\frac{y^2}{36}=1$

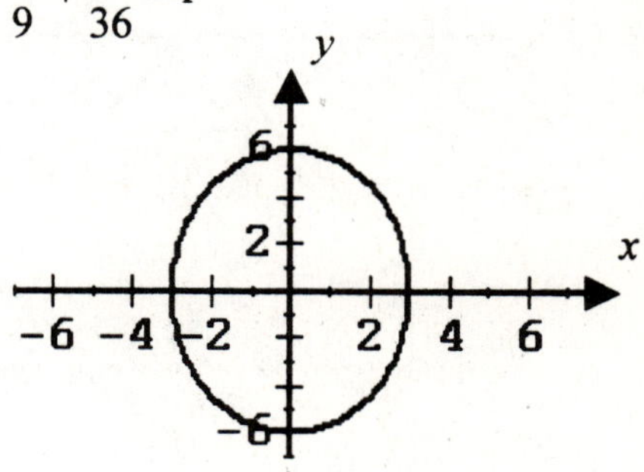

7. $\frac{(x+3)^2}{25}+\frac{(y-1)^2}{36}=1$

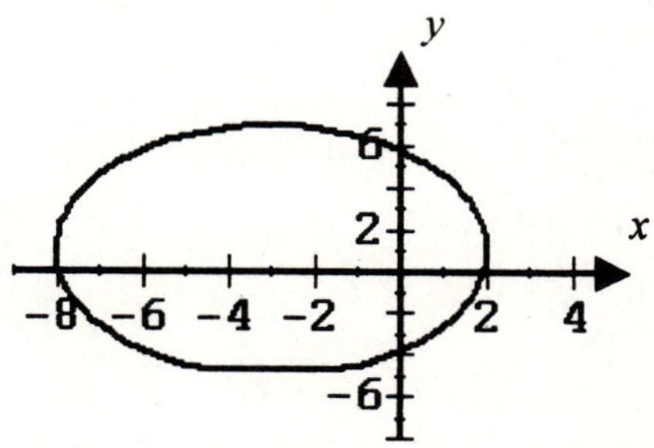

8. $25y^2 - 9x^2 = 225$

$$\frac{y^2}{9} - \frac{x^2}{25} = 1$$

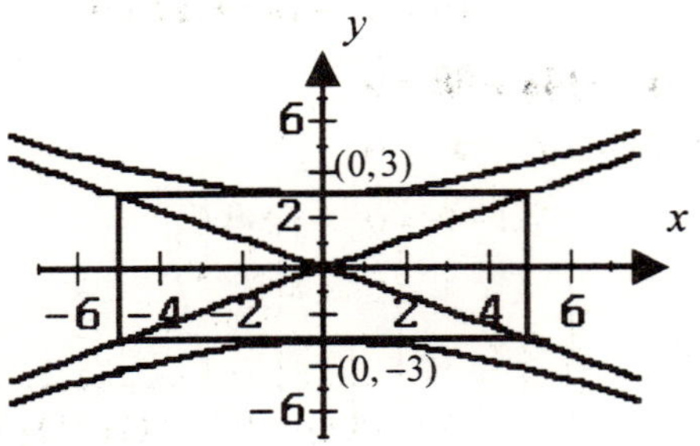

9. $\dfrac{(x-2)^2}{4} - \dfrac{(y+1)^2}{9} = 1$

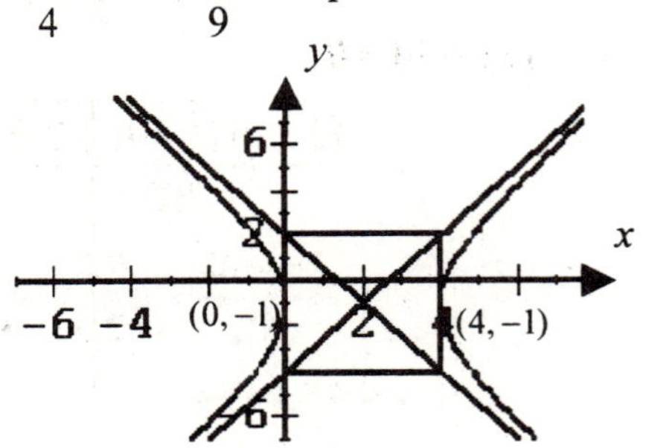

10. $x^2 + y^2 - 25$

$3x + 4y = 0 \Rightarrow y = \dfrac{-3x}{4}$, substitute into first equation.

$$x^2 + \left(\frac{-3x}{4}\right)^2 = 25$$

$$x^2 + \frac{9x^2}{16} = 25$$

$$16x^2 + 9x^2 = 400$$

$$x^2 = 16$$

$$x = \pm 4$$

$$y = \frac{-3(\pm 4)}{4} = \mp 3$$

$(4, -3)$, $(-4, 3)$ is the solution.

11. $y = x^2 + 1 \Rightarrow x^2 y - 1$, substitute into second equation.

$$4y^2 = 4 - x^2$$

$$4y^2 = 4 - (y - 1)$$

$$4y^2 = 4 - y + 1$$

$$4y^2 + y - 5 = 0$$

$$(y-1)(4y+5) = 0$$

$$y - 1 = 0,\ 4y + 5 = 0$$

$$y = 1,\ y = -\frac{5}{4}$$

$$x^2 = y - 1 = \begin{cases} 1 - 1 = 0 \\ -\dfrac{5}{4} - 1 = -\dfrac{9}{4} \end{cases}$$

$$x^2 = 0$$

$$x = 0$$

$(0, 1)$ is the solution.

10.1 Exercises

1. Subtract the value of the point and use the absolute value: $|-2-4| = |-6| = 6$.

3. The center is $(1, -2)$ and the radius is 3.

5. $d = \sqrt{(x_2 - x_1)^2 + (y_2 - y_1)^2}$

$$d = \sqrt{(1-2)^2 + (6-4)^2} = \sqrt{5}$$

7. $d = \sqrt{(x_2 - x_1)^2 + (y_2 - y_1)^2}$

$$d = \sqrt{\left(\frac{1}{2} - \frac{3}{4}\right)^2 + \left(\frac{5}{2} - \frac{3}{2}\right)^2} = \frac{\sqrt{17}}{4}$$

9. $d = \sqrt{(x_2 - x_1)^2 + (y_2 - y_1)^2}$

$$d = \sqrt{(3-(-2))^2 + (9-(-3))^2} = 13$$

11. $d=\sqrt{(x_2-x_1)^2+(y_2-y_1)^2}$
$d=\sqrt{(4-0)^2+(1-(-3))^2}=\sqrt{32}$
$d=4\sqrt{2}$

13. $d=\sqrt{(x_2-x_1)^2+(y_2-y_1)^2}$
$d=\sqrt{\left(\frac{7}{3}-\frac{1}{3}\right)^2+\left(\frac{1}{5}-\frac{3}{5}\right)^2}$
$d=\sqrt{\frac{104}{25}}$
$d=\frac{2\sqrt{26}}{5}$

15. $d=\sqrt{(x_2-x_1)^2+(y_2-y_1)^2}$
$d=\sqrt{(1.3-(-5.7))^2+(2.6-1.6)^2}$
$d=\sqrt{50}$
$d=5\sqrt{2}$

17. $d=\sqrt{(x_2-x_1)^2+(y_2-y_1)^2}$
$10=\sqrt{(y-6)^2+(1-7)^2}$
$100=y^2-4y+40$
$y^2-4y-60=0$
$(y-10)(y+6)=0$
$y-10=0,\ y+6=0$
$y=10,\ y=-6$

19. $d=\sqrt{(x_2-x_1)^2+(y_2-y_1)^2}$
$2.5=\sqrt{(0-1.5)^2+(y-2)^2}$
$6.25=2.25+y^2-4y+4$
$y^2-4y=0$
$y(y-4)=0$
$y=0,\ y-4=0$
$y=0,\ y=4$

21. $d=\sqrt{(x_2-x_1)^2+(y_2-y_1)^2}$
$\sqrt{10}=\sqrt{(x-7)^2+(6-3)^2}$
$10=x^2-14x+49+9$
$x^2-14x+48=0$
$(x-6)(x-8)=0$
$x-6=0,\ x-8=0$
$x=6,\ x=8$

23. $d=\sqrt{(x_2-x_1)^2+(y_2-y_1)^2}$
$4=\sqrt{(4-6)^2+(y-6)^2}$
$16=4+y^2-12y+36$
$y^2-12y+24=0$
$y=\frac{12\pm\sqrt{12^2-4(1)(24)}}{2(1)}$
$y=\frac{12\pm\sqrt{48}}{2}\approx\begin{cases}9.5\\2.5\end{cases}$
$y\approx 9.5$ miles

25. $(x-h)^2+(y-k)^2=r^2$
$(x-(-1))^2+(y-(-7))^2=\sqrt{5}^2$
$(x+1)^2+(y+7)^2=5$

27. $(x-h)^2+(y-k)^2=r^2$
$(x-(-3.5))^2+(y-0)^2=6^2$
$(x+3.5)^2+y^2=36$

29. $(x-h)^2+(y-k)^2=r^2$
$\left(x-\frac{7}{4}\right)^2+(y-0)^2=\left(\frac{1}{3}\right)^2$
$\left(x-\frac{7}{4}\right)^2+y^2=\frac{1}{9}$

31. $x^2 + y^2 = 25$
$C(0,0),\ r = 5$

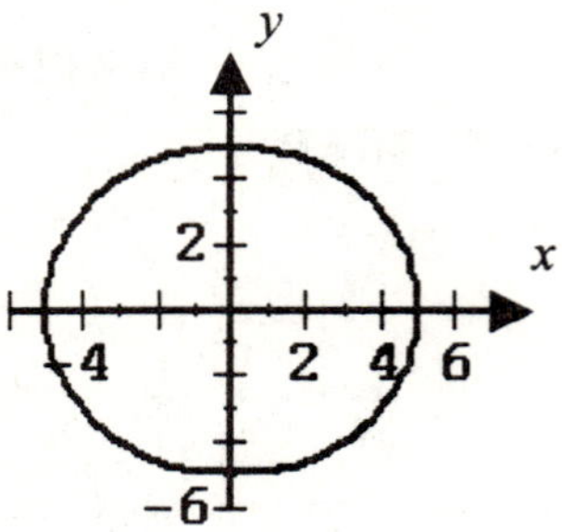

33. $(x-3)^2 + (y-2)^2 = 4 = 2^2$
$C(3,2),\ r = 2$

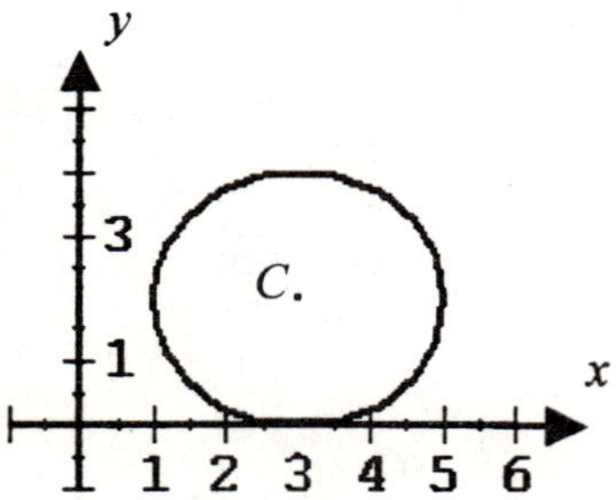

35. $(x+2)^2 + (y-3)^2 = 25 = 5^2$
$C(-2,3),\ r = 5$

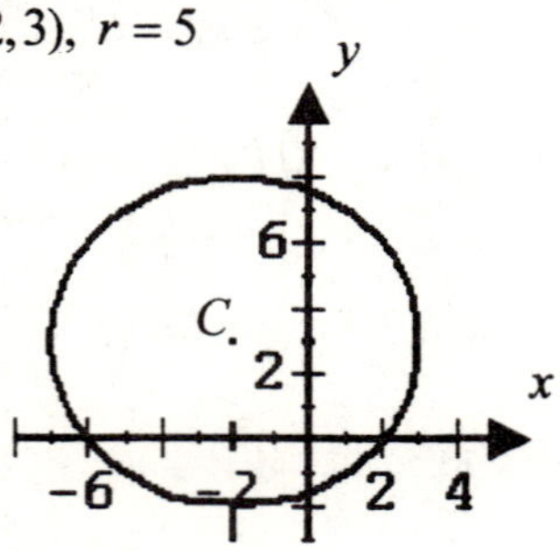

37. $$x^2 + y^2 + 6x - 4y - 3 = 0$$
$$x^2 + 6x + 9 + y^2 - 4y + 4 = 3 + 9 + 4$$
$$(x+3)^2 + (y-2)^2 = 16 = 4^2$$
$$C(-3,2),\ r = 4$$

39. $$x^2 + y^2 - 12x + 2y - 12 = 0$$
$$x^2 - 12x + 36 + y^2 + 2y + 1 = 12 + 36 + 1$$
$$(x-6)^2 + (y+1)^2 = 49 = 7^2$$
$$C(6,-1),\ r = 7$$

41. $$x^2 + y^2 + 3x - 2 = 0$$
$$x^2 + 3x + y^2 = 2$$
$$x^2 + 3x + \frac{9}{4} + y^2 = 2 + \frac{9}{4} = \frac{17}{4}$$
$$\left(x + \frac{3}{2}\right)^2 + y^2 = \left(\frac{\sqrt{17}}{2}\right)^2$$
$$C\left(-\frac{3}{2}, 0\right),\ r = \frac{\sqrt{17}}{2}$$

43. $(x-42.7)^2 + (y-29.7)^2 = 25.1^2 = 630.01$

45. $(x-5.32)^2 + (y+6.54)^2 = 47.28$

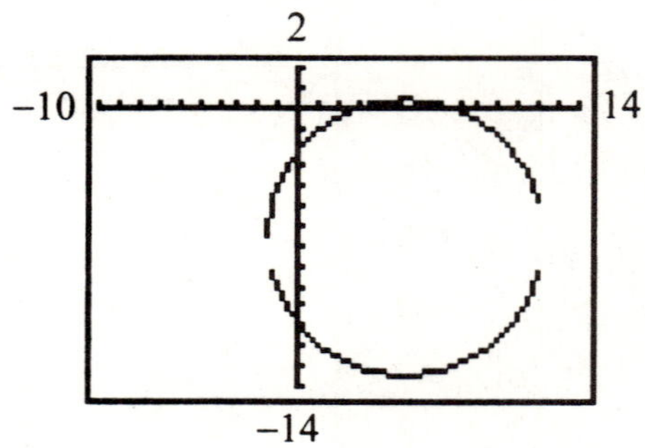

Cumulative Review Problems

47. $$9 + \frac{3}{x} = \frac{2}{x^2}$$
$$9x^2 + 3x - 2 = 0$$
$$(3x+2)(3x-1) = 0$$
$$3x + 2 = 0,\ 3x - 1 = 0$$
$$x = -\frac{2}{3},\ x = \frac{1}{3}$$

49. $5x^2 - 6x - 7 = 0$

$$x = \frac{-(-6) \pm \sqrt{(-6)^2 - 4(5)(-7)}}{2(5)}$$

$$x = \frac{6 \pm \sqrt{176}}{10} = \frac{6 \pm \sqrt{16(11)}}{10} = \frac{6 \pm 4\sqrt{11}}{10}$$

$$x = \frac{3 \pm 2\sqrt{11}}{5}$$

51. $V = Ah$

$$V = 20\text{ mi}^2(150\text{ ft})\left(\frac{5280^2\text{ ft}^2}{\text{mi}^2}\right)$$

$$V = 8.364 \times 10^{10}\text{ ft}^3$$

10.2 Exercises

1. The graph of $y = x^2$ is symmetric about the <u>y-axis</u>. The graph of $x = y^2$ is symmetric about the <u>x-axis</u>.

3. Since $y = 2(x-3)+4$ is in standard form, $y = a(x-h)^2 + k$, the vertex is $(h, k) = (3, 4)$.

5. $y = -4x^2$
$V(0,0)$
y-intercept = $(0,0)$

x	−2	−1	0	1	2
y	−16	−4	0	−4	−16

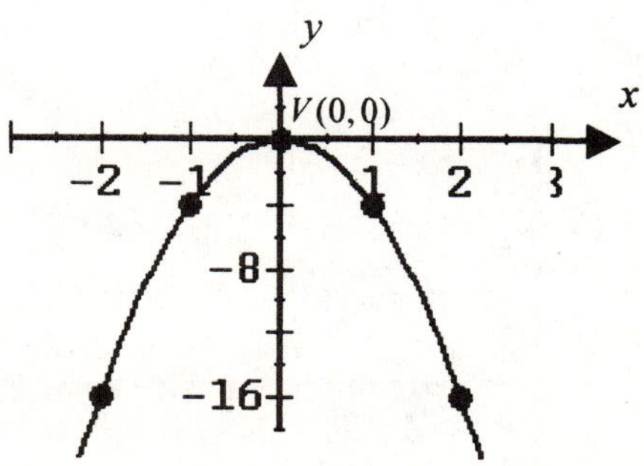

7. $y = x^2 - 6$
$V(0,-6)$
y-intercept: $(0,-6)$

x	−3	−2	0	2	3
y	3	−2	−6	−2	6

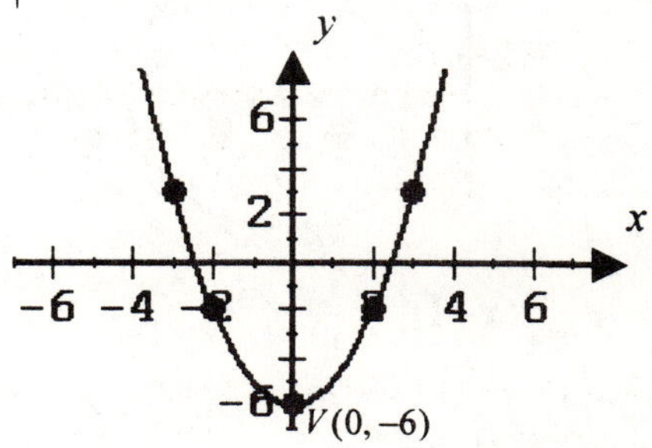

9. $y = -2x^2 + 4$
$V(0,4)$
y-intercept: $(0,4)$

x	−2	−1	0	1	2
y	−4	2	4	2	−4

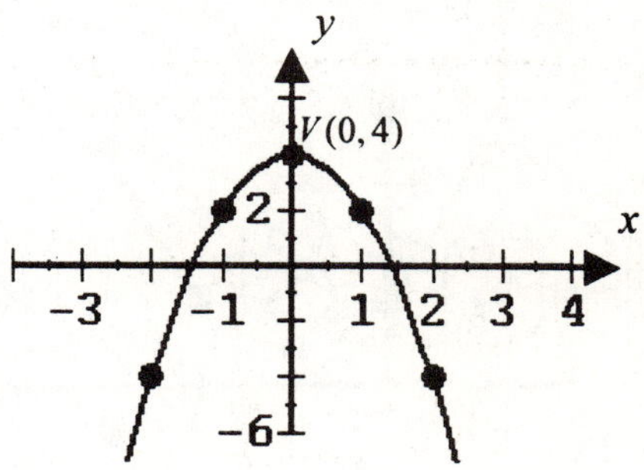

11. $y = (x-3)^2 - 2$
$V(3,-2)$
y-intercept: $(0,7)$

x	1	2	3	4	5
y	2	−1	−2	−1	2

(continued)

11. (continued)

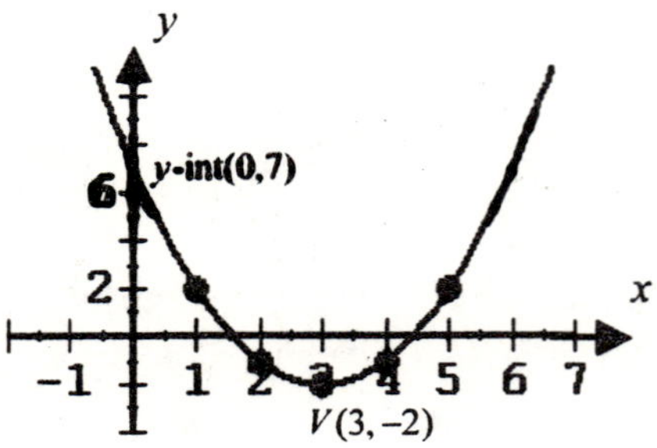

13. $y = 2(x-1)^2 + \frac{3}{2}$

$V\left(1, \frac{3}{2}\right)$

y-intercept: $\left(0, \frac{7}{2}\right)$

x	−1	0	1	2	3
y	$\frac{19}{2}$	$\frac{7}{2}$	$\frac{3}{2}$	$\frac{7}{2}$	$\frac{19}{2}$

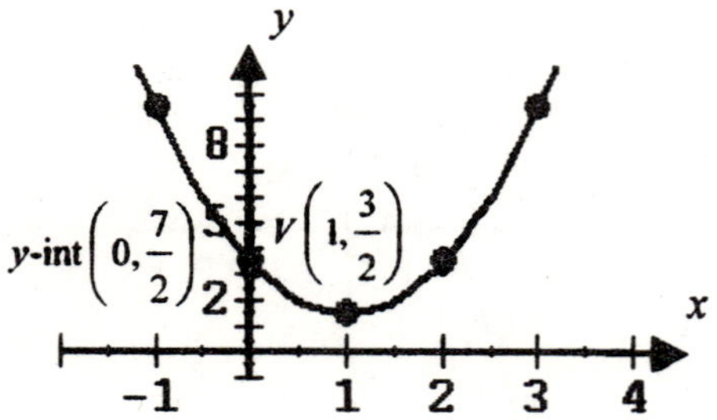

15. $y = -4\left(x + \frac{3}{2}\right)^2 + 5$

$V\left(-\frac{3}{2}, 5\right)$

y-intercept: $(0, -4)$

x	−3	−2	$-\frac{3}{2}$	−1	0
y	−4	4	5	4	−4

(continued)

15. (continued)

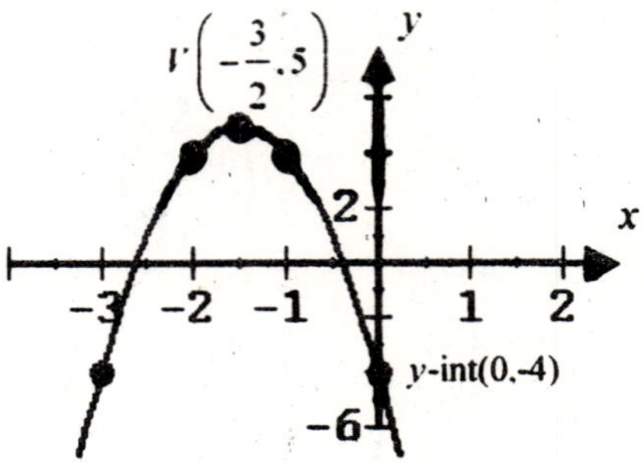

17. $x = \frac{1}{4}y^2 - 2$

$V(-2, 0)$

x-intercept: $(-2, 0)$

x	−2	0	0	2	2
y	0	−2	2	−4	4

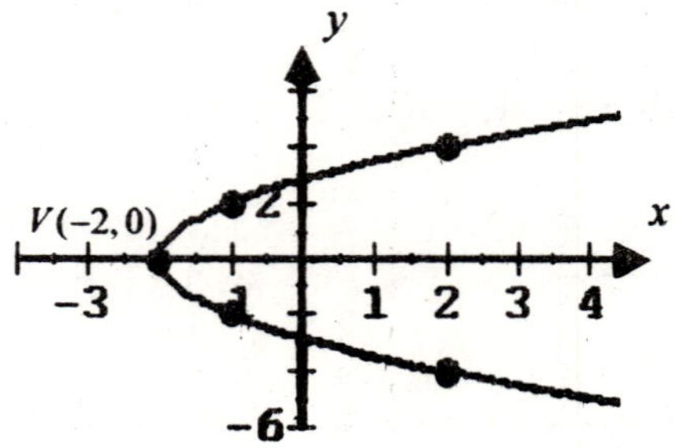

19. $x = (y-2)^2 + 3$

$V(3, 2)$

x-intercept: $(7, 0)$

x	3	4	4	7	7
y	2	1	3	4	0

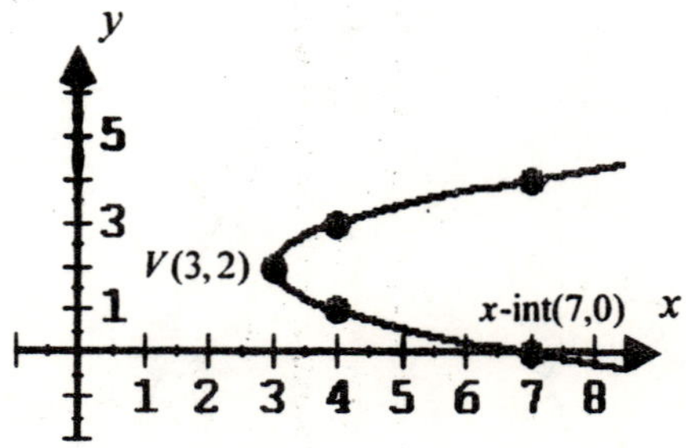

21. $x = -3(y+1) - 2$

$V(-2, -1)$

x-intercept: $(-5, 0)$

x	-2	-5	-5
y	-1	0	-2

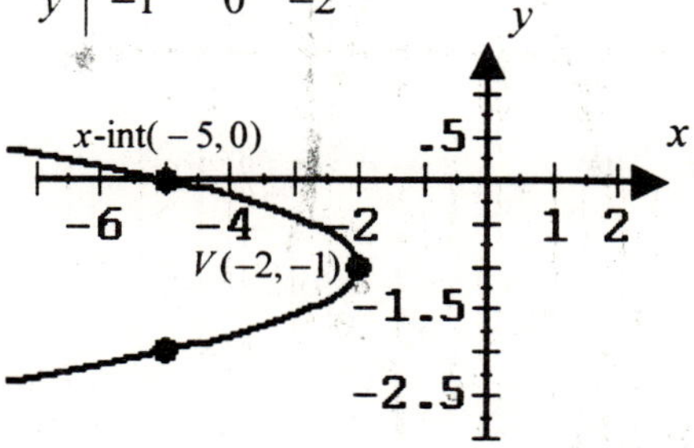

23. $x = \frac{1}{2}(y+1)^2 + 2$

$V(2, -1)$

x-intercept: $(2.5, 0)$

x	2	2.5	4	4
y	-1	0	1	-3

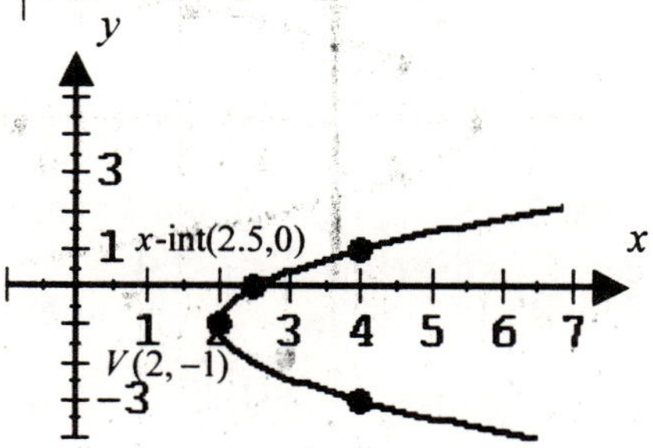

25. $x = -(y-2)^2 + \frac{1}{2}$

$V\left(\frac{1}{2}, 2\right)$

x-intercept: $\left(-\frac{7}{2}, 0\right)$

x	$-\frac{7}{2}$	$\frac{1}{2}$
y	0	2

(continued)

25. (continued)

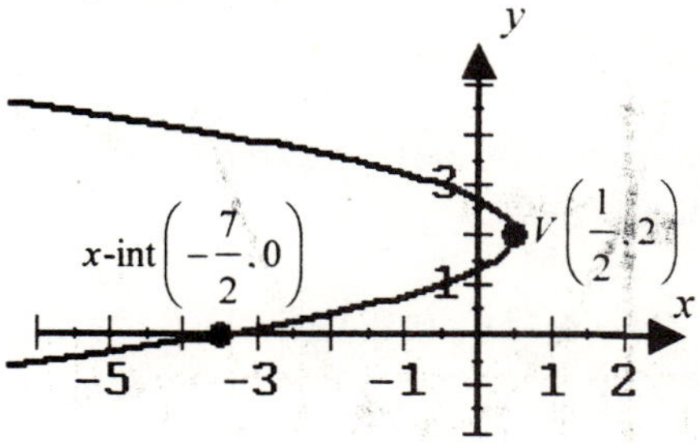

27. $y = x^2 + 12x + 25$

$y = x^2 + 12x + 36 - 11$

$y = (x+6)^2 - 11$

(a) Vertical

(b) Opens upward

(c) $V(-6, -11)$

29. $y = -2x^2 + 4x - 3$

$y = -2(x^2 - 2x + 1) - 3 + 2$

$y = -2(x-1)^2 - 1$

(a) Vertical

(b) Downward

(c) $V(1, -1)$

31. $x = y^2 + 8y + 9$

$x = y^2 + 8y + 16 - 7$

$x = (y+4)^2 - 7$

(a) Horizontal

(b) Opens right

(c) $V(-7, -4)$

33. $y = ax^2$, $8 = a(16)^2$, $a = \frac{1}{32}$

$y = \frac{1}{32}x^2$

35. $a=\frac{1}{4p}=\frac{1}{32}\Rightarrow p=8$

The distance from (0,0) to the focus point is 8 inches.

37. $y=2x^2+6.48x-0.1312$

$y=2(x^2+3.24x+2.6244)-0.1312-5.2488$

$y=2(x+1.62)^2-5.38$

$V(-1.62,-5.38)$

y-intercept: -0.1312

$$\frac{-6.48\pm\sqrt{6.48^2-4(2)(-0.1312)}}{2(2)}$$

$$=\begin{cases}0.020121947\\-3.260121947\end{cases}=x\text{-intercepts}$$

39. $P=-2x^2+200x+47,000$

$P=-2(x^2-100x+2500)+47,000+5000$

$P=-2(x-50)^2+52,000$

The maximum profit is \$52,000 and the number of items needed is 50.

41. $S=650d-2d^2$

$S=-2(d^2-325d+26,406.25)+52,812.5$

$S=-2(d-162.5)^2+52,812.5$

The maximum sensitivity is 52,812.5 and the dosage is 162.5 milligrams.

Cumulative Review Problems

43. $\sqrt{50x^3}=\sqrt{25x^2\cdot 2x}$

$=5x\sqrt{2x}$ for $x\geq 0$

45. $\sqrt{98x}+x\sqrt{8}-3\sqrt{50x}$

$=\sqrt{49\cdot 2x}+x\sqrt{4\cdot 2}-3\sqrt{25\cdot 2x}$

$=7\sqrt{2x}+2x\sqrt{2}-15\sqrt{2x}$

$=2x\sqrt{2}-8\sqrt{2x}$

47. $d=rt$

$d=40t=\frac{40}{60}(56-15)=27\frac{1}{3}$

Matthew lives $27\frac{1}{3}$ miles from work.

49. $8(1050)(0.88)=7392$

Sir George can expect 7392 blooms if there is heavy rainfall this year.

10.3 Exercises

1. $\frac{(x+2)^2}{4}+\frac{(y-3)}{9}=1$ is in the form $\frac{(x-h)^2}{a^2}+\frac{(y-k)^2}{b^2}=1$ where (h,k) is the center of the ellipse. Therefore, the center of the ellipse is $(-2,3)$.

3. $\frac{x^2}{36}+\frac{y^2}{4}=1\Rightarrow\frac{x^2}{6^2}+\frac{y^2}{2^2}=1$

$a=6,\ b=2$

Intercepts: $(\pm 6,0)$, $(0,\pm 2)$

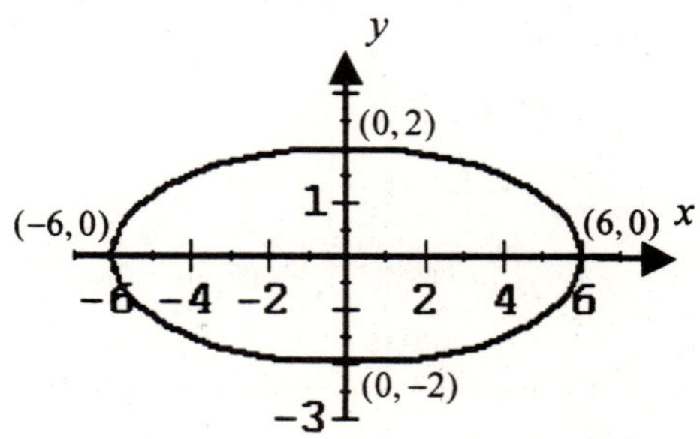

5. $\dfrac{x^2}{81}+\dfrac{y^2}{100}=1 \Rightarrow \dfrac{x^2}{9^2}+\dfrac{y^2}{10^2}=1$

$a=9,\ b=10$

Intercepts: $(\pm 9,0),\ (0,\pm 10)$

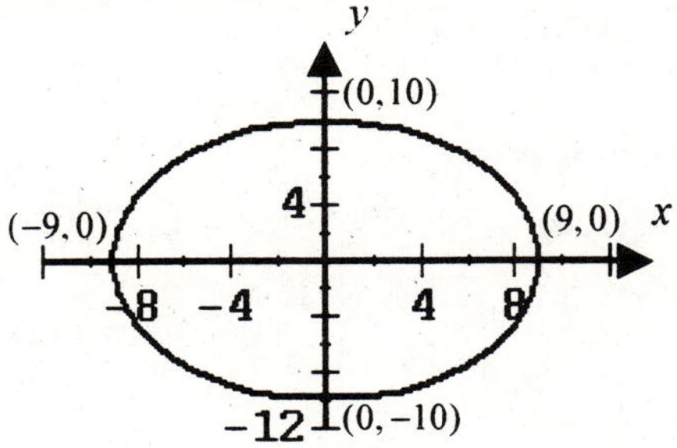

7. $4x^2+y^2-36=0$

$\dfrac{x^2}{9}+\dfrac{y^2}{36}=1 \Rightarrow \dfrac{x^2}{3^2}+\dfrac{y^2}{6^2}=1$

$a=3,\ b=6$

Intercepts: $(\pm 3,0),\ (0,\pm 6)$

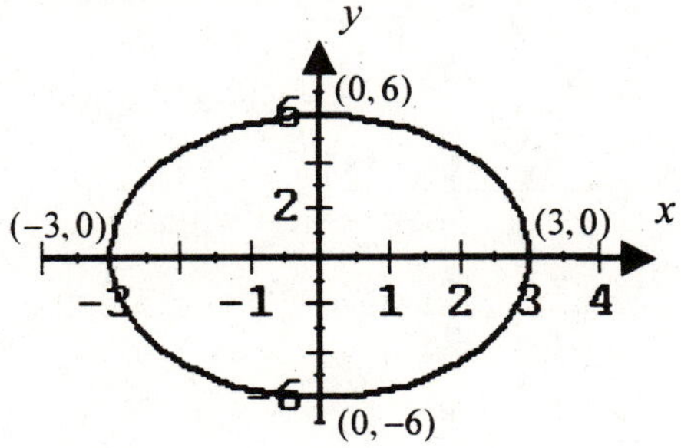

9. $x^2+9y^2=81 \Rightarrow \dfrac{x^2}{81}+\dfrac{y^2}{9}=1$

$\dfrac{x^2}{9^2}+\dfrac{y^2}{3^2}=1$

$a=9,\ b=3$

Intercepts: $(\pm 9,0),\ (0,\pm 3)$

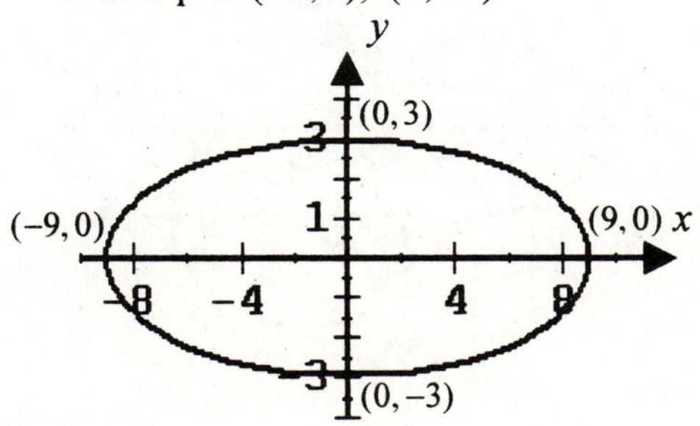

11. $x^2+12y^2=36$

$\dfrac{x^2}{36}+\dfrac{y^2}{3}=1 \Rightarrow \dfrac{x^2}{6^2}+\dfrac{y^2}{\sqrt{3}^2}=1$

$a=6,\ b=\sqrt{3}$

Intercepts: $(\pm 6,0),\ (0,\pm\sqrt{3})$

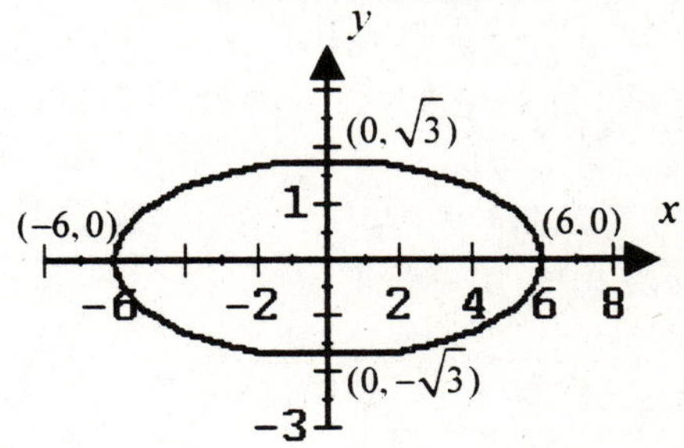

13. $\dfrac{x^2}{\frac{9}{4}}+\dfrac{y^2}{\frac{25}{4}}=1 \Rightarrow \dfrac{x^2}{\left(\frac{3}{2}\right)^2}+\dfrac{y^2}{\left(\frac{5}{2}\right)^2}=1$

$a=\dfrac{3}{2},\ b=\dfrac{5}{2}$

Intercepts: $\left(\pm\dfrac{3}{2},0\right),\ \left(0,\pm\dfrac{5}{2}\right)$

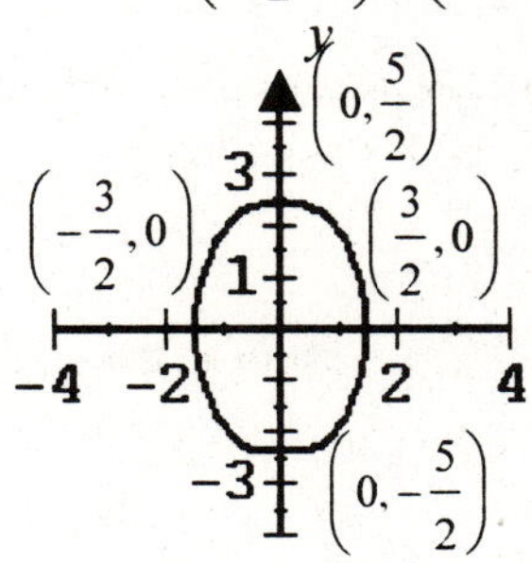

15. $121x^2+64y^2=7744$

$\dfrac{x^2}{64}+\dfrac{y^2}{121}=1$

$\dfrac{x^2}{8^2}+\dfrac{y^2}{11^2}=1$

$a=8,\ b=11,\ C(0,0)$

Intercepts: $(0,\pm 11),(\pm 8,0)$

(continued)

15. (continued)

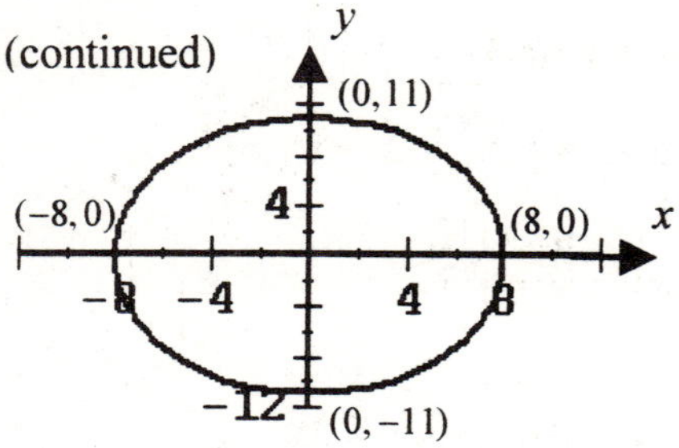

17. $\frac{(x-h)^2}{a^2}+\frac{(y-k)^2}{b^2}=1$

$C(0,0)\Rightarrow(h,k)=(0,0)$

$x\text{-int}(9,0)\Rightarrow a=9$

$y\text{-int}(0,-2)\Rightarrow b=2$

$\frac{x^2}{9^2}+\frac{y^2}{2^2}=1$

$\frac{x^2}{81}+\frac{y^2}{4}=1$

19. $\frac{(x-h)^2}{a^2}+\frac{(y-k)^2}{b^2}=1$

$C(0,0)\Rightarrow(h,k)=(0,0)$

$x\text{-int}(9,0)\Rightarrow a=9$

$y\text{-int}(0,3\sqrt{2})\Rightarrow b=3\sqrt{2}$

$\frac{x^2}{9^2}+\frac{y^2}{(3\sqrt{2})^2}=1$

$\frac{x^2}{81}+\frac{y^2}{18}=1$

21. $\frac{(x-h)^2}{a^2}+\frac{(y-k)^2}{b^2}=1$

$C(0,0)\Rightarrow(h,k)=(0,0)$

$x\text{-int}(30,0)\Rightarrow a=30$

$y\text{-int}(0,18)\Rightarrow b=18$

$\frac{x^2}{30^2}+\frac{y^2}{18^2}=1$

$\frac{x^2}{900}+\frac{y^2}{324}=1$

23. $\frac{(x-h)^2}{a^2}+\frac{(y-k)^2}{b^2}=1$

$\frac{(x-5)^2}{9}+\frac{(y-2)^2}{1}=1$

$\frac{(x-5)^2}{3^2}+\frac{(y-2)^2}{1^2}=1$

$a=3,\ b=1$

$C(5,2)$

Vertices: $(2,2),\ (5,1)$
$(8,2),\ (5,3)$

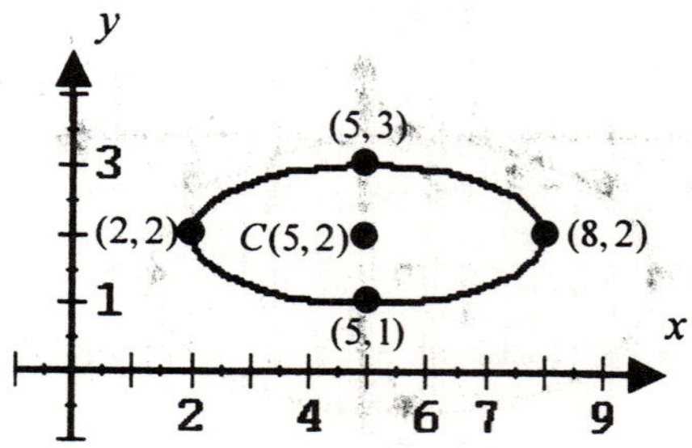

25. $\frac{(x-h)^2}{a^2}+\frac{(y-k)^2}{b^2}=1$

$\frac{(x+2)^2}{49}+\frac{y^2}{25}=1$

$\frac{(x+2)^2}{7^2}+\frac{y^2}{5^2}=1$

$a=7,\ b=5$

$C(-2,0)$

Vertices: $(-9,0),\ (-2,5),$
$(-2,-5),\ (5,0)$

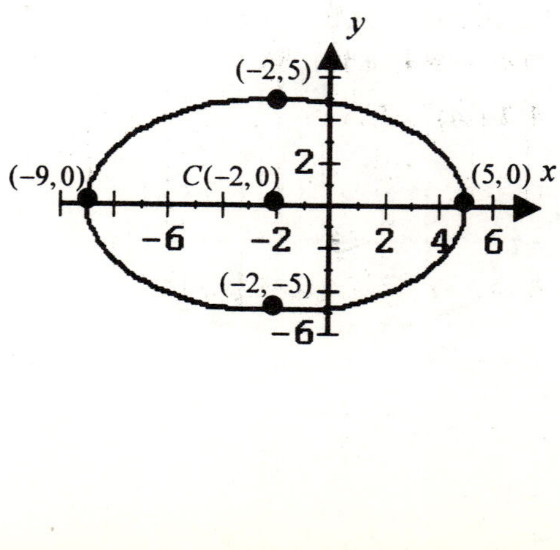

27. $\dfrac{(x-h)^2}{a^2}+\dfrac{(y-k)^2}{b^2}=1$

$\dfrac{(x+1)^2}{36}+\dfrac{(y+4)^2}{16}=1$

$\dfrac{(x+1)^2}{6^2}+\dfrac{(y+4)^2}{4^2}=1$

$a=6,\ b=4$

$C(-1,-4)$

Vertices: $(-7,-4),\ (-1,0),$
$(5,-4),\ (-1,-8)$

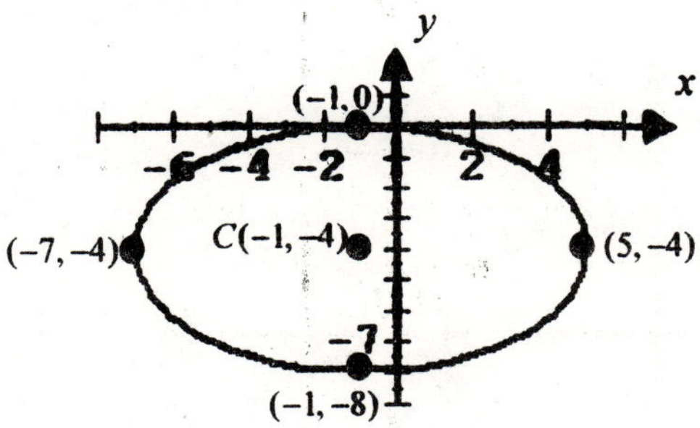

29. $\left(\dfrac{2+6}{2},\dfrac{3+3}{2}\right)=(4,3)$

$a=|6-4|=2$

$b=|7-3|=4$

$\dfrac{(x-4)^2}{2^2}+\dfrac{(y-3)^2}{4^2}=1$

$\dfrac{(x-4)^2}{4}+\dfrac{(y-3)^2}{16}=1$

31. $\dfrac{(-4+5)^2}{4}+\dfrac{(4+a)^2}{9}=1$

$9(1)^2+4(4+a)^2=36$

$4(4+a)^2=27$

$(4+a)^2=\dfrac{27}{4}$

$4+a=\pm\dfrac{\sqrt{27}}{2}$ (continued)

31. (continued)

$a=-4\pm\dfrac{\sqrt{9\cdot 3}}{2}=-4\pm\dfrac{3\sqrt{3}}{2}$

$a=\dfrac{-8\pm 3\sqrt{3}}{2}$

33. $\dfrac{(0-3.6)^2}{14.98}+\dfrac{(y-5.3)^2}{28.98}=1$

$(y-5.3)^2=28.98\left(1-\dfrac{12.96}{14.98}\right)$

$y=\pm\sqrt{28.98\left(1-\dfrac{12.967}{14.98}\right)}+5.3$

$y=\begin{cases}7.2768\\3.3232\end{cases}$

$\dfrac{(x-3.6)^2}{14.98}+\dfrac{(0-5.3)^2}{28.98}=1$

$(x-3.6)^2=14.98\left(1-\dfrac{28.09}{28.98}\right)$

$x=\pm\sqrt{14.98\left(1-\dfrac{28.09}{28.98}\right)}+3.6$

$x=\begin{cases}4.2783\\2.9217\end{cases}$

x-intercepts: $(4.2783,0),\ (2.9217,0)$

y-intercepts: $(0,7.2768),\ (0,3.3232)$

35. $A=\pi ab$

$A=3.1416\left(\dfrac{185}{2}\right)\left(\dfrac{154}{2}\right)$

$A=22{,}376.0$ square meters

Cumulative Review Problems

37. $\dfrac{5}{\sqrt{2x}-\sqrt{y}}\cdot\dfrac{\sqrt{2x}+\sqrt{y}}{\sqrt{2x}+\sqrt{y}}=\dfrac{5(\sqrt{2x}+\sqrt{y})}{2x-y}$

39. $$m = \frac{y_2 - y_1}{x_2 - x_1} = \frac{8-(-2)}{-2-0} = -5$$
$$y - y_1 = m(x - x_1)$$
$$y-(-2) = -5(x-0)$$
$$y = -5x - 2$$

41. $$\frac{5+3i}{2i} \cdot \frac{i}{i} = \frac{5i+3i^2}{2i^2} = \frac{5i-3}{-2} = \frac{3-5i}{2}$$

10.4 Exercises

1. $\dfrac{x^2}{a^2} - \dfrac{y^2}{b^2} = 1$

3. This is a horizontal hyperbola with vertices at $(4,0)$ and $(-4,0)$. Draw a fundamental rectangle with corners at $(4,2)$, $(4,-2)$, $(-4,2)$, $(-4,-2)$. Extend the diagonals through the rectangle as asymptotes of the hyperbola. Construct each branch of the hyperbola passing through the vertex and approaching the asymptotes.

5. $\dfrac{x^2}{4} - \dfrac{y^2}{25} = 1 \Rightarrow \dfrac{x^2}{2^2} - \dfrac{y^2}{5^2} = 1$

$a = 2,\ b = 5$

$V(\pm 2, 0),\ y_{\text{asymptote}} = \pm\dfrac{5}{2}x$

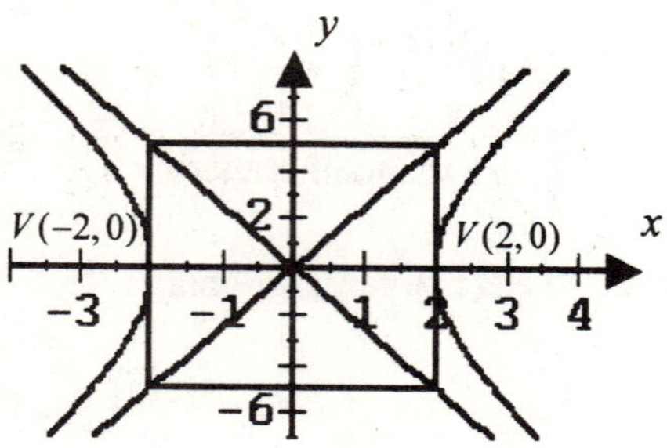

7. $\dfrac{y^2}{36} - \dfrac{x^2}{49} = 1 \Rightarrow \dfrac{y^2}{6^2} - \dfrac{x^2}{7^2} = 1$

$a = 7,\ b = 6,\ V(0, \pm 6)$

$y_{\text{asymptote}} = \pm\dfrac{6}{7}x$

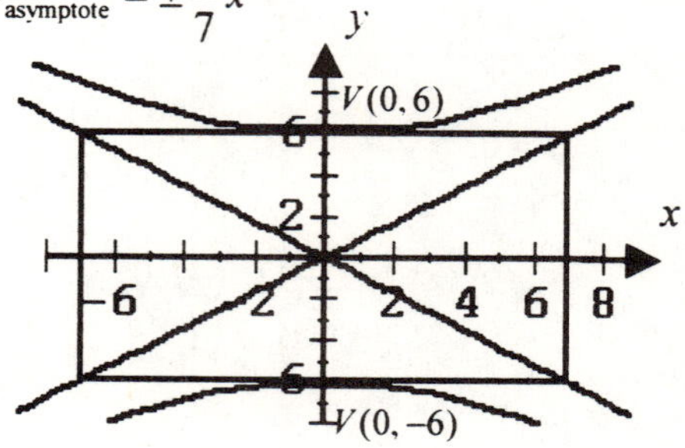

9. $4x^2 - y^2 = 64,\ \dfrac{x^2}{16} - \dfrac{y^2}{64} = 1$

$\dfrac{x^2}{4^2} - \dfrac{y^2}{8^2} = 1,\ a = 4,\ b = 8$

$V(\pm 4, 0),\ y_{\text{asymptote}} = \pm\dfrac{8}{4}x$

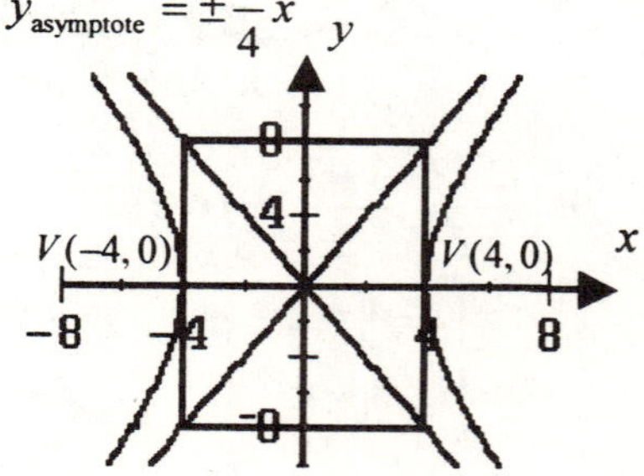

11. $8x^2 - y^2 = 16 \Rightarrow \dfrac{x^2}{2} - \dfrac{y^2}{16} = 1$

$\dfrac{x^2}{\sqrt{2}^2} - \dfrac{y^2}{4^2} = 1,\ a = \sqrt{2},\ b = 4$

$V(\pm\sqrt{2}, 0),\ y_{\text{asymptote}} = \pm\dfrac{4}{\sqrt{2}}x$

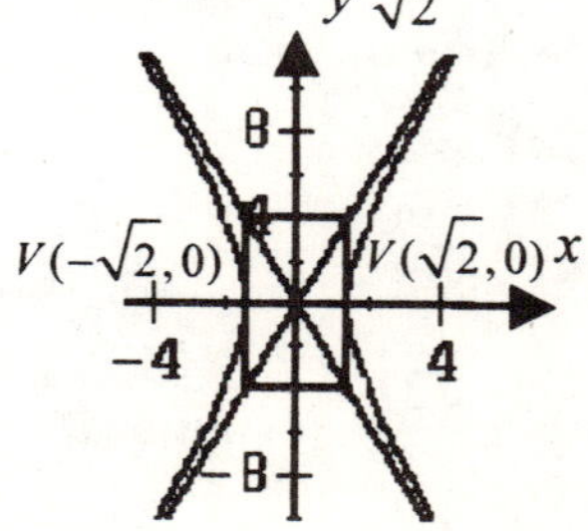

13. $2y^2 - 3x^2 = 54 \Rightarrow \frac{y^2}{27} - \frac{x^2}{18} = 1$

$$\frac{y^2}{(3\sqrt{3})^2} - \frac{x^2}{(3\sqrt{2})^2} = 1$$

$$a = 3\sqrt{2}$$

$$b = 3\sqrt{3}$$

$$V(0, \pm 3\sqrt{3})$$

$$y_{\text{asymptote}} = \pm \frac{\sqrt{3}}{\sqrt{2}} x$$

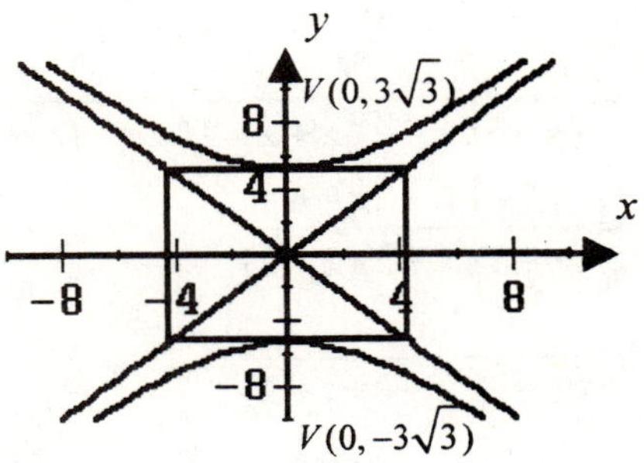

15. $V(\pm 3, 0)$

$$a = 3$$

$$a^2 = 9$$

$$y_{\text{asymptote}} = \pm \frac{4}{3} x$$

$$b = 4$$

$$b^2 = 16$$

$$\frac{x^2}{9} - \frac{y^2}{16} = 1$$

17. $V(0, \pm 7) \Rightarrow b = 7,\ b^2 = 49$

$$y_{\text{asymptote}} = \pm \frac{7}{3} x$$

$$a = 3$$

$$a^2 = 9$$

$$\frac{y^2}{49} - \frac{x^2}{9} = 1$$

19. $\frac{x^2}{a^2} - \frac{y^2}{b^2} = 1$, from graph, $a = 120$.

$$y_{\text{asymptote}} = \frac{3}{1} x$$

$$\frac{b}{a} = \frac{3}{1}$$

$$\frac{b}{120} = \frac{3}{1}$$

$$b = 360$$

$$\frac{x^2}{120^2} - \frac{y^2}{360^2} = 1$$

$$\frac{x^2}{14,400} - \frac{y^2}{129,600} = 1$$

21. $\frac{(x-6)^2}{25} - \frac{(y-4)^2}{49} = 1$

$$\frac{(x-6)^2}{5^2} - \frac{(y-4)^2}{7^2} = 1$$

$C(6,4)$

$a = 5,\ b = 7$

$V(6 \pm 5, 4) = (11,4),\ (1,4)$

$$y_{\text{asymptote}} = \pm \frac{7}{5}(x-6) + 4$$

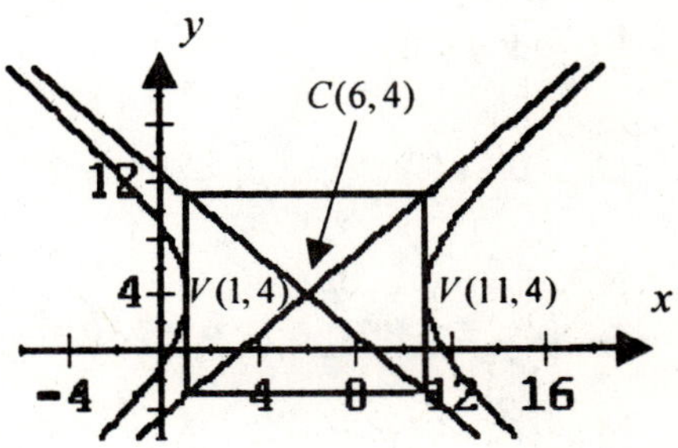

23. $\dfrac{(y+2)^2}{36}-\dfrac{(x+1)^2}{81}=1$

$\dfrac{(y+2)^2}{6^2}-\dfrac{(x+1)^2}{9^2}=1$

$C(-1,-2)$

$a=6,\ b=9$

$V(-1,-2\pm 6)=(-1,4),\ (-1,-8)$

$y_{asymptote}=\pm\dfrac{6}{9}(x-(-1))+(-2)$

$y_{asymptote}=\pm\dfrac{2}{3}(x+1)-2$

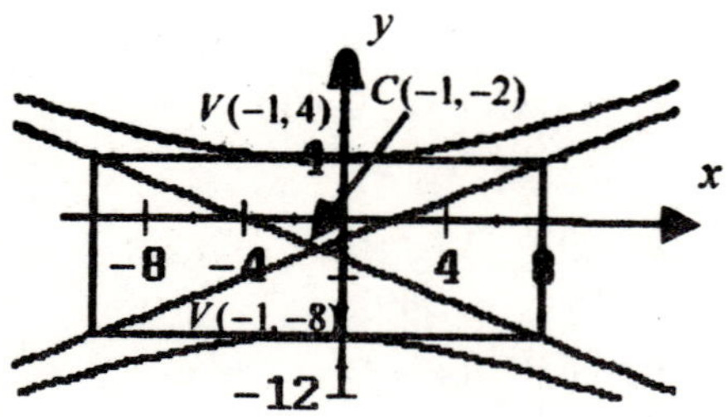

25. $\dfrac{(x+6)^2}{7}-\dfrac{y^2}{3}=1$

$\dfrac{(x-(-6))^2}{\sqrt{7}^2}-\dfrac{(y-0)^2}{\sqrt{3}^2}=1$

$C(-6,0)$

$V(-6\pm\sqrt{7},0)$

27. $C\left(5,\dfrac{0+14}{2}\right)=(5,7)$

$y_{asymptote}=\dfrac{7}{5}x=\dfrac{b}{a}x\Rightarrow a=5,\ b=7$

$\dfrac{(y-7)^2}{7^2}-\dfrac{(x-5)^2}{5^2}=1$

$\dfrac{(y-7)^2}{49}-\dfrac{(x-5)^2}{25}=1$

29. $8x^2-y^2=16\big|_{x=3.5}$

$8(3.5)^2-y^2=16$

$y^2=8(3.5)^2-16$

$y=\pm\sqrt{8(3.5)^2-16}=\pm 9.055385138$

Cumulative Review Problems

31. $12x^2+x-6=(4x+3)(3x-2)$

33. $\dfrac{3}{x^2-5x+6}+\dfrac{2}{x^2-4}$

$=\dfrac{3(x+2)}{(x-3)(x-2)(x+2)}+\dfrac{2(x-3)}{(x-2)(x+2)(x-3)}$

$=\dfrac{3x+6+2x-6}{(x-3)(x-2)(x+2)}$

$=\dfrac{5x}{(x-3)(x-2)(x+2)}$

35. $\dfrac{x}{9}=\dfrac{49}{14}$

$14x=441$

$x=\dfrac{441}{14}$

$x=31.5$ ft

37. $\dfrac{(2.1\times 10^9+I)\text{ pencils}}{274\times 10^6\text{ people}}=\dfrac{10\text{ pencils}}{\text{person}}$

$I=10(274\times 10^6)-2.1\times 10^9$

$I=640{,}000{,}000$ pencils were imported

10.5 Exercises

1. $y^2 = 4x$

   $y = x + 1$, substitute into first equation

   $(x+1)^2 = 4x$

   $x^2 - 2x + 1 = 0$

   $(x-1)^2 = 0$

   $x - 1 = 0$

   $x = 1$

   $y = x + 1 = 2$

   $(1, 2)$ is the solution.

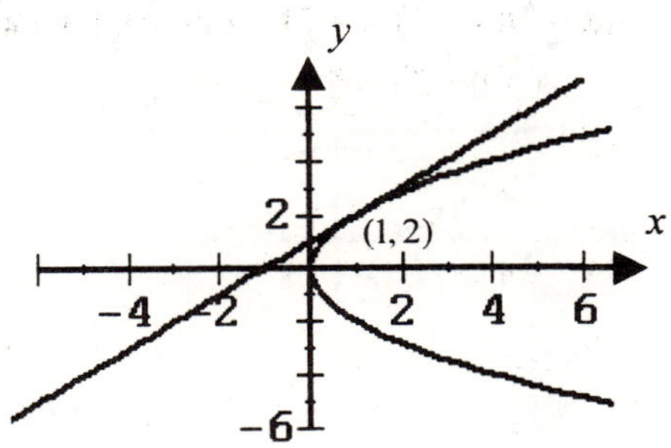

3. $y - 4x = 0 \Rightarrow y = 4x$

   $4x^2 + y^2 = 20$

   $4x^2 + (4x)^2 = 20$

   $4x^2 + 16x^2 = 20x^2 = 20$

   $x^2 = 1,\ x = \pm 1$

   $y = 4x = 4(\pm 1) = \pm 4$

   $(1, 4),\ (-1, -4)$ is the solution

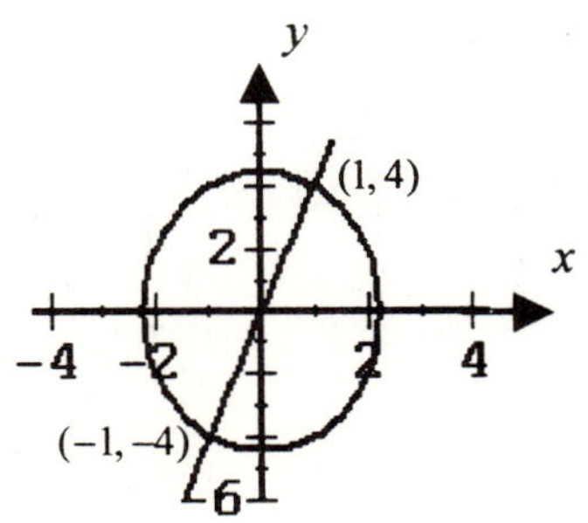

5. $\dfrac{x^2}{1} - \dfrac{y^2}{3} = 1$

   $x + y = 1$

   $y = 1 - x$

   $3x^2 - (1-x)^2 = 3$

   $3x^2 - 1 + 2x - x^2 = 3$

   $2x^2 + 2x - 4 = 0$

   $x^2 + x - 2 = 0$

   $(x+2)(x-1) = 0$

   $x + 2 = 0,\ x - 1 = 0$

   $x = -2,\ x = 1$

   $y = 1 - x = \begin{cases} 1-(-2) = 3 \\ 1 - 1 = 0 \end{cases}$

   $(-2, 3),\ (1, 0)$ is the solution

7. $x^2 + y^2 - 25 = 0$

   $3y = x + 5$

   $x = 3y - 5$

   $(3y-5)^2 + y^2 - 25 = 0$

   $9y^2 - 30y + 25 + y^2 - 25 = 0$

   $10y^2 - 30y = 0$

   $10y(y-3) = 0$

   $10y = 0,\ y - 3 = 0$

   $y = 0,\ y = 3$

   $x = 3y - 5 = \begin{cases} 3(0) - 5 = -5 \\ 3(3) - 5 = 4 \end{cases}$

   $(-5, 0),\ (4, 3)$ is the solution

9. $x^2+2y^2=4$

$y=-x+2$

$x^2+2(-x+2)^2=4$

$x^2+2x^2-8x+8=4$

$3x^2-8x+4=0$

$(3x-2)(x-2)=0$

$3x-2=0,\ x-2=2$

$x=\frac{2}{3},\ x=2$

$y=-x+2=\begin{cases}-\frac{2}{3}+2=\frac{4}{3}\\ -2+2=0\end{cases}$

$\left(\frac{2}{3},\frac{4}{3}\right)$, $(2,0)$ is the solution

11. $\frac{x^2}{4}-\frac{y^2}{4}=1$

$x+y-4=0 \Rightarrow x=4-y$

$(4-y)^2-y^2=4$

$16-8y+y^2-y^2=4$

$8y=12$

$y=\frac{3}{2}$

$x=4-y=4-\frac{3}{2}=\frac{5}{2}$

$\left(\frac{5}{2},\frac{3}{2}\right)$ is the solution

13. $2x^2-5y^2=-2 \xrightarrow{\times 2} 4x^2-10y^2=-4$

$3x^2+2y^2=35 \xrightarrow{\times 5} 15x^2+10y^2=175$

$19x^2 \quad =171$

$x=\pm 3$

(continued)

13. (continued)

$y^2=\frac{35-3x^2}{2}$

$y=\pm\sqrt{\frac{35-3x^2}{2}}\Bigg|_{x=\pm 3}$

$y=\pm\sqrt{\frac{35-3(\pm 3)^2}{2}}$

$y=\pm 2$

$(3,\pm 2)$, $(-3,\pm 2)$ is the solution

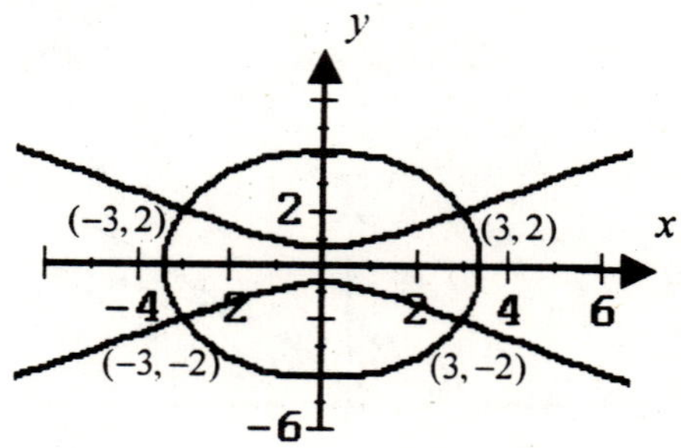

15. $2x^2+5y^2=42 \xrightarrow{\times -3} -6x^2-15y^2=-126$

$3x^2+4y^2=35 \xrightarrow{\times 2} \quad 6x^2+8y^2=70$

$-7y^2=-56$

$y^2=8$

$y=\pm 2\sqrt{x}$

$x^2=\frac{42-5y^2}{2}$

$x^2=\frac{42-5(8)}{2}$

$x^2=1$

$x=\pm 1$

$(1,\pm 2\sqrt{2}),(-1,\pm 2\sqrt{2})$ is the solution

17. $x^2+2y^2=8 \rightarrow x^2+2y^2=8$

$x^2-\ y^2=1\xrightarrow{\times -1}\underline{-x^2+y^2=-1}$

$3y^2=7$

$y=\pm\frac{\sqrt{21}}{3}$

$x^2=1+y^2=1+\frac{7}{3}=\frac{10}{3}$

$x=\pm\frac{\sqrt{30}}{3}$

$\left(\frac{\sqrt{30}}{3},\pm\frac{\sqrt{21}}{3}\right),\left(-\frac{\sqrt{30}}{3},\pm\frac{\sqrt{21}}{3}\right)$

is the solution

19. $x^2+y^2=7\rightarrow\ x^2+y^2=7$

$\frac{x^2}{3}-\frac{y^2}{9}=1\rightarrow \underline{3x^2-y^2=9}$

$4x^2\qquad=16$

$x^2=4$

$x=\pm 2$

$y^2=7-x^2=7-4=3$

$y=\pm\sqrt{3}$

$(2,\pm\sqrt{3}),\ (-2,\pm\sqrt{3})$ is the solution

21. $xy=3\Rightarrow x=\frac{3}{y}$

$3y=3x+6=3\cdot\frac{3}{y}+6$

$3y^2-6y-9=0\Rightarrow y^2-2y-3=0$

(continued)

21. (continued)

$(y-3)(y+1)=0$

$y-3=0,\ y+1=0$

$y=3,\ y=-1$

$x=\frac{3}{y}=\begin{cases}\frac{3}{3}=1\\ \frac{3}{-1}=-3\end{cases}$

$(1,3),\ (-3,-1)$ is the solution

23. $xy=8$

$y=x+2,$ substitute into first equation

$x(x+2)=8$

$x^2+2x-8=0$

$(x+4)(x-2)=0$

$x+4=0,\ x-2=0$

$x=-4,\ x=2$

$y=x+2=\begin{cases}-4+2=-2\\ 2+2=4\end{cases}$

$(-4,-2),\ (2,4)$ is the solution

25. $x+y=5\Rightarrow x=5-y$

$x^2+y^2=4\Rightarrow(5-y)^2+y^2=4$

$25-10y+y^2+y^2=4$

$2y^2-10y+21=0$

$y=\frac{-(-10)\pm\sqrt{(-10)^2-4(2)(21)}}{2(2)}$

$y=\frac{10\pm\sqrt{-68}}{4}$

No real solution

27. $540 = xy \Rightarrow y = \dfrac{540}{x}$

$$39^2 = x^2 + y^2 = x^2 + \left(\frac{540}{x}\right)$$

$$1521 = x^2 + \frac{291,600}{x^2}$$

$$x^4 - 1521x^2 + 291,600 = 0$$

$$(x^2 - 1296)(x^2 - 225) = 0$$

$$x^2 = 1296,\ x^2 = 225$$

$$x = \sqrt{1296} = 36,\ x = \sqrt{225} = 15$$

$$y = \frac{540}{x} = \begin{cases} \dfrac{540}{36} = 15 \\ \dfrac{540}{15} = 36 \end{cases}$$

(36,15), (15,36) is the solution

The dimensions of the rectangle are 15 meters by 36 meters.

29. $x^2 + y^2 = 16,000,000 \Rightarrow$

$$y^2 = 16,000,000 - x^2$$

$$25,000,000x^2 - 9,000,000y^2$$
$$= 2.25 \times 10^{14}$$

$$25x^2 - 9y^2 = 2.25 \times 10^8$$

$$25x^2 - 9(16,000,000 - x^2) = 2.25 \times 10^8$$

$$34x^2 = 369,000,000$$

$$x^2 = 10,852,941.18$$

$$x = 3294.380242$$

$$y^2 = 16,000,000 - 10,852,941.18$$

$$y^2 = 5,147,058.824$$

$$y = 2268.713032$$

The hyperbola intersects the circle when $(x, y) \approx (3290, 2270)$

Cumulative Review Problems

31. $\dfrac{6x^4 - 24x^3 - 30x^2}{3x^3 - 21x^2 + 30x} = \dfrac{6x^2(x^2 - 4x - 5)}{3x(x^2 - 7x + 10)}$

$$= \frac{2x(x-5)(x+1)}{(x-5)(x-2)}$$

$$= \frac{2x(x+1)}{x-2}$$

33. $d = 55(5) = v(11)$

$v = 25$ mph

Putting Your Skills to Work

1. Let station A be at (0,0) in an xy coordinate system. The coordinates of station $B$are (50,30). The epicenter is located on the intersection of the circles $x^2 + y^2 = 40^2$ and

$(x-50)^2 + (y-30)^2 = 20^2$.

Graphing the two circles shows two points of intersection in the first quadrant, $x, y \geq 0$. Solve as follows

$$x^2 - 100x + 50^2 + y^2 - 60y + 30^2 = 20^2$$

$$40^2 - 100x + 50^2 - 60y + 30^2 = 20^2$$

$$-100x + 4600 - 60y = 0$$

$$-100x + 4600 - 60\sqrt{1600 - x^2} = 0$$

$$3\sqrt{1600 - x^2} = -5x + 230$$

$$9(1600 - x^2) = 25x^2 - 2300x + 52,900$$

$34x^2 - 2300x + 38,500 = 0$, solving with the quadratic formula gives

$$x = 30.40619117,\ x = 37.24086766$$

$$y = \sqrt{1600 - x^2} = \begin{cases} 25.98968139 \\ 14.5985539 \end{cases}$$

The epicenter is approximately 30.4 miles east and 26.0 miles north of

1. (continued)
station *A* or approximately 37.2 miles east and 14.6 miles north of station *A*.

2. The epicenter lies on the intersection of the two circles, $x^2 + y^2 = 50^2$ and $(x-60)^2 + (y-20)^2 = 30^2$. The points of intersection may be found algebraically or using a graphing calculator, which gives

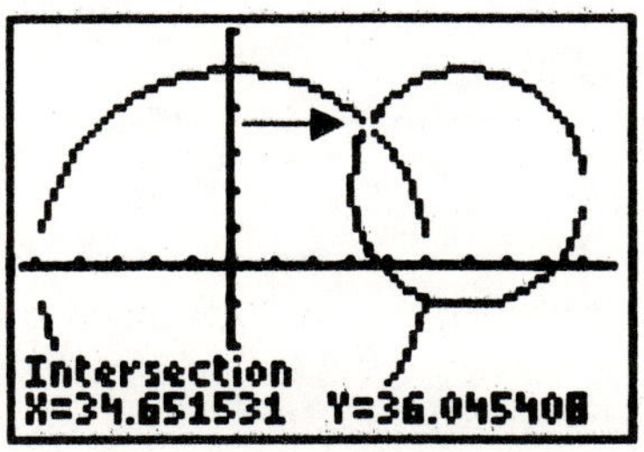

and

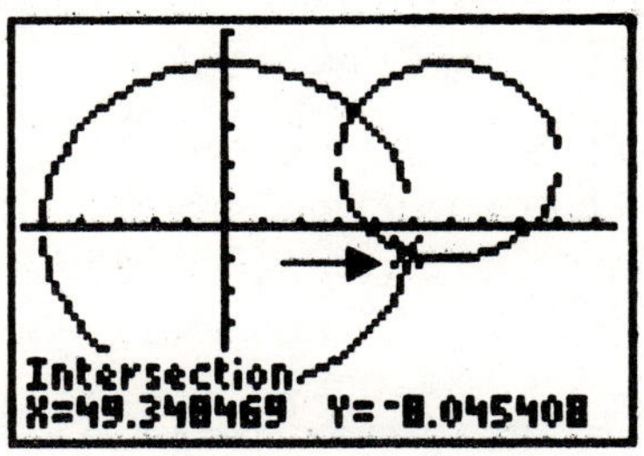

The epicenter is approximately 34.7 miles east and 36.0 miles north of station *A* or approximately 49.3 miles east and 8.0 miles south of station *A*.

3. The epicenter lies on the intersection of the three circles
$x^2 + y^2 = 30^2$
$(x-40)^2 + (y-30)^2 = 25^2$
$(x-30)^2 + (y+20)^2 = 27.55^2$.
From a graphing calculator,

(continued)

3. (continued)

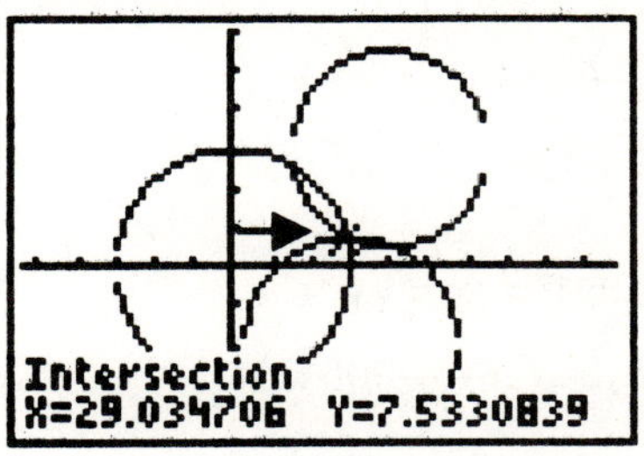

The epicenter is 29.0 miles east and 7.5 miles north of station *A*.

4. The epicenter lies on the intersection of the three circles
$x^2 + y^2 = 50^2$
$(x-40)^2 + (y-30)^2 = 40^2$
$(x-10)^2 + (y-15)^2 = 45.92^2$.
From a graphing calculator,

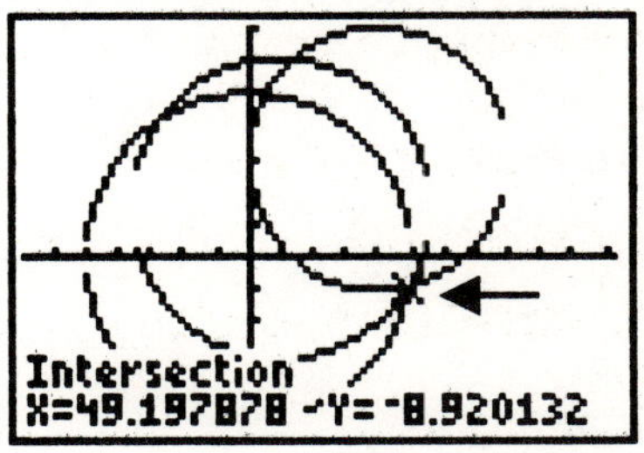

The epicenter is 49.2 miles east and 8.9 miles south of station *A*.

Chapter 10 Review Problems

1. $d = \sqrt{(x_2 - x_1)^2 + (y_2 - y_1)^2}$
$d = \sqrt{(7.5 - 10.5)^2 + (-4 - (-6))^2} = \sqrt{13}$

2. $d = \sqrt{(x_2 - x_1)^2 + (y_2 - y_1)^2}$
$d = \sqrt{(-2 - (-7))^2 + (-1 - 3)^2} = \sqrt{41}$

3. $$(x-h)^2+(y-k)^2=r^2$$
$$(x-(-6))^2+(y-3)^2=\sqrt{15}^2$$
$$(x+6)^2+(y-3)^2=15$$

4. $$(x-h)^2+(y-k)^2=r^2$$
$$(x-0)^2+(y-(-7))^2=5^2$$
$$x^2+(y+7)^2=25$$

5. $$x^2+y^2-6x-8y+3=0$$
$$x^2-6x+9+y^2-8y+16=-3+9+16$$
$$(x-3)^2+(y-4)^2=22=\sqrt{22}^2$$
$$C(3,4),\ r=\sqrt{22}$$

6. $x^2+y^2-10x+12y+52=0$
$$x^2-10x+25+y^2+12y+36=-52+25+36$$
$$(x-5)^2+(y+6)^2=9$$
$$(x-5)^2+(y+6)^2=3^2$$
$$C(5,-6),\ r=3$$

7. $x=\frac{1}{3}y$

x	y	
0	0	$V(0,0)$
3	3	y-intercept: $(0,0)$
3	−3	

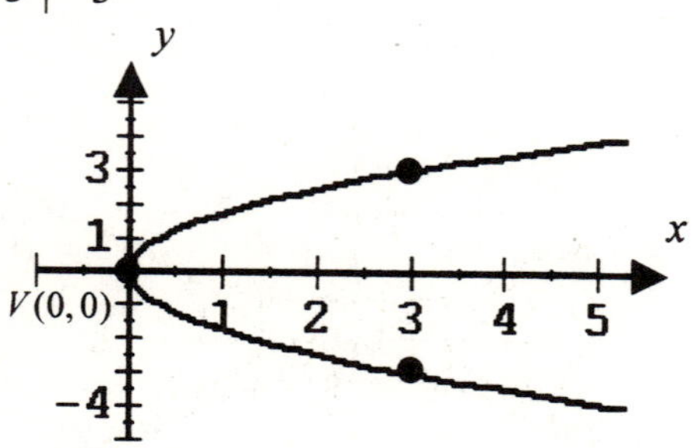

8. $x=\frac{1}{2}(y-2)^2+4$

x	y	
4	2	$V(4,2)$
6	0	y-intercept: $(6,0)$
6	4	

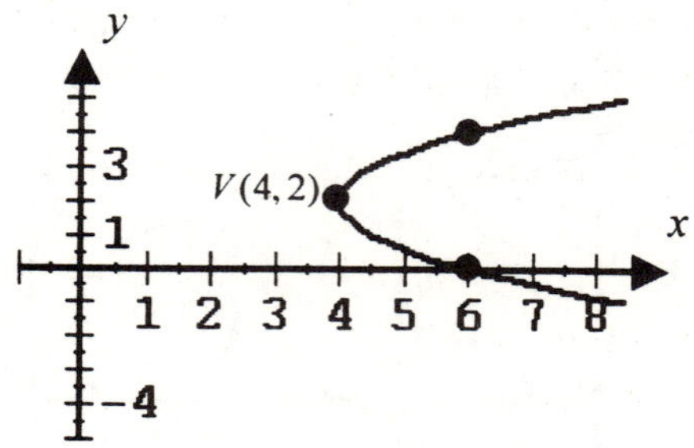

9. $y=-2(x+1)^2$

x	y	
−1	0	$V(-1,0)$
0	−2	y-intercept: $(0,-2)$
−2	−2	

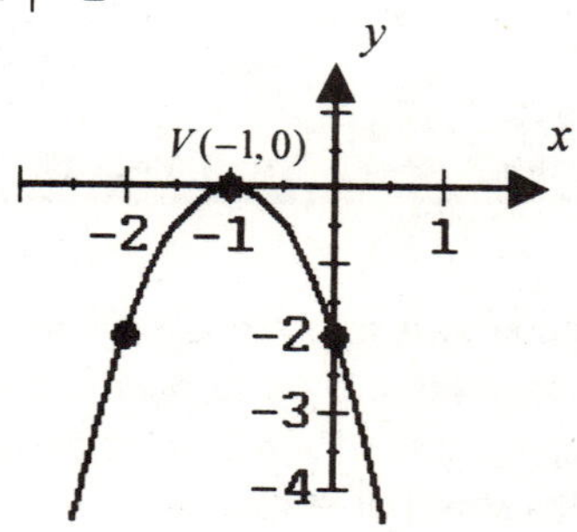

10. $x+8y=y^2+10$
$$x=y^2-8y+16-6$$
$$x=(y-4)^2-6$$
$$V(-6,4)$$
Opens to right.

11. $x^2 + 6x = y - 4$

$y = x^2 + 6x + 9 - 9 + 4$

$y = (x+3)^2 - 5, \; V(-3,-5)$

Opens upward.

12. $\frac{x^2}{\frac{1}{4}} + \frac{y^2}{1} = 1 \Rightarrow \frac{x^2}{\left(\frac{1}{2}\right)^2} + \frac{y^2}{1^2} = 1$

$a = \frac{1}{2}, \; b = 1, \; C(0,0)$

Vertices: $(0,1), \; (0,-1), \; \left(-\frac{1}{2},0\right), \; \left(\frac{1}{2},0\right)$

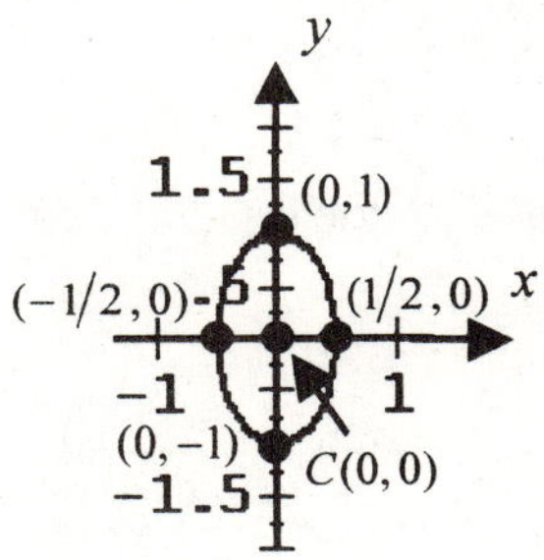

13. $16x^2 + y^2 - 32 = 0 \Rightarrow \frac{x^2}{2} + \frac{y^2}{32} = 1$

$\frac{x^2}{\sqrt{2}^2} + \frac{y^2}{(4\sqrt{2})^2} = 1$

$a = \sqrt{2}, \; b = 4\sqrt{2}$

$C(0,0)$

Vertices: $(0, 4\sqrt{2}), \; (0, -4\sqrt{2})$
$(-\sqrt{2}, 0), \; (\sqrt{2}, 0)$

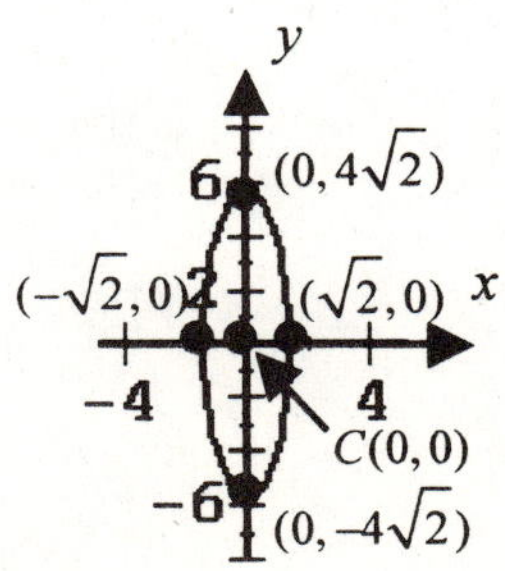

14. $\frac{(x+5)^2}{4} + \frac{(y+3)^2}{25} = 1$

$\frac{(x-(-5))^2}{2^2} + \frac{(y-(-3))^2}{5^2} = 1$

$C(-5,-3)$

$a = 2, \; b = 5$

Vertices: $(-3,-3), \; (-7,-3)$
$(-5,2), \; (-5,-8)$

15. $\frac{(x+1)^2}{9} + \frac{(y-2)^2}{16} = 1$

$\frac{(x-(-1))^2}{3^2} + \frac{(y-2)^2}{4^2} = 1$

$C(-1,2)$

$a = 3, \; b = 4$

Vertices: $(2,2), \; (-4,2)$
$(-1,6), \; (-1,-2)$

16. $x^2 - 4y^2 - 16 = 0 \Rightarrow \frac{x^2}{16} - \frac{y^2}{4} = 1$

$\frac{x^2}{4^2} - \frac{y^2}{2^2} = 1$

$C(0,0)$

$a = 4, \; b = 2$

Vertices: $(-4,0), \; (4,0)$

$y_{\text{asymptote}} = \pm\frac{2}{4}x = \pm\frac{1}{2}x$

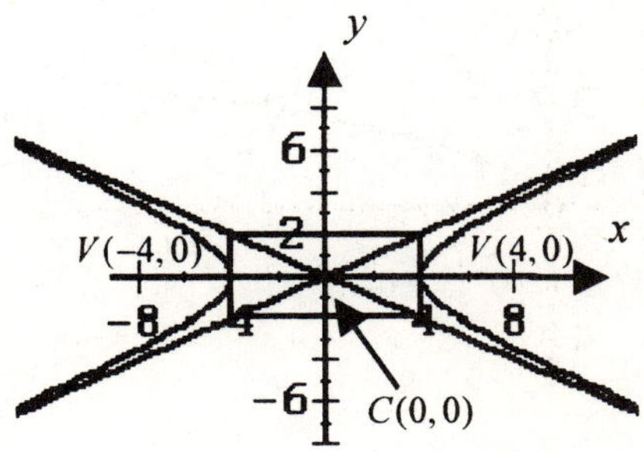

17. $9y^2 - 25x^2 = 225 \Rightarrow \frac{y^2}{25} - \frac{x^2}{9} = 1$

$\frac{y^2}{5^2} - \frac{x^2}{3^2} = 1$

$a = 5,\ b = 3$

$C(0,0)$

Vertices: $(0,5),\ (0,-5)$

$y_{\text{asymptote}} = \pm\frac{5}{3}x$

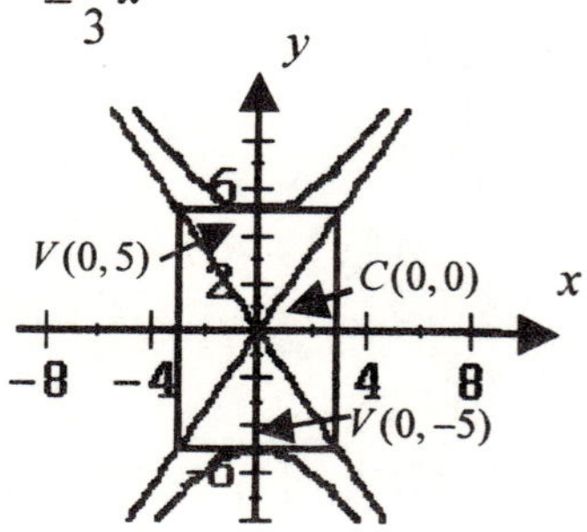

18. $\frac{(x-2)^2}{4} - \frac{(y+3)^2}{25} = 1$

$\frac{(x-2)^2}{2^2} - \frac{(y-(-3))^2}{5^2} = 1$

$C(2,-3)$

$a = 2,\ b = 5$

Vertices: $(0,-3),\ (4,-3)$

19. $9(y-2)^2 - (x+5)^2 - 9 = 0$

$\frac{(y-2)^2}{1^2} - \frac{(x-(-5))^2}{3^3} = 0$

$C(-5,2)$

$a = 3,\ b = 1$

Vertices: $(-5,3),\ (-5,1)$

20. $x^2 + y = 9 \rightarrow x^2 + y = 9$

$y - x = 3 \xrightarrow{\times -1} \underline{x - y = -3}$

$x^2 - x = 6$

(continued)

20. (continued)

$x^2 - x - 6 = 0$

$(x+3)(x-2) = 0$

$x + 3 = 0,\ x - 2 = 0$

$x = -3,\ x = 2$

$y = x + 3 = \begin{cases} -3 + 3 = 0 \\ 2 + 3 = 5 \end{cases}$

$(-3,0),\ (2,5)$ is the solution.

21. $y^2 + x^2 = 3$

$x - 2y = 1 \Rightarrow x = 2y + 1$

$y^2 + (2y+1)^2 = 3$

$y^2 + 4y^2 + 4y + 1 = 3$

$5y^2 + 4y - 2 = 0$

Solving with the quadratic formula gives

$y = \frac{-2 \pm \sqrt{14}}{5}$

$x = 2y + 1 = \begin{cases} 2\left(\frac{-2+\sqrt{14}}{5}\right) + 1 \\ 2\left(\frac{-2\sqrt{14}}{5}\right) + 1 \end{cases}$

$x = \begin{cases} \frac{1+2\sqrt{14}}{5} \\ \frac{1-2\sqrt{14}}{5} \end{cases}$

$\left(\frac{1-2\sqrt{14}}{5}, \frac{-2-\sqrt{14}}{5}\right)$ and

$\left(\frac{1+2\sqrt{14}}{5}, \frac{-2+\sqrt{14}}{5}\right)$ is the solution.

22. $2x^2+y^2=17 \Rightarrow y^2=17-2x^2$
$$x^2+2y^2=22$$
$$x^2+2(17-2x^2)=22$$
$$x^2+34-4x^2=22$$
$$3x^2=12$$
$$x^2=4$$
$$x=\pm 2$$
$$y=\pm\sqrt{17-2x^2}$$
$$y=\pm\sqrt{17-2(4)}$$
$$y=\pm 9$$
$(2,\pm 3)$, $(-2,\pm 3)$ is the solution.

23. $xy=-2 \Rightarrow y=\dfrac{-2}{x}$
$$x^2+y^2=5 \Rightarrow x^2+\left(\frac{-2}{x}\right)^2=5$$
$$x^4+4=5x^2$$
$$x^4-5x^2+4=0$$
$$(x^2-4)(x^2-1)=0$$
$$x^2-4=0,\ x^2-1=0$$
$$x^2=4,\ x^2=1$$
$$x=\pm 2,\ x=\pm 1$$
$$y=\frac{-2}{x}=\begin{cases}\dfrac{-2}{1}=-1\\[4pt] \dfrac{-2}{-2}=1\\[4pt] \dfrac{-2}{1}=-2\\[4pt] \dfrac{-2}{-1}=2\end{cases}$$
$(2,-1)$, $(-2,1)$, $(1,-2)$, $(-1,2)$ is the solution.

24. $3x^2-4y^2=12$
$$7x^2-y^2=8 \Rightarrow y^2=7x^2-8$$
$$3x^2-4(7x^2-8)=12$$
$$3x^2-28x^2+32=12$$
$$25x^2=20 \Rightarrow x^2=\frac{20}{25}$$
$$y^2=7x^2-8=7\cdot\frac{20}{25}-8=-\frac{12}{5}$$
No real solution.

25. $y=x^2+1 \Rightarrow x^2=y-1$
$$x^2+y^2-8y+7=0$$
$$y-1+y^2-8y+7=0$$
$$y^2-7y+6=0$$
$$(y-1)(y-6)=0$$
$$y-1=0,\ y-6=0$$
$$y=1,\ y=6$$
$$x^2=y-1=\begin{cases}1-1=0\\ 6-1=5\end{cases}$$
$$x=\begin{cases}0\\ \pm\sqrt{5}\end{cases}$$
$(0,1)$, $(\sqrt{5},6)$, $(-\sqrt{5},6)$ is the solution.

26. $2x^2+y^2=18$
$$xy=4 \Rightarrow y=\frac{4}{x}$$
$$2x^2+\left(\frac{4}{x}\right)^2=18$$
$$2x^4+16=18x^2$$
$$x^4-9x^2+8=0$$
$$(x^2-8)(x^2-1)=0$$
$$x^2-8=0,\ x^2-1=0$$

(continued)

26. (continued)

$x^2 = 8,\ x^2 = 1$

$x = \pm 2\sqrt{2},\ x = \pm 1$

$$y = \frac{4}{x} = \begin{cases} \dfrac{4}{2\sqrt{2}} = \sqrt{2} \\ \dfrac{4}{-2\sqrt{2}} = -\sqrt{2} \\ \dfrac{4}{1} = 4 \\ \dfrac{4}{-1} = -4 \end{cases}$$

$(2\sqrt{2}, \sqrt{2}),\ (-2\sqrt{2}, -\sqrt{2}),$
$(1, 4),\ (-1, -4)$ is the solution.

27. $y^2 - 2x^2 = 2 \xrightarrow{\times -2} -2y^2 + 4x^2 = -4$

$2y^2 - 3x^2 = 5 \rightarrow \underline{\ 2y^2 - 3x^2 = 5\ }$

$x^2 = 1$

$x = \pm 1$

$y^2 = 2x^2 + 2 = 2(1) + 2 = 4$

$y = \pm 2$

$(1, \pm 2),\ (-1, \pm 2)$ is the solution.

28. $y^2 = \frac{1}{2}x$

$y = x - 1 \Rightarrow x = y + 1$

$y^2 = \frac{1}{2}(y+1)$

$2y^2 - y - 1 = 0$

$(2y+1)(y-1) = 0$

$2y + 1 = 0,\ y - 1 = 0$

$y = -\frac{1}{2},\ y = 1$

(continued)

28. (continued)

$$x = y + 1 = \begin{cases} -\dfrac{1}{2} + 1 = \dfrac{1}{2} \\ 1 + 1 = 2 \end{cases}$$

$\left(\frac{1}{2}, -\frac{1}{2}\right)$, (2,1) is the solution.

29. $y^2 = 2x$

$y = \frac{1}{2}x + 1 \Rightarrow x = 2y - 2$

$y^2 = 2(2y - 2) = 4y - 4$

$y^2 - 4y + 4 = 0$

$(y - 2)^2 = 2$

$y = 2$

$x = 2y - 2 = 2(2) - 2 = 4 - 2 = 2$

$x = 2$

$(2, 2)$ is the solution.

30. $y^2 = 4px$

$5^2 = 4p(4)$

$p = \frac{25}{16} = 1.5625$

The receiver should be placed 1.56 feet from the center of the dish.

31. $y^2 = 4px$

$y^2 = 4(2)x = 8x$

$y^2 = 8x$

$2.5^2 = 8x$

$x = 0.78125$

The searchlight should be 0.78 feet deep.

Chapter 10 Test

1. $d = \sqrt{(-2-(-6))^2 + (5-(-8))^2}$

$d = \sqrt{185}$

2. $y^2 - 6y - x + 13 = 0$

   $x - 13 + 9 = y^2 - 6y + 9$

   $x = (y-3)^2 + 4$

   Parabola: $V(4,3)$

   x-int: $(13,0)$

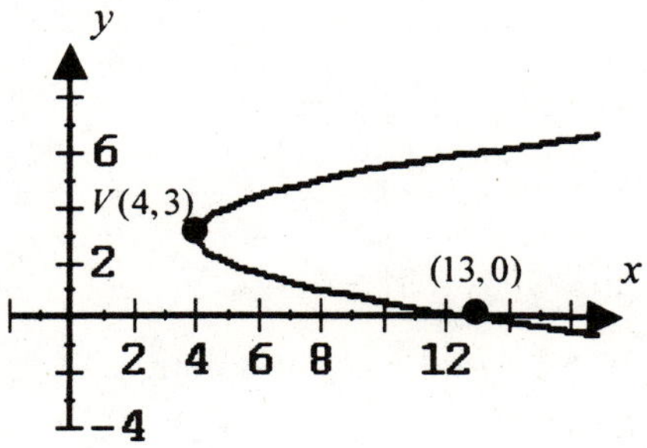

3. $x^2 + y^2 + 6x - 4y + 9 = 0$

   $x^2 + 6x + 9 + y^2 - 4y + 4 = -9 + 9 + 4$

   $(x+3)^2 + (y-2)^2 = 4 = 2^2$

   Circle: $C(-3,0)$, $r = 2$

   x-int: $(-3,0)$

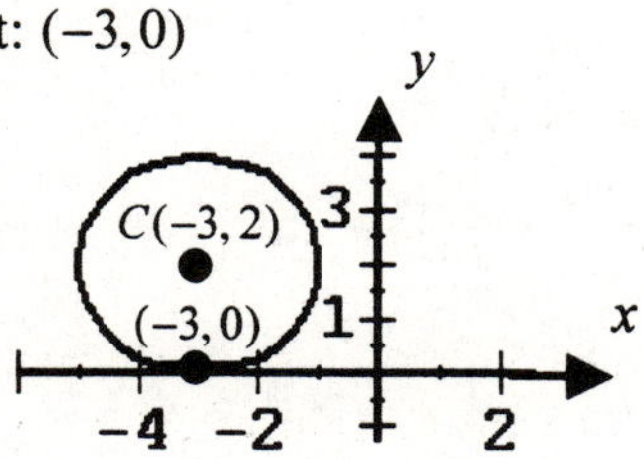

4. $\dfrac{x^2}{25} + \dfrac{y^2}{1} = 1 \Rightarrow \dfrac{x^2}{5^2} + \dfrac{y^2}{1^2} = 1$

   Ellipse: $C(0,0)$

   $a = 5,\ b = 1$

   Vertices: $(0,5), (-5,0), (0,1), (0,-1)$

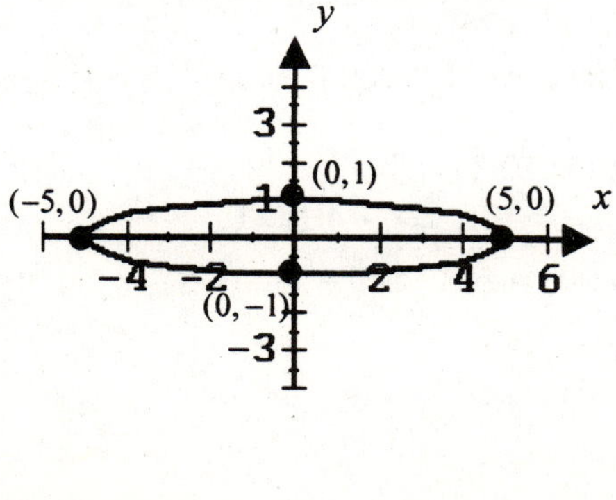

5. $\dfrac{x^2}{10} - \dfrac{y^2}{9} = 1$, hyperbola

   $C(0,0)$, $V(\pm\sqrt{10},0)$

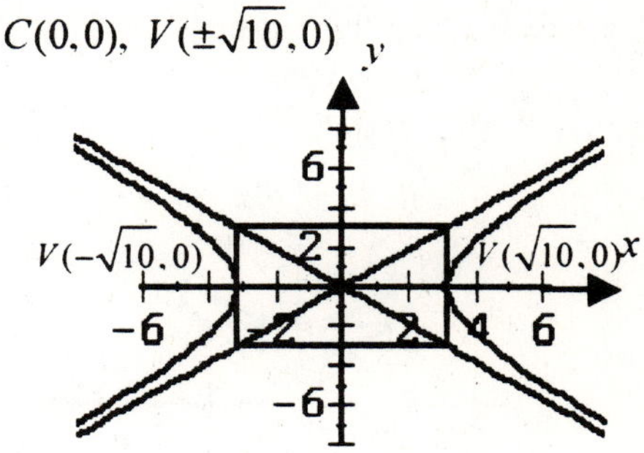

6. $y = -2(x+3)^2 + 4$

   Parabola: $C(-3,4)$

   y-intercept: $(0,-14)$

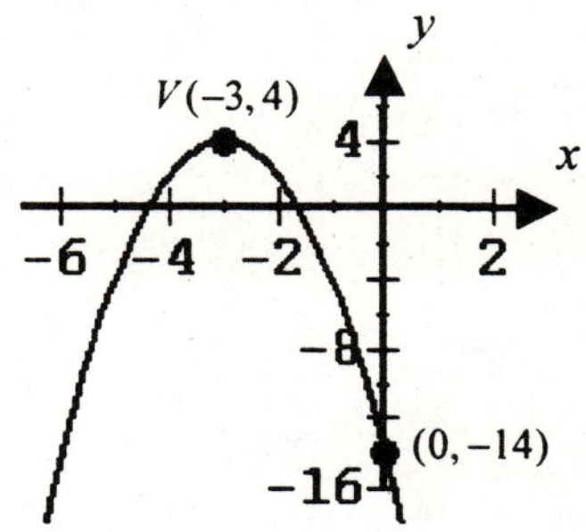

7. $\dfrac{(x+2)^2}{16} + \dfrac{(y-5)^2}{4} = 1$

   $\dfrac{(x-(-2))^2}{4^2} + \dfrac{(y-5)^2}{2^2} = 1$

   Ellipse: $C(-2,5)$

   $a = 2, b = 2$

   Vertices: $(-2,7), (-2,3), (-6,5), (2,5)$

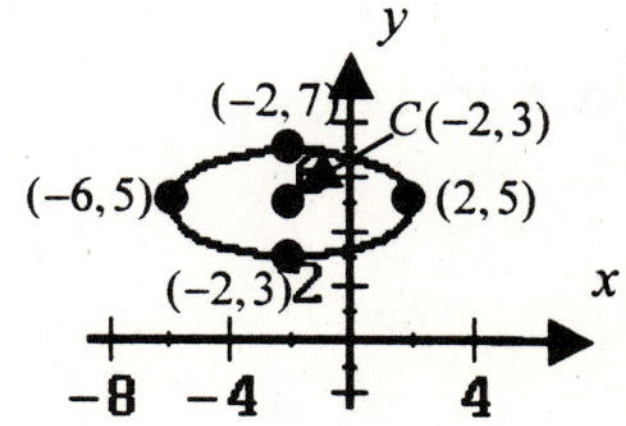

8. $7y^2 - 7x^2 = 28 \Rightarrow \frac{y^2}{2^2} - \frac{x^2}{2^2} = 1$

Hyperbola: $C(0,0)$, Vertices: $(0,2)$, $(0,-2)$

$a = 2,\ b = 2,\ y_{\text{asymptote}} = \pm\frac{2}{2}x = \pm x$

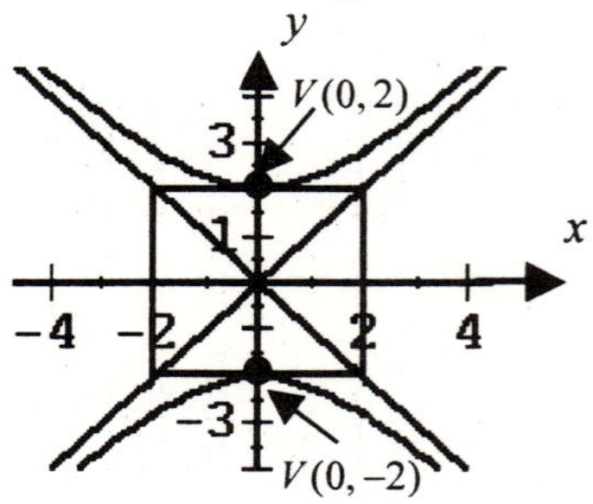

9. $(x-h)^2 + (y-k)^2 = r^2$

$(x-3)^2 + (y-(-5))^2 = \sqrt{8}^2$

$(x-3)^2 + (y+5)^2 = 8$

10. $a = 1,\ b = 3,\ (h,k) = (-4,-2)$

$\frac{(x-h)^2}{a^2} + \frac{(y-k)^2}{b^2} = 1$

$\frac{(x-(-4))^2}{1^2} + \frac{(y-(-2))^2}{3^2} = 1$

$\frac{(x+4)^2}{1} + \frac{(y+2)^2}{9} = 1$

11. $x = (y-k)^2 + h,\ (h,k) = (-7,3)$

$x = (y-3)^2 + (-7)$

$x = (y-3)^2 - 7$

12. $(h,k) = (6,7),\ a = 3,\ b = 14 - 7 = 7$

$\frac{(y-k)^2}{b^2} - \frac{(x-h)^2}{a^2} = 1$

$\frac{(y-7)^2}{7^2} - \frac{(x-6)^2}{3^2} = 1$

$\frac{(y-7)^2}{49} - \frac{(x-6)^2}{9} = 1$

13. $-2x + y = 5 \Rightarrow y = 2x + 5$

$x^2 + y^2 - 25 = 0$

$x^2 + (2x+5)^2 - 25 = 0$

$x^2 + 4x^2 + 20x + 25 - 25 = 0$

$5x^2 + 20x = 0$

$5x(x+4) = 0$

$5x = 0,\ x + 4 = 0$

$x = 0,\ x = -4$

$y = 2x + 5 = 2(0) + 5 = 5$

$y = 2(-4) + 5 = -3$

$(0,5)$, $(-4,-3)$ is the solution.

14. $x^2 + y^2 = 9$

$y = x - 3$

$x^2 + (x-3)^2 = 9$

$x^2 + x^2 - 6x + 9 = 9$

$x^2 - 3x = 0$

$x(x-3) = 0$

$x = 0,\ x = 3$

$y = x - 3 = 0 - 3 = -3$

$y = x - 3 = 3 - 3 = 0$

$(0,-3)$, $(3,0)$ is the solution.

15. $4x^2 + y^2 - 4 = 0 \Rightarrow y^2 = 4 - 4x^2$

$9x^2 - 4y^2 - 9 = 0$

$9x^2 - 4(4 - 4x^2) - 9 = 0$

$9x^2 - 16 + 16x^2 - 9 = 0$

$25x^2 = 25$

$x^2 = 1$

$x = \pm 1$

$y = 4 - 4x^2 = 4 - 4(1) = 0$

$(1,0)$, $(-1,0)$ is the solution.

16. $2x^2 + y^2 = 9$

$xy = 3 \Rightarrow y = -\frac{3}{x}$

$2x^2 + \left(-\frac{3}{x}\right)^2 = 9$

$2x^4 + 9 = 9x^2$

$2x^4 - 9x^2 + 9 = 0$

$(2x^2 - 3)(x^2 - 3) = 0$

$2x^2 - 3 = 0,\ x^2 - 3 = 0$

$x^2 = \frac{3}{2},\ x^2 = 3$

$x = \pm\sqrt{\frac{3}{2}\cdot\frac{2}{2}} = \pm\frac{\sqrt{6}}{2},\ x = \pm\sqrt{3}$

$y = -\frac{3}{x} = -\frac{3}{\frac{\sqrt{6}}{2}} = -\frac{6}{\sqrt{6}}\cdot\frac{\sqrt{6}}{\sqrt{6}} = -\sqrt{6}$

$y = -\frac{3}{x} = -\frac{3}{-\frac{\sqrt{6}}{2}} = \sqrt{6}$

$y = -\frac{3}{x} = -\frac{3}{\sqrt{3}}\cdot\frac{\sqrt{3}}{\sqrt{3}} = -\sqrt{3}$

$y = -\frac{3}{x} = -\frac{3}{-\sqrt{3}}\cdot\frac{\sqrt{3}}{\sqrt{3}} = \sqrt{3}$

$(\sqrt{3}, -\sqrt{3}),\ (-\sqrt{3}, \sqrt{3}),$

$\left(\frac{\sqrt{6}}{2}, -\sqrt{6}\right),\ \left(-\frac{\sqrt{6}}{2}, \sqrt{6}\right)$ is the solution.

Cumulative Test for Chapters 1-10

1. $2\{x - 3[x - 2(x+1)]\}$

$= 2\{x - 3[x - 2x - 2]\}$

$= 2\{x - 3[-x - 2]\} = 2\{x + 3x + 6\}$

$= 2\{4x + 6\} = 8x + 12$

2. $3(4-6)^3 + \sqrt{25} = 3(-2)^3 + \sqrt{25}$

$= 3(-8) + 5$

$= -24 + 5$

$= -19$

3. $A = 3bt + prt$

$prt = A - 3bt$

$p = \frac{A - 3bt}{rt}$

4. $m = \frac{4-2}{-\frac{1}{2} - \left(-\frac{1}{2}\right)} = \frac{2}{0}$, no slope

5. $25x^2 - 40xy + 16y^2 = (5x - 4y)^2$

6. $\frac{3}{x-4} + \frac{6}{x^2 - 16}$

$= \frac{3(x+4)}{(x-4)(x+4)} + \frac{6}{(x-4)(x+4)}$

$= \frac{3x + 12 + 6}{(x-4)(x+4)}$

$= \frac{3x + 18}{(x-4)(x+4)}$

7. $\frac{3}{2x+3} = \frac{1}{2x-3} + \frac{2}{4x^2 - 9}$

$\frac{3}{(2x+3)} = \frac{1}{2x-3} + \frac{2}{(2x+3)(2x-3)}$

$3(2x - 3) = 2x + 3 + 2$

$6x - 9 = 2x + 5$

$4x = 14$

$x = \frac{7}{2}$

8. $3x-2y-9z=9$
$x-y+z=8$
$2x+3y-z=-2$
Switch the first and second equation.
$x-y+z=8$
$3x-2y-9z=9$
$2x+3y-z=-2$
Multiply the first equation by -3 and add to the second equation and multiply the first equation by -2 and add to the third equation
$x-y+z=8$
$y-12z=-15$
$5y-3z=-18$
Multiply the second equation by -5 and add to the third equation
$x-y+z=8$
$y-12z=-15$
$57z=57$
$z=1$
$y=12z-15=12(1)-15$
$y=-3$
$x-y+z=8$
$x-(-3)+1=8$
$x=4$
$(4,-3,1)$ is the solution.

9. $(\sqrt{2}+\sqrt{3})(2\sqrt{6}-\sqrt{3})$
$=2\sqrt{12}-\sqrt{6}+2\sqrt{18}-3$
$=2\sqrt{4\cdot 3}+2\sqrt{9\cdot 2}-\sqrt{6}-3$
$=4\sqrt{3}+6\sqrt{2}-\sqrt{6}-3$

10. $\sqrt{8x}+3x\sqrt{50}-4x\sqrt{32}$
$=\sqrt{4\cdot 2x}+3x\sqrt{25\cdot 2}-4x\sqrt{16\cdot 2}$
$=2\sqrt{2x}+15x\sqrt{2}-16x\sqrt{2}$
$=2\sqrt{2x}-x\sqrt{2}$

11. $2x+(4x-1)>6-x$
$2x+4x-1>6-x$
$6x-1>6-x$
$7x>7$
$x>1$

12. $\frac{6(x-4)}{5}\ge\frac{3(x+2)}{4}$
$24(x-4)\ge 15(x+2)$
$24x-96\ge 15x+30$
$9x\ge 126$
$x\ge 14$

13. $d=\sqrt{(x_2-x_1)^2+(y_2-y_1)^2}$
$d=\sqrt{(-3-6)^2+(-4-(-1))^2}=\sqrt{90}$
$d=3\sqrt{10}$

14. $y=-\frac{1}{2}(x+2)^2-3$
Parabola: $C(-2,-3)$
y-intercept: $(0,-5)$

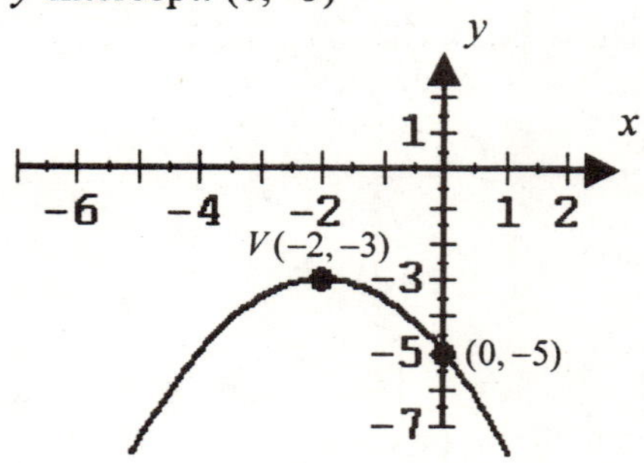

15. $25x^2+25y^2=125$
$x^2+y^2=5=\sqrt{5}^2$
Circle: $C(0,0)$, $r=\sqrt{5}$

(continued)

15. (continued)

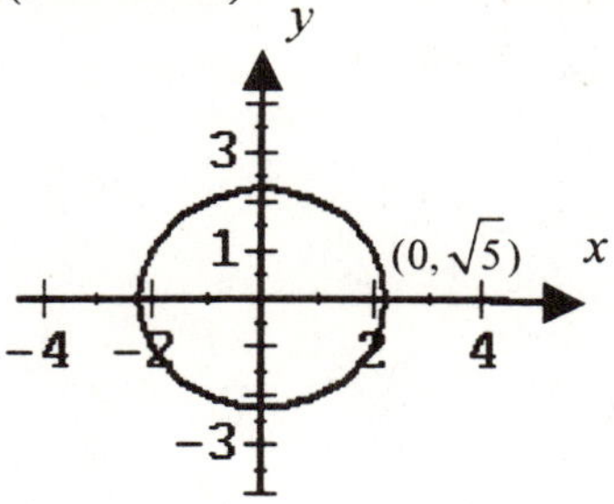

16. $16x^2 - 4y^2 = 64 \Rightarrow \frac{x^2}{4} - \frac{y^2}{16} = 1$

$\frac{x^2}{2^2} - \frac{y^2}{4^2} = 1$

Hyperbola: $C(0,0)$

$a = 2,\ b = 4$

Vertices: $(-2,0),\ (2,0)$

$y_{\text{asymptote}} = \pm\frac{4}{2}x = \pm 2x$

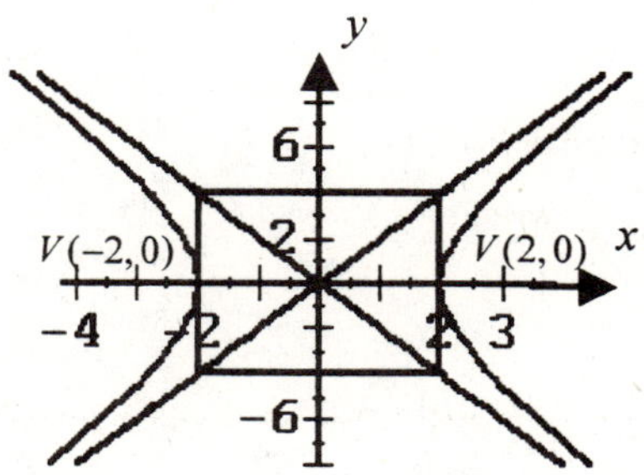

17. $\frac{(x-2)^2}{25} + \frac{(y-3)^2}{16} = 1$

$\frac{(x-2)^2}{5^2} + \frac{(y-3)^2}{4^2} = 1$

Ellipse: $C(2,3)$

$a = 5,\ b = 3$

Vertices: $(2,7),(-3,3),(2,-1),(7,3)$

(continued)

17. (continued)

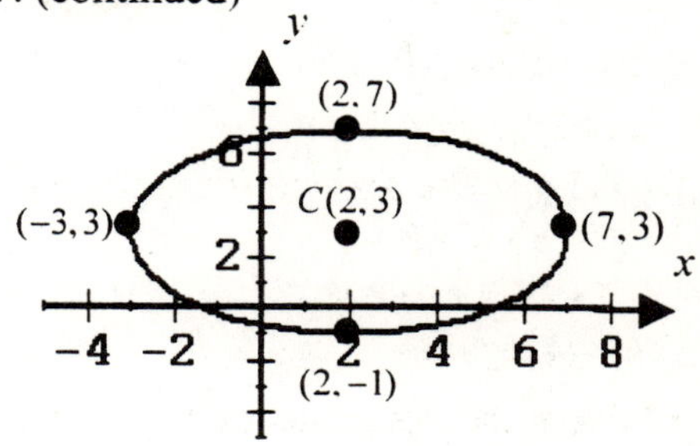

18.

$$y = 2x^2$$
$$y = 2x + 4$$
$$2x^2 = 2x + 4$$
$$x^2 - x - 2 = 0$$
$$(x-2)(x+1) = 0$$
$$x - 2 = 0,\ x + 1 = 0$$
$$x = 2,\ x = -1$$
$$y = 2x^2 = 2(2)^2 = 8$$
$$y = 2(-1)^2 = 2$$

$(2,8),\ (-1,2)$ is the solution.

19. $x^2 + 2y^2 = 16 \Rightarrow x^2 = 16 - 2y^2$

$4x^2 - y^2 = 24$

$4(16 - 2y^2) - y^2 = 24$

$64 - 8y^2 - y^2 = 24$

$9y^2 = 40$

$y = \pm\frac{2\sqrt{10}}{3}$

$x^2 = 16 - 2y^2 = 16 - 2\left(\frac{40}{9}\right) = \frac{64}{9}$

$x = \pm\frac{8}{3}$

$\left(\frac{8}{3}, \pm\frac{2\sqrt{10}}{3}\right),\ \left(-\frac{8}{3}, \pm\frac{2\sqrt{10}}{3}\right)$ is the solution.

20. $x^2 + y^2 = 25$

$x - 2y = -5 \Rightarrow x = 2y - 5$

$(2y-5)^2 + y^2 = 25$

$4y^2 - 20y + 25 + y^2 = 25$

$5y^2 - 20y = 0$

$5y(y-4) = 0$

$5y = 0,\ y - 4 = 0$

$y = 0,\ y = 4$

$x = 2y - 5 = 2(0) - 5 = -5$

$x - 2y - 5 = 2(4) - 5 = 3$

$(3,4),\ (-5,0)$ is the solution

21. $xy = -15 \Rightarrow y = -\dfrac{15}{x}$

$4x + 3y = 3$

$4x + 3\left(-\dfrac{15}{x}\right) = 3$

$4x^2 - 45 = 3x$

$4x^2 - 3x - 45 = 0$

$(x+3)(4x-15) = 0$

$x + 3 = 0,\ 4x - 15 = 0$

$x = -3,\ x = \dfrac{15}{4}$

$y = -\dfrac{15}{x} = -\dfrac{15}{-3} = 5$

$y = -\dfrac{15}{x} = -\dfrac{15}{\frac{15}{4}} = -4$

$(-3,5),\ \left(\dfrac{15}{4}, -4\right)$ is the solution

Chapter 11

Pretest Chapter 11

1. $f(x) = 2x - 6$

   (a) $f(-3) = 2(-3) - 6 = -12$

   (b) $f(a) = 2a - 6$

   (c) $f(2a) = 2(2a) - 6 = 4a - 6$

   (d) $f(a+2) = 2(a+2) - 6 = 2a - 2$

2. $f(x) = 5x^2 + 2x - 3$

   (a) $f(-2) = 5(-2)^2 + 2(-2) - 3 = 13$

   (b) $f(a) = 5a^2 + 2a - 3$

   (c) $f(a+1) = 5(a+1)^2 + 2(a+1) - 3$

   $= 5(a^2 + 2a + 1) + 2a + 2 - 3$

   $= 5a^2 + 10a + 5 + 2a - 1$

   $= 5a^2 + 12a + 4$

3. $f(x) = \dfrac{3x}{x+2}$

   (a) $f(a) + f(a-2)$

   $$= \frac{3a}{a+2} + \frac{3(a-2)}{a-2+2} = \frac{3a^2 + 3(a+2)(a-2)}{a(a+2)}$$

   $$= \frac{3a^2 + 3a^2 - 12}{a(a+2)} = \frac{6a^2 - 12}{a(a+2)} = \frac{6(a^2 - 2)}{a(a+2)}$$

   (b) $f(3a) - f(3) = \dfrac{3(3a)}{3a+2} - \dfrac{3(3)}{3+2}$

   $$= \frac{9a}{3a+2} - \frac{9}{5}$$

   $$= \frac{45a - 9(3a+2)}{5(3a+2)}$$

   $$= \frac{45a - 27a - 18}{5(3a+2)}$$

   $$= \frac{18(a-1)}{5(3a+2)}$$

4. Graph passes vertical line test and therefore represents a function.

5. Graph does not pass vertical line test and hence does not represent a function.

6. $f(x) = |x|$, $s(x) = |x - 3|$

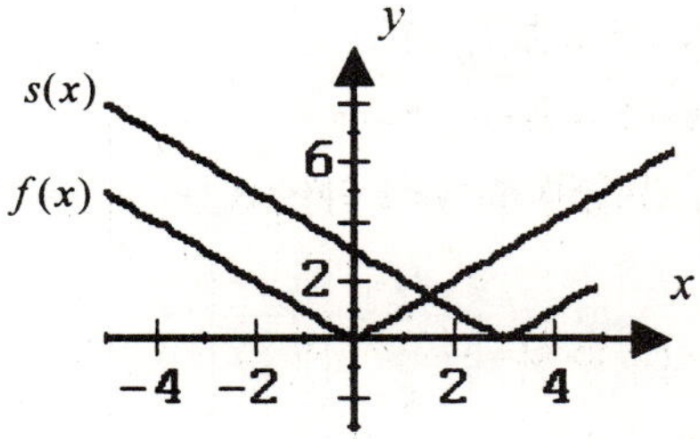

7. $f(x) = x^2$, $h(x) = (x+2)^2 + 3$

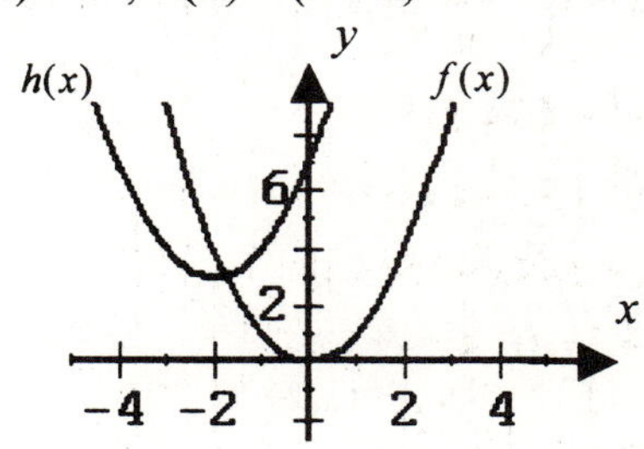

8. $f(x) = \dfrac{2}{x+6}$, $g(x) = -3x + 1$

   (a) $(fg)(x) = f(x)g(x)$

   $$= \frac{2}{x+6}(-3x+1)$$

   $$= \frac{-6x+2}{x+6}$$

   (b) $(fg)(-4) = \dfrac{-6(-4)+2}{-4+6} = 13$

   (c) $f[g(x)] = f(-3x+1)$

   $$= \frac{2}{-3x+1+6} = \frac{2}{-3x+7}, x \neq \frac{7}{3}$$

9. $f(x)=3x-4,\ g(x)=-2x^2-6x+3$

   (a) $(f+g)(x)=f(x)+g(x)$
   $$=3x-4-2x^3-6x+3$$
   $$=-2x^3-3x-1$$

   (b) $(f+g)(2)=-2(2)^3-3(2)-1=-23$

   (c) $f[g(x)]=f(-2x^3-6x+3)$
   $$=3(-2x^3-6x+3)-4$$
   $$=-6x^3-18x+9-4$$
   $$=-6x^3-18x+5$$

10. $f(x)=6x^2-5x-4,\ g(x)=3x-4$

   (a) $\left(\dfrac{f}{g}\right)(x)=\dfrac{f(x)}{g(x)}=\dfrac{6x^2-5x-4}{3x-4}$
   $$=\frac{(3x-4)(2x+1)}{(3x-4)}$$
   $$=2x+1,\ x\neq\frac{4}{3}$$

   (b) $\left(\dfrac{f}{g}\right)(-1)=2(-1)+1=-1$

   (c) $(f\circ g)(x)=f[g(x)]=f(3x-4)$
   $$=6(3x-4)^2-5(3x-4)-4$$
   $$=6(9x^2-24x+16)-15x+20-4$$
   $$=54x^2-144x+96+16$$
   $$=54x^2-159x+112$$

   (d) $(g\circ f)(x)=g[f(x)]$
   $$=g(6x^2-5x-4)$$
   $$=3(6x^2-5x-4)-4$$
   $$=18x^2-15x-12-4$$
   $$=18x^2-15x-16$$

11. Graph passes vertical line test and horizontal line test. Graph represents a one-to-one function.

12. Graph passes vertical line test but does not pass horizontal line test. Graph represents a function but not a one-to-one function.

13. Since not two ordered pairs have the same second coordinate, A is a one-to-one function.

14. $F=\{(7,1),(6,3),(2,-1),(-1,5)\}$
   $F^{-1}=\{(1,7),(3,6),(-1,2),(5,-1)\}$

15. $g(x)=3-5x$
   $$y=3-5x$$
   $$x=3-5y$$
   $$y=\frac{3-x}{5}$$
   $$g^{-1}(x)=\frac{3-x}{5}$$

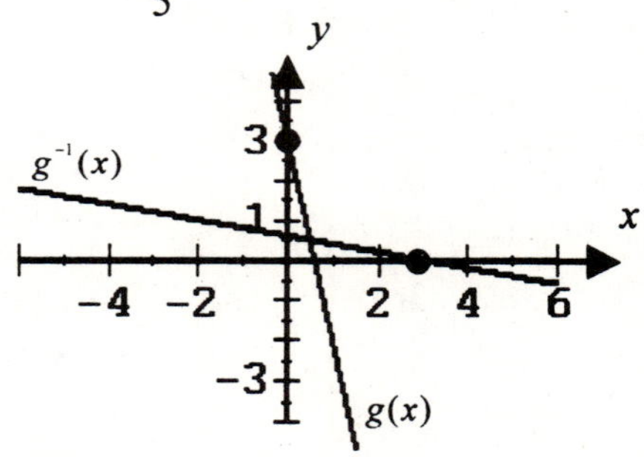

11.1 Exercises

1. $f(x)=3x-5$
   $$f\left(-\frac{2}{3}\right)=3\left(-\frac{2}{3}\right)-5=-7$$

3. $f(x)=3x-5$
   $$f(a-4)=3(a-4)-5=3a-12-5$$
   $$=3a-17$$

5. $g(x)=\frac{1}{2}x-3$

$g(4)+g(a)=\frac{1}{2}(4)-3+\frac{1}{2}a-3$

$=\frac{1}{2}a-4$

7. $g(x)=\frac{1}{2}x-3$

$g(2a)=\frac{1}{2}(2a)-3=a-3$

9. $g(x)=\frac{1}{2}x-3$

$g(2a-4)=\frac{1}{2}(2a-4)-3=a-2-3$

$=a-5$

11. $g(x)=\frac{1}{2}x-3$

$g(a^2)-g\left(\frac{2}{5}\right)=\frac{1}{2}a^2-3-\left(\frac{1}{2}\cdot\frac{2}{5}-3\right)$

$=\frac{1}{2}a^2-3-\frac{1}{5}+3$

$=\frac{1}{2}a^2-\frac{1}{5}$

13. $p(x)=3x^2+4x-2$

$p(-2)=3(-2)^2+4(-2)-2=2$

15. $p(x)=3x^2+4x-2$

$p\left(\frac{1}{2}\right)=3\left(\frac{1}{2}\right)^2+4\left(\frac{1}{2}\right)-2=\frac{3}{4}$

17. $p(x)=3x^2+4x-2$

$p(a+1)=3(a+1)^2+4(a+1)-2$

$=3a^2+6a+3+4a+4-2$

$=3a^2+10a+5$

19. $p(x)=3x^2+4x-2$

$p\left(-\frac{a^2}{2}\right)=3\left(-\frac{a^2}{2}\right)^2+4\left(-\frac{a^2}{2}\right)-2$

$=\frac{3a^4}{4}-2a^2-2$

21. $h(x)=\sqrt{x+5}$

$h(-1)=\sqrt{-1+5}=\sqrt{4}=2$

23. $h(x)=\sqrt{x+5}$

$h(3)=\sqrt{3+5}=\sqrt{8}=2\sqrt{2}$

25. $h(x)=\sqrt{x+5}$

$h(a^2-1)=\sqrt{a^2-1+5}=\sqrt{a^2+4}$

27. $h(x)=\sqrt{x+5}$

$h(3a)=\sqrt{3a+5}$

29. $h(x)=\sqrt{x+5}$

$h(4a-1)=\sqrt{4a-1+5}=\sqrt{4a+4}$

$=2\sqrt{a+1}$

31. $h(x)=\sqrt{x+5}$

$h(b^2+b)=\sqrt{b^2+b+5}$

33. $r(x)=\frac{7}{x-3}$

$r(7)=\frac{7}{7-3}$

$r(7)=\frac{7}{4}$

35. $r(x)=\frac{7}{x-3}$

$r(1.5)=\frac{7}{1.5-3}=-4.\bar{6}$

37. $r(x) = \dfrac{7}{x-3}$

$r(a^2) = \dfrac{7}{a^2-3}$

39. $r(x) = \dfrac{7}{x-3}$

$r(a+2) = \dfrac{7}{a+2-3} = \dfrac{7}{a-1}$

41. $r(x) = \dfrac{7}{x-3}$

$r\left(\dfrac{1}{2}\right) + r(8) = \dfrac{7}{\frac{1}{2}-3} + \dfrac{7}{8-3}$

$= -\dfrac{7}{5}$

43. $f(x) = 5-2x$

$\dfrac{f(x+h)-f(x)}{h}$

$= \dfrac{5-2(x+h)-(5-2x)}{h}$

$= \dfrac{5-2x-2h-5+2x}{h}$

$= -2$

45. $f(x) = 2x^2$

$\dfrac{f(x+h)-f(x)}{h}$

$= \dfrac{2(x+h)^2-2x^2}{h}$

$= \dfrac{2x^2+4xh+2h^2-2x^2}{h}$

$= \dfrac{2h(2x+h)}{h}$

$= 2(2x+2h)$

$= 4x+2h$

47. $P = 2.5w^2$

(a) $P(w) = 2.5w^2$

(b) $P(20) = 2.5(20)^2 = 1000$ kilowatts

(c) $P = 2.5(20+e)^2 = 2.5(400+40e+e^2)$

$P(e) = 2.5e^2 + 100e + 1000$

(d) $P(2) = 2.5(2)^2 + 100(2) + 1000$

$= 1210$ kilowatts

49. The function values associated with $p(x)-13$ would be the function values of $p(x)$ decreased by 13.

$p(3)-13 \approx 39-13 = 26$

51. $f(x) = 3x^2 - 4.6x + 1.23$

$f(3.56a) = 3(3.56a)^2 - 4.6(3.56a) + 1.23$

$= 38.021a^2 - 16.376a + 1.23$

53. $f(x) = 3x^2 - 4.6x + 1.23$

$f(a-0.152)$

$= 3(a-0.152)^2 - 4.6(a-0.152) + 1.23$

$= 3a^2 - 0.912a + 0.069312 - 4.6a + 0.6992 + 1.23$

$= 3a^2 - 5.512a + 1.999$

55. $A(x) = \left(\dfrac{x}{4}\right)^2 + \left(\dfrac{20-x}{4}\right)^2$

$= \dfrac{x^2+400-40x+x^2}{16}$

$= \dfrac{2x^2-40x+400}{16}$

$A(2) = \dfrac{2(2)^2-40(2)+400}{16} = 20.5$

$A(5) = \dfrac{2(5)^2-40(5)+400}{16} = 15.625$

$A(8) = \dfrac{2(8)^2-40(8)+400}{16} = 13$

Cumulative Review Problems

57. $\frac{7}{6}+\frac{5}{x}=\frac{3}{2x}$
$7x+30=9$
$7x=-21$
$x=-3$

59. $\frac{V_{\text{Earth}}}{V_{\text{Mercury}}}=\frac{\frac{4}{3}\pi\left(\frac{7927}{2}\right)^3}{\frac{4}{3}\pi\left(\frac{3031}{2}\right)^3}$
$V_{\text{Earth}}=17.88828747...(V_{\text{Mercury}})$

The volume of the Earth is approximately 17.9 times greater than the volume of Mercury.

11.2 Exercises

1. No, $f(x+2)$ means substitute $x+2$ for x in the function, $f(x)$. $f(x)+f(2)$ means evaluate $f(x)$ and $f(2)$ and then add the two results. One example is
$f(x)=2x+1$
$f(x+2)=2(x+2)+1=2x+5$
$f(x)+f(2)=2x+1+2(2)+1=2x+6$

3. To obtain the graph of $f(x)+k$, shift the graph of $f(x)$ up k units.

5. Graph fails vertical line test and does not represent a function.

7. Graph passes vertical line test and does represent a function.

9. Graph passes vertical line test and does represent a function.

11. Graph fails vertical line test and does not represent a function.

13. Graph fails vertical line test and does not represent a function.

For Exercises 15, 17, and 19:

x	$f(x)=x^2$
−2	4
−1	1
0	0
1	1
2	4

15. $f(x)=x^2$, $h(x)=x^2+4$
Shift $f(x)$ up 4 units.

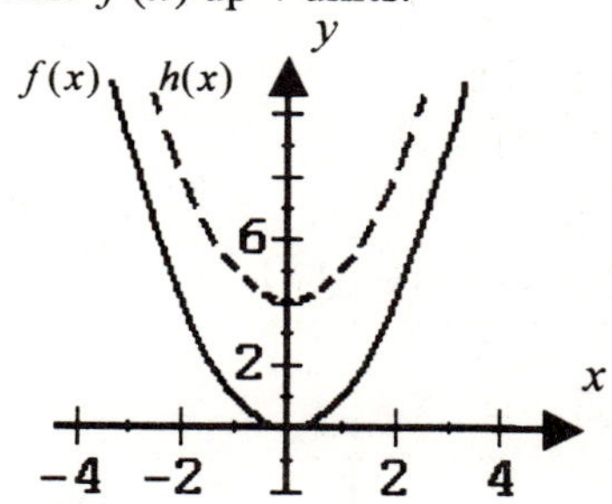

17. $f(x)=x^2$, $p(x)=(x+1)^2$
Shift $f(x)$ left 1 unit.

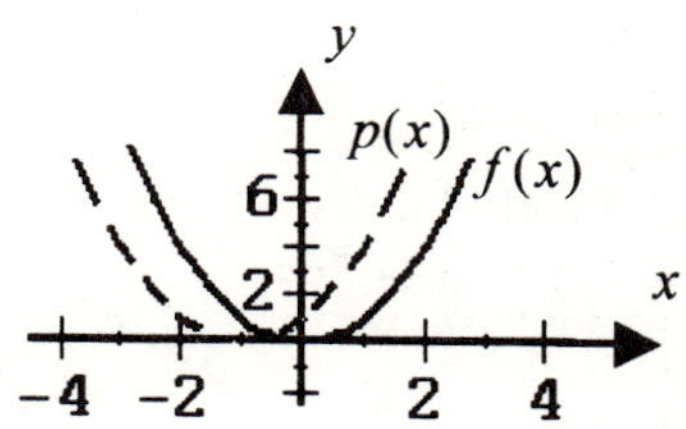

19. $f(x)=x^2$, $g(x)=(x-2)^2+1$
    Shift $f(x)$ right 2 and up 1 units.

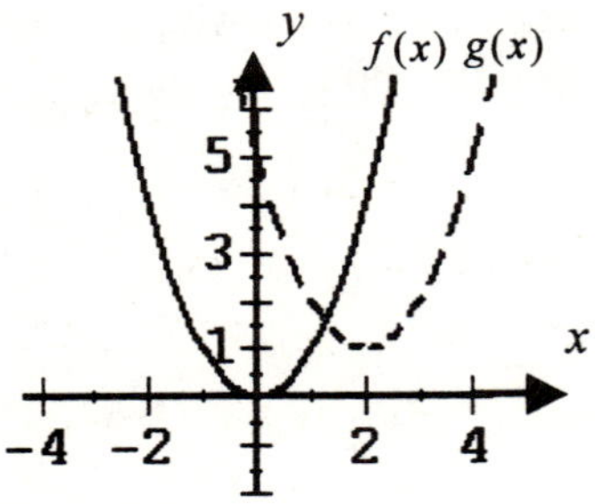

For Exercises 21, 23, and 25:

x	$f(x)=\|x\|$
−2	2
−1	1
0	0
1	1
2	2

21. $f(x)=|x|$, $r(x)=|x|-1$
    Shift $f(x)$ down 1 unit.

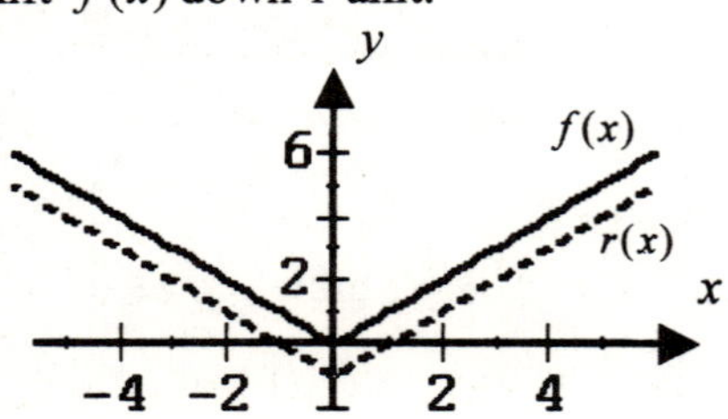

23. $f(x)=|x|$, $s(x)=|x+4|$
    Shift $f(x)$ left 4 units.

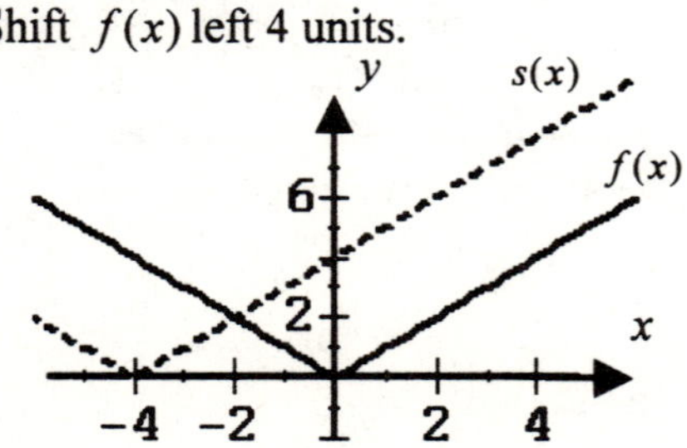

25. $f(x)=|x|$, $t(x)=|x-3|-4$
    Shift right 3 and down 4 units.

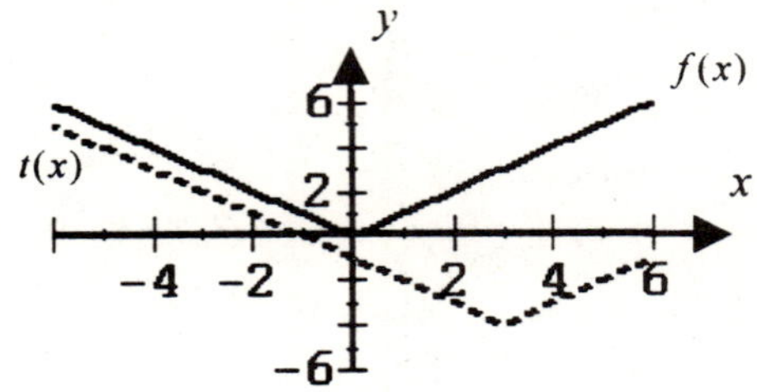

27. $f(x)=x^3$, $j(x)=(x-3)^3+3$
    Shift $f(x)$ right 3 and up 3 units.

x	$f(x)=x^3$
−2	−8
−1	−1
0	0
1	1
2	8

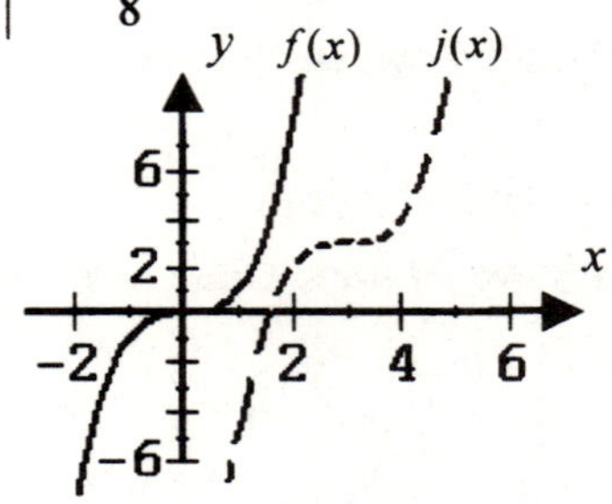

29. $f(x)=\dfrac{2}{x}$, $g(x)=\dfrac{2}{x}+3$
    Shift $f(x)$ up 3 units.

x	$f(x)=\dfrac{2}{x}$
−2	−1
−1	−2
0	undefined
1	2
2	1

(continued)

29. (continued)

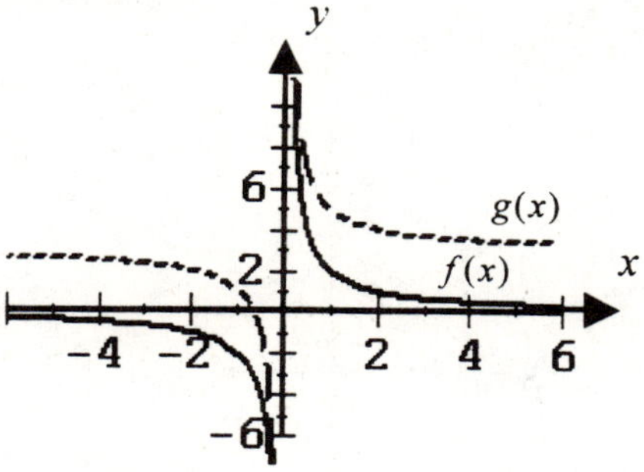

31. $f(x)=x^4$, $f(x)=(x-3.2)^4-2.6$

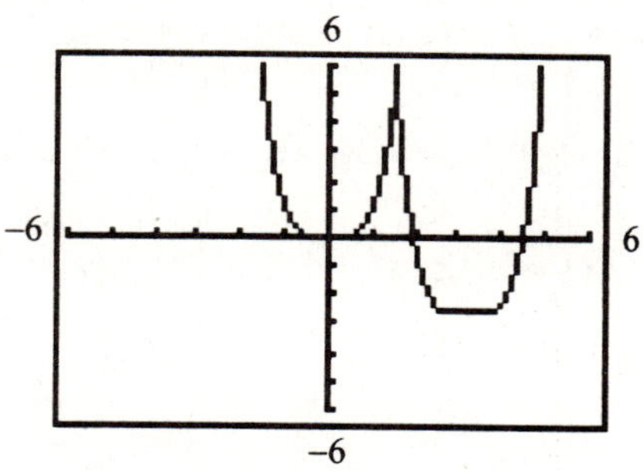

Cumulative Review Problems

33. $\sqrt{12}+3\sqrt{50}-4\sqrt{27}$
$=\sqrt{4\cdot 3}+3\sqrt{25\cdot 2}-4\sqrt{9\cdot 3}$
$=2\sqrt{3}+15\sqrt{2}-12\sqrt{3}$
$=15\sqrt{2}-10\sqrt{3}$

35. $\dfrac{2\sqrt{3}+\sqrt{5}}{\sqrt{3}-2\sqrt{5}}\cdot\dfrac{\sqrt{3}+2\sqrt{5}}{\sqrt{3}+2\sqrt{5}}$
$=\dfrac{6+4\sqrt{15}+\sqrt{15}+10}{3-20}$
$=-\dfrac{16+5\sqrt{15}}{17}$

37. $y=kx^2$, $36=k(3)^2\Rightarrow k=4$
$y=4x^2\big|_{x=5}=4(5)^2=100$

11.3 Exercises

1. $f(x)=-2x+3$, $g(x)=2+4x$
(a) $(f+g)(x)=f(x)+g(x)$
$=-2x+3+2+4x$
$=2x+5$
(b) $(f-g)(x)=f(x)-g(x)$
$=-2x+3-(2+4x)$
$=-2x+3-2-4x$
$=-6x+1$
(c) $(f+g)(2)=2(2)+5=9$
(d) $(f-g)(-1)=-6(-1)+1=7$

3. $f(x)=x^3-4x+5$, $g(x)=2x-1$
(a) $(f+g)(x)=f(x)+g(x)$
$=2x^2-4x+5+2x-1$
$=2x^2-2x+4$
(b) $(f-g)(x)=f(x)-g(x)$
$=2x^2-4x+5-(2x-1)$
$=2x^2-6x+6$
(c) $(f+g)(2)=2(2)^2-2(2)+4=8$
(d) $(f-g)(-1)=2(-1)^2-6(-1)+6=14$

5. $f(x)=x^3-\frac{1}{2}x^2+x$, $g(x)=x^2-\frac{x}{4}-5$
(a) $(f+g)(x)=f(x)+g(x)$
$=x^3-\frac{1}{2}x^2+x+x^2-\frac{x}{4}-5$
$=x^3+\frac{1}{2}x^2+\frac{3x}{4}-5$
(b)

(continued)

5. (continued)

(b) $(f-g)(x) = f(x) - g(x)$

$= x^3 - \frac{1}{2}x^2 + x - (x^2 - \frac{x}{4} - 5)$

$= x^3 - \frac{1}{2}x^2 + x - x^2 + \frac{x}{4} + 5$

$= x^3 - \frac{3}{2}x^2 + \frac{5x}{4} + 5$

(c) $(f+g)(2) = (2)^3 + \frac{1}{2}(2)^2 + \frac{3(2)}{4} - 5$

$= \frac{13}{2}$

(d) $(f-g)(-1) = (-1)^3 - \frac{3}{2}(-1)^2 + \frac{5(-1)}{4} + 5$

$= \frac{5}{4}$

7. $f(x) = 3\sqrt{3-x},\ g(x) = -5\sqrt{3-x}$

(a) $f(x) + g(x) = f(x) + g(x)$

$= 3\sqrt{3-x} + (-5\sqrt{3-x})$

$= -2\sqrt{3-x}$

(b) $f(x) - g(x) = 3\sqrt{3-x} - (-5\sqrt{3-x})$

$= 8\sqrt{3-x}$

(c) $(f+g)(2) = -2\sqrt{3-1} = -2$

(d) $(f-g)(-1) = 8\sqrt{3-(-1)} = 16$

9. $f(x) = x^2 - 3x + 2,\ g(x) = 1 - x$

(a) $(fg)(x) = f(x)g(x)$

$= (x^2 - 3x + 2)(1 - x)$

$= x^2 - 3x + 2 - x^3 + 3x^2 - 2x$

$= -x^3 + 4x^2 - 5x + 2$

(b) $(fg)(-3)$

$= -(-3)^3 + 4(-3)^2 - 5(-3) + 2$

$= 80$

11. $f(x) = \frac{2}{x^2},\ g(x) = x^2 - x$

(a) $(fg)(x) = f(x)g(x) = \frac{2}{x^2}(x^2 - x)$

$= 2 - \frac{2}{x} = \frac{2(x-1)}{x}$

(b) $(fg)(-3) = \frac{2(-3-1)}{-3} = \frac{8}{3}$

13. $f(x) = \sqrt{-2x+1},\ g(x) = -3x$

(a) $(fg)(x) = f(x)g(x) = \sqrt{-2x+1}(-3x)$

$= -3x\sqrt{-2x+1}$

(b) $(fg)(-3) = -3(-3)\sqrt{-2(-3)+1} = 9\sqrt{7}$

15. $f(x) = 3x,\ g(x) = 4x - 1$

(a) $\left(\frac{f}{g}\right)(x) = \frac{f(x)}{g(x)} = \frac{3x}{4x-1},\ x \neq \frac{1}{4}$

(b) $\left(\frac{f}{g}\right)(2) = \frac{3(2)}{4(2)-1} = \frac{6}{7}$

17. $f(x) = x^2 - 1,\ g(x) = x - 1$

(a) $\left(\frac{f}{g}\right)(x) = \frac{f(x)}{g(x)} = \frac{x^2-1}{x-1}$

$= \frac{(x-1)(x+1)}{x-1}$

$= x + 1,\ x \neq 1$

(b) $\left(\frac{f}{g}\right)(2) = 2 + 1 = 3$

19. $f(x) = x^2 + 10x + 25,\ g(x) = x + 5$

(a) $\left(\frac{f}{g}\right)(x) = \frac{f(x)}{g(x)} = \frac{x^2 + 10x + 25}{x+5}$

$= \frac{(x+5)(x+5)}{(x+5)}$

$= x + 5,\ x \neq -5$

19. (b) $\left(\frac{f}{g}\right)(2) = 2 + 5 = 7$

21. $f(x) = 4x - 1,\ g(x) = 4x^2 + 7x - 2$

(a) $\left(\frac{f}{g}\right)(x) = \frac{f(x)}{g(x)} = \frac{4x-1}{4x^2+7x-2}$

$= \frac{(4x-1)}{(4x-1)(x+2)}$

$= \frac{1}{x+2},\ x \neq -2, \frac{1}{4}$

(b) $\left(\frac{f}{g}\right)(2) = \frac{1}{2+2} = \frac{1}{4}$

23. $f(x) = 3x + 2,\ g(x) = x^2 - 2x$

$(f - g)(x) = f(x) - g(x)$

$= 3x + 2 - (x^2 - 2x)$

$= -x^2 + 5x + 2$

25. $f(x) = 3x + 2,\ g(x) = x^2 - 2x$

$(fg)(x) = f(x)g(x)$

$= (3x + 2)(x^2 - 2x)$

$= 3x^3 - 6x^2 + 2x^2 - 4x$

$= 3x^3 - 4x^2 - 4x$

27. $(fg)(x) = 3x^3 - 4x^2 - 4x$

$(fg)(-1) = 3(-1)^3 - 4(-1)^2 - 4(-1) = -3$

29. $f(x) = 3x + 2,\ h(x) = \frac{x-2}{3}$

$\left(\frac{f}{h}\right)(x) = \frac{f(x)}{h(x)}$

$= \frac{3x+2}{\frac{x-2}{3}}$

$= \frac{9x+6}{x-2},\ x \neq 2$

31. $f(x) = 2 - 3x,\ g(x) = 2x + 5$

$f[g(x)] = f[2x + 5]$

$= 2 - 3(2x + 5)$

$= 2 - 6x - 15$

$= -6x - 13$

33. $f(x) = 3x^2,\ g(x) = x - 4$

$f[g(x)] = f[x - 4]$

$= 3(x - 4)^2$

$= 3(x^2 - 8x + 16)$

$= 3x^2 - 24x + 48$

35. $f(x) = 4 - 3x,\ g(x) = 2x^2 - 1$

$f[g(x)] = f[2x^2 - 1]$

$= 4 - 3(2x^2 - 1)$

$= 4 - 6x^2 + 3$

$= -6x^2 + 7$

37. $f(x) = \frac{3}{x+1},\ g(x) = 2x - 1$

$f[g(x)] = f(2x - 1)$

$= \frac{3}{2x-1+1}$

$= \frac{3}{2x},\ x \neq 0$

39. $f(x) = |x + 3|,\ g(x) = 2x - 1$

$f[g(x)] = f[2x - 1]$

$= |2x - 1 + 3|$

$= |2x + 2|$ or

$= |2(x + 1)|$

$= |2||x + 1|$

$= 2|x + 1|$

41. $f(x) = x^2 + 2,\ g(x) = 3x + 5$

$$\begin{aligned} f[g(x)] &= f(3x+5) \\ &= (3x+5)^2 + 2 \\ &= 9x^2 + 30x + 25 + 2 \\ &= 9x^2 + 30x + 27 \end{aligned}$$

43. $f(x) = x^2 + 2,\ g(x) = 3x + 5$

$$\begin{aligned} g[f(x)] &= g[x^2 + 2] \\ &= 3(x^2 + 2) + 5 \\ &= 3x^2 + 6 + 5 \\ &= 3x^2 + 11 \end{aligned}$$

45. From Exercise 43,

$$\begin{aligned} g[f(x)] &= 3x^2 + 11 \\ g[f(3)] &= 3(3)^2 + 11 \\ &= 38 \end{aligned}$$

47. $p(x) = \sqrt{x-1},\ f(x) = x^2 + 2$

$$\begin{aligned} (p \circ f)(x) &= p[f(x)] \\ &= p[x^2 + 2] \\ &= \sqrt{x^2 + 2 - 1} \\ &= \sqrt{x^2 + 1} \end{aligned}$$

49. $g(x) = 3x + 5,\ h(x) = \dfrac{1}{x}$

$$\begin{aligned} (g \circ h)(\sqrt{2}) &= g\left[h(\sqrt{2})\right] \\ &= g\left[\frac{1}{\sqrt{2}}\right] \\ &= 3 \cdot \frac{1}{\sqrt{2}} \cdot \frac{\sqrt{2}}{\sqrt{2}} + 5 \\ &= \frac{3\sqrt{2}}{2} + 5 \end{aligned}$$

51. $p(x) = \sqrt{x-1},\ f(x) = x^2 + 2$

$$\begin{aligned} (p \circ f)(-5) &= p[f(-5)] \\ &= p\left[(-5)^2 + 2\right] \\ &= p(27) \\ &= \sqrt{27 - 1} \\ &= \sqrt{26} \end{aligned}$$

53. $$\begin{aligned} K[C(F)] &= K\left[\frac{5F - 160}{9}\right] \\ &= \frac{5F - 160}{9} + 273 \\ &= \frac{5F - 160 + 9(273)}{9} \\ &= \frac{5F + 2297}{9} \end{aligned}$$

55. $$\begin{aligned} v[r(h)] &= v[3.5h] \\ &= 31.4(3.5h)^2 \\ &= 384.65h^2 \end{aligned}$$

$$384.65h^2\Big|_{h=8} = 24{,}617.6 \text{ ft}^3$$

Cumulative Review Problems

57. $$\begin{aligned} 6ab - 3ac + 2b - c &= 3a(2b - c) + (2b - c) \\ &= (3a + 1)(2b - c) \end{aligned}$$

59. $3x^2 - 7x + 2 = (3x - 1)(x - 2)$

61. $x =$ number of 60-sec commercials

$$\begin{aligned} 60x + 30(20 - x) &= 14(60) \\ 60x + 600 - 30x &= 840 \\ 30x &= 240 \\ x &= 8 \text{ 60-sec commercials} \\ 20 - x &= 12 \text{ 30-sec commercials} \end{aligned}$$

11.4 Exercises

1. A one-to-one function is a function in which no ordered pairs <u>have the same second coordinate</u>.

3. The graphs of a function f and its inverse f^{-1} are symmetric about the line $\underline{y = x}$.

5. $B = \{(0,1),(1,0),(10,0)\}$ is not one-to-one since two ordered pairs, $(1,0)$ and $(10,0)$, have the same second coordinate.

7. $F = \left\{\left(\frac{2}{3},2\right),\left(3,-\frac{4}{5}\right),\left(-\frac{2}{3},-2\right),\left(-3,\frac{4}{5}\right)\right\}$ is a one-to-one function since no two ordered pairs have the same second coordinate.

9. $E = \left\{(1,3),\left(\frac{1}{2},-5\right),(-1,-3),\left(-5,\frac{1}{2}\right)\right\}$ is a one-to-one function since no two ordered pairs have the same second coordinate.

11. Graph of function passes horizontal line test and therefore, function is one-to-one.

13. Graph of function does not pass horizontal line test and therefore, function is not one-to-one.

15. Graph of function passes horizontal line test and therefore, function is one-to-one.

17. Yes, it passes the vertical line test. No, it does not pass the horizontal line test.

19. $J = \{(8,2),(1,1),(0,0),(-8,-2)\}$

$J^{-1} = \{(2,8),(1,1),(0,0),(-2,-8)\}$

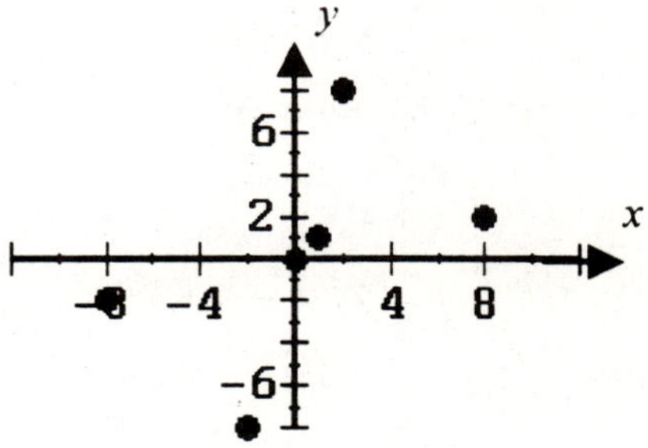

21.
$$f(x) = 4x - 5,\ f(x) \to y$$
$$y = 4x - 5,\ x \leftrightarrow y$$
$$x = 4y - 5$$
$$4y = x + 5$$
$$y = \frac{x+5}{4}$$
$$f^{-1}(x) = \frac{x+5}{4}$$

23.
$$f(x) = x^3 - 2,\ f(x) \to y$$
$$y = x^3 - 2,\ x \leftrightarrow y$$
$$x = y^3 - 2$$
$$y^3 = x + 2$$
$$y = \sqrt[3]{x+2}$$
$$f^{-1}(x) = \sqrt[3]{x+2}$$

25.
$$f(x) = -\frac{4}{x},\ f(x) \to y$$
$$y = -\frac{4}{x},\ x \leftrightarrow y$$
$$y = -\frac{4}{x}$$
$$f^{-1}(x) = -\frac{4}{x}$$

27. $f(x) = -\dfrac{3}{x-2},\ f(x) \to y$

$y = -\dfrac{3}{x-2},\ x \leftrightarrow y$

$x = -\dfrac{3}{y-2}$

$y - 2 = -\dfrac{3}{x}$

$y = 2 - \dfrac{3}{x}$

$f^{-1}(x) = 2 - \dfrac{3}{x} = \dfrac{2x-3}{x}$

29. No. $f(x) = 2x^2 + 3$ is vertical parabola and fails the horizontal line test; it is not one-to-one and therefore, does not have an inverse.

31. $g(x) = 2x + 5,\ g(x) \to y$

$y = 2x + 5,\ x \leftrightarrow y$

$x = 2y + 5$

$2y = x - 5$

$y = \dfrac{x-5}{2}$

$g^{-1}(x) = \dfrac{x-5}{2}$

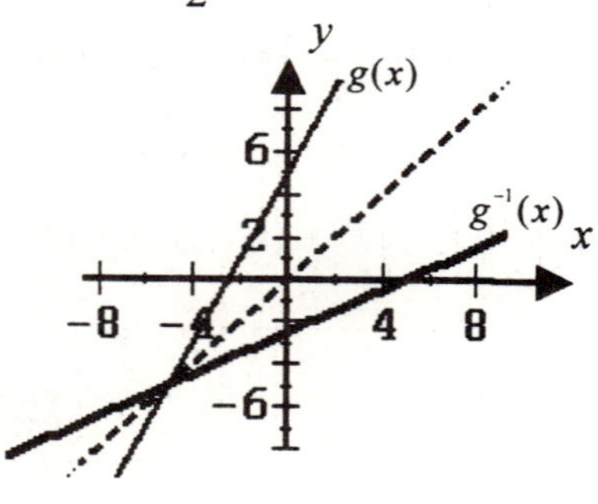

$g(x)$ and $g^{-1}(x)$ are symmetrical about the line $y = x$.

33. $h(x) = \dfrac{1}{2}x - 2,\ h(x) \to y$

$y = \dfrac{1}{2}x - 2,\ x \leftrightarrow y$

$x = \dfrac{1}{2}y - 2$

$2x = y - 4$

$y = 2x + 4$

$h^{-1}(x) = 2x + 4$

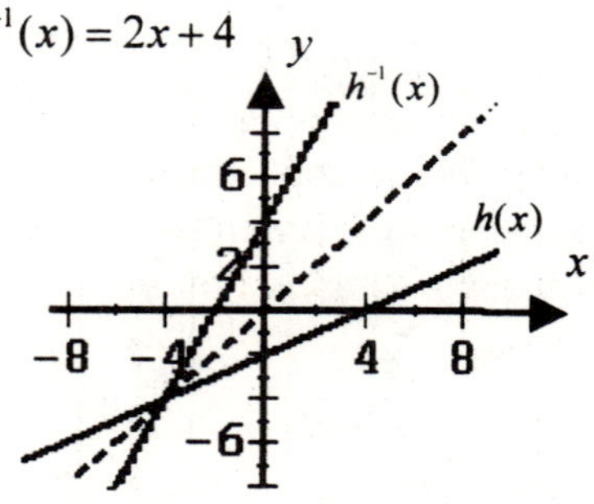

$h(x)$ and $h^{-1}(x)$ are symmetrical about the line $y = x$.

35. $k(x) = 3 - 2x,\ k(x) \to y$

$y = 3 - 2x,\ x \leftrightarrow y$

$x = 3 - 2y$

$2y = 3 - x$

$y = \dfrac{3-x}{2}$

$k^{-1}(x) = \dfrac{3-x}{2}$

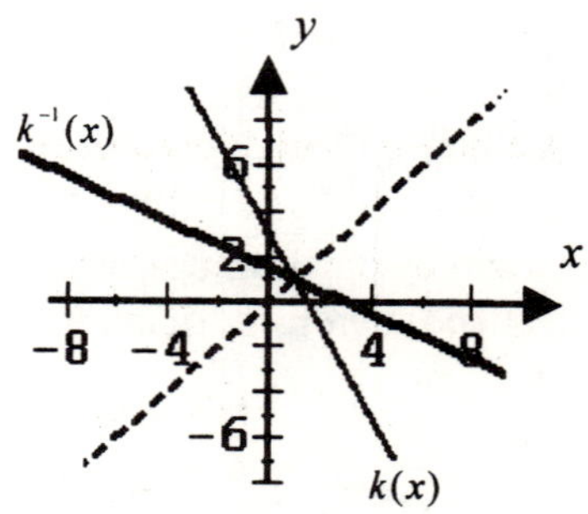

37. $f(x) = 0.0063x - 5,\ f(x) \to y$

$y = 0.0063x - 5,\ x \leftrightarrow y$

$x = 0.0063y - 5$

$0.0063y = x + 5$

$y = \dfrac{x+5}{0.0063}$

$f^{-1}(x) = \dfrac{x+5}{0.0063}$

The inverse function tells how many Spanish pesetas are given by the bank for *x* dollars. No, because of the bank fee the inverse function will not work for Manuela's transaction.

39. $f\left[f^{-1}(x)\right] = f\left(\frac{1}{2}x - \frac{3}{4}\right)$

$= 2\left(\frac{1}{2}x - \frac{3}{4}\right) + \frac{3}{2}$

$= x - \frac{3}{2} + \frac{3}{2}$

$= x$

$f^{-1}\left[f(x)\right] = f^{-1}\left(2x + \frac{3}{2}\right)$

$= \frac{1}{2}\left(2x + \frac{3}{2}\right) - \frac{3}{4}$

$= x + \frac{3}{4} - \frac{3}{4}$

$= x$

Cumulative Review Problems

41. $x^{2/3} + 7x^{1/3} + 12 = 0$

$(x^{1/3} + 4)(x^{1/3} + 3) = 0$

$x^{1/3} + 4 = 0,\ x^{1/3} + 3 = 0$

$x^{1/3} = -4,\quad x^{1/3} = -3$

$x = -64,\quad x = -27$

43. ratio $= \dfrac{4.6}{4.2} = \dfrac{23}{21}$

45. $\dfrac{1}{16}(12{,}800{,}000) = 800{,}000$ people

Putting Your Skills to Work

1. $p(x) = x^3 - 80x^2 + 1900x + 2000$

x	0	10	20	30
y	2000	14,000	16,000	14,000

x	40	50	60
y	14,000	22,000	44,000

2. 1960 corresponds to $x = 20$ and 1980 corresponds to $x = 40$. From the graph, the smallest number of cubic feet of lumber produced during this period appears to be sometime in 1975 or 1976.

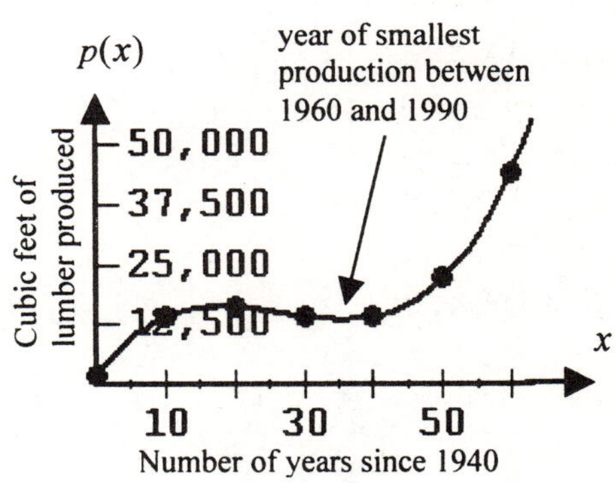

3. $p(x) = -x^4 + 15x^3 - 66x^2 + 80x + 50$

x	0	1	2	3	4	5	6	7	8
$p(x)$	50	78	50	20	18	50	98	120	50

From the table, 78 per hour at 12 noon and 120 per hour at 6 P.M.

4. From the graph, the hourly demand for Big Macs is lowest at $x = 3.5$ which corresponds to 2:30 P.M.

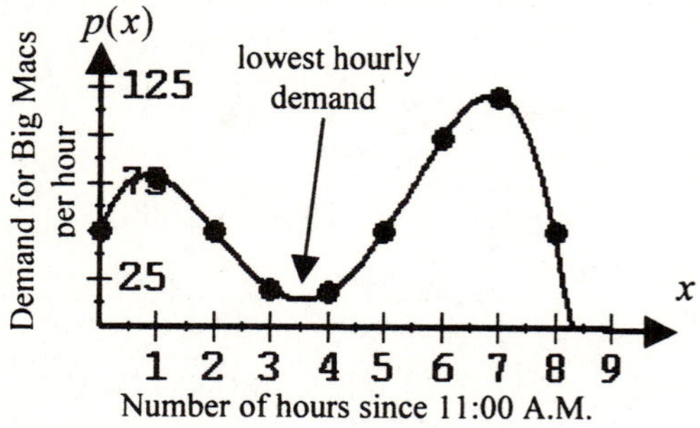

Chapter 11 Review Problems

1. $f(x) = \frac{1}{2}x + 3$

$$f(a-1) = \frac{1}{2}(a-1) + 3 = \frac{1}{2}a + \frac{5}{2}$$

2. $f(x) = \frac{1}{2}x + 3$

$$f(a+2) = \frac{1}{2}(a+2) + 3 = \frac{1}{2}a + 4$$

3. $f(x) = \frac{1}{2}x + 3$

$$\begin{aligned} f(a-1) - f(a) &= \frac{1}{2}(a-1) + 3 - \left(\frac{1}{2}a + 3\right) \\ &= \frac{1}{2}a - \frac{1}{2} + 3 - \frac{1}{2}a - 3 \\ &= -\frac{1}{2} \end{aligned}$$

4. $f(x) = \frac{1}{2}x + 3$

$$\begin{aligned} f(a+2) - f(a) &= \frac{1}{2}(a+2) + 3 - \left(\frac{1}{2}a + 3\right) \\ &= \frac{1}{2}a + 1 + 3 - \frac{1}{2}a - 3 \\ &= 1 \end{aligned}$$

5. $f(x) = \frac{1}{2}x + 3$

$$\begin{aligned} f(2a+3) &= \frac{1}{2}(2a+3) + 3 \\ &= a + \frac{3}{2} + 3 \\ &= a + \frac{9}{2} \end{aligned}$$

6. $f(x) = \frac{1}{2}x + 3$

$$\begin{aligned} f(2a-3) &= \frac{1}{2}(2a-3) + 3 \\ &= a - \frac{3}{2} + 3 \\ &= a + \frac{3}{2} \end{aligned}$$

7. $p(x) = -2x^2 + 3x - 1$

$$\begin{aligned} p(-3) &= -2(-3)^2 + 3(-3) - 1 \\ &= -28 \end{aligned}$$

8. $p(x) = -2x^2 + 3x - 1$

$$p(4) = -2(4)^2 + 3(4) - 1 = -21$$

9. $p(x) = -2x^2 + 3x - 1$

$$\begin{aligned} & p(2a) + p(-2) \\ &= -2(2a)^2 + 3(2a) - 1 + (-2(-2)^2 + 3(-2) - 1) \\ &= -8a^2 + 6a - 1 + (-8 - 6 - 1) \\ &= -8a^2 + 6a - 16 \end{aligned}$$

10. $p(x) = -2x^2 + 3x - 1$

$$\begin{aligned} & p(3a) + p(3) \\ &= -2(3a)^2 + 3(3a) - 1 + (-2(3)^2 + 3(3) - 1) \\ &= -18a^2 + 9a - 1 + (-18 + 9 - 1) \\ &= -18a^2 + 9a - 11 \end{aligned}$$

11. $p(x) = -2x^2 + 3x - 1$
$$\begin{aligned} p(a+2) &= -2(a+2)^2 + 3(a+2) - 1 \\ &= -2(a^2 + 4a + 4) + 3(a+2) - 1 \\ &= -2a^2 - 8a - 8 + 3a + 6 - 1 \\ &= -2a^2 - 5a - 3 \end{aligned}$$

12. $p(x) = -2x^2 + 3x - 1$
$$\begin{aligned} p(a-3) &= -2(a-3)^2 + 3(a-3) - 1 \\ &= -2(a^2 - 6a + 9) + 3a - 9 - 1 \\ &= -2a^2 + 12a - 18 + 3a - 10 \\ &= -2a^2 + 15a - 28 \end{aligned}$$

13. $h(x) = |2x - 1|$
$$h(8a) = |2(8a) - 1| = |16a - 1|$$

14. $h(x) = |2x - 1|$
$$h(7a) = |2(7a) - 1| = |14a - 1|$$

15. $h(x) = |2x - 1|$
$$\begin{aligned} h\left(\frac{1}{4}a\right) &= \left|2\left(\frac{1}{4}a\right) - 1\right| \\ &= \left|\frac{1}{2}a - 1\right| \end{aligned}$$

16. $h(x) = |2x - 1|$
$$h\left(\frac{3}{2}a\right) = \left|2\left(\frac{3}{2}a\right) - 1\right| = |3a - 1|$$

17. $h(x) = |2x - 1|$
$$h(a-5) = |2(a-5) - 1| = |2a - 11|$$

18. $h(x) = |2x - 1|$
$$h(a+4) = |2(a+4) - 1| = |2a + 7|$$

19. $r(x) = \dfrac{3x}{x+4},\ x \neq -4$
$$r(5) = \frac{3(5)}{5+4} = \frac{15}{9} = \frac{5}{3}$$

20. $r(x) = \dfrac{3x}{x+4},\ x \neq -4$
$$\begin{aligned} r(-6) &= \frac{3(-6)}{-6+4} \\ &= \frac{-18}{-2} = 9 \end{aligned}$$

21. $r(x) = \dfrac{3x}{x+4},\ x \neq -4$
$$\begin{aligned} r(a+3) &= \frac{3(a+3)}{a+3+4} \\ &= \frac{3a+9}{a+7} \end{aligned}$$

22. $r(x) = \dfrac{3x}{x+4},\ x \neq -4$
$$\begin{aligned} r(a-2) &= \frac{3(a-2)}{a-2+4} \\ &= \frac{3a-6}{a+2} \end{aligned}$$

23. $r(x) = \dfrac{3x}{x+4},\ x \neq -4$
$$\begin{aligned} r(3) + r(a) &= \frac{3(3)}{3+4} + \frac{3(a)}{a+4} \\ &= \frac{9}{7} + \frac{3a}{a+4} \\ &= \frac{9(a+4) + 7(3a)}{7(a+4)} \\ &= \frac{9a + 36 + 21a}{7a + 28} \\ &= \frac{30a + 36}{7a + 28} \end{aligned}$$

24. $$r(x)=\frac{3x}{x+4},\ x\neq -4$$

$$r(a)+r(-2)=\frac{3a}{a+4}+\frac{3(-2)}{-2+4}$$
$$=\frac{3a}{a+4}+\frac{-6}{2}$$
$$=\frac{3a}{a+4}-3$$
$$=\frac{3a-3(a+4)}{a+4}$$
$$=\frac{-12}{a+4}$$

25. $f(x)=7x-4$

$$\frac{f(x+h)-f(x)}{h}=\frac{7(x+h)-4-(7x-4)}{h}$$
$$=\frac{7x+7h-4-7x+4}{h}$$
$$=7$$

26. $f(x)=6x-5$

$$\frac{f(x+h)-f(x)}{h}=\frac{6(x+h)-5-(6x-5)}{h}$$
$$=\frac{6x+6h-5-6x+5}{h}$$
$$=6$$

27. $f(x)=2x^2-5x$

$$\frac{f(x+h)-f(x)}{h}$$
$$=\frac{2(x+h)^2-5(x+h)-(2x^2-5x)}{h}$$
$$=\frac{2x^2+4xh+2h^2-5x-5h-2x^2+5x}{h}$$
$$=\frac{4xh+2h^2-5h}{h}$$
$$=4x+2h-5$$

28. $f(x)=2x-3x^2$

$$\frac{f(x+h)-f(x)}{h}$$
$$=\frac{2(x+h)-3(x+h)^2-(2x-3x^2)}{h}$$
$$=\frac{2x+2h-3x^2-6xh-3h^2-2x+3x^2}{h}$$
$$=\frac{2h-6xh-3h^2}{h}$$
$$=2-6x-3h$$
$$=-6x-3h+2$$

29. (a) Yes, the graph passes the vertical line test and therefore represents a function.
(b) Yes, the graph passes the horizontal line test and therefore represents a one-to-one function.

30. (a) No, the graph fails the vertical line test and therefore does not represent a function.
(b) No, unless the graph represents a function first it cannot represent a one-to-one function.

31. (a) Yes, the graph passes the vertical line test and therefore represents a function.
(b) No, the graph fails the horizontal line test and therefore does not represent a one-to-one function.

32. (a) Yes, the graph passes the vertical line test and therefore represents a function.
(b) No, the graph fails the horizontal line test and therefore does not represent a one-to-one function.

33. (a) No, the graph fails the vertical line test and therefore does not represent a function.

33. (b) No, unless the graph represents a function first it cannot represent a one-to-one function.

34. (a) Yes, the graph passes the vertical line test and therefore represents a function.
(b) Yes, the graph passes the horizontal line test and therefore represents a one-to-one function.

35. $f(x) = x^2$
$g(x) = (x+2)^2 + 4$ is $f(x)$ shifted left 2 units and up 4 units.

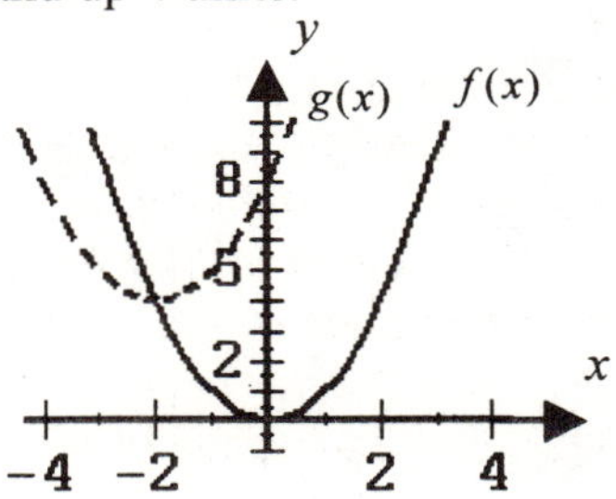

36. $f(x) = |x|$
$g(x) = |x+3|$ is $f(x)$ shifted left 3 units.

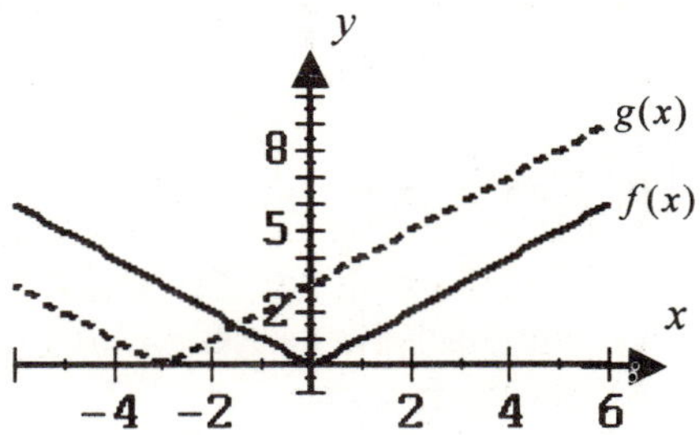

37. $f(x) = |x|$
$g(x) = |x-4|$ is $f(x)$ shifted right 2 units.

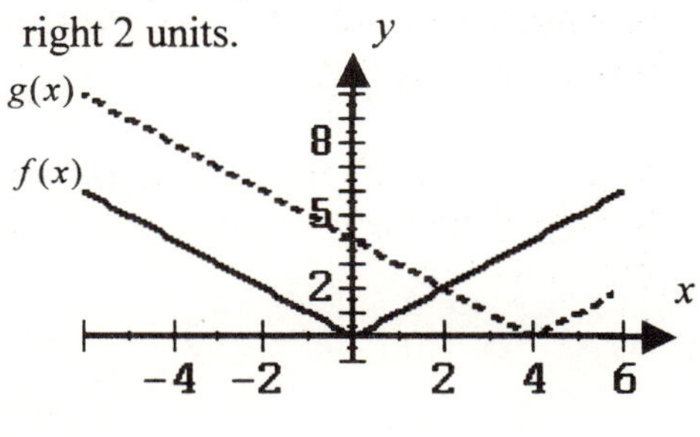

38. $f(x) = |x|$
$h(x) = |x| + 3$ is $f(x)$ shifted up 3 units.

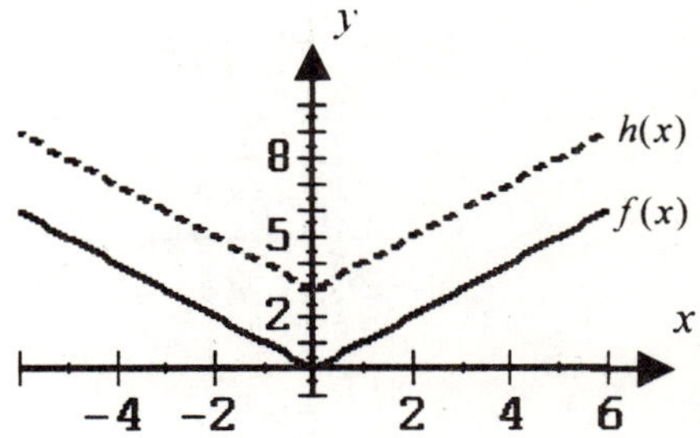

39. $f(x) = |x|$
$h(x) = |x| - 2$ is $f(x)$ shifted down 2 units.

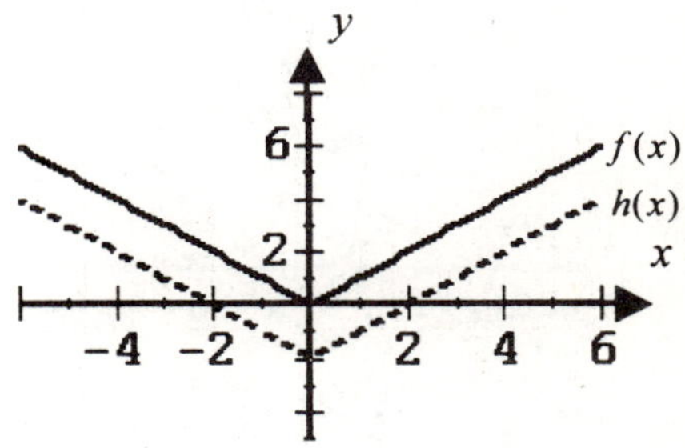

40. $f(x) = x^3$
$r(x) = (x+3)^3 + 1$ is $f(x)$ shifted left 3 units and up 1unit.

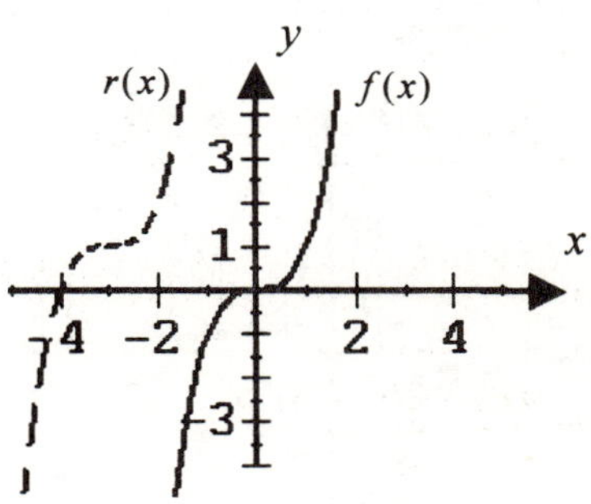

41. $f(x)=x^3$

$r(x)=(x-1)^3+5$ is $f(x)$ shifted right 1 unit and up 5 units.

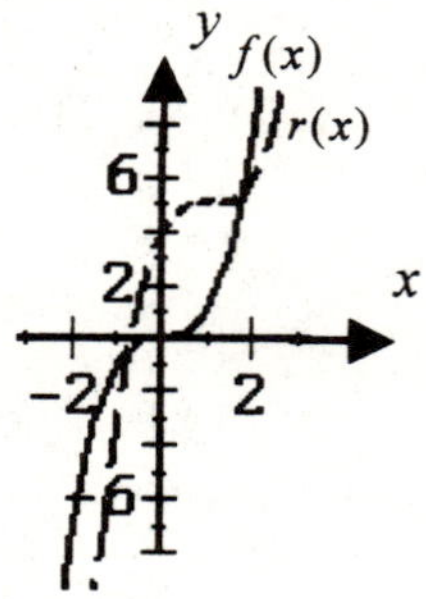

42. $f(x)=\frac{2}{x},\ x\neq 0$

$r(x)=\frac{2}{x+3}-2,\ x\neq -3$ is $f(x)$ shifted left 3 units and down 2 units.

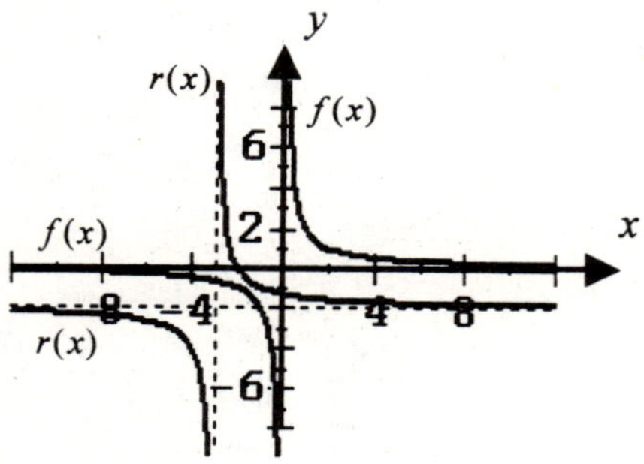

43. $f(x)=\frac{4}{x},\ x\neq 0$

$r(x)=\frac{4}{x+2},\ x\neq 2$ is $f(x)$ shifted left 2 units.

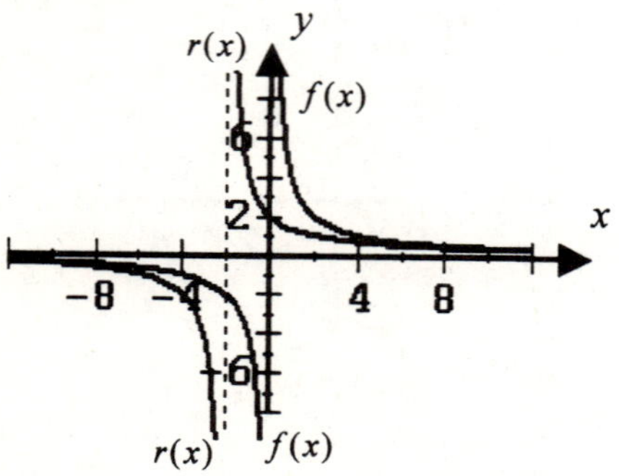

In Exercises 44-63.

$$f(x)=3x+5;\ g(x)=\frac{2}{x},\ x\neq 0$$

$$s(x)=\sqrt{x-2},\ x\geq 2;\ h(x)=\frac{x+1}{x-4},\ x\neq 4$$

$$p(x)=2x^2-3x+4;\ t(x)=-\frac{1}{2}x-3$$

44. $(f+p)(x)=f(x)+p(x)$
$=3x+5+2x^2-3x+4$
$=2x^2+9$

45. $(f+t)(x)=f(x)+t(x)$
$=3x+5+\left(-\frac{1}{2}x-3\right)$
$=\frac{5}{2}x+2$

46. $(t-f)(x)=t(x)-f(x)$
$=-\frac{1}{2}x-3-(3x+5)$
$=-\frac{7}{2}x-8$

47. $(p-f)(x)=p(x)-f(x)$
$=2x^2-3x+4-(3x+5)$
$=2x^2-3x+4-3x-5$
$=2x^2-6x-1$

48. From Exercise 43,
$(p-f)(x)=2x^2-6x-1$
$(p-f)(2)=2(2)^2-6(2)-1$
$=2(4)-12-1$
$=8-13$
$=-5$

49. From Exercise 46.

$$(t-f)(x) = -\frac{7}{2}x - 8$$

$$(t-f)(-3) = -\frac{7}{2}(-3) - 8 = \frac{5}{2}$$

50. $(fg)(x) = f(x)g(x)$

$$= (3x+5)\left(\frac{2}{x}\right) = \frac{6x+10}{x},\ x \neq 0$$

51. $(tp)(x) = t(x)p(x)$

$$= \left(-\frac{1}{2}x - 3\right)(2x^2 - 3x + 4)$$

$$= -x^3 + \frac{3x^2}{2} - 2x - 6x^2 + 9x - 12$$

$$= -x^3 - \frac{9}{2}x^2 + 7x - 12$$

52. $\left(\frac{g}{h}\right)(x) = \frac{g(x)}{h(x)} = \frac{\frac{2}{x}}{\frac{x+1}{x+4}}$

$$= \frac{2}{x} \cdot \frac{x-4}{x+1}$$

$$= \frac{2x-8}{x^2+1},\ x \neq 0, 4, -1$$

53. $\left(\frac{g}{f}\right)(x) = \frac{g(x)}{f(x)} = \frac{\frac{2}{x}}{3x+5}$

$$= \frac{2}{x} \cdot \frac{1}{3x+5}$$

$$= \frac{2}{3x^2+5x},\ x \neq 0, -\frac{5}{3}$$

54. From Exercise 52.

$$\left(\frac{g}{h}\right)(x) = \frac{2x-8}{x^2+x},\ x \neq -1, 0, 4$$

$$\left(\frac{g}{h}\right)(-2) = \frac{2(-2)-8}{(-2)^2+(-2)} = -6$$

55. From Exercise 53.

$$\left(\frac{g}{f}\right)(x) = \frac{2}{3x^2+5x},\ x \neq 0, -\frac{5}{3}$$

$$\left(\frac{g}{f}\right)(-3) = \frac{2}{3(-3)^2+5(-3)} = \frac{1}{6}$$

56. $f[t(x)] = f\left[-\frac{1}{2}x - 3\right]$

$$= 3\left(-\frac{1}{2}x - 3\right) + 5$$

$$= -\frac{3}{2}x - 4$$

57. $h[f(x)] = h(3x+5)$

$$= \frac{3x+5+1}{3x+5-4}$$

$$= \frac{3x+6}{3x+1},\ x \neq -\frac{1}{3}$$

58. $s[p(x)] = s\left[2x^2 - 3x + 4\right]$

$$= \sqrt{2x^2 - 3x + 4 - 2}$$

$$= \sqrt{2x^2 - 3x + 2}$$

59. $s[t(x)] = s\left[-\frac{1}{2}x - 3\right]$

$$= \sqrt{-\frac{1}{2}x - 3 - 2}$$

$$= \sqrt{-\frac{1}{2}x - 5},\ x \leq -10$$

60. From Exercise 58,

$$s[p(x)]=\sqrt{2x^2-3x+2}$$

$$s[p(2)]=\sqrt{2(2)^2-3(2)+2}=2$$

61. From Exercise 59,

$$s[t(x)]=\sqrt{-\frac{1}{2}x-5},\ x\le 10$$

$$s[t(-18)]=\sqrt{-\frac{1}{2}(-18)-5}=2$$

62. $f[g(x)]=f\left[\frac{2}{x}\right],\ x\ne 0$

$$=3\left(\frac{2}{x}\right)+5=\frac{6}{x}+5$$

$$=\frac{6+5x}{x}$$

$$g[f(x)]=g[3x+5]$$

$$=\frac{2}{3x+5}$$

$$f[g(x)]\ne g[f(x)]$$

63. $p[g(x)]=p\left[\frac{2}{x}\right]$

$$=2\left(\frac{2}{x}\right)^2-3\left(\frac{2}{x}\right)+4$$

$$=\frac{8-6x+4x^2}{x^2}$$

$$g[p(x)]=g\left[2x^2-3x+4\right]$$

$$=\frac{2}{2x^2-3x+4}$$

$$p[g(x)]\ne g[p(x)]$$

64. $B=\{(3,7),(7,3),(0,8),(0,-8)\}$

(a) $D=\{0,3,7\}$

(continued)

64. (continued)

(b) $R=\{-8,3,7,8\}$

(c) No, the set does not define a function since two of the ordered pairs have the same first coordinate.

(d) No, since the set does not define a function it cannot define a one-to-one function.

65. $A=\{(100,10),(200,20),(300,30),(400,10)\}$

(a) $D=\{100,200,300,400\}$

(b) $R=\{10,20,30\}$

(c) Yes, the set defines a function since no two of the ordered pairs have the same first coordinate.

(d) No, the set does not define a one-to-one function since two of the ordered pairs have the same second coordinate.

66. $D=\left\{\left(\frac{1}{2},2\right),\left(\frac{1}{4},4\right),\left(-\frac{1}{3},-3\right),\left(4,\frac{1}{4}\right)\right\}$

(a) domain $=\left\{\frac{1}{2},\frac{1}{4},-\frac{1}{3},4\right\}$

(b) $R=\left\{2,4,-3,\frac{1}{4}\right\}$

(c) Yes, the set defines a function since no two of the ordered pairs have the same first coordinate.

(d) Yes, the set defines a one-to-one function since it is a function and no two ordered pairs have the same second coordinate.

67. $C=\{(12,6),(0,6),(0,-1),(-6,-12)\}$

(a) $D=\{12,0,-6\}$

(b) $R=\{-1,-12,6\}$

(continued)

67. (continued)

(c) No, the set does not define a function since two of the ordered pairs have the same first coordinate.

(d) No, since the set does not define a function it cannot define a one-to-one function.

68. $E = \{(0,1),(1,2),(2,9),(-1,-2)\}$

(a) $D = \{-1,0,1,2\}$

(b) $R = \{-2,1,2,9\}$

(c) Yes, the set defines a function since no two of the ordered pairs have the same first coordinate.

(d) Yes, the set defines a one-to-one function since it is a function and no two ordered pairs have the same second coordinate.

69. $F = \{(3,7),(2,1),(0,-3),(1,1)\}$

(a) $D = \{0,1,2,3\}$

(b) $R = \{-3,1,7\}$

(c) Yes, the set defines a function since no two of the ordered pairs have the same first coordinate.

(d) No, the set does not define a one-to-one function since two of the ordered pairs have the same second coordinate.

70. $A = \left\{\left(3,\frac{1}{3}\right),\left(-2,-\frac{1}{2}\right),\left(-4,-\frac{1}{4}\right),\left(5,\frac{1}{5}\right)\right\}$

$A^{-1} = \left\{\left(\frac{1}{3},3\right),\left(-\frac{1}{2},-2\right),\left(-\frac{1}{4},-4\right),\left(\frac{1}{5},5\right)\right\}$

71. $B = \{(1,10),(3,7),(12,15),(10,1)\}$

$B^{-1} = \{(10,1),(7,3),(15,12),(1,10)\}$

72. $f(x) = -\frac{3}{4}x + 2,\ f(x) \to y$

$y = -\frac{3}{4}x + 2,\ x \leftrightarrow y$

$x = -\frac{3}{4}y + 2$

$y = -\frac{4}{3}x + \frac{8}{3},\ y \to f^{-1}(x)$

$f^{-1}(x) = -\frac{4}{3}x + \frac{8}{3}$

73. $g(x) = -8 - 4x,\ g(x) \to y$

$y = -8 - 4x,\ x \leftrightarrow y$

$x = -8 - 4y$

$y = -\frac{1}{4}x - 2,\ y \to g^{-1}(x)$

$g^{-1}(x) = -\frac{1}{4}x - 2$

74. $h(x) = \frac{x+2}{3},\ h(x) \to y$

$y = \frac{x+2}{3},\ x \leftrightarrow y$

$x = \frac{y+2}{3}$

$y = 3x - 2,\ y \to h^{-1}(x)$

$h^{-1}(x) = 3x - 2$

75. $j(x) = \frac{1}{x-3},\ j(x) \to y$

$y = \frac{1}{x-3},\ x \leftrightarrow y$

$x = \frac{1}{y-3}$

$y = \frac{1}{x} + 3,\ y \to j^{-1}(x)$

$j^{-1}(x) = \frac{1}{x} + 3$

76. $p(x) = \sqrt[3]{x+1},\ p(x) \to y$

$y = \sqrt[3]{x+1},\ x \leftrightarrow y$

$x = \sqrt[3]{y+1}$

$y = x^3 - 1,\ y \to p^{-1}(x)$

$p^{-1}(x) = x^3 - 1$

77. $r(x) = x^3 + 2,\ r(x) \to y$

$y = x^3 + 2,\ x \leftrightarrow y$

$x = y^3 + 2$

$y = \sqrt[3]{x-2},\ y \to r^{-1}(x)$

$r^{-1}(x) = \sqrt[3]{x-2}$

78. $f(x) = \dfrac{-x-2}{3},\ f(x) \to y$

$y = \dfrac{-x-2}{3},\ x \leftrightarrow y$

$x = \dfrac{-y-2}{3}$

$y = -3x - 2,\ y \to f^{-1}(x)$

$f^{-1}(x) = -3x - 2$

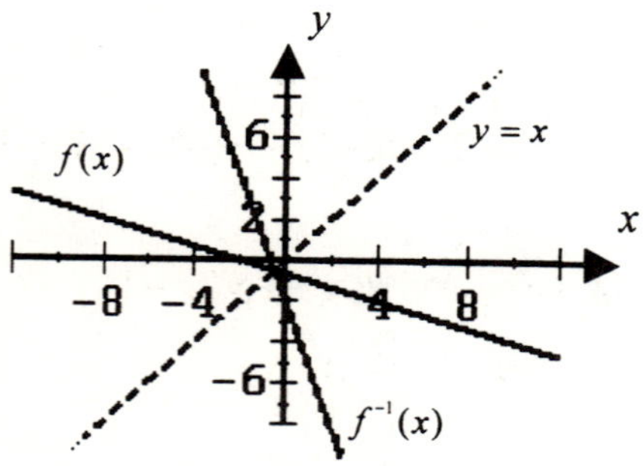

79. $f(x) = -\dfrac{3}{4}x + 1,\ f(x) \to y$

$y = -\dfrac{3}{4}x + 1,\ x \leftrightarrow y$

(continued)

79. (continued)

$x = -\dfrac{3}{4}y + 1$

$y = -\dfrac{4}{3}x + \dfrac{4}{3},\ y \to f^{-1}(x)$

$f^{-1}(x) = -\dfrac{4}{3}x + \dfrac{4}{3}$

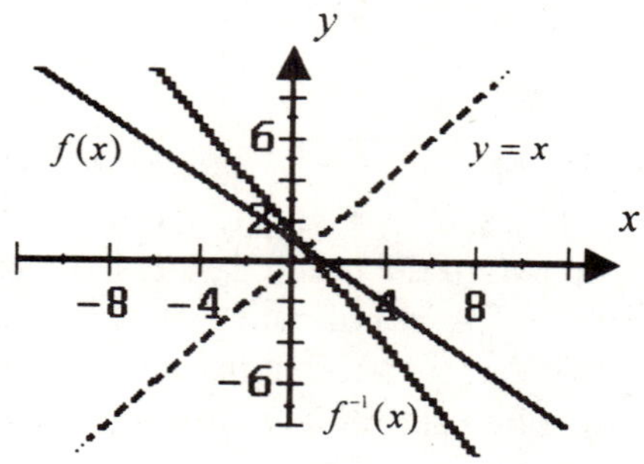

Chapter 11 Test

1. $f(x) = \dfrac{3}{4}x - 2$

$f(-8) = \dfrac{3}{4}(-8) - 2 = -8$

2. $f(x) = \dfrac{3}{4}x - 2$

$f(2a) = \dfrac{3}{4}(2a) - 2 = \dfrac{3}{2}a - 2$

3. $f(x) = \dfrac{3}{4}x - 2$

$$\begin{aligned} f(a) - f(2) &= \frac{3}{4}a - 2 - \left(\frac{3}{4}(2) - 2\right) \\ &= \frac{3}{4}a - 2 - \frac{3}{2} + 2 \\ &= \frac{3}{4}a - \frac{3}{2} \end{aligned}$$

4. $f(x) = 3x^2 - 2x + 4$
$f(-6) = 3(-6)^2 - 2(-6) + 4 = 124$

5. $f(x) = 3x^2 - 2x + 4$
$$f(a+1) = 3(a+1)^2 - 2(a+1) + 4$$
$$= 3a^2 + 6a + 3 - 2a - 2 + 4$$
$$= 3a^2 + 4a + 5$$

6. $f(x) = 3x^2 - 2x + 4$
$$f(a) + f(1) = 3a^2 - 2a + 4 + 3(1)^2 - 2(1) + 4$$
$$= 3a^2 - 2a + 9$$

7. $f(x) = 3x^2 - 2x + 4$
$$f(-2a) - 2 = 3(-2a)^2 - 2(-2a) + 4 - 2$$
$$= 12a^2 + 4a + 2$$

8. (a) Graph passes vertical line test and therefore represents a function.
(b) Graph fails horizontal line test and does not represent a one-to-one function.

9. (a) Graph passes vertical line test and therefore represents a function.
(b) Graph passes horizontal line test and therefore represents a one-to-one function.

10. $f(x) = x^2$
$g(x) = (x-1)^2 + 3$ is $f(x)$ shifted right 1 unit and up 3 units.

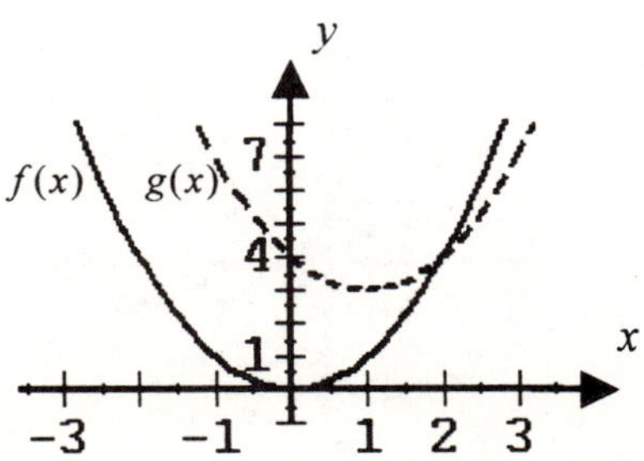

11. $f(x) = |x|$
$g(x) = |x+1| + 2$ is $f(x)$ shifted left 1 unit and up 2 units.

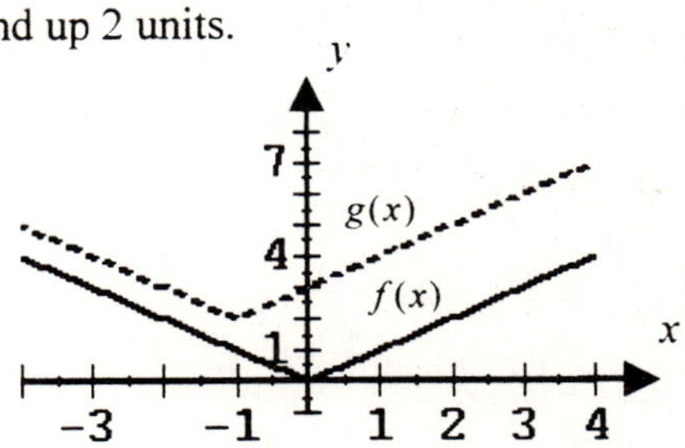

12. $f(x) = 3x^2 - x - 6,\ g(x) = -2x^2 + 5x + 7$
(a)
$$(f+g)(x) = f(x) + g(x)$$
$$= 3x^2 - x - 6 + (-2x^2 + 5x + 7)$$
$$= 3x^2 - x - 6 - 2x^2 + 5x + 7$$
$$= x^2 + 4x + 1$$
(b)
$$(f-g)(x) = f(x) - g(x)$$
$$= 3x^2 - x - 6 - (-2x^2 + 5x + 7)$$
$$= 3x^2 - x - 6 + 2x^2 - 5x - 7$$
$$= 5x^2 - 6x - 13$$
(c) From (b)
$$(f-g)(x) = 5x^2 - 6x - 13$$
$$(f-g)(-2) = 5(-2)^2 - 6(-2) - 13 = 19$$

13. $f(x) = \frac{3}{x},\ x \neq 0;\ g(x) = 2x - 1$
(a) $(fg)(x) = f(x)g(x)$
$$= \frac{3}{x}(2x-1)$$
$$= \frac{6x-3}{x},\ x \neq 0$$

(continued)

13. (continued)

(b) $\left(\dfrac{f}{g}\right)(x) = \dfrac{f(x)}{g(x)}$

$= \dfrac{\frac{3}{x}}{2x-1}$

$= \dfrac{3}{2x^2 - x},\ x \neq 0, \dfrac{1}{2}$

(c) $g[f(x)] = g\left[\dfrac{3}{x}\right]$

$= 2 \cdot \dfrac{3}{x} - 1$

$= \dfrac{6}{x} - 1,\ x \neq 0$

14. $f(x) = \dfrac{1}{2}x - 3,\ g(x) = 4x + 5$

(a) $(f \circ g)(x) = f[g(x)]$

$= f[4x+5]$

$= \dfrac{1}{2}(4x+5) - 3$

$= 2x - \dfrac{1}{2}$

(b) $(g \circ f)(x) = g[f(x)]$

$= g\left[\dfrac{1}{2}x - 3\right]$

$= 4\left(\dfrac{1}{2}x - 3\right) + 5$

$= 2x - 7$

(c) $f[f(x)] = f\left(\dfrac{1}{2}x - 3\right)$

$= \dfrac{1}{2}\left(\dfrac{1}{2}x - 3\right) - 3$

$= \dfrac{1}{4}x - \dfrac{9}{2}$

15. $B = \{(1,8), (8,1), (9,10), (-10,9)\}$

(a) Yes, the function is one-to-one since no two ordered pairs have the same second coordinate.

(b) $B^{-1} = \{(8,1), (1,8), (10,9), (9,-10)\}$

16. $A = \{(1,5), (2,1), (4,-7), (0,7)\}$

(a) Yes, the function is one-to-one since no two ordered pairs have the same second coordinate.

(b) $A^{-1} = \{(5,1), (1,2), (-7,4), (7,0)\}$

17. $f(x) = \dfrac{1}{2}x - \dfrac{1}{5},\ f(x) \to y$

$y = \dfrac{1}{2}x - \dfrac{1}{5},\ x \leftrightarrow y$

$x = \dfrac{1}{2}y - \dfrac{1}{5}$

$y = 2x + \dfrac{2}{5},\ y \to f^{-1}(x)$

$f^{-1}(x) = 2x + \dfrac{2}{5}$

18. $f(x) = -3x - 2,\ f(x) \to y$

$y = -3x + 2,\ x \leftrightarrow y$

$x = -3y + 2$

$y = -\dfrac{1}{3}x + \dfrac{2}{3},\ y \to f^{-1}(x)$

$f^{-1}(x) = -\dfrac{1}{3}x + \dfrac{2}{3}$

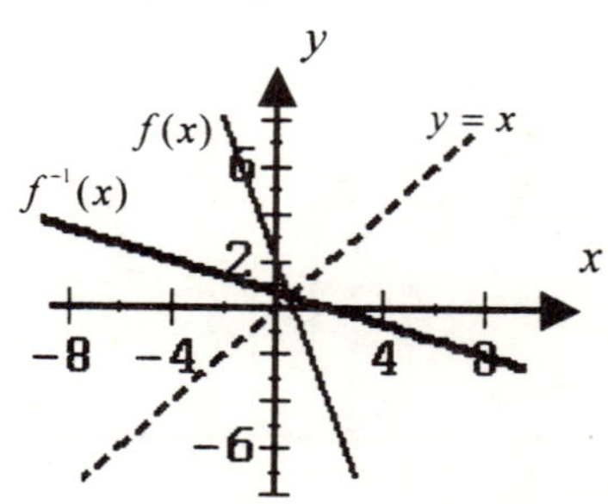

19. $f(x)=\frac{3}{7}x+\frac{1}{2}, f^{-1}(x)=\frac{14x-7}{6}$

$$f^{-1}[f(x)]=f^{-1}\left[\frac{3}{7}x+\frac{1}{2}\right]$$
$$=\frac{14\left[\frac{3}{7}x+\frac{1}{2}\right]-7}{6}$$
$$=\frac{6x+7-7}{6}=\frac{6x}{6}$$
$$=x$$

Cumulative Test for Chapters 1-11

1. $3x\{2y-3[x+2(x+2y)]\}$
$$=3x\{2y-3[x+2x+4y]\}$$
$$=3x\{2y-3[3x+4y]\}$$
$$=3x\{2y-9x-12y\}$$
$$=3x\{-10y-9x\}$$
$$=-30xy-27x^2$$

2. $2x^2-3xy+y^2\big|_{x=2,y=-3}$
$$=2(2)^2-3(2)(-3)+(-3)^2=35$$

3. $\frac{1}{2}(x-2)=\frac{1}{3}(x+10)-2x$
$$3(x-2)=2(x+10)-12x$$
$$3x-6=2x+20-12x$$
$$13x=26$$
$$x=2$$

4. $16x^4-1=(4x^2+1)(4x^2-1)$
$$=(4x^2+1)(2x+1)(2x-1)$$

5. $(3x+1)(2x-3)(x+4)$
$$=(6x^2-7x-3)(x+4)$$
$$=6x^3-7x^2-3x+24x^2-28x-12$$
$$=6x^3+17x^2-31x-12$$

6. $\frac{3x}{x^2-4}=\frac{2}{x+2}+\frac{4}{2-x}$
$$\frac{3x}{(x+2)(x-2)}=\frac{2}{x+2}-\frac{4}{x-2}$$
$$3x=2(x-2)-4(x+2)$$
$$3x=2x-4-4x-8$$
$$5x=-12$$
$$x=-\frac{12}{5}$$

7. $y-y_1=m(x-x_1)$
$$y-(-1)=-3(x-2)$$
$$y+1=-3x+6$$
$$y=-3x+5$$

8. $3x+2y=5 \Rightarrow y=\frac{5-3x}{2}$
$$7x+5y=11$$
$$7x+5\cdot\frac{5-3x}{2}=11$$
$$14x+25-15x=22$$
$$-x=-3$$
$$x=3$$
$$y=\frac{5-3x}{2}=\frac{5-3(3)}{2}=-2$$
(3, −2) is the solution.

9. $\sqrt{18x^5y^6z^3}=\sqrt{9x^4y^6z^2\cdot 2xz}$
$$=3x^2y^3z\sqrt{2xz}$$

10. $(\sqrt{2}+\sqrt{3})(2\sqrt{2}-4\sqrt{3})$
$=2(2)-4\sqrt{6}+2\sqrt{6}-4(3)$
$=4-2\sqrt{6}-12$
$=-2\sqrt{6}-8$

11. $d=\sqrt{(x_2-x_1)^2+(y_2-y_1)^2}$
$d=\sqrt{(-3-6)^2+(-4-(-1))^2}=\sqrt{90}$
$d=3\sqrt{10}$

12. $12x^2-11x+2=(4x-1)(3x-2)$

13. $x^4-10x^2+9=(x^2-9)(x^2-1)$
$=(x+3)(x-3)(x+1)(x-1)$

14. $(x-h)^2+(y-k)^2=r^2$
$(x-(-3))^2+(y-6)^2=14^2$
$(x+3)^2+(y-6)^2=196$

15. $f(x)=3x^2-2x+1$
(a) $f(-2)=3(-2)^2-2(-2)+1=17$
(b) $f(a-2)=3(a-2)^2-2(a-2)+1$
$=3a^2-12a+12-2a+4+1$
$=3a^2-14a+17$
(c) $f(a)+f(-2)=3a^2-2a+1+17$
$=3a^2-2a+18$

16. $f(x)=x^3$, $g(x)=(x+2)^3+4$ is $f(x)$ shifted right 2 units and up 4 units.

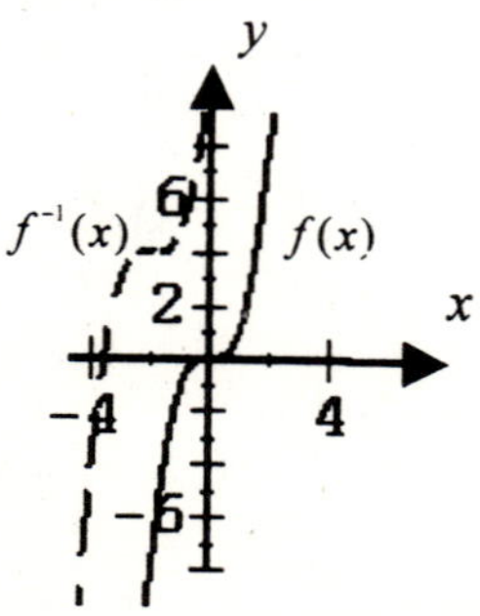

17. $f(x)=2x^2-5x-6,\ g(x)-5x+3$

(a) $(fg)(x)=f(x)g(x)$
$=(2x^2-5x-6)(5x+3)$
$=10x^3+6x^2-25x^2-15x-30x+18$
$=10x^3-19x^2-45x+18$

(b) $\left(\dfrac{f}{g}\right)(x)=\dfrac{f(x)}{g(x)}$
$=\dfrac{2x^2-5x-6}{5x+3},\ x\neq-\dfrac{3}{5}$

(c) $f[g(x)]=f[5x+3]$
$=2(5x+3)^2-5(5x+3)-6$
$=50x^2+60x+18-25x-15-6$
$=50x^2+35x-3$

18. $A=\{(3,6),(1,8),(2,7),(4,4)\}$
(a) Yes, A is a function; no two ordered pairs have the same first coordinate.
(b) Yes, A is a one-to-one function since no two ordered pairs have the same second coordinate.
(c) $A^{-1}=\{(6,3),(8,1),(7,2),(4,4)\}$

19. $f(x)=7x-3,\ f(x)\to y$
$y=7x-3,\ x\leftrightarrow y$
$x=7y-3$
$y=\dfrac{x+3}{7},\ y\to f^{-1}(x)$
$f^{-1}(x)=\dfrac{x+3}{7}$

20. $f(x) = 5x^3 - 3x^2 - 6$

(a) $f(5) = 5(5)^3 - 3(5)^2 - 6 = 544$

(b) $f(-3) = 5(-3)^3 - 3(-3)^2 - 6 = -168$

(c) $f(2a) = 5(2a)^3 - 3(2a)^2 - 6$

$= 40a^3 - 12a^{-2} - 6$

21. (a) $f(x) = -\frac{2}{3}x + 2,\ f(x) \rightarrow y$

$y = -\frac{2}{3}x + 2,\ x \leftrightarrow y$

$x = -\frac{2}{3}y + 2$

$y = -\frac{3}{2}x + 3,\ y \rightarrow f^{-1}(x)$

$f^{-1}(x) = -\frac{3}{2}x + 3$

(b)

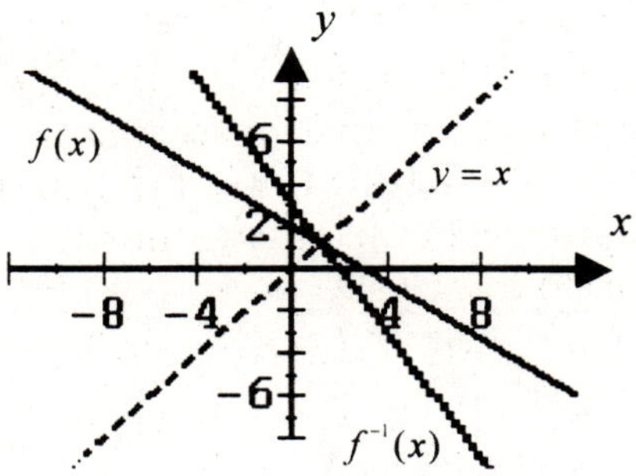

22. From Exercise 21.

$f\left[f^{-1}(x)\right] = f\left[-\frac{3}{2}x + 3\right]$

$= -\frac{2}{3}\left[-\frac{3}{2}x + 3\right] + 2$

$= x - 2 + 2$

$= x$

Chapter 12

Pretest Chapter 12

1. $f(x)=2^{-x}$

x	$y=f(x)=2^{-x}$
-2	4
-1	2
0	1
1	$\frac{1}{2}$
2	$\frac{1}{4}$

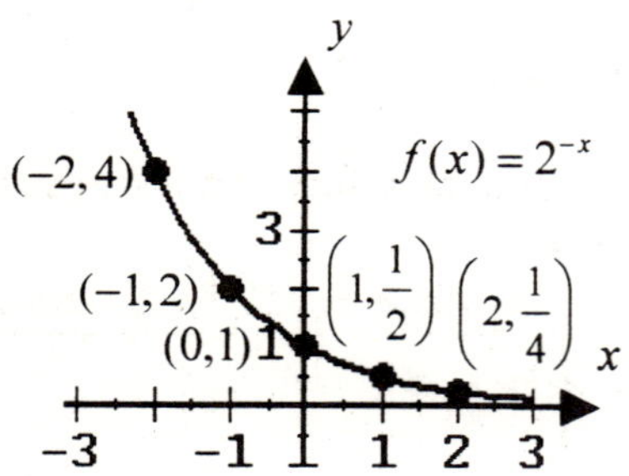

2. $3^{2x-1}=27$

$$3^{2x-1}=3^3$$
$$2x-1=3$$
$$2x=4$$
$$x=2$$

3. $A=P(1+r)^t$

$A=10{,}000(1+0.12)^4=15{,}735.1936$

In 4 years Nancy will have \$15,735.19.

4. $\frac{1}{49}=7^{-2}$

$$\log_7\left(\frac{1}{49}\right)=-2$$

5. $\log_5 x=3$

$$x=5^3=125$$

6. Let $N=\log_{10}(10{,}000)$

$$10^N=10{,}000$$
$$10^N=10^4$$
$$N=4$$

7. $\log_5\left(\frac{x^2y^5}{z^3}\right)=\log_5(x^2y^5)-\log_5 z^3$

$$=\log_5 x^2+\log_5 y^5-\log_5 z^3$$
$$=2\log_5 x+5\log_5 y-3\log_5 z$$

8. $\frac{1}{2}\log_4 x-3\log_4 w=\log_4 x^{1/2}-\log_4 w^3$

$$=\log_4\sqrt{x}-\log_4 w^3$$
$$=\log_4\left(\frac{\sqrt{4}}{w^3}\right)$$

9. $\log_3 x+\log_3 2=4$

$$\log_3(2x)=4$$
$$2x=3^4$$
$$2x=81$$
$$x=\frac{81}{2}$$

10. $\log x=3.9170$

$$x=10^{3.9170}=8260.3795$$

11. $\ln 4.79=1.5665$

12. $\log_6 5.02=\frac{\ln 5.02}{\ln 6}=0.9005$

13. $\log 0.7523 = -0.1236$

14. $\ln x = 22.976$
$$e^{\ln x} = e^{22.976}$$
$$x = 9.5137 \times 10^9$$

15. $\log x - \log(x+3) = -1$
$$\log \frac{x}{x+3} = -1$$
$$\frac{x}{x+3} = 10^{-1}$$
$$10x = x+3$$
$$9x = 3$$
$$x = \frac{3}{9} = \frac{1}{3}$$
check:
$$\log \frac{1}{3} - \log\left(\frac{1}{3}+3\right) \stackrel{?}{=} -1$$
$$\log\left(\frac{\frac{1}{3}}{\frac{1}{3}+3}\right) \stackrel{?}{=} -1$$
$$\log \frac{1}{10} = \log 1 - \log 10 = 0 - 1 = -1$$

16. $4^{2x+1} = 9$
$$\log 4^{2x+1} = \log 9$$
$$(2x+1)\log 4 = \log 9$$
$$2x+1 = \frac{\log 9}{\log 4}$$
$$x = \frac{1}{2}\left(\frac{\log 9}{\log 4} - 1\right) \approx 0.2925$$

17. $A = P(1+r)^t$
$$7000 = 2000(1+0.06)^t$$
(continued)

17. (continued)
$$(1.06)^t = \frac{7}{2}$$
$$\log(1.06)^t = \log \frac{7}{2}$$
$$t \log(1.06) = \log \frac{7}{2}$$
$$t = \frac{\log \frac{7}{2}}{\log(1.06)} = 21.49968154$$

It would take about 21 years for \$2000 to grow to \$7000 and 6% compounded yearly.

12.1 Exercises

1. The exponential function is an equation of the form
<u>$f(x) = b^x$ where $b > 0$, $b \neq 0$, and x is a real number.</u>

3. $f(x) = 3^x$

x	$y = f(x) = 3^x$
-2	$\frac{1}{9}$
-1	$\frac{1}{3}$
0	1
1	3
2	9

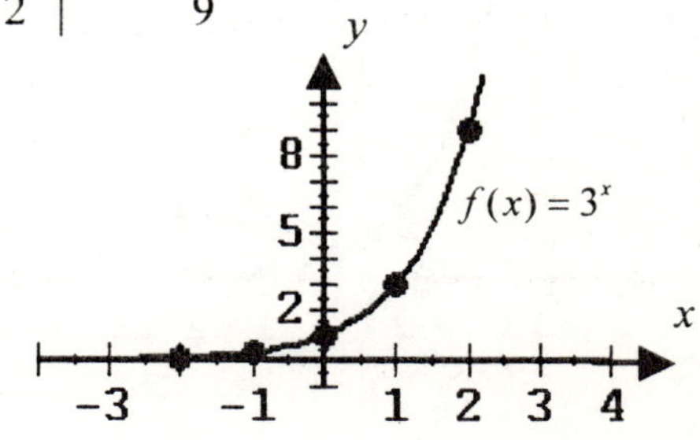

5. $f(x) = 2^{-x}$

x	$y = f(x) = 2^{-x}$
-2	4
-1	2
0	1
1	$\frac{1}{2}$
2	$\frac{1}{4}$

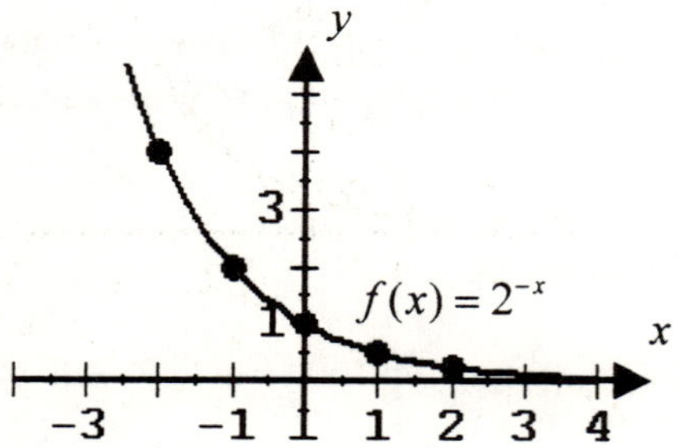

7. $f(x) = 3^{-x}$

x	$y = f(x) = 3^{-x}$
-2	9
-1	3
0	1
1	$\frac{1}{3}$
2	$\frac{1}{9}$

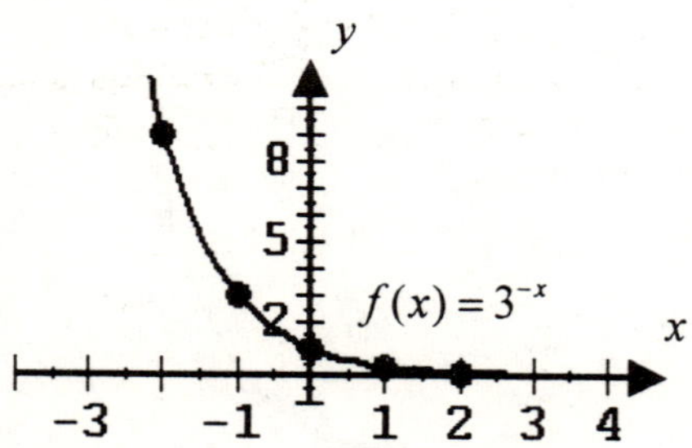

9. $f(x) = 2^{x+5}$

x	$y = f(x) = 2^{x+5}$
-7	$\frac{1}{4}$
-6	$\frac{1}{2}$
-5	1
-4	2
-3	4

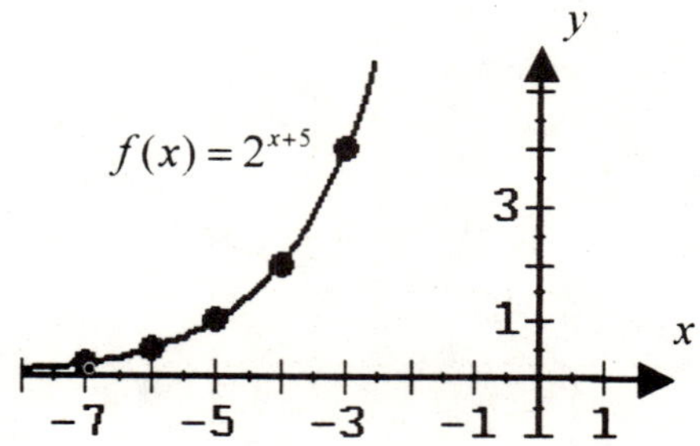

11. $f(x) = 3^{x-4}$

x	$y = f(x) = 3^{x-4}$
6	9
5	3
4	1
3	$\frac{1}{3}$
2	$\frac{1}{9}$

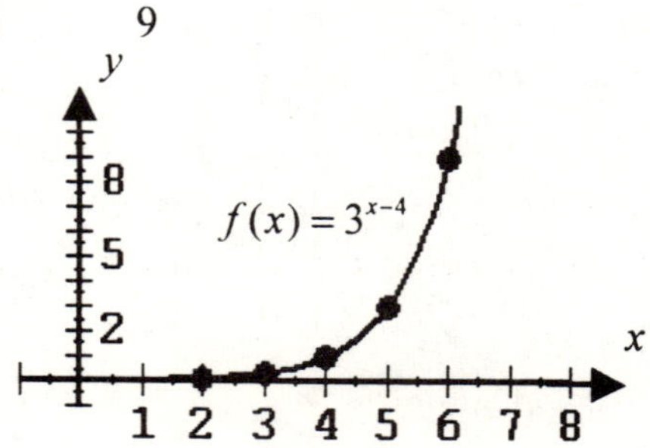

13. $f(x) = 2^x + 2$

x	$y = f(x) = 2^x + 2$
-2	$\frac{9}{4}$
-1	$\frac{5}{2}$
0	3
1	4
2	6

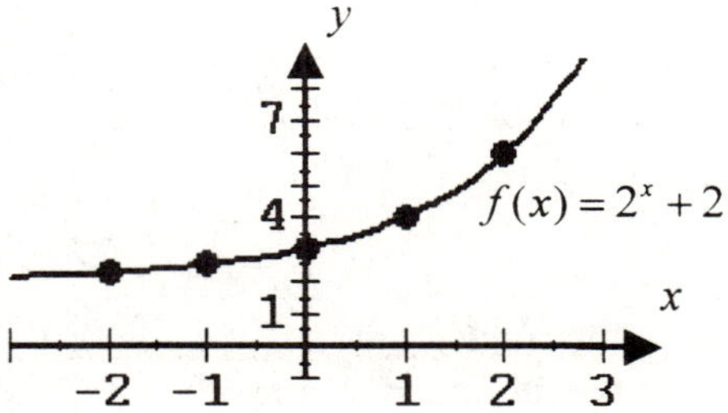

15. $f(x) = e^{x-1}$

x	$y = e^{x-1}$
-2	0.05
-1	0.14
0	0.37
1	1
2	2.7

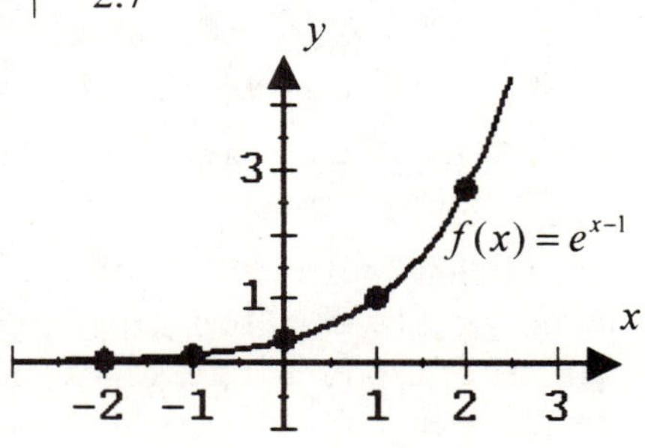

17. $f(x) = 2e^x$

x	$y = f(x) = 2e^x$
-2	0.27
-1	0.74
0	2
1	5.44
2	14.8

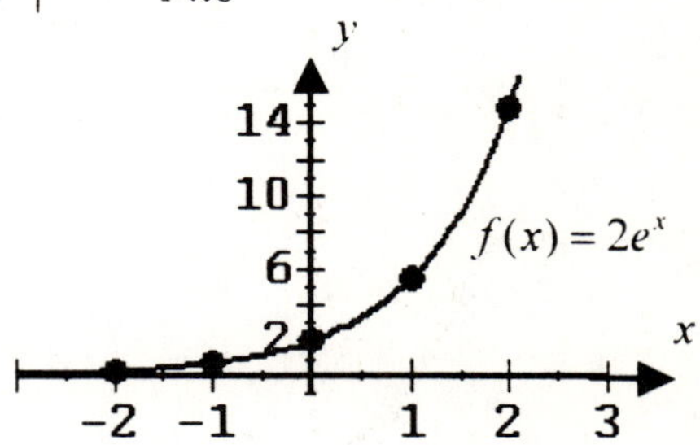

19. $f(x) = e^{1-x}$

x	$y = f(x) = e^{1-x}$
-2	20.1
-1	7.39
0	2.72
1	0.37
2	0.14

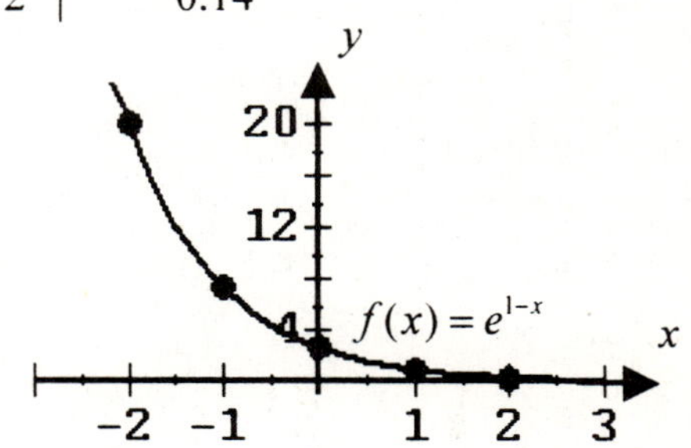

21. $2^x = 4$

$2^x = 2^2$

$x = 2$

23. $2^x = 1$
$2^x = 2^0$
$x = 0$

25. $2^x = \frac{1}{8}$
$2^x = \frac{1}{2^3} = 2^{-3}$
$x = -3$

27. $3^x = 81$
$3^x = 3^4$
$x = 4$

29. $3^x = 1$
$3^x = 3^0$
$x = 0$

31. $3^{-x} = \frac{1}{9} = \frac{1}{3^2}$
$3^{-x} = 3^{-2}$
$-x = -2$
$x = 2$

33. $4^x = 256$
$4^x = 4^4$
$x = 4$

35. $7^{x-1} = 49$
$7^{x-1} = 7^2$
$x - 1 = 2$
$x = 3$

37. $8^{3x-1} = 64 = 8^2$
$3x - 1 = 2$
$3x = 3$
$x = 1$

39. $A = P\left(1 + \frac{r}{n}\right)^{nt}$

$A = 2000\left(1 + \frac{0.063}{1}\right)^{1(3)} = 2402.314094$

Alicia will have \$2402.31 after 3 years.

41. $A = P\left(1 + \frac{r}{n}\right)^{nt}$

$A = 2000\left(1 + \frac{0.14}{4}\right)^{4(5)} = 3979.577727$

$A = 2000\left(1 + \frac{0.14}{12}\right)^{12(5)} = 4011.219586$

She will have \$3979.58 if it is compounded quarterly and \$4011.22 if it is compounded monthly.

43. $B(t) = 4000(2^t)$
$B(3) = 4000(2^3) = 32{,}000$
$B(9) = 4000(2^9) = 2{,}048{,}000$

At the end of 3 hours there will be 32,000 bacteria in the culture and at the end of 9 hours there will be 2,048,000 bacteria in the culture.

45. $f(t) = (1 - 0.08)^t = 0.92^t$
$f(5) = 0.92^5 = 0.65908...$
$f(25) = 0.92^{25} = 0.1243... > 0.10$

In 5 years approximately 65.9% of the homeowners will still be using septic tanks. At the present rate, 12.4% of the homeowners will still be using septic tanks in 25 years so the goal of less than 10% will not be achieved.

47. $A = Ce^{-0.0004297t}$
$A = 6e^{-0.0004297(1000)} \approx 3.91$

There will be approximately 3.91 mg of radium in the container after 1000 years.

49. $P = 14.7e^{-0.21d}$
$P = 14.7e^{-0.21(2)} \approx 9.66$

The man will experience a pressure of approximately 9.66 lb/in^2.

51. $N = 0.00472e^{0.11596t}$
$N = 0.00472e^{0.11596(80)} \approx 50.4$
$N = 0.00472e^{0.11596(90)} \approx 160.8$
$\frac{160.8 - 50.4}{50.4} \approx 2.19$

About 50.4 million stocks were traded in 1980 and about 160.8 million stocks were traded in 1990, an increase of about 219%.

53. From the graph the world's population reached three billion people sometime in 1955.

55. $5.68(1.017)^{10} \approx 6.7$. The world's population would be about 6.7 billion people in 2005.

57. $f(x) = \frac{e^x + e^{-x}}{2}$

X	Y1
-1	1.5431
-.5	1.1276
0	1
.5	1.1276
1	1.5431
1.5	2.3524
2	3.7622

Y1 = cosh(X)

57. (continued)

$f(x) = \frac{e^x + e^{-x}}{2} = \cosh x$

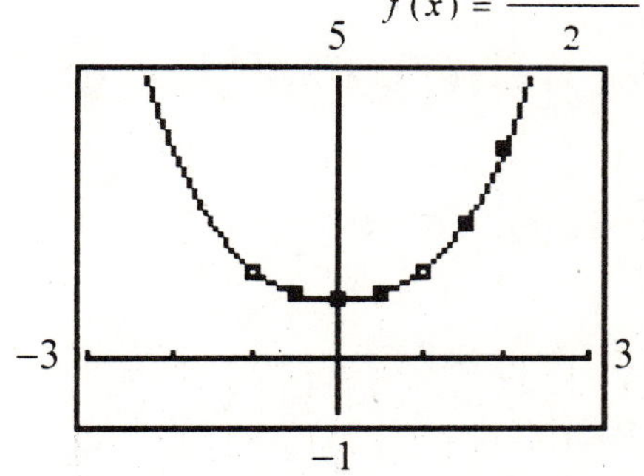

Cumulative Review Problems

59. $5 - 2(3 - x) = 2(2x + 5) + 1$
$5 - 6 + 2x = 4x + 10 + 1$
$2x = -12$
$x = -6$

12.2 Exercises

1. A logarithm is an <u>exponent</u>.

3. In the equation $y = \log_b x$, the domain (the permitted values of x) is $\underline{x > 0.}$

5. $81 = 3^4 \Leftrightarrow \log_3 81 = 4$

7. $36 = 6^2 \Leftrightarrow \log_6 36 = 2$

9. $\frac{1}{25} = 5^{-2} \Leftrightarrow \log_5 \frac{1}{25} = -2$

11. $\frac{1}{32} = 2^{-5} \Leftrightarrow \log_2 \frac{1}{32} = -5$

13. $y = e^5 \Leftrightarrow \log_e y = 5$

15. $2 = \log_3 9 \Leftrightarrow 3^2 = 9$

17. $0 = \log_5 1 \Leftrightarrow 5^0 = 1$

19. $\frac{1}{2} = \log_{16} 4 \Leftrightarrow 16^{1/2} = 4$

21. $-2 = \log_{10}(0.01) \Leftrightarrow 10^{-2} = 0.01$

23. $-4 = \log_3\left(\frac{1}{81}\right) \Leftrightarrow 3^{-4} = \frac{1}{81}$

25. $\frac{2}{3} = \log_e x \Leftrightarrow e^{2/3} = x$

27. $\log_2 x = 4 \Leftrightarrow 2^4 = x$
$x = 16$

29. $\log_{10} x = -3 \Leftrightarrow 10^{-3} = x$
$x = \frac{1}{1000}$

31. $\log_4 64 = y \Leftrightarrow 4^y = 64 = 4^3$
$y = 3$

33. $\log_8\left(\frac{1}{64}\right) = y \Leftrightarrow 8^y = \frac{1}{64} = \frac{1}{8^2} = 8^{-2}$
$y = -2$

35. $\log_a 144 = 2 \Leftrightarrow a^2 = 144 = 12^2$
$a = 12$

37. $\log_a 1000 = 3 \Leftrightarrow a^3 = 1000 = 10^3$
$a = 10$

39. $\log_{25} 5 = w \Leftrightarrow 25^w = 5$
$(5^2)^w = 5^{2w} = 5^1$
$2w = 1$
$w = \frac{1}{2}$

41. $\log_3\left(\frac{1}{3}\right) = w$
$3^w = \frac{1}{3} = \frac{1}{3^1} = 3^{-1}$
$w = -1$

43. $\log_{15} w = 0 \Leftrightarrow 15^0 = w$
$w = 1$

45. $\log_w 81 = -2 \Leftrightarrow w^{-2} = 81$
$w^2 = \frac{1}{81}$
$w = \pm\frac{1}{9}$, pick + since $w > 0$
$w = \frac{1}{9}$

47. $\log_{10}(0.001) = x \Leftrightarrow 10^x = 0.001 = 10^{-3}$
$x = -3$

49. $\log_2 128 = x \Leftrightarrow 2^x = 128 = 2^7$
$x = 7$

51. $\log_{23} 1 = x \Leftrightarrow 23^x = 1 = 23^0$
$x = 0$

53. $\log_6 \sqrt{6} = x \Leftrightarrow 6^x = \sqrt{6} = 6^{1/2}$
$x = \frac{1}{2}$

55. $\log_2 64 = x \Leftrightarrow 2^x = 64 = 2^6$
$x = 6$

57. $\log_4 x = y$
$4^y = x$

(continued)

57. (continued)

$x = 4^y$	y
$\frac{1}{16}$	−2
$\frac{1}{4}$	−1
1	0
4	1
16	2

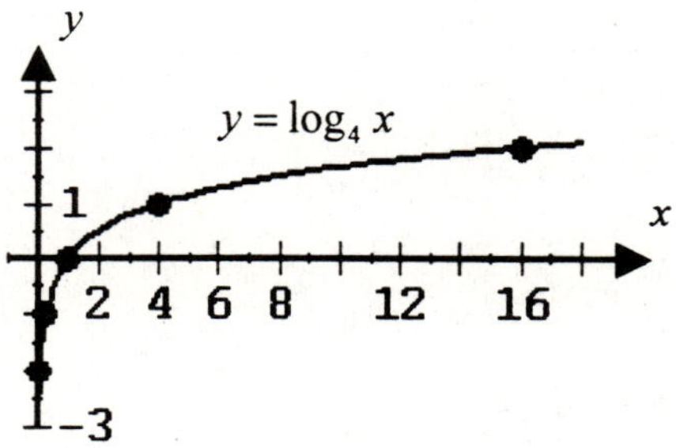

59. $\log_{1/3} x = y \Leftrightarrow (1/3)^y = x$

$x = (1/3)^y$	y
9	−2
3	−1
1	0
1/3	1
1/9	2

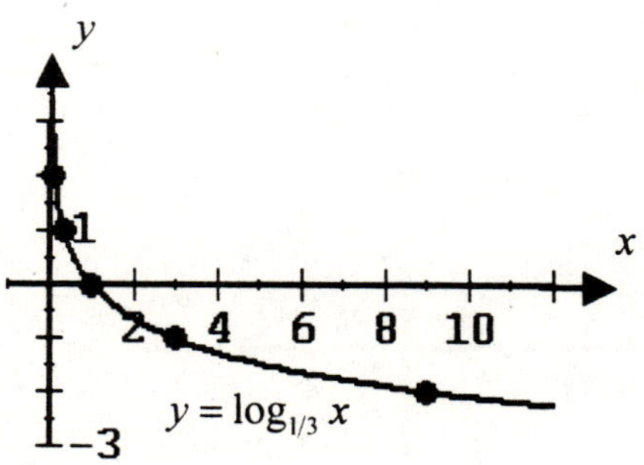

61. $\log_{10} x = y \Leftrightarrow 10^y = x$

$x = 10^y$	y
1/100	−2
1/10	−1
1	0
10	1
100	2

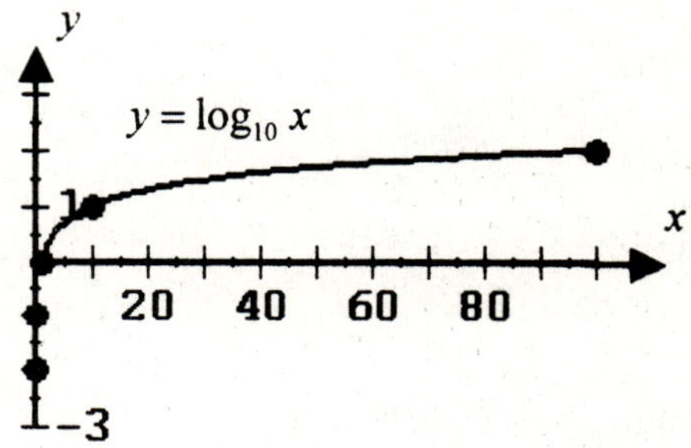

63. $f(x) = \log_3 x,\ f(x) \to y$

$y = \log_3 x \Leftrightarrow 3^y = x$

$x = 3^y$	y
1/9	−2
1/3	−1
1	0
3	1
9	2

$f^{-1}(x) = 3^x,\ f^{-1}(x) \to y$

$y = 3^x$

x	$y = 3^x$
−2	1/9
−1	1/3
0	0
1	3
2	9

(continued)

63. (continued)

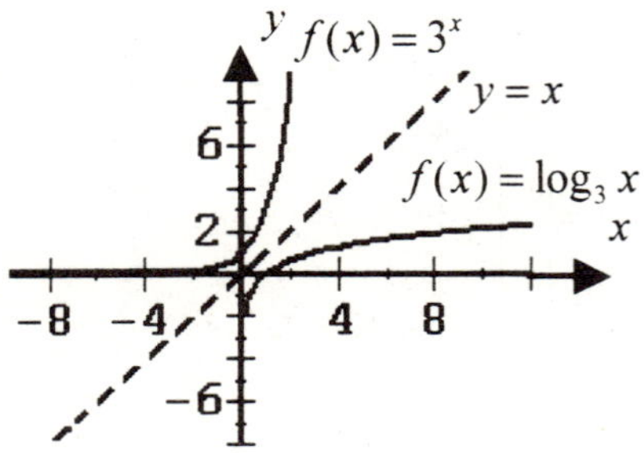

65. $\text{pH}=-\log_{10}\left[\text{H}^+\right]$

$\text{pH}=-\log_{10}\left[10^{-2.5}\right]$

$-\text{pH}=\log_{10}\left[10^{-2.5}\right]$

$10^{-\text{pH}}=10^{-2.5}$

$-\text{pH}=-2.5$

$\text{pH}=2.5$

The pH is 2.5.

67. $\text{pH}=-\log_{10}\left[\text{H}^+\right]$

$8=-\log_{10}\left[\text{H}^+\right]$

$-8=\log_{10}\left[\text{H}^+\right]$

$10^{-8}=\text{H}^+$

The concentration of hydrogen ions in the solution is 10^{-8}.

69. $\text{pH}=-\log_{10}\left[\text{H}^+\right]$

$9.25=-\log_{10}\left[\text{H}^+\right]$

$-9.25=\log_{10}\left[\text{H}^+\right]$

$\text{H}^+=10^{-9.25}=5.623\times10^{-10}$

The concentration of hydrogen ions in the solution is 5.623×10^{-10}.

71. $N=1200+(2500)(\log_{10} d)$

$\log_{10} d=\dfrac{N-1200}{2500}$

$\log_{10} 100{,}000=\dfrac{N-1200}{2500}$

$10^{\frac{N-1200}{2500}}=100{,}000=10^5$

$\dfrac{N-1200}{2500}=5$

$N-1200=12{,}500$

$N=13{,}700$

13,700 sets of software were sold.

73. $N=1200+(2500)(\log_{10} d)$

$\log_{10} d=\dfrac{18{,}700-1200}{2500}$

$\log_{10} d=7$

$d=10^7$

$d=10{,}000{,}000$

They should spend \$10,000,000.

75. $5^{\log_5 4}=y\Leftrightarrow \log_5 4=\log_5 y\rightarrow y=4$

Cumulative Review Problems

77. $y=-\dfrac{2}{3}x+5$,

x	y
0	5
3	3

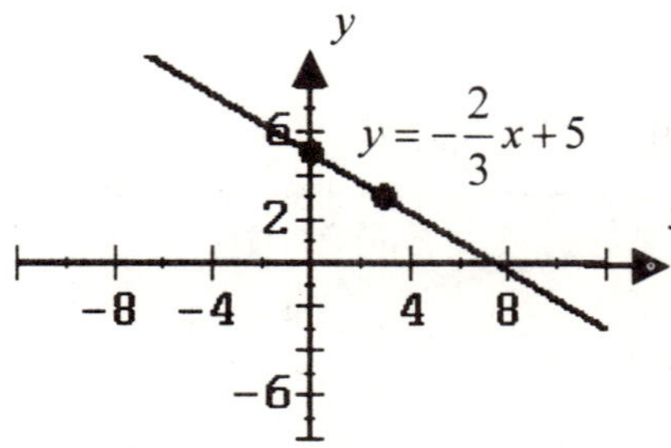

79. $m = \frac{y_2 - y_1}{x_2 - x_1} = \frac{3-2}{-6-(-1)} = -\frac{1}{5}$

81. (a) $C(t) = P(1.04)^t$
$C(5) = 4400(1.04)^5 = \$5353.27$
(b) $C(t) = P(1.04)^t$
$C(10) = 16,500(1.04)^{10} = \$24,424.03$

83. (a) $C(t) = P(1.04)^t$
$C(5) = 4400(1.04)^5$
$C(5) = \$5353.27$
(b) $C(10) = 16,500(1.04)^{10}$
$C(10) = \$24,424.03$

12.3 Exercises

1. $\log_7 MN = \log_7 M + \log_7 N$

3. $\log_5(7 \cdot 11) = \log_5 7 + \log_5 11$

5. $\log_b 9f = \log_b 9 + \log_b f$

7. $\log_9\left(\frac{2}{7}\right) = \log_9 2 - \log_9 7$

9. $\log_a\left(\frac{G}{7}\right) = \log_a G - \log_a 7$

11. $\log_a\left(\frac{E}{F}\right) = \log_a E - \log_a F$

13. $\log_8 a^7 = 7\log_8 a$

15. $\log_b A^{-2} = -2\log_b A$

17. $\log_5 \sqrt{w} = \log_5 w^{1/2}$
$= \frac{1}{2}\log_5 w$

19. $\log_5 \sqrt{x}y^3 = \log_5 x^{1/2} + \log_5 y^3$
$= \frac{1}{2}\log_5 x + 3\log_5 y$

21. $\log_{13}\left(\frac{5B}{A^2}\right) = \log_{13}(5B) - \log_{13} A^2$
$= \log_{13} 5 + \log_{13} B - 2\log_{13} A$

23. $\log_2\left(\frac{5xy^4}{\sqrt{z}}\right)$
$= \log_2(5xy^4) - \log_2 \sqrt{z}$
$= \log_2 5 + \log_2 x + \log_2 y^4 - \log_2 z^{1/2}$
$= \log_2 5 + \log_2 x + 4\log_2 y - \frac{1}{2}\log_2 z$

25. $\log_b \sqrt[3]{\frac{x}{y^2 z}}$
$= \log_b\left(\frac{x}{y^2 z}\right)^{1/3}$
$= \frac{1}{3}\log_b\left(\frac{x}{y^2 z}\right)$
$= \frac{1}{3}\left[\log_b x - \log_b(y^2 z)\right]$
$= \frac{1}{3}\left[\log_b x - \log_b y^2 - \log_b z\right]$
$= \frac{1}{3}[\log_b x - 2\log_b y - \log_b z]$
$= \frac{1}{3}\log_b x - \frac{2}{3}\log_b y - \frac{1}{3}\log_b z$

27. $\log_4 13 + \log_4 y + \log_4 3$
$= \log_4(13 \cdot y \cdot 3) = \log_4(39y)$

29. $5\log_3 x - \log_3 7 = \log_3 x^5 - \log_3 7$
$= \log_3 \frac{x^5}{7}$

31. $\frac{2}{3}\log_b x + \frac{1}{2}\log_b y - 3\log_b z$
$= \log_b x^{2/3} + \log_b y^{1/2} - \log_b z^3$
$= \log_b \sqrt[3]{x^2} + \log_b \sqrt{y} - \log_b z^3$
$= \log_b \frac{\sqrt[3]{x^2}\sqrt{y}}{z^3}$

33. $\log_3 3 = 1$

35. $\log_e e = 1$

37. $\log_9 1 = 0$

39. $\log_5 5 + \log_5 1 = 1 + 0 = 1$

41. $\log_8 x = \log_8 7 \Rightarrow x = 7$

43. $\log_5(2x+7) = \log_5(29)$
$2x + 7 = 29$
$2x = 22$
$x = 11$

45. $\log_3 1 = x \Rightarrow x = 0$

47. $\log_7 7 = x \Rightarrow x = 1$

49. $\log_{10} x + \log_{10} 25 = 2$
$\log_{10}(25x) = 2 \Leftrightarrow 10^2 = 25x$
$25x = 100 \rightarrow x = 4$

51. $\log_2 7 = \log_2 x - \log_2 3$
$\log_2 7 = \log_2 \frac{x}{3}$
$\frac{x}{3} = 7$
$x = 21$

53. $3\log_5 x = \log_5 8$
$\log_5 x^3 = \log_5 8$
$x^3 = 8$
$x = 2$

55. $\log_e x - \log_e 2 = 2$
$\log_e \frac{x}{2} = 2 \Leftrightarrow e^2 = \frac{x}{2}$
$x = 2e^2$

57. $\log_6(5x+21) - \log_6(x+3) = 1$
$\log_6 \frac{5x+21}{x+3} = 1 \Rightarrow \frac{5x+21}{x+3} = 6$
$5x + 21 = 6x + 18$
$x = 3$

59. $5^{\log_5 4} + 3^{\log_3 2} = 4 + 2 = 6$

61. Let $\log_b M = x,\ \log_b N = y \Leftrightarrow$
$b^x = M,\ b^y = N$
$\frac{M}{N} = \frac{b^x}{b^y}$
$\frac{M}{N} = b^{x-y}$
$\log_b \frac{M}{N} = \log_b b^{x-y}$
$\log_b \frac{M}{N} = x - y$
$\log_b \frac{M}{N} = \log_b M - \log_b N$

Cumulative Review Problems

63. $A = \pi r^2 = \pi(4)^2 \approx 50.27 \text{ m}^2$

65. $$\begin{aligned} 5x+3y=9 &\xrightarrow{\times 2} 10x+6y=18 \\ 7x-2y=25 &\xrightarrow{\times 3} \underline{21x-6y=75} \\ & \quad 31x \qquad = 93 \end{aligned}$$

$x = 3$

$5(3)+3y=9$

$3y=-6$

$y=-2$

$(3,-2)$ is the solution.

67. $\dfrac{9.30\times10^8 - 8.01\times10^8}{8.01\times10^8} = 0.161048...$

$\dfrac{E-9.3\times10^8}{9.3\times10^8} = 0.161048...$

$E \approx 1.08\times10^9$

Emissions increased 16.1% from 1996 to 2000. Emissions in 2004 will be 1.08×10^9 metric tons.

69. $\dfrac{84+80+92+95+S}{5} = 88$

$351+S=440$

$S=89$

12.4 Exercises

1. $\log 5.13 \approx 0.7101173651$

3. $\log 25.6 \approx 1.408239965$

5. $\log 356 \approx 2.551449998$

7. $\log 125{,}000 \approx 5.096910013$

9. $\log 0.0123 \approx -1.910094889$

11. Error. You cannot take the log of a negative number.

13. $\log x = 2.016$
$x = 10^{2.016} \approx 103.7528416$

15. $\log x = 1.7860$
$x = 10^{1.7860} \approx 61.09420249$

17. $\log x = 3.9304$
$x = 10^{3.9304} \approx 8519.223264$

19. $\log x = 6.4683$
$x = 10^{6.4683} \approx 2939679.609$

21. $\log x = -3.3893$
$x = 10^{-3.3893} \approx 0.000408037$

23. $\log x = 2.0030$
$x = 10^{2.0030} \approx 100.6391669$

25. $\text{antilog}(7.6215) \approx 41831168.87$

27. $\text{antilog}(-1.0826) \approx 0.0826799109$

29. $\ln 5.62 \approx 1.726331664$

31. $\ln 107 \approx 4.672828834$

33. $\ln 136{,}000 \approx 11.82041016$

35. $\ln 0.00579 \approx -5.151622987$

37. $\ln x = 0.95$
$x = e^{0.95} \approx 2.585709659$

39. $\ln x = 2.4$
$x = e^{2.4} \approx 11.02317638$

41. $\ln x = -0.13$
$x = e^{-0.13} \approx 0.8780954309$

43. $\ln x = -2.7$
$x = e^{-2.7} \approx 0.0672055127$

45. $\text{antilog}_e(6.1582) \approx 472.5766708$

47. $\text{antilog}_e(-2.1298) \approx 0.1188610637$

49. $\log_3 9.2 = \dfrac{\log 9.2}{\log 3} \approx 2.020006063$

51. $\log_7(7.35) = \dfrac{\log 7.35}{\log 7} \approx 1.025073184$

53. $\log_6 0.127 = \dfrac{\log 0.127}{\log 6} \approx -1.151699337$

55. $\log_{11} 128 = \dfrac{\log 128}{\log 11} \approx 2.023453784$

57. $\log_4 0.07733 = \dfrac{\ln 0.07733}{\ln 4} \approx -1.846414$

59. $\log_{18} 98{,}265 = \dfrac{\ln 98{,}265}{\ln 18} \approx 3.97714348$

61. $\ln 1537 \approx 7.337587744$

63. $\text{antilog}_e(-1.874) \approx 0.1535083985$

65. $\log x = 8.5634$
$x = 10^{8.5634} \approx 3.65931672 \times 10^8$

67. $\log_4 x = 0.8645 \Leftrightarrow x = 4^{0.8645}$
$x \approx 3.314979618$

69. $y = \log_6 x$

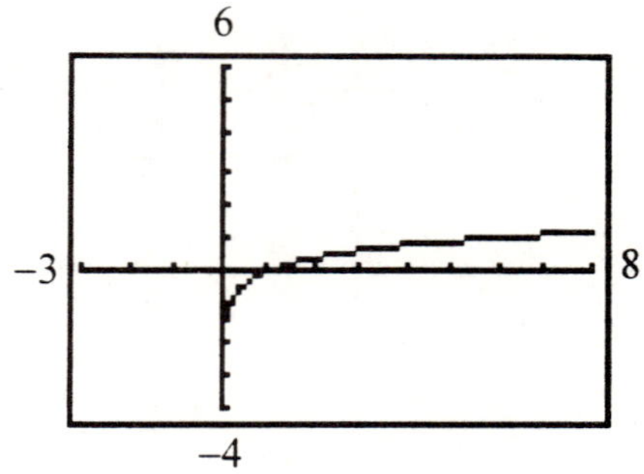

71. $y = \log_{0.2} x$

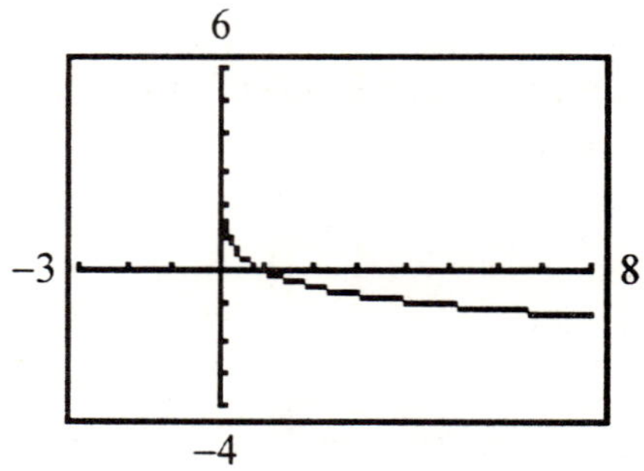

73. $N = 32.82 + 1.0249 \ln x$
$N = 32.82 + 1.0249 \ln 10 \approx 35.18$ in 2000
$N = 32.82 + 1.0249 \ln 20 \approx 35.89$ in 2010
$\dfrac{35.89 - 35.18}{35.18} \approx 0.02$, a 2% increase

75. $R = \log x$
$R = \log 184{,}000$
$R \approx 5.26$

77. $R = \log x$
$5.4 = \log x \Leftrightarrow x = 10^{5.4} \approx 251188.6432$

The shock wave is about 251,000 times greater than the smallest detectable shock wave.

Cumulative Review Problems

79. $2y^2+4y-3=0$

$$y=\frac{-4\pm\sqrt{4^2-4(2)(-3)}}{2(2)}=\frac{-4\pm\sqrt{40}}{4}$$

$$y=\frac{-4\pm\sqrt{4\cdot 10}}{4}=\frac{-4\pm 2\sqrt{10}}{4}$$

$$y=\frac{-2\pm\sqrt{10}}{2}$$

81. $(x-2)^2+(y+1)^2=3^2=9$

83. Let x, y, z, w, s be the distances between adjacent exits, then:

$$x+y+z+w+s=36$$
$$x+y=12\rightarrow \text{first eqn}$$
$$y+z=15$$
$$z+w=12\rightarrow \text{first eqn}$$
$$w+s=15$$
$$12+12+s=36$$
$$s=12$$
$$w+s=w+12=15$$
$$w=3$$
$$z+w=z+3=12$$
$$z=9$$
$$y+z=y+9=15$$
$$y=6$$
$$x+y=x+6=12$$
$$x=6$$

distance between Exit 1 and Exit 4:
$x+y+z=6+6+9=21$ miles
distance between Exit 1 and Exit 5:
$x+y+z+w$
$=6+6+9+3=12+12=24$ miles

12.5 Exercises

1. $\log_8(x+12)+\log_8 4=2$

$$\log_8(4x+48)=2 \Leftrightarrow 4x+48=8^2$$
$$4x+48=64$$
$$4x=16,\ x=4$$

check: $\log_8(4+12)+\log_8 4\stackrel{?}{=}2$
$$\log_8 16+\log_8 4\stackrel{?}{=}2$$
$$\log_8 64\stackrel{?}{=}2$$
$$2=2$$

3. $\log_5(x-2)+\log_5 3=2$

$$\log_5(3x-6)=2 \Leftrightarrow 3x-6=5^2=25$$
$$3x=31,\ x=\frac{31}{3}$$

check: $\log_5\left(\frac{31}{3}-2\right)+\log_5 3\stackrel{?}{=}2$
$$\log_5\left(3\left(\frac{31}{3}-2\right)\right)\stackrel{?}{=}2$$
$$\log_5 25\stackrel{?}{=}2$$
$$2=2$$

5. $\log_3(2x-1)=2-\log_3 4$

$$\log_3(2x-1)+\log_3 4=2$$
$$\log_3(8x-4)=2 \Leftrightarrow 8x-4=3^2=9$$
$$8x=13,\ x=\frac{13}{8}$$

check: $\log_3\left(2\cdot\frac{13}{8}-1\right)\stackrel{?}{=}2-\log_3 4$
$$\log_3\left(\frac{9}{4}\right)\stackrel{?}{=}2-\log_3 4$$
$$\log_3 9-\log_3 4\stackrel{?}{=}2-\log_3 4$$
$$2-\log_3 4=2-\log_3 4$$

7. $$\log(30x+40) = 2+\log(x-1)$$
$$\log(30x+40)-\log(x-1) = 2$$
$$\log\left(\frac{30x+40}{x-1}\right) = 2 \Leftrightarrow \frac{30x+40}{x-1} = 10^2$$
$$30x+40 = 100x-100$$
$$70x = 140$$
$$x = 2$$
check: $\log(30(2)+40) \overset{?}{=} 2+\log(2-1)$
$$\log 100 \overset{?}{=} 2+\log(1)$$
$$2 \overset{?}{=} 2+\log(1) = 2+0$$
$$2 = 2$$

9. $$2+\log_6(x-1) = \log_6(12x)$$
$$\log_6(x-1)-\log_6(12x) = -2$$
$$\log_6\left(\frac{x-1}{12x}\right) = -2 \Leftrightarrow \frac{x-1}{12x} = 6^{-2}$$
$$36x-36 = 12x$$
$$24x = 36$$
$$x = \frac{3}{2}$$
check: $2+\log_6\left(\frac{3}{2}-1\right) \overset{?}{=} \log_6\left(12\cdot\frac{3}{2}\right)$
$$2+\log_6\frac{1}{2} \overset{?}{=} \log_6\frac{36}{2}$$
$$2+\log_6 1-\log_6 2 \overset{?}{=} \log_6\frac{36}{2}$$
$$2+0-\log_6 2 \overset{?}{=} \log_6 36-\log_6 2$$
$$2-\log_6 2 = 2-\log_6 2$$

11. $\log(x+20)-\log x = 2$
$$\log(x+20)-\log x = 2$$
$$\log\left(\frac{x+20}{x}\right) = 2 \Leftrightarrow \frac{x+20}{x} = 10^2$$
$$100x = x+20$$
$$99x = 20,\ x = \frac{20}{99}$$
check: $\log\left(\frac{20}{99}+20\right)-\log\left(\frac{20}{99}\right) \overset{?}{=} 2$
$$\log\left(\frac{2000}{99}\right)-\log\left(\frac{20}{99}\right) \overset{?}{=} 2$$
$$\log\left(\frac{100\cdot 20}{99}\right)-\log\left(\frac{20}{99}\right) \overset{?}{=} 2$$
$$\log 100+\log\left(\frac{20}{99}\right)-\log\left(\frac{20}{99}\right) \overset{?}{=} 2$$
$$2 = 2$$

13. $2\log_3 4 = \log_3 x-\log_3(x-1)$
$$\log_3 4^2 = \log_3\left(\frac{x}{x-1}\right)$$
$$\frac{x}{x-1} = 4^2 = 16$$
$$x = 16x-16$$
$$15x = 16$$
$$x = \frac{16}{15}$$
check: $2\log_3 4 \overset{?}{=} \log_3\left(\frac{16}{15}\right)-\log_3\left(\frac{16}{15}-1\right)$
$$\log_3 4^2 \overset{?}{=} \log_3\left(\frac{16}{15}\right)-\log_3\left(\frac{1}{15}\right)$$
$$\log_3 4^2 \overset{?}{=} \log_3\left(\frac{\frac{16}{15}}{\left(\frac{1}{15}\right)}\right) = \log_3 16$$
$$\log_3 16 = \log_3 16$$

15. $1+\log(x-2)=\log(6x)$

$\log(6x)-\log(x-2)=1$

$\log\dfrac{6x}{x-2}=1 \Leftrightarrow \dfrac{6x}{x-2}=10^1$

$6x=10x-20$

$4x=20$

$x=5$

check: $1+\log(5-2)\overset{?}{=}\log(6\cdot 5)=\log 30$

$1+\log 3\overset{?}{=}\log(10\cdot 3)$

$1+\log 3\overset{?}{=}\log 10+\log 3$

$1+\log 3=1+\log 3$

17. $\log_2(x+5)-2=\log_2 x$

$\log_2(x+5)-\log_2 x=2$

$\log_2\dfrac{x+5}{x}=2 \Leftrightarrow \dfrac{x+5}{x}=2^2=4$

$x+5=4x$

$3x=5$

$x=\dfrac{5}{3}$

check: $\log_2\left(\dfrac{5}{3}+5\right)-2\overset{?}{=}\log_2\dfrac{5}{3}$

$\log_2\dfrac{20}{3}-2\overset{?}{=}\log_2\dfrac{5}{3}$

$\log_2\left(4\cdot\dfrac{5}{3}\right)-2\overset{?}{=}\log_2\dfrac{5}{3}$

$\log_2 4+\log_2\dfrac{5}{3}-2\overset{?}{=}\log_2\dfrac{5}{3}$

$2+\log_2\dfrac{5}{3}-2\overset{?}{=}\log_2\dfrac{5}{3}$

$\log_2\dfrac{5}{3}=\log_2\dfrac{5}{3}$

19. $2\log_7 x=\log_7(x+4)+\log_7 2$

$\log_7 x^2-\log_7(x+4)=\log_7 2$

$\log_7\dfrac{x^2}{(x+4)}=\log_7 2$

$\dfrac{x^2}{(x+4)}=2$

$x^2-2x-8=0$

$(x-4)(x+2)=0$

$x-4=0,\ x+2=0$

$x=4,\qquad x=-2$

$x=4$, $x=-2$, reject, gives $\log_7$(negative)

check: $2\log_7 4\overset{?}{=}\log_7(4+4)+\log_7 2$

$\log_7 4^2\overset{?}{=}\log_7(8)+\log_7 2$

$\log_7 16\overset{?}{=}\log_7(8\cdot 2)$

$\log_7 16=\log_7 16$

21. $\ln 10-\ln x=\ln(x-3)$

$\ln\dfrac{10}{x}=\ln(x-3)$

$\dfrac{10}{x}=x-3$

$x^2-3x-10=0$

$(x-5)(x+2)=0$

$x-5=0,\ x+2=0$

$x=5,\qquad x=-2$

$x=5$, $x=-2$, reject, gives ln(negative)

check: $\ln 10-\ln 5\overset{?}{=}\ln(5-3)$

$\ln\dfrac{10}{5}\overset{?}{=}\ln 2$

$\ln 2=\ln 2$

23.
$$8^{x-1} = 11$$
$$\log 8^{x-1} = \log 11$$
$$(x-1)\log 8 = \log 11$$
$$x-1 = \frac{\log 11}{\log 8}$$
$$x = 1 + \frac{\log 11}{\log 8}$$
$$x = \frac{\log 8}{\log 8} + \frac{\log 11}{\log 8}$$
$$x = \frac{\log 8 + \log 11}{\log 8}$$

25.
$$2^{3x+4} = 17$$
$$\log 2^{3x+4} = \log 17$$
$$(3x+4)\log 2 = \log 17$$
$$3x+4 = \frac{\log 17}{\log 2}$$
$$3x = \frac{\log 17}{\log 2} - 4 \cdot \frac{\log 2}{\log 2}$$
$$x = \frac{\log 17}{3\log 2} - 4 \cdot \frac{\log 2}{3\log 2}$$
$$x = \frac{\log 17 - 4\log 2}{3\log 2}$$

27.
$$6^{4x-1} = 225$$
$$\log 6^{4x-1} = \log 225$$
$$(4x-1)\log 6 = \log 225$$
$$4x-1 = \frac{\log 225}{\log 6}$$
$$4x = 1 + \frac{\log 225}{\log 6}$$
$$4x = \frac{\log 6 + \log 225}{\log 6}$$
$$x = \frac{\log 6 + \log 225}{4\log 6} \approx 1.006$$

29.
$$5^x = 4^{x+1}$$
$$\log 5^x = \log 4^{x+1}$$
$$x\log 5 = (x+1)\log 4$$
$$x\log 5 = x\log 4 + \log 4$$
$$x(\log 5 - \log 4) = \log 4$$
$$x = \frac{\log 4}{\log 5 - \log 4}$$
$$x \approx 6.213$$

31.
$$e^{x-1} = 28$$
$$\ln e^{x-1} = \ln 28$$
$$(x-1)\ln e = \ln 28$$
$$x = 1 + \ln 28$$
$$x \approx 4.332$$

33.
$$88 = e^{2x+1}$$
$$\ln 88 = \ln e^{2x+1}$$
$$\ln 88 = (2x+1)\ln e$$
$$\ln 88 = (2x+1)(1)$$
$$\ln 88 = 2x+1$$
$$2x = \ln 88 - 1$$
$$x = \frac{\ln 88 - 1}{2}$$
$$x \approx 1.739$$

35.
$$A = P(1+r)^t$$
$$5000 = 1500(1+0.08)^t$$
$$1.08^t = \frac{10}{3}$$
$$\ln 1.08^t = \ln \frac{10}{3}$$
$$t = \frac{\ln \frac{10}{3}}{\ln 1.08} = 15.64392564...$$

It will take approximately 16 years.

37. $$A = P(1+r)^t$$
$$3P = P(1+0.06)^t$$
$$1.06^t = 3$$
$$\ln 1.06^t = \ln 3$$
$$t \ln 1.06 = \ln 3$$
$$t = \frac{\ln 3}{\ln 1.06}$$
$$t = 18.85417668...$$

It will take approximately 19 years.

39. $$A = P(1+r)^t$$
$$6500 = 5000(1+r)^6$$
$$1.3 = (1+r)^6$$
$$\ln 1.3 = \ln(1+r)^6 = 6\ln(1+r)$$
$$\ln(1+r) = \frac{\ln 1.3}{6}$$
$$e^{\ln(1+r)} = e^{\frac{\ln 1.3}{6}}$$
$$1+r = e^{\frac{\ln 1.3}{6}}$$
$$r = e^{\frac{\ln 1.3}{6}} - 1$$
$$r = 0.0446975079...$$

The rate is approximately 4.5%.

41. $$A = A_0 e^{rt}$$
$$9 = 6e^{0.02t},\ 1.5 = e^{0.02t}$$
$$\ln 1.5 = \ln e^{0.02t}$$
$$\ln 1.5 = 0.02t$$
$$t = \frac{\ln 1.5}{0.02} = 20.27325541...$$

It would take approximately 20 years.

43. $$A = A_0 e^{rt}$$
$$4A_0 = A_0 e^{0.02t}.\ 4 = e^{0.02t}$$
$$\ln 4 = \ln e^{0.02t}$$
$$\ln 4 = 0.02t$$
$$t = \frac{\ln 4}{0.02} = 69.31471806...$$

It would take approximately 69 years.

45. $$N = 20{,}800(1.264)^x$$
$$N = 20{,}800(1.264)^{13}$$
$$N = 437{,}295.79$$

In 2003 there will be approximately 437,000 employees.

47. $$N = 20{,}800(1.264)^x$$
$$274{,}000 = 20{,}800(1.264)^x$$
$$1.264^x = \frac{274{,}000}{20{,}800}$$
$$\ln 1.264^x = \ln\frac{274{,}000}{20{,}800}$$
$$x = \frac{\ln\frac{274{,}000}{20{,}800}}{\ln 1.264} = 11.00461354...$$

The number of employees will reach 274,000 sometime in 2001.

49. $A = A_0e^{rt}$

$3.5 = 3e^{0.03t}$

$e^{0.03t} = \frac{7}{6}$

$\ln e^{0.03t} = \ln\frac{7}{6}$

$0.03t = \ln\frac{7}{6}$

$t = \frac{\ln\frac{7}{6}}{0.03} = 5.138355994...$

There will be 3.5 million people in approximately 5 years.

51. $A = A_0e^{rt}$

$1800 = 200e^{0.04t}$

$e^{0.04t} = 9$

$0.04t = \ln 9$

$t = \frac{\ln 9}{0.04} = 54.93061443...$

It will take approximately 55 hours.

53. $A = A_0e^{rt}$

$A = 24{,}500e^{0.05(13)} = 46930.75031...$

By the end of 2010 approximately 46,931 people will be infected.

55. $R = \log\left(\frac{I}{I_0}\right)$

$6.8 = \log\left(\frac{I_N}{I_0}\right) = \log I_N - \log I_0$

$7.2 = \log\left(\frac{I_J}{I_0}\right) = \log I_J - \log I_0$

(continued)

55. (continued)

Subtracting the two equations gives

$-0.4 = \log I_N - \log I_J = \log\frac{I_N}{I_J}$

$10^{\log\frac{I_N}{I_J}} = 10^{-0.4}$

$\frac{I_N}{I_J} = 10^{-0.4}$

$I_J = 10^{0.4}I_N \approx 2.5I_N$

The Japan earthquake was about 2.5 times as intense as the Northridge earthquake.

57. $R = \log\left(\frac{I}{I_0}\right)$

$8.3 = \log\left(\frac{I_S}{I_0}\right) = \log I_S - \log I_0$

$6.8 = \log\left(\frac{I_J}{I_0}\right) = \log I_J - \log I_0$

Subtracting the two equations gives

$1.5 = \log I_S - \log I_J = \log\frac{I_S}{I_J}$

$10^{\log\frac{I_S}{I_J}} = 10^{1.5}$

$\frac{I_S}{I_J} = 10^{1.5}$

$I_S = 10^{1.5}I_J \approx 31.6I_J$

The intensity of the San Francisco earthquake was approximately 31.6 times greater than the intensity of the Japanese earthquake.

59. Graph $y_1 = 300e^{0.12x}$ and $y_2 = 750 + 100x$ on the same screen and use the *intersect* feature to solve.

(continued)

59. (continued)

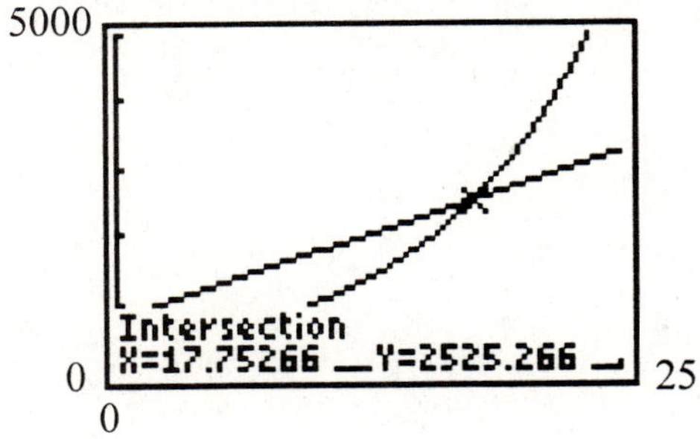

Populations will be equal in ≈ 17.8 years.

Cumulative Review Problems

61. $\sqrt{98x^3y^2} = \sqrt{49x^2y^2 \cdot 2x} = 7xy\sqrt{2x}$

63. $\dfrac{40,500}{1.08} = 37,500$

65. $|3x-5| = 7$

$3x-5=7$ or $3x-5=-7$

$3x=12 \qquad 3x=-2$

$x=4 \qquad x=-\dfrac{2}{3}$

Putting Your Skills to Work

1. $T = C + (T_0 - C)e^{-kt}$

$140 = 72 + (220-72)e^{-k(30)}$

$68 = 148e^{-k(30)}$

$e^{k(30)} = \dfrac{148}{68}$

$\ln e^{k(30)} = \ln\dfrac{148}{68}$

$k(30) = \ln\dfrac{148}{68}$

$k = \dfrac{\ln\frac{148}{68}}{30} \approx 0.026$

2. $T = C + (T_0 - C)e^{-kt}$

$T = 72 + (220-72)e^{-0.26t}$

3. $T = 72 + (220-72)e^{-0.26t}$

$90 = 72 + (220-72)e^{-0.026t}$

$18 = 148e^{-0.026t}$

$e^{0.026t} = \dfrac{148}{18}$

$0.026t = \ln\dfrac{148}{18}$

$t = \dfrac{\ln\frac{148}{18}}{0.026} = 81.03232753...$

It will take approximately 81 minutes for the pie to cool.

4. $T = C + (T_0 - C)e^{-kt}$

ceramic: $T = 70 + (190-70)e^{-0.13722(20)}$

$T = 77.7°\text{F}$

foam: $T = 70 + (190-70)e^{-0.05(20)}$

$T = 114.1°\text{F}$

cardboard: $T = 70 + (190-70)e^{-0.08(20)}$

$T = 94.2°\text{F}$

5. ceramic: $125 = 70 + (190-70)e^{-0.13722t}$

$125 = 70 + (190-70)e^{-0.13722t}$

$e^{0.13722t} = \dfrac{120}{55}$

$0.13722t = \ln\dfrac{120}{55}$

$t = \dfrac{\ln\frac{120}{55}}{0.13722} \approx 5.7$ minutes

foam: $125 = 70 + (190-70)e^{-0.05t}$

(continued)

5. (continued)

$$125 = 70 + (190 - 70)e^{-0.05t}$$

$$e^{0.05t} = \frac{120}{55}$$

$$0.05t = \ln\frac{120}{55}$$

$$t = \frac{\ln\frac{120}{55}}{0.05} \approx 15.6 \text{ minutes}$$

cardboard: $125 = 70 + (190 - 70)e^{-0.08t}$

$$125 = 70 + (190 - 70)e^{-0.08t}$$

$$e^{0.08t} = \frac{120}{55}$$

$$0.08t = \ln\frac{120}{55}$$

$$t = \frac{\ln\frac{120}{55}}{0.08} \approx 9.8 \text{ minutes}$$

6. Coffee cools the slowest in a foam container and fastest in a ceramic container.

Chapter 12 Review Problems

1. $f(x) = 4^{3+x}$

x	$y = f(x) = 4^{3+x}$
–5	1/16
–4	1/4
–3	1
–2	4
–1	16

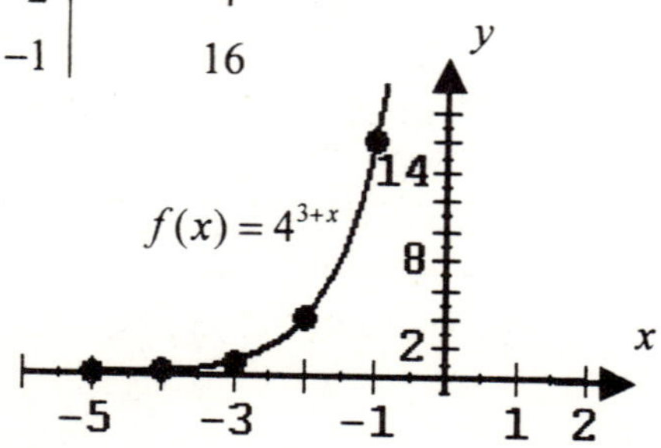

2. $f(x) = e^{x-3}$

x	$y = f(x) = e^{x-3}$
1	0.14
2	0.37
3	1
4	2.72
5	7.39

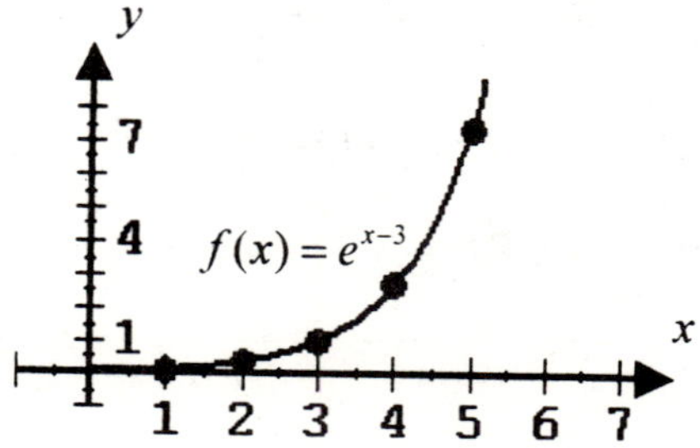

3. $5^{x+2} = 125 = 5^3$

$$x + 2 = 3$$

$$x = 1$$

4. $-3 = \log_{10}(0.001) \Leftrightarrow 10^{-3} = 0.001$

5. $\frac{1}{32} = 2^{-5} \Leftrightarrow \log_2 \frac{1}{32} = -5$

6. $\log_w 16 = 4 \Leftrightarrow w^4 = 16 = 2^4,\ w = 2$

7. $\log_3 x = -2 \Leftrightarrow 3^{-2} = x,\ x = \frac{1}{9}$

8. $\log_8 x = 0 \Leftrightarrow 8^0 = x,\ x = 1$

9. $\log_7 w = -1 \Leftrightarrow 7^{-1} = w,\ w = \frac{1}{7}$

10. $\log_w 27 = 3 \Leftrightarrow w^3 = 27 = 3^3,\ w = 3$

11. $\log_{10} w = -3 \Leftrightarrow 10^{-3} = w,\ w = 0.001$

12. $\log_{10} 1000 = x \Leftrightarrow 10^x = 1000 = 10^3,\ x = 3$

13. $\log_2 64 = x \Leftrightarrow 2^x = 64 = 2^6,\ x = 6$

14. $\log_2 \frac{1}{4} = x \Leftrightarrow 2^x = \frac{1}{4} = 2^{-2},\ x = -2$

15. $\log_5 125 = x \Leftrightarrow 5^x = 125 = 5^3,\ x = 3$

16. $\log_3 x = y \Leftrightarrow 3^y = x$

$x = 3^y$	y
1/9	−2
1/3	−1
1	0
3	1
9	2

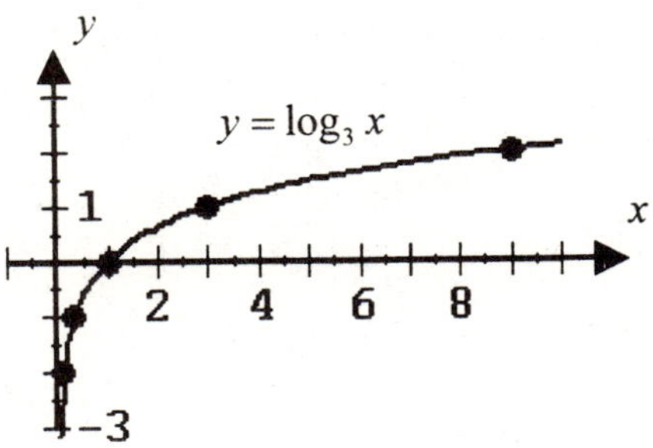

17. $\log_2\left(\frac{5x}{\sqrt{w}}\right) = \log_2(5x) - \log_2 \sqrt{w}$
$= \log_2(5x) - \log_2 w^{1/2}$
$= \log_2 5 + \log_2 x - \frac{1}{2}\log_2 w$

18. $\log_2 x^3\sqrt{y} = \log_2 x^3 + \log_2 \sqrt{y}$
$= 3\log_2 x + \log_2 y^{1/2}$
$= 3\log_2 x + \frac{1}{2}\log_2 y$

19. $\log_3 x + \log_3 w^{1/2} - \log_3 2$
$= \log_3 x + \log_3 \sqrt{w} - \log_3 2$
$= \log_3(x\sqrt{w}) - \log_3 2$
$= \log_3 \frac{x\sqrt{w}}{2}$

20. $4\log_8 w - \frac{1}{3}\log_8 z = \log_8 w^4 - \log_8 z^{1/3}$
$= \log_8 w^4 - \log_8 \sqrt[3]{z}$
$= \log_8 \frac{w^4}{\sqrt[3]{z}}$

21. $\log_e e^6 = 6\log_e e$
$= 6(1)$
$= 6$

22. $\log_5 100 - \log_5 x = \log_5 4$
$\log_5 \frac{100}{x} = \log_5 4$
$\frac{100}{x} = 4$
$x = 25$

23. $\log_8 x + \log_8 3 = \log_8 75$
$\log_8(3x) = \log_8 75$
$3x = 75$
$x = 25$

24. $\log 23.8 = 1.376576957$

25. $\log 0.0817 = -1.087777943$

26. $\ln 3.92 = 1.366091654$

27. $\ln 803 = 6.688354714$

28. $\log n = 1.1367 \Leftrightarrow n = 10^{1.1367}$
$$n = 13.69935122$$

29. $\ln n = 1.7 \Leftrightarrow n = e^{1.7}$
$$n = 5.473947392$$

30. $\log_8 2.81 = \dfrac{\ln 2.81}{\ln 8}$
$$\log_8 2.81 = 0.4968567101$$

31. $\log_7(x+3) + \log_7 5 = 2$
$$\log_7(5(x+3)) = 2$$
$$5(x+3) = 7^2$$
$$5x + 15 = 49$$
$$5x = 34$$
$$x = \frac{34}{5}$$
check: $\log_7\left(\dfrac{34}{5}+3\right) + \log_7 5 \stackrel{?}{=} 2$
$$\log_7\left(\frac{49}{5}\right) + \log_7 5 \stackrel{?}{=} 2$$
$$\log_7 49 - \log_7 5 + \log_7 5 \stackrel{?}{=} 2$$
$$\log_7 7^2 - \log_7 5 + \log_7 5 \stackrel{?}{=} 2$$
$$2\log_7 7 - \log_7 5 + \log_7 5 \stackrel{?}{=} 2$$
$$2 = 2$$

32. $\log_3(2x+3) = \log_3 2 - 3$
$$\log_3(2x+3) - \log_3 2 = -3$$
$$\log_3 \frac{2x+3}{2} = -3$$
$$\frac{2x+3}{2} = 3^{-3} = \frac{1}{3^3} = \frac{1}{27}$$
$$27(2x+3) = 2$$
$$54x + 81 = 2$$
$$54x = -79$$
$$x = \frac{-79}{54}$$
check: $\log_3\left(2\cdot\dfrac{-79}{54}+3\right) \stackrel{?}{=} \log_3 2 - 3$
$$\log_3\left(\frac{2}{27}\right) \stackrel{?}{=} \log_3 2 - 3$$
$$\log_3 2 - \log_3 27 \stackrel{?}{=} \log_3 2 - 3$$
$$\log_3 2 - \log_3 3^3 \stackrel{?}{=} \log_3 2 - 3$$
$$\log_3 2 - 3\log_3 3 \stackrel{?}{=} \log_3 2 - 3$$
$$\log_3 2 - 3\log_3 3 \stackrel{?}{=} \log_3 2 - 3$$
$$\log_3 2 - 3(1) \stackrel{?}{=} \log_3 2 - 3$$
$$\log_3 2 - 3 = \log_3 2 - 3$$

33. $\log_5(x+1) - \log_5 8 = \log_5 x$
$$\log_5(x+1) - \log_5 x = \log_5 8$$
$$\log_5 \frac{x+1}{x} = \log_5 8$$
$$\frac{x+1}{x} = 8$$
$$x + 1 = 8x$$
$$7x = 1$$
$$x = \frac{1}{7}$$

(continued)

33. (continued)

$$\text{check: } \log_5\left(\frac{1}{7}+1\right)-\log_5 8 \stackrel{?}{=} \log_5\left(\frac{1}{7}\right)$$

$$\log_5\left(\frac{8}{7}\right)-\log_5 8 \stackrel{?}{=} \log_5\left(\frac{1}{7}\right)$$

$$\log_5\left(\frac{1}{7}\cdot 8\right)-\log_5 8 \stackrel{?}{=} \log_5\left(\frac{1}{7}\right)$$

$$\log_5\left(\frac{1}{7}\right)+\log_5 8-\log_5 8 \stackrel{?}{=} \log_5\left(\frac{1}{7}\right)$$

$$\log_5\left(\frac{1}{7}\right)=\log_5\left(\frac{1}{7}\right)$$

34. $2\log_3(x+3)-\log_3(x+1)=3\log_3 2$

$$\log_3(x+3)^2-\log_3(x+1)=\log_3 2^3$$

$$\log_3\left(\frac{(x+3)^2}{(x+1)}\right)=\log_3 8$$

$$\frac{(x+3)^2}{(x+1)}=8$$

$$(x+3)^2=8(x+1)$$

$$x^2+6x+9=8x+8$$

$$x^2-2x+1=0$$

$$(x-1)^2=0$$

$$x-1=0$$

$$x=1$$

(continued)

34. (continued)

$$\text{check: } 2\log_3(1+3)-\log_3(1+1) \stackrel{?}{=} 3\log_3 2$$

$$2\log_3(4)-\log_3(2) \stackrel{?}{=} 3\log_3 2$$

$$\log_3(4^2)-\log_3(2) \stackrel{?}{=} 3\log_3 2$$

$$\log_3(16)-\log_3(2) \stackrel{?}{=} 3\log_3 2$$

$$\log_3\frac{16}{2} \stackrel{?}{=} 3\log_3 2$$

$$\log_3 8 \stackrel{?}{=} 3\log_3 2$$

$$\log_3 2^3 \stackrel{?}{=} 3\log_3 2$$

$$3\log_3 2=3\log_3 2$$

35. $\log_2(x-2)+\log_2(x+5)=3$

$$\log_2\big((x-2)(x+5)\big)=3$$

$$(x-2)(x+5)=2^3=8$$

$$x^2+3x-10=8$$

$$x^2+3x-18=0$$

$$(x+6)(x-3)=0$$

$x=3$, $x=-6$, reject -6 since it gives $\log_2(\text{negative})$.

$$\text{check: } \log_2(3-2)+\log_2(3+5) \stackrel{?}{=} 3$$

$$\log_2(1)+\log_2(8) \stackrel{?}{=} 3$$

$$0+\log_2(2^3) \stackrel{?}{=} 3$$

$$3\log_2 2 \stackrel{?}{=} 3$$

$$3=3$$

36. $\log_5(x+1)+\log_5(x-3)=1$

$$\log_5\big((x+1)(x-3)\big)=1$$

$$(x+1)(x-3)=5^1=5$$

$$x^2-2x-3=5$$

(continued)

36. (continued)

$$x^2-2x-8=0$$
$$(x-4)(x+2)=0$$
$x=4$, $x=-2$, reject -2 since it gives $\log_5(\text{negative})$

check: $\log_5(4+1)+\log_5(4-3)\stackrel{?}{=}1$
$$\log_5(5)+\log_5(1)\stackrel{?}{=}1$$
$$1+0\stackrel{?}{=}1$$
$$1=1$$

37. $\log(2t+1)+\log(4t-1)=2\log 3$

$$\log\left((2t+3)(4t-1)\right)=\log 3^2=\log 9$$
$$(2t+3)(4t-1)=9$$
$$8t^2+10t-3=9$$
$$8t^2+10t-12=0$$
$$4t^2+5t-6=0$$
$$(4t-3)(t+2)=0$$
$t=\frac{3}{4}$, $t=-2$, reject -2 since it gives $\log(\text{negative})$

check:
$$\log\left(2\cdot\frac{3}{4}+3\right)+\log\left(4\cdot\frac{3}{4}-1\right)\stackrel{?}{=}2\log 3$$
$$\log\left(\frac{9}{2}\right)+\log(2)\stackrel{?}{=}2\log 3$$
$$\log\left(\frac{9}{2}\cdot 2\right)\stackrel{?}{=}2\log 3$$
$$\log(9)\stackrel{?}{=}2\log 3$$
$$\log(3^2)\stackrel{?}{=}2\log 3$$
$$2\log 3=2\log 3$$

38. $\log(2t+4)-\log(3t+1)=\log 6$

$$\log\left(\frac{2t+4}{3t+1}\right)=\log 6$$
$$\frac{2t+4}{3t+1}=6$$
$$2t+4=18t+6$$
$$16t=-2$$
$$t=-\frac{1}{8}$$

check: $\log\left(2\cdot\frac{-1}{8}+4\right)-\log\left(3\cdot\frac{-1}{8}+1\right)\stackrel{?}{=}\log 6$
$$\log\left(\frac{15}{4}\right)-\log\left(\frac{5}{8}\right)\stackrel{?}{=}\log 6$$
$$\log\left(\frac{\frac{15}{4}}{\frac{5}{8}}\right)\stackrel{?}{=}\log 6$$
$$\log\left(\frac{\frac{15}{4}}{\frac{5}{8}}\right)\stackrel{?}{=}\log 6$$
$$\log 6=\log 6$$

39. $3^x=14$

$$\log 3^x=\log 14$$
$$x\log 3=\log 14$$
$$x=\frac{\log 14}{\log 3}$$

40. $5^x=4^{x+2}$

$$\log 5^x=\log 4^{x+2}$$
$$x\log 5=(x+2)\log 4=x\log 4+2\log 4$$
$$x(\log 5-\log 4)=2\log 4$$
$$x=\frac{2\log 4}{\log 5-\log 4}$$

41. $$16e^{x+1} = 56$$
$$\ln(16e^{x+1}) = \ln 56$$
$$\ln 16 + \ln e^{x+1} = \ln 56$$
$$(x+1)\ln e = \ln 56 - \ln 16 = \ln\frac{56}{16}$$
$$x+1 = \ln 3.5$$
$$x = -1 + \ln 3.5$$

42. $$e^{2x} = 30.6$$
$$\ln e^{2x} = \ln 30.6$$
$$2x\ln 3 = \ln 30.6$$
$$2x = \ln 30.6$$
$$x = \frac{\ln 30.6}{2}$$

43. $$2^{3x+1} = 5^x$$
$$\ln 2^{3x+1} = \ln 5^x$$
$$(3x+1)\ln 2 = x\ln 5$$
$$3x\ln 2 + \ln 2 = x\ln 5$$
$$x(3\ln 2 - \ln 5) = -\ln 2$$
$$x = \frac{\ln 2}{\ln 5 - 3\ln 2}$$
$$x \approx -1.4748$$

44. $$3^{x+1} = 7$$
$$\ln 3^{x+1} = \ln 7$$
$$(x+1)\ln 3 = \ln 7$$
$$x = -1 + \frac{\ln 7}{\ln 3}$$
$$x \approx 0.7712$$

45. $$e^{3x-4} = 20$$
$$\ln e^{3x-4} = \ln 20$$
$$(3x-4)\ln e = \ln 20$$
$$3x - 4 = \ln 20$$
$$x = \frac{\ln 20 + 4}{3} \approx 2.3319$$

46. $$1.03^x = 20$$
$$\ln 1.03^x = \ln 20$$
$$x\ln 1.03 = \ln 20$$
$$x = \frac{\ln 20}{\ln 1.03}$$
$$x \approx 101.3482$$

47. $$A = P(1+r)^t$$
$$2P = P(1+0.08)^t$$
$$2 = 1.08^t$$
$$\ln 2 = \ln 1.08^t$$
$$\ln 2 = t\ln 1.08$$
$$t = \frac{\ln 2}{\ln 1.08}$$
$t \approx 9$ years to double money in account

48. $$A = P(1+r)^t$$
$$A = 5000(1+0.06)^4$$
$A = \$6312.38$ in account after 4 years

49. $$A = P(1+r)^t$$
$$20{,}000 = 12{,}000(1+0.07)^t$$
$$\frac{5}{3} = 1.07^t$$
$$\ln 1.07^t = \ln\frac{5}{3}$$
$$t\ln 1.07 = \ln\frac{5}{3}$$
$$t = \frac{\ln\frac{5}{3}}{\ln 1.07}$$
$$t = 7.550041795...$$
$$t \approx 8$$

It would take approximately 8 years for \$12,000 to grow to \$20,000.

50. $A = P(1+r)^t$

$$A_{\text{Robert}} + 500 = A_{\text{Brother}}$$
$$3500(1+0.05)^t + 500 = 3500(1+0.06)^t$$
$$7(1.06)^t - 7(1.05)^t = 1$$
$$(1.06)^t - (1.05)^t = \frac{1}{7}$$

Solve with a graphing calculator. Graph $y_1 = 1.06^x - 1.05^x$ and $y_2 = \frac{1}{7}$ and use the *intersect* feature.

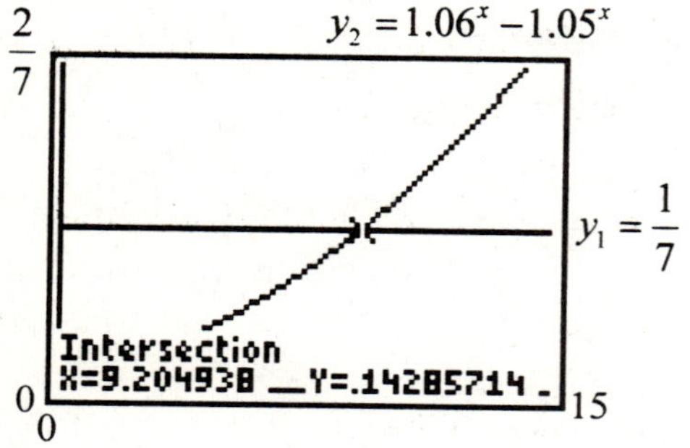

It will take approximately 9 years for Robert's amount to be $500 less than his brother's amount.

51. $A = A_0 e^{rt}$

$$10 = 6e^{0.02t}$$
$$\ln 1.\overline{6} = \ln e^{0.02t}$$
$$\ln 1.\overline{6} = 0.02t \ln e$$
$$t = \frac{\ln 1.\overline{6}}{0.02} = 25.54128119...$$

It will take approximately 26 years.

52. $A = A_0 e^{rt}$

$$16 = 7e^{0.02t}$$
$$\ln \frac{16}{7} = \ln e^{0.02t}$$

(continued)

52. (continued)

$$\ln \frac{16}{7} = 0.02t \ln e$$
$$t = \frac{\ln \frac{16}{7}}{0.02} = 41.33392866...$$

It will take approximately 41 years.

53. $A = A_0 e^{rt}$

$$2600 = 2000e^{0.03t}$$
$$\ln 1.3 = \ln e^{0.03t}$$
$$\ln 1.3 = 0.03t$$
$$t = \frac{\ln 1.3}{0.03} = 8.745475482...$$

It will take approximately 9 years.

54. $A = A_0 e^{rt}$

$$95{,}000 = 40{,}000e^{0.08t}$$
$$2.375 = e^{0.08t}$$
$$\ln 2.375 = \ln e^{0.08t}$$
$$0.08t = \ln 2.375$$
$$t = \frac{\ln 2.375}{0.08} = 10.81246797...$$

It will take approximately 11 years.

55. $M = \log\left(\frac{I}{I_0}\right)$

$$8.4 = \log\left(\frac{I_A}{I_0}\right) = \log I_A - \log I_0$$
$$-\left(6.7 = \log\left(\frac{I_T}{I_0}\right) = \log I_T - \log I_0\right)$$
$$1.7 = \log I_A - \log I_T$$

(continued)

55. (continued)

$$\log \frac{I_A}{I_T} = 1.7$$

$$10^{\log \frac{I_A}{I_T}} = 10^{1.7}$$

$$\frac{I_A}{I_T} = 50.11872336...$$

The Alaska earthquake was about 50.1 times more intense than the Turkey earthquake.

56. $W = p_0 V_0 \ln\left(\frac{V_1}{V_0}\right)$

(a) $W = 40(15)\ln\left(\frac{24}{15}\right) \approx 282$

(b) $100 = p_0(8)\ln\left(\frac{40}{8}\right)$

$$p_0 = \frac{100}{(8)\ln\left(\frac{40}{8}\right)} \approx 7.77$$

Chapter 12 Test

1. $f(x) = 3^{4-x}$

x	$y = f(x) = 3^{4-x}$
3	3
4	1

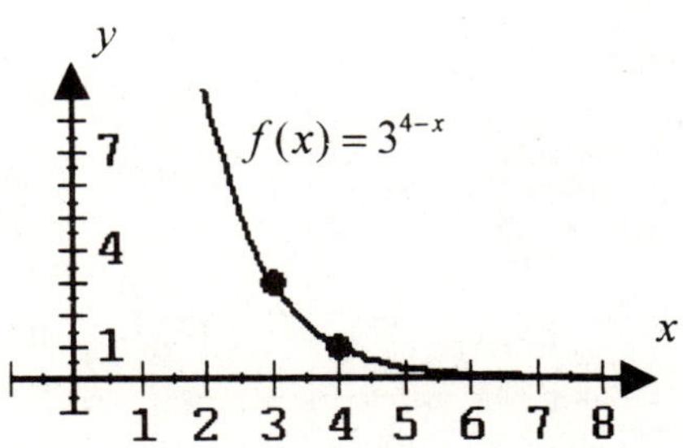

2. $4^{x+3} = 64 = 4^3$

$$x + 3 = 3$$

$$x = 0$$

3. $\log_w 125 = 3 \Leftrightarrow w^3 = 125 = 5^3, \; w = 5$

4. $\log_8 x = -2 \Leftrightarrow x = 8^{-2} = \frac{1}{64}$

5. $\log_8 x + \log_8 w - \frac{1}{4}\log_8 3$

$$= \log_8(xw) - \log_8 3^{1/4}$$

$$= \log_8(xw) - \log_8 \sqrt[4]{3}$$

$$= \log_8 \frac{xw}{\sqrt[4]{3}}$$

6. $\ln 5.99 = 1.7901$

7. $\log 23.6 = 1.3729$

8. $\log_3 1.62 = \frac{\log 1.62}{\log 3} = 0.4391$

9. $\log x = 3.7284 \Leftrightarrow x = 10^{3.7284}$

$$x = 5350.569382$$

10. $\ln x = 0.14 \Leftrightarrow x = e^{0.14} = 1.150273799$

11. $\log_8(x+3) - \log_8 2x = \log_8 4$

$$\log_8\left(\frac{x+3}{2x}\right) = \log_8 4$$

$$\frac{x+3}{2x} = 4$$

$$x + 3 = 8x$$

$$7x = 3$$

$$x = \frac{3}{7}$$

(continued)

11. (continued)

$$\text{check}: \log_8\left(\frac{3}{7}+3\right)-\log_8\left(2\cdot\frac{3}{7}\right)\stackrel{?}{=}\log_8 4$$

$$\log_8\left(\frac{24}{7}\right)-\log_8\left(\frac{6}{7}\right)\stackrel{?}{=}\log_8 4$$

$$\log_8\left(\frac{\frac{24}{7}}{\frac{6}{7}}\right)\stackrel{?}{=}\log_8 4$$

$$\log_8 4=\log_8 4$$

12. $\log_8 2x+\log_8 6=2$

$$\log_8((2x)(6))=2 \Leftrightarrow 12x=8^2$$

$$12x=64$$

$$x=\frac{16}{3}$$

$$\log_8 2\cdot\frac{16}{3}+\log_8 6\stackrel{?}{=}2$$

$$\log_8(2\cdot\frac{16}{3}\cdot 6)\stackrel{?}{=}2$$

$$\log_8 64\stackrel{?}{=}2$$

$$\log_8 8^2\stackrel{?}{=}2$$

$$2\log_8 8\stackrel{?}{=}2$$

$$2=2$$

13.

$$29=16e^{3x+1}$$

$$e^{3x+1}=0.25$$

$$\ln e^{3x+1}=\ln 0.25$$

$$(3x+1)\ln e=\ln 0.25$$

$$3x+1=\ln 0.25$$

$$3x=-1+\ln 0.25$$

$$x=\frac{-1+\ln 0.25}{3}$$

14.

$$5^{3x+6}=17$$

$$\ln 5^{3x+6}=\ln 17$$

$$(3x+6)\ln 5=\ln 17$$

$$3x+6=\frac{\ln 17}{\ln 5}$$

$$x=\frac{-6+\frac{\ln 17}{\ln 5}}{3}\approx -1.4132$$

15. $A=P(1+r)^t$

$$A=2000(1+0.08)^5=2938.656154...$$

Henry will have \$2938.66

16. $A=P(1+r)^t$

$$2P=P(1+0.05)^t$$

$$2=(1.05)^t$$

$$\ln 2=\ln 1.05^t$$

$$\ln 2=t\ln 1.05$$

$$t=\frac{\ln 2}{\ln 1.05}=14.20669908...$$

Barb can double her money in approximately 14 years.

Cumulative Test for Chapters 1-12

1. $2(-3)+12\div(-2)+3\sqrt{36}$

$$=-6+(-6)+3(6)=6$$

2. $H=3bx-2ay$

$$3bx=H+2ay$$

$$x=\frac{H+2ay}{3b}$$

3. $y=-\frac{2}{3}x+4$

(continued)

3. (continued)

x	y
0	4
6	0

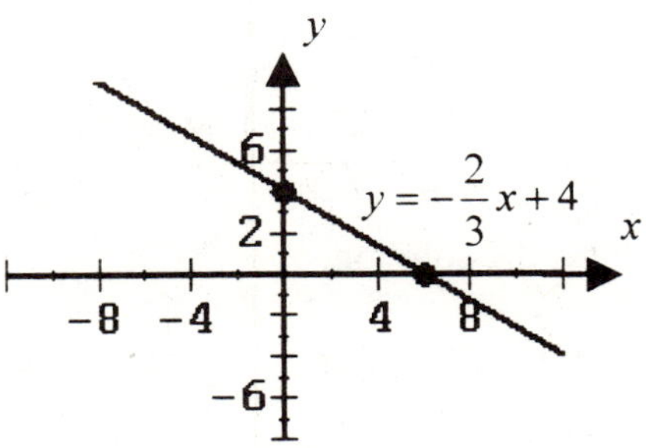

4. $5ax+5ay-7wx-7wy$
$=5a(x+y)-7w(x+y)$
$=(x+y)(5a-7w)$

5.
$3x-y+\ z=6$
$2x-y+2z=7$
$x+y\ +z=2$

Switch first and third equation
$x+y\ +z=2$
$3x-y+\ z=6$
$2x-y+2z=7$

Add -3 time first equation to second equation and -2 times first equation to third equation
$x+y\ +z=2$
$-4y\ -2z=0$
$-3y\qquad =3\Rightarrow y=-1$, from second equation, $-4(-1)-2z=0$
$2z=4\Rightarrow z=2$, from first equation,
$x+(-1)+2=2\Rightarrow x=1$

$(1,-1,2)$ is the solution.

6. $(5\sqrt{2}+\sqrt{3})(\sqrt{5}-2\sqrt{6})$
$=5\sqrt{10}-10\sqrt{12}+\sqrt{15}-2\sqrt{18}$
$=5\sqrt{10}+\sqrt{15}-10\sqrt{4\cdot 3}-2\sqrt{9\cdot 2}$
$=5\sqrt{10}+\sqrt{15}-20\sqrt{3}-6\sqrt{2}$

7. $x^4-5x^2-6=0$
$(x^2-6)(x^2+1)=0$
$x^2-6=0,\ x^2+1=0$
$x^2=6,\ x^2=-1$
$x=\pm\sqrt{6},\ x=\pm i$

8. $2x-y=4\Rightarrow y=2x-4$
$4x-y^2=0\Rightarrow 4x-(2x-4)^2=0$
$4x-4x^2+16x-4=0$
$x^2-5x+4=0$
$(x-4)(x-1)=0$
$x=4,\ x=1$
$y=2(4)-4=4,$
$y=2(1)-4=-2$
$x=4,\ y=4;\ x=1,\ y=-2$ is the solution.

9. $2x-3=\sqrt{7x-3}$
$4x^2-12x+9=7x-3$
$4x^2-19x+12=0$
$(x-4)(4x-3)=0$
$x=4,\ x=\frac{3}{4}$ which does not check
$x=4$ is the solution.

10. $\frac{5}{\sqrt[3]{2xy^2}}\cdot\frac{\sqrt[3]{4x^2y}}{\sqrt[3]{4x^2y}}$
$=\frac{5\sqrt[3]{4x^2y}}{2xy}$

11. $f(x) = 2^{3-2x}$

x	$y = f(x) = 2^{3-2x}$
1	2
2	1/2
3	1/8
4	1/32

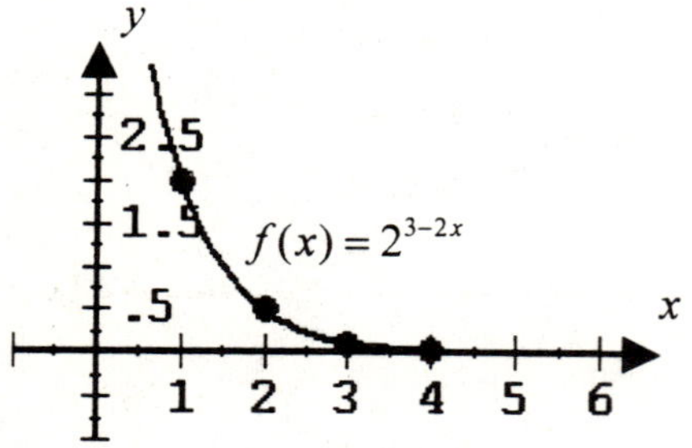

12. $\log_x\left(\frac{1}{64}\right) = 3 \Leftrightarrow x^3 = \frac{1}{64} = \left(\frac{1}{4}\right)^3$

$$x = \frac{1}{4}$$

13. $5^{2x-1} = 25 = 5^2$

$$2x - 1 = 2$$
$$2x = 3$$
$$x = \frac{3}{2}$$

14. $\log 7.67 = 0.8847953639$

15. $\log_x = 1.8209 \Leftrightarrow x = 10^{1.8209}$

$$x \approx 66.20640403$$

16. $\log_3 7 = \frac{\ln 7}{\ln 3}$

$$= 1.771243749$$

17. $\ln x = 1.9638 \Leftrightarrow x = e^{1.9638}$

$$x \approx 7.1263558$$

18. $\log_9 x = 1 - \log_9(x - 8)$

$$\log_9 x + \log_9(x-8) = 1$$
$$\log_9(x(x-8)) = 1 \Leftrightarrow x(x-8) = 9^1$$
$$x^2 - 8x - 9 = 0$$
$$(x-9)(x+1) = 0$$

$x = 9$, $x = -1$ which must be rejected since it gives $\log_9$(negative)

$x = 9$ is the solution.

19. $\log_5 x = \log_5 2 + \log_5(x^2 - 3)$

$$\log_5 x - \log_5(x^2 - 3) = \log_5 2$$
$$\log_5\left(\frac{x}{x^2 - 3}\right) = \log_5 2$$
$$\frac{x}{x^2 - 3} = 2$$
$$2x^2 - 6 = x$$
$$2x^2 - x - 6 = 0$$
$$(x-2)(2x+3) = 0$$

$x = 2$, $x = -\frac{3}{2}$ which must be rejected since it gives $\log_5$(negative)

$x = 2$ is the solution.

20. $3^{x+2} = 5$

$$\ln 3^{x+2} = \ln 5$$
$$(x+2)\ln 3 = \ln 5$$
$$x + 2 = \frac{\ln 5}{\ln 3}$$
$$x = -2 + \frac{\ln 5}{\ln 3} \approx -0.535$$

21. $33 = 66e^{2x} \Rightarrow e^{2x} = 0.5 \Rightarrow \ln e^{2x} = \ln 0.5$

$$\ln e^{2x} = \ln 0.5$$
$$2x = \ln 0.5,\ x = \frac{\ln 0.5}{2}$$

22. $A = P(1+r)^t$

$A = 3000(1+0.09)^4 = 4234.74483...$

They will have \$4234.74.

Practice Final Examination

1. $(4-3)^2 + \sqrt{9} \div (-3) + 4$

$= 1^2 + 3 \div (-3) + 4$

$= 1 + (-1) + 4$

$= 4$

2. $5a - 2ab - 3a^2 - 6a - 8ab + 2a^2$

$= -a^2 - 10ab - a$

3. $-2x + 3y\{7 - 2[x - (4x + y)]\}$

$= -2x + 3y\{7 - 2[x - 4x - y]\}$

$= -2x + 3y\{7 - 2[-3x - y]\}$

$= -2x + 3y\{7 + 6x + 2y\}$

$= -2x + 21y + 18xy + 6y^2$

4. $2x^2 - 3xy - 4y\big|_{x=-2, y=3}$

$= 2(-2)^2 - 3(-2)(3) - 4(3)$

$= 2(4) + 6(3) - 12$

$= 8 + 18 - 12$

$= 26 - 12$

-14

5. $F = \frac{9}{5}C + 32 = \frac{9}{5}(-35) + 32 = -31°$

6. $\frac{1}{3}y - 4 = \frac{1}{2}y + 1$

$\frac{1}{6}y = -5$

$y = -30$

7. $A = \frac{1}{2}a(b+c)$

$2A = ab + ac$

$ab = 2A - ac$

$b = \frac{2A - ac}{a}$

8. $5x + 3 - (4x - 2) \le 6x - 8$

$5x + 3 - 4x + 2 \le 6x - 8$

$-5x \le -13$

$x \ge 2.6$

9. $P = 2L + 2W = 1760$

$L + W = 880$

$2W - 200 + W = 880$

$3W = 1080$

$W = 360$

$L = 2W - 200 = 520$

The width is 360 meters and the length is 520 meters.

10. Let $x =$ amount invested at 14%

$0.14x + 0.12(4000 - x) = 508$

$0.14x + 480 - 0.12x = 508$

$0.02x = 28$

$x = 1400$

$4000 - x = 2600$

\$1400 was invested at 14% and \$2500 at 12%.

11. $7x-2y=-14$

x	y
0	7
−2	0

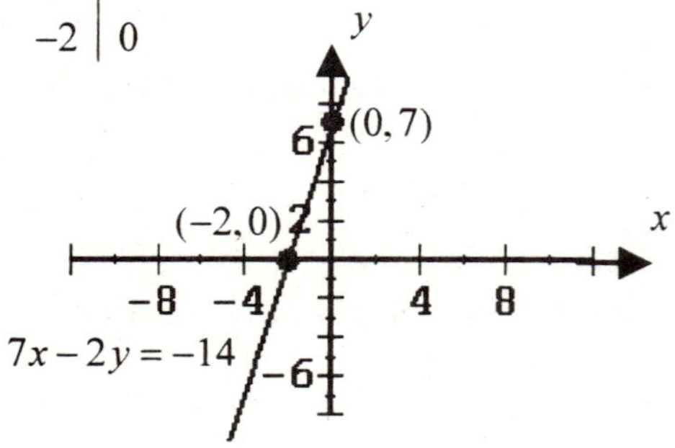

12. $3x-4y\le 6$

Test point: $(0,0)$

$3(0)-4(0)\le 6$

$0\le 6$, True

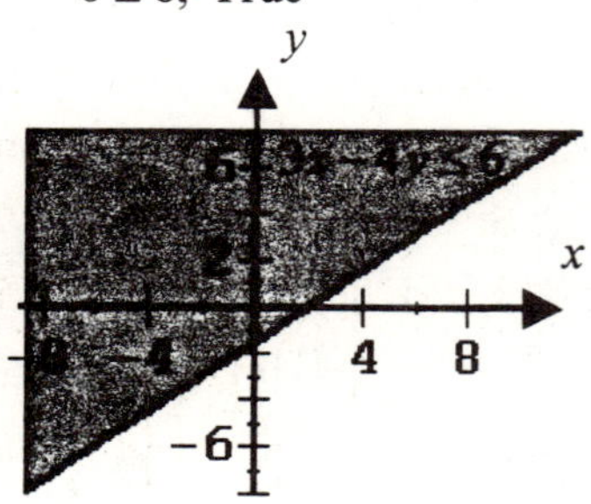

13. $m=\dfrac{y_2-y_1}{x_2-x_1}=\dfrac{-3-5}{-2-1}=\dfrac{8}{3}$

14. $3x+y=8\Rightarrow y=-3x+8,\ m=-3$

$m_{\parallel}=-3,\ y-4=-3(x-(-1))$

$y-4=-3x-3$

$y=-3x+1$

15. $f(x)=3x^2-4x-3$

$f(3)=3(3^2)-4(3)-3$

$f(3)=12$

16. $f(x)=3x^2-4x-3$

$f(-2)=3((-2)^2)-4(-2)-3$

$f(-2)=17$

17. $f(x)=-\dfrac{1}{2}x^2$

x	$y=f(x)=-\dfrac{1}{2}x^2$
−2	−2
0	0
2	−2

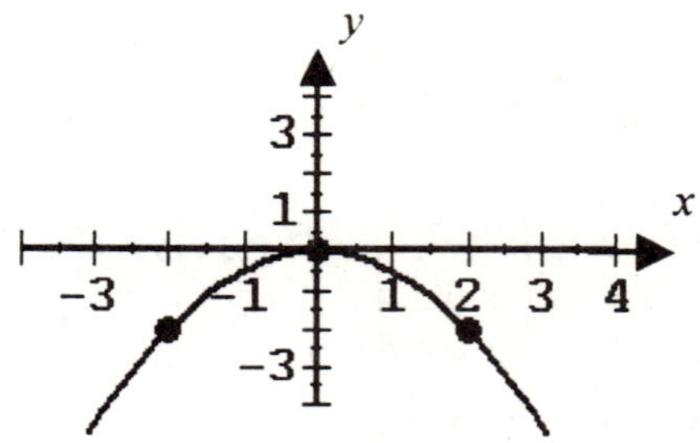

18. $\dfrac{1}{2}x+\dfrac{2}{3}y=1\xrightarrow{\times 6}3x+4y=6$

$\dfrac{1}{3}x+y=-1\rightarrow y=-1-\dfrac{1}{3}x$

$3x+4\left(-1-\dfrac{1}{3}x\right)=6$

$3x-4-\dfrac{4}{3}x=6$

$9x-12-4x=18$

$5x=30$

$x=6$

$y=-1-\dfrac{1}{3}x=-1-\dfrac{1}{3}(6)=-3$

$x=6,\ y=-3$ is the solution.

19. $4x-3y=12\xrightarrow{\times 4}16x-12y=48$

$3x-4y=2\xrightarrow{\times -3}-9x+12y=-6$

$7x=42$

$x=6$

$4(6)-3y=12$

$-3y=-12$

$y=4$

$x=6,\ y=4$ is the solution.

20. Solve the system.

$2x+3y-z=16$

$x-y+3z=-9$

$5x+2y-z=15$

Multiply first equation by 3 and add to the second equation

$6x+9y-3z=48$

$x-y+3z=-9$

$7x+8y=39$

Multiply third equation by 3 and add to the second equation

$x-y+3z=-9$

$15x+6y-3z=45$

$16x+5y=36$

Now solve the system

$7x+8y=39\xrightarrow{\times 5}35x+40y=195$

$16x+5y=36\xrightarrow{\times -8}-128x-40y=-288$

$-93x=-93$

$x=1$

$16x+5y=36\Rightarrow 16(1)+5y=36$

$5y=20$

$y=4$

$x-y+3z=-9\Rightarrow 1-4+3z=-9$

$z=-2$

$x=1,\ y=4,\ z=-2$ is the solution.

21. Solve the system.

$y+z=2$

$x+z=5$

$x+y=5$

Subtract the second equation from the first.

$y+z=2$

$-x-z=-5$

$-x+y=-3$, add third equation

$x+y=5$

$2y=2$

$y=1$, from $x+y=4\Rightarrow x+1=5,\ x=4$

From $x+z=5\Rightarrow 4+z=5,\ z=1$

$x=4,\ y=1,\ z=1$ is the solution.

22. $3y\ge 8x-12$ | $2x+3y\le -6$

Test point: $(0,0)$ | Test point: $(0,0)$

$3(0)\ge 8(0)-12$ | $2(0)+3(0)\le -6$

$0\ge -12$, true | $0\le -6$, false

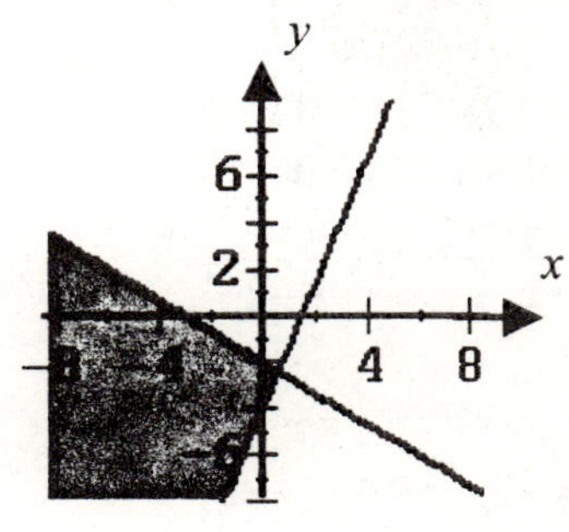

23. $(-3x^2y)(-6x^3y^4)$

$=(-3)(-6)x^{2+3}y^{1+4}$

$=18x^5y^5$

24. $(3x-2)(2x^2-4x+3)$

$=6x^3-12x^2+9x-4x^2+8x-6$

$=6x^3-16x^2+17x-6$

25. $(25x^3+9x+2)\div(5x+1)$

$$\begin{array}{r} 5x^2-x+2 \\ 5x+1\overline{)25x^3+0x^2+9x+2} \\ \underline{25x^3+5x^2} \qquad\quad \\ -5x^2+9x \quad \\ \underline{-5x^2-\ x} \quad \\ 10x+2 \\ \underline{10x+2} \\ 0 \end{array}$$

$(25x^3+9x+2)\div(5x+1)=5x^2-x+2$

26. $9x^2-30x+25=(3x-5)^2$

27. x^3+2x^2-4x-8
$=x^2(x+2)-4(x+2)$
$=(x+2)(x^2-4)=(x+2)(x+2)(x-2)$

28. $2x^3+15x^2-8x=x(2x^2+15x-8)$
$=x(2x-1)(x+8)$

29. $x^2+15x+54=0$
$(x+9)(x+6)=0$
$x+9=0,\quad x+6=0$
$x=-9,\quad x=-6$

30. $\dfrac{9x^3-x}{3x^2-8x-3}=\dfrac{x(9x^2-1)}{(3x+1)(x-3)}$
$=\dfrac{x(3x+1)(3x-1)}{(3x+1)(x-3)}$
$=\dfrac{x(3x-1)}{(x-3)}$

31. $\dfrac{x^2-9}{2x^2+7x+3}\div\dfrac{x^2-3x}{2x^2+11x+5}$
$=\dfrac{(x+3)(x-3)}{(2x+1)(x+3)}\cdot\dfrac{(2x+1)(x+5)}{x(x-3)}$
$=\dfrac{(x+5)}{x}$

32. $\dfrac{3x}{x+5}-\dfrac{2}{x^2+7x+10}$
$=\dfrac{3x(x+2)}{(x+5)(x+2)}-\dfrac{2}{(x+5)(x+2)}$
$=\dfrac{3x(x+2)-2}{(x+5)(x+2)}$
$=\dfrac{3x^2+6x-2}{(x+5)(x+2)}$

33. $\dfrac{\frac{3}{2x+1}+2}{1-\frac{2}{4x^2-1}}$
$=\dfrac{3(2x-1)+2(2x-1)(2x+1)}{(2x-1)(2x+1)-2}$
$=\dfrac{6x-3+2(4x^2-1)}{4x^2-1-2}$
$=\dfrac{8x^2+6x-5}{4x^2-3}$

34. $\dfrac{x-1}{x^2-4}=\dfrac{2}{x+2}+\dfrac{4}{x-2}$
$\dfrac{x-1}{(x+2)(x-2)}=\dfrac{2}{x+2}+\dfrac{4}{x-2}$
$x-1=2(x-2)+4(x+2)$
$x-1=2x-4+4x+8$
$5x=-5$
$x=-1$

35. $\dfrac{5x^{-4}y^{-2}}{15x^{-1/2}y^{3}}=\dfrac{1}{3x^{7/2}y^{5}}$

36. $\sqrt[3]{40x^4y^7}=\sqrt[3]{8x^3y^6\cdot 5x^1y}=2xy^2\sqrt[3]{5xy}$

37. $5\sqrt{2}-3\sqrt{50}+4\sqrt{98}$
$=5\sqrt{2}-3\sqrt{25\cdot 2}+4\sqrt{49\cdot 2}$
$=5\sqrt{2}-15\sqrt{2}+28\sqrt{2}$
$=18\sqrt{2}$

38. $\dfrac{2\sqrt{3}+1}{3\sqrt{3}-\sqrt{2}}\cdot\dfrac{3\sqrt{3}+\sqrt{2}}{3\sqrt{3}+\sqrt{2}}$
$=\dfrac{18+2\sqrt{6}+3\sqrt{3}+\sqrt{2}}{27-2}$
$=\dfrac{18+2\sqrt{6}+3\sqrt{3}+\sqrt{2}}{25}$

39. $i^3+\sqrt{-25}+\sqrt{-16}=-i+5i+4i=8i$

40. $\sqrt{x+7}=x+7$
$x+7=x^2+10x+25$
$x^2+9x+18=0$
$(x+6)(x+3)=0$
$x+6=0,\ x+3=0$
$x=-6,\ x=-3$
check: $\sqrt{-6+7}\overset{?}{=}-6+5$
$\sqrt{1}\overset{?}{=}-1$
$1\neq -1$
$\sqrt{-3+7}\overset{?}{=}-3+5$
$\sqrt{4}\overset{?}{=}-3+5$
$2=2$
$x=-3$ is the solution.

41. $y=kx^2$
$15=k(2)^2\Rightarrow k=\dfrac{15}{4}\Rightarrow y=\dfrac{15}{4}x^2$
$y=\dfrac{15}{4}(3)^2=33.75$

42. $5x(x+1)=1+6x$
$5x^2+5x=1+6x$
$5x^2-x-1=0$. use quadratic formula
$x=\dfrac{-(-1)\pm\sqrt{(-1)^2-4(5)(-1)}}{2(5)}$
$x=\dfrac{1\pm\sqrt{21}}{10}$

43. $5x^2-9x=-12x\Rightarrow 5x^2+3x=0$
$x(5x+3)=0$
$x=0,\ x=-\dfrac{3}{5}$

44. $x^{2/3}+5x^{1/3}-14=0$, let $x^{1/3}=w,\ x^{2/3}=w^2$
$w^2+5w-14=0$
$(w-2)(w+7)=0$
$w-2=0,\ w+7=0$
$w=2,\quad w=-7$
$x^{1/3}=2,\quad x^{1/3}=-7$
$x=8,\quad x=-343$

45. $3x^2-11x-4\geq 0$
$3x^2-11x-4=0$
$(3x+1)(x-4)=0$
$3x+1=0$ or $\quad x-4=0$
$x=-\dfrac{1}{3}$ or $\quad x=4$

(continued)

45. (continued)

Region I: Test $x = -1$

$3(-1)^2 - 11(-1) - 4 = -4 < 0$

Region II: Test $x = 0$

$3(0)^2 - 11(0) - 4 = -4 < 0$

Region III: Test $x = 5$

$3(5)^2 - 11(5) - 4 = 16 > 0$

$x \le -\frac{1}{3}$ or $x \ge 4$

46. $f(x) = -x^2 - 4x + 5$

parabola, opening downward

$f(0) = -0^2 - 4(0) + 5 = 5 \Rightarrow$ y-int: $(0,5)$

$-x^2 - 4x + 5 = 0 \Rightarrow x^2 + 4x - 5 = 0$

$(x+5)(x-1) = 0 \Rightarrow x = -5,\ x = 1$

x-int: $(-5,0)$, $(1,0)$

$-\frac{b}{2a} = -\frac{-4}{2(-1)} = -2$

$f(-2) = -(-2)^2 - 4(-2) + 5 = 9$

$V(-2,9)$

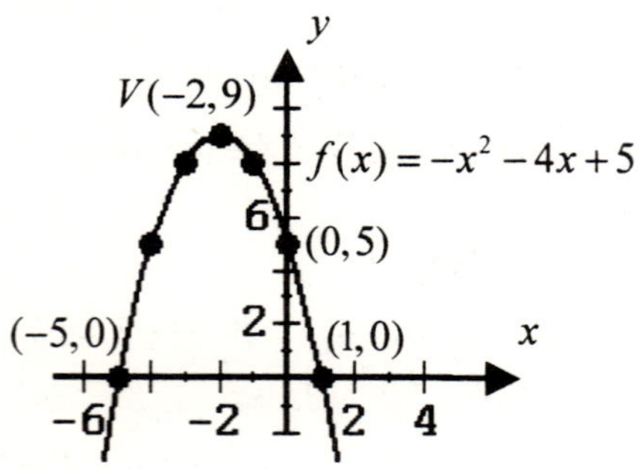

47. $L = 3W + 1$

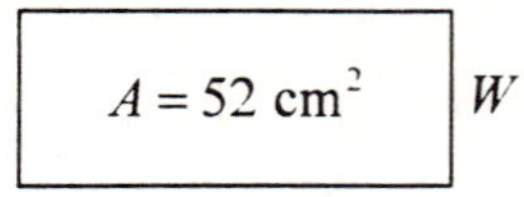

$A = LW = (3W+1)W = 52$

$3W^2 + W - 52 = 0$

$(W-4)(3W+13) = 0$

$W = 4$ or $W = -\frac{13}{3}$ reject, $W > 0$

$L = 3W + 1 = 13$

The width is 4 cm and the length is 13 cm.

48. $\left|\frac{2}{3}x - 4\right| = 2$

$\frac{2}{3}x - 4 = 2$ or $\frac{2}{3}x - 4 = -2$

$2x - 12 = 62 \qquad x - 12 = -6$

$2x = 18 \qquad 2x = 6$

$x = 9 \qquad x = 3$

49. $|2x - 5| < 10$

$-10 < 2x - 5 < 10$

$-5 < 2x < 15$

$-\frac{5}{2} < x < \frac{15}{2}$

50. $x^2 + y^2 + 6x - 4y = -9$

$x^2 + 6x + 9 + y^2 - 4y + 4 = -9 + 9 + 4 = 4$

$(x+3)^2 + (y-2)^2 = 2^2$

$C(-3,2),\ r = 2$

51. $\frac{x^2}{16}+\frac{y^2}{25}=1$

$\frac{x^2}{4^2}+\frac{y^2}{5^2}=1$

ellipse: $C(0,0)$

$a=4,\ b=5$

x-int: $(\pm 4,0)$

y-int: $(0,\pm 5)$

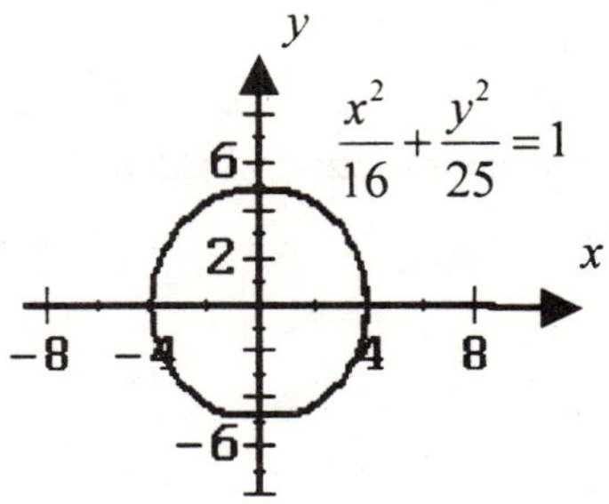

52. $\frac{x^2}{4}-\frac{y^2}{9}=1 \Rightarrow \frac{x^2}{2^2}-\frac{y^2}{3^2}=1$, hyperbola

$C(0,0),\ a=2,\ b=3$

x-int: $(\pm 2,0)$

$y_{\text{asymptote}}=\pm\frac{3}{2}x$

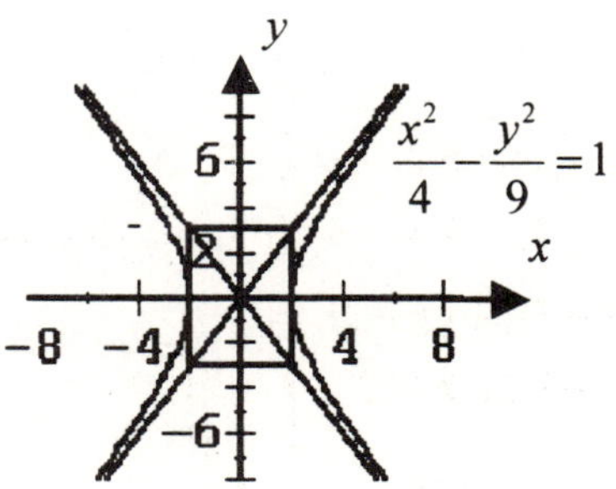

53. $x=(y-3)^2+5$

parabola opening right, $V(5,3)$

$x=(0-3)^2+5=14$, x-int: $(14,0)$

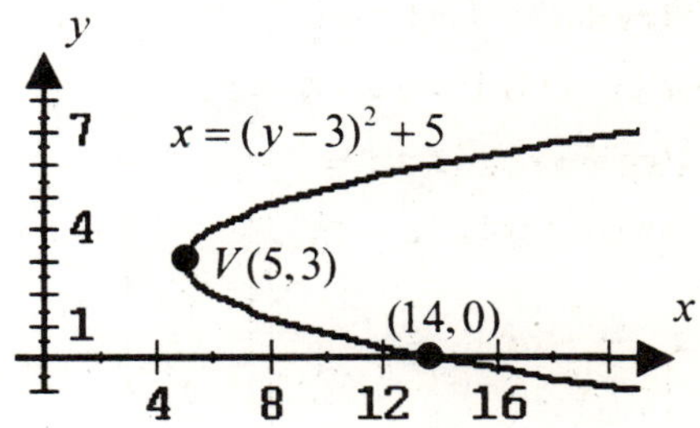

54.
$$x^2+y^2=16$$
$$x^2-y=4 \Rightarrow y=x^2-4$$
$$x^2+(x^2-4)^2=16$$
$$x^2+x^4-8x^2+16=16$$
$$x^4-7x^2=0$$
$$x^2(x^2-7)=0$$
$$x^2=0,\ x^2=7$$
$$x=0,\ x=\pm\sqrt{7}$$

$y=x^2-4=0^2-4=-4$

$y=x^2-4=(\pm\sqrt{7})^2-4=3$

$(0,-4),\ (\pm\sqrt{7},3)$ is the solution.

55. $f(x)=3x^2-2x+5$

(a) $f(-1)=3(-1)^2-2(-1)+5=10$

(b) $f(a)=3a^2-2a+5$

(c) $f(a+2)=3(a+2)^2-2(a+2)+5$

$=3a^2+12a+12-2a-4+5$

$=3a^2+10a+13$

56. $f(x)=5x^2-3,\ g(x)=-4x-2$

$$\begin{aligned}f[g(x)]&=f(-4x-2)\\&=5(-4x-2)^2-3\\&=5(16x^2+16x+4)-3\\&=80x^2+80x+20-3\\&=80x^2+80x+17\end{aligned}$$

57. $f(x)=\frac{1}{2}x-7,\ f(x)\to y$

$y=\frac{1}{2}x-7,\ x\leftrightarrow y$

$x=\frac{1}{2}y-7\Rightarrow y=2x+14,\ y\to f^{-1}(x)$

$f^{-1}(x)=2x+14$

58. $f(x)=|x|$

$f(x+2)\to f(x)$ shifted left 2 units

$f(x)-3\to f(x)$ shifted down 3 units

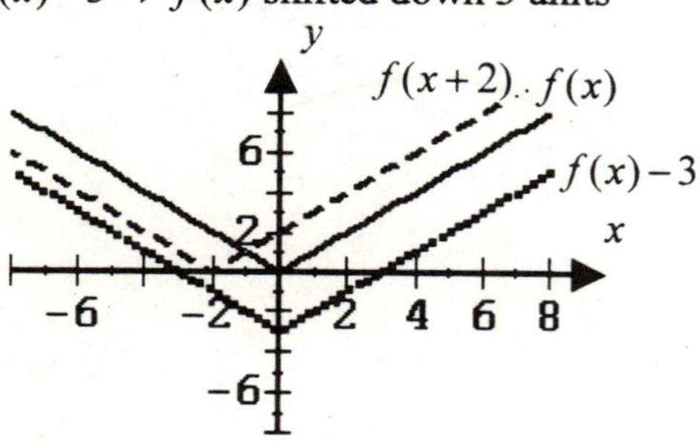

59. $f(x)=2^{1-x}$

x	$y=f(x)=2^{1-x}$
−1	4
0	2
1	1

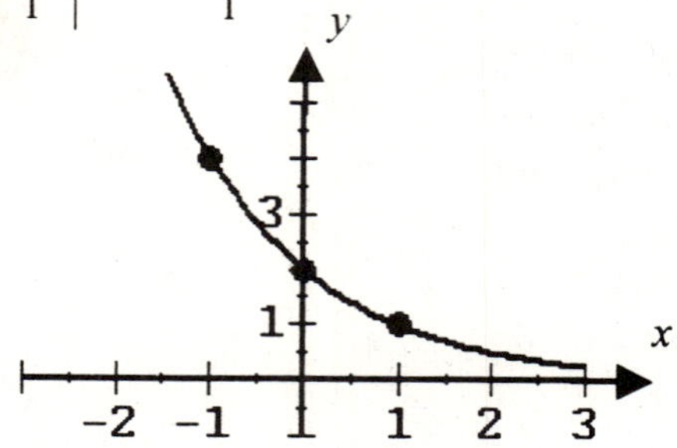

60. $\log_5 x=-4\Leftrightarrow x=5^{-4}=\frac{1}{5^4}=\frac{1}{625}$

61. $\log_4(3x+1)=3\Leftrightarrow 3x+1=4^3=64$

$$\begin{aligned}3x+1&=64\\3x&=63\\x&=21\end{aligned}$$

62. $\log_{10}0.01=y\Leftrightarrow 10^y=0.01=10^{-2}$

$y=-2$

63.
$$\begin{aligned}\log_2 6+\log_2 x&=4+\log_2(x-5)\\\log_2(x-5)-\log_2 x&=\log_2 6-4\\\log_2\frac{x-5}{x}&=\log_2 6-4\\\frac{x-5}{x}&=2^{\log_2 6-4}\\\frac{x-5}{x}&=2^{\log_2 6}\cdot 2^{-4}\\\frac{x-5}{x}&=\frac{6}{16}=\frac{3}{8}\\8x-40&=3x\\5x&=40\\x&=8\end{aligned}$$